M. Mehring

Principles of
High Resolution
NMR in Solids

Second, Revised and Enlarged Edition

Springer-Verlag
Berlin Heidelberg New York 1983

Professor Dr. Michael Mehring
Physikalisches Institut, Teilinstitut 2
Universität Stuttgart
Pfaffenwaldring 57, D-7000 Stuttgart 80

Second, revised and enlarged edition of NMR – Basic Principles and Progress, Vol. 11

ISBN-13: 978-3-642-68758-7   e-ISBN-13: 978-3-642-68756-3
DOI: 10.1007/978-3-642-68756-3

Library of Congress Cataloging in Publication Data
Mehring, M., 1937– Principles of high-resolution NMR in solids.
Rev. ed. of: High resolution NMR spectroscopy in solids. 1976.
Includes bibliographical references.
1. Solids – Spectra. 2. Nuclear magnetic resonance spectroscopy.
I. Title. II. Title: Principles of high-resolution N.M.R. in solids.
QC176.8.06M43 1982   538′.362   82-10827

Typesetting, printing, and binding: Universitätsdruckerei H. Stürtz AG, Würzburg
2152/3140-543210

Preface

The field of Nuclear Magnetic Resonance (NMR) has developed at a fascinating pace during the last decade. It always has been an extremely valuable tool to the organic chemist by supplying molecular "finger print" spectra at the atomic level. Unfortunately the high resolution achievable in liquid solutions could not be obtained in solids and physicists and physical chemists had to live with unresolved lines open to a wealth of curve fitting procedures and a vast amount of speculations.

High resolution NMR in solids seemed to be a paradoxon. Broad structureless lines are usually encountered when dealing with NMR in solids. Only with the recent advent of multiple pulse, magic angle, cross-polarization, two-dimensional and multiple-quantum spectroscopy and other techniques during the last decade it became possible to resolve finer details of nuclear spin interactions in solids.

I have felt that graduate students, researchers and others beginning to get involved with these techniques needed a book which treats the principles, theoretical foundations and applications of these rather sophisticated experimental techniques. Therefore I wrote a monograph on the subject in 1976. Very soon new ideas led to the developement of "two-dimensional spectroscopy" and "multiple-quantum spectroscopy", topics which were not covered in the first edition of my book. Moreover an exponential growth of literature appeared in this area of research leaving the beginner in an awkward situation of tracing back from a current article to the roots of the experiment. I therefore felt a second enlarged edition was necessary. Springer-Verlag demanded to publish it in the same series and to keep the chapters of the first edition. Only some chapters covering new aspects were supposed to be added. Some revisions of the old chapters, however, were also necessary. Then Springer-Verlag decided to publish it as a separate book. It therefore contains most parts of the first edition and some new chapters. Parts of these were written in Dortmund, parts on the train when commuting between Münster, Dortmund and Stuttgart. The final version was finished in Stuttgart where I moved in spring 1982. Like in the first edition this book contains some material which has never been published separately.

Prerequisite to reading this monograph is some familiarity with the principles of magnetic resonance as can be found in the fundamental books by A. Abragam, M. Goldmann and C.P. Slichter. Additional reading of the monograph written by U. Haeberlen is highly recommended. I have tried very hard to cover the whole current literature in this rapidly expanding field; however, I am aware that I have certainly missed important contributions. Among these are multiple-pulse experiments applied to liquids. Of those who suffer from this, I herewith beg pardon.

Among my friends and colleagues I am particularly indebted to O. Kanert, A. Pines and J.S. Waugh for their criticisms, discussions and comments. Among these my friend A. Pines has encouraged and stimulated me continously. I gratefully acknowledge the kind hospitality of John S. Waugh during my stay at the Massachusetts Institute of Technology in 1969–1971, where I was introduced to these fascinating experiments. My friend O. Kanert has continously encouraged me during the time of writing. I am very much obliged to my co-workers who supplied some of the material covered in this book. There are numerous scientists from whose discussion I have benefited greatly in the past. Among these I am particularly indebted to R.G. Griffin, the late R.W. Vaughan and U. Haeberlen.

I also gratefully acknowledge the patience and endurance of Mrs. R. Beck and Mrs. E. Winkler who typed and edited the additional chapters.

Finally I want to apologize to my wife Sabine and the children for spoiling many evenings and sunny weekends by working on this monograph. Their patience and endurance is gratefully acknowledged here.

This book would never have been completed without the patience, support and constant urge by Dr. F.L. Boschke and Mrs. A. Heinrich from Springer-Verlag.

Stuttgart, September 1, 1982 Michael Mehring

Contents

1 Introduction

Manipulation and Dilution
Tools for Ruling Abundant Species

Spin engineering has brought about a wealth of techniques to overcome the natural line broadening mechanisms in solids, such as dipole-dipole and quadrupole interactions. We are going to review in this monograph the different techniques involved and we shall discuss the results obtained. For the convenience of the unbiased reader let us first take a look at some representative results.

As is well known to the chemist, the NMR spectrum of a liquid consists of numerous sharp lines typically with less than 1 Hz linewidth, due to magnetic field inhomogeneities or spin relaxation [1]. In order to supply a reference to this concept of "High Resolution NMR", Fig. 1.1 displays as a representative example the spectrum of ethyl alcohol [2]. Neither *manipulation* nor *dilution* is indicated in order to obtain the NMR spectrum of this compound in the

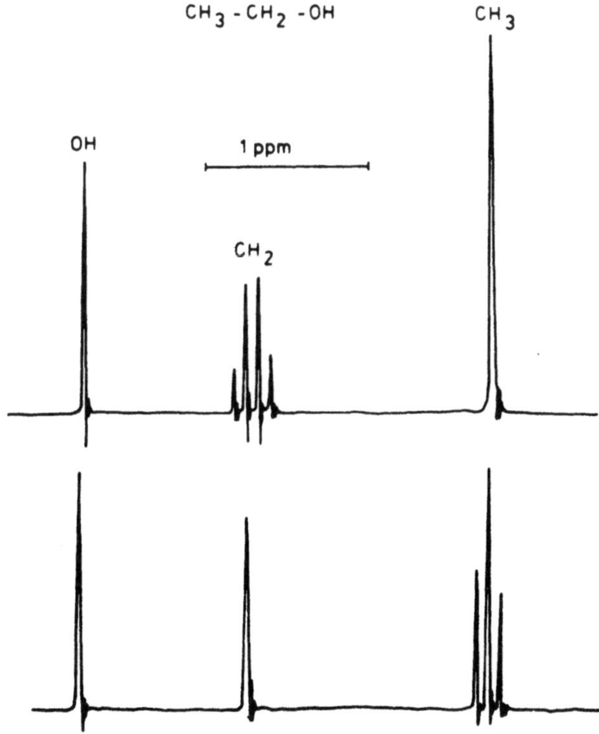

Fig. 1.1. Highly resolved proton spectrum of ethanol using spin decoupling. Top: While recording the methyl group line, irradiation was performed on the methylene resonance. Bottom: While recording the methylene group line, irradiation was performed on the methyl group resonance [2]

liquid state. It may be obtained in a rather standard fashion by taking simply the NMR spectrum of the liquid sample. However, also high resolution NMR spectroscopists like to manipulate on their spectra as is demonstrated in Fig. 1.1. We note in advance that in general "High Resolution NMR in Solids" has not lived up to this state of the art. The interaction Hamiltonian in a liquid sample is represented by isotropic chemical shift and scalar spin-spin interactions. All possible anisotropic interactions, namely chemical shift anisotropy, dipole-dipole interaction, quadrupole interaction etc. are averaged to zero due to the rapid isotropic molecular motion i.e., nature performs some manipulations in this case.

In the solid state, however, all these anisotropic interactions are retained and may be used to monitor the symmetry properties and the electronic state of the solid [1]. Unfortunately in many cases (like ^{19}F and ^{1}H) the dipole-dipole interaction is overwhelming at ordinary magnetic field strength (1–6 Tesla). This results in a more or less bell shaped, structureless line shape, from which very little information can be extracted about the local symmetry and electronic configuration, In this sense the goal of high resolution NMR in solids can be formulated as designing methods to repress the "unwanted" dipolar interaction considerably, leaving chemical shift anisotropies, scalar interaction etc. more or less unaffected.

The natural way of achieving this goal would be by *dilution* of the spins, since the dipolar interaction is proportional to r^{-3}, where r is the distance between the spins. This, however, leads to "High Resolution" only in that case, where no other heteronuclear spins are present. In favourable cases such as in a dilute spin system with small gyromagnetic ratio and large chemical shift anisotropy, already the ordinary NMR spectrum yields a "High Resolution Spectrum" in the sense that anisotropic shift interactions are observable. If in addition considerable molecular motion is present, highly resolved spectral lines can be observed in a solid, as was demonstrated by Andrew and co-workers [3] in the case of solid P_4S_3 whose spectrum is shown in Fig. 1.2.

In this monograph, however, we are dealing with the case where the natural line broadening due to dipolar and quadrupolar interactions masks other weak interactions, such as shift interactions and scalar couplings. The first attempt to overcome this obstacle was made independently by E.R. Andrew et al. [4]

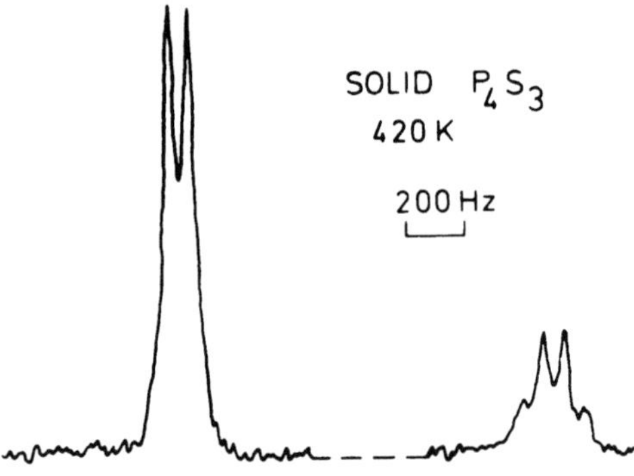

Fig. 1.2. ^{31}P NMR spectrum in solid P_4S_3 at 420 K (melting point 446 K) by Andrew, Hinshaw and Jasinski [3]. The spectrum is strongly narrowed by molecular motion in the solid. The AB_3 type fine structure is represented by a chemical shift separation between the doublet (basal nuclei) and the quartet (apical nuclei) of 185 ±2 ppm. The coupling constant is $J = 70 \pm 3$ Hz

and by I. Lowe [5] by using a specimen rotation method. The whole sample is rapidly rotated, in this method, about an axis tilted by the "magic angle" ϑ_m $= 54°44'8''12'''$ $(\tan \vartheta_m = \sqrt{2})$ with respect to the static magnetic field B_0. It can be shown that the average dipolar interaction vanishes in this case. A representative example is shown in Fig. 1.3. The reader will realize that shielding

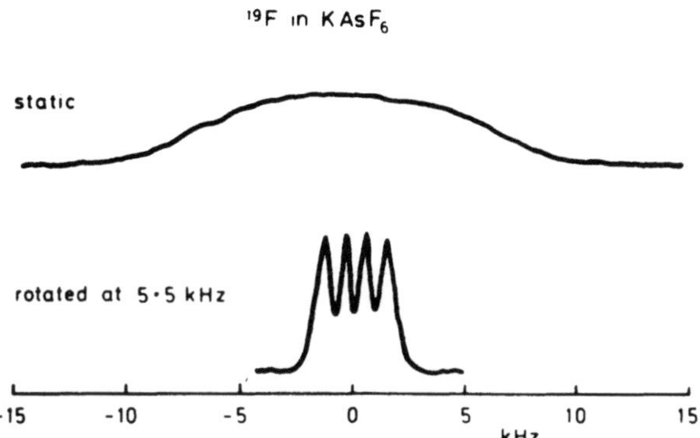

Fig. 1.3. Application of the magic angle specimen rotation method of ^{19}F in polycrystalline KAsF$_6$ by Andrew, Farnell and Gledhill [6b]. The upper spectrum shows the ordinary NMR line without sample rotation. Dipolar coupling among the spins governs the line width. The lower spectrum corresponds to a magic angle rotation of the sample with 5.5 KHz, displaying a quartet structure due to electron-coupled interaction between ^{19}F and ^{75}As nuclei $(I = 3/2)$. The coupling constant is $J = 905$ Hz

anisotropies are also averaged to zero and only the isotropic shift is retained. On the other hand, homonuclear and heteronuclear dipolar interactions vanish, since this technique is not spin specific. Excellent review articles have been written on this subject by E.R. Andrew [6]. Therefore we are not treating this subject in detail; however, we shall touch upon it occasionally. One of our main purposes is to review the recently developed techniques of *manipulation*, which operate in spin space and which are capable of reducing dipolar as well as quadrupolar interactions considerably. The first useful multiple pulse cycle which has successfully been applied in this sense, was the four-pulse cycle of J.W. Waugh, L.M. Huber and U. Haeberlen [7], often referred to as WAHUHA experiment. Patent holders: J.S. Waugh and U. Haeberlen, U.S. patent No. 3,530,374. Figure 1.4 represents the "High Resolution" spectrum of ^{19}F in C$_6$F$_{12}$ obtained with such a four-pulse experiment in comparison with the ordinary NMR spectrum [8]. The great potential of these techniques for the resolution of weak spin interactions is demonstrated clearly. The reason why no such anisotropy is observed in Fig. 1.4 is because nature supplies enough motion at the applied temperature, to average anisotropic shielding, leaving only the scalar interactions. Since this method operates in spin space, all anisotropic spin interactions which are linear in the spin variable are retained and have been extensively studied in powder samples as well as in single crystals [9–12]. Modifications of this basic multiple-pulse line narrowing

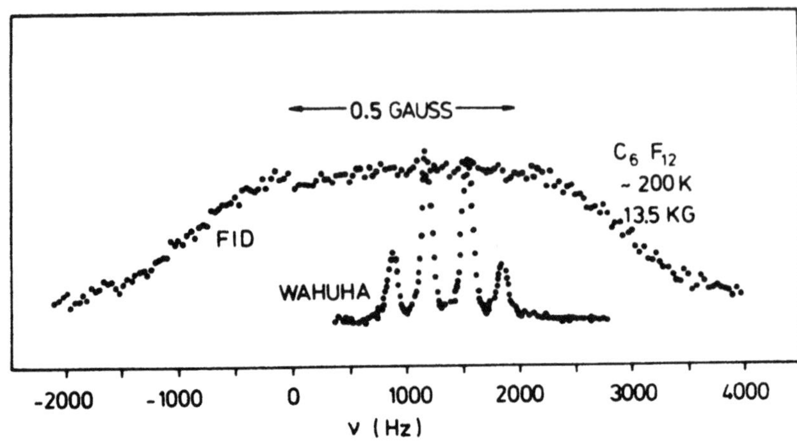

Fig. 1.4. Application of the multiple pulse method to ^{19}F in polycrystalline C_6F_{12} at 200 K by Ellett, Haeberlen and Waugh [8]. Upper curve: Normal NMR spectrum governed by dipolar coupling among the ^{19}F nuclei. Lower curve AB type spectrum ($J = 310$ Hz, $\delta = 17.5$ ppm) after removal of the dipolar interaction by applying a multiple-pulse sequence (WAHUHA). Chemical shift anisotropy is removed by isotropic molecular motion in the solid, but is observed at a lower temperature

experiment have been developed which in some cases are capable of extremely high resolution [13].

Dilution of spin systems is often done on purpose, however, nature supplies a wealth of diluted spin systems with very low natural abundance, such as ^{13}C, ^{15}N, ^{43}Ca etc. in a surrounding of abundant spins like ^{1}H and ^{19}F. In order to obtain a high signal to noise, high resolution spectrum, the rare spins are first polarized by the virtue of spin order transfer from the abundant spins. During the subsequent observation of the rare spins, the abundant spins are decoupled, in order to repress the broadening due to heteronuclear dipolar interaction. This technique was first applied by A. Pines, M. Gibby, and J.S. Waugh and is

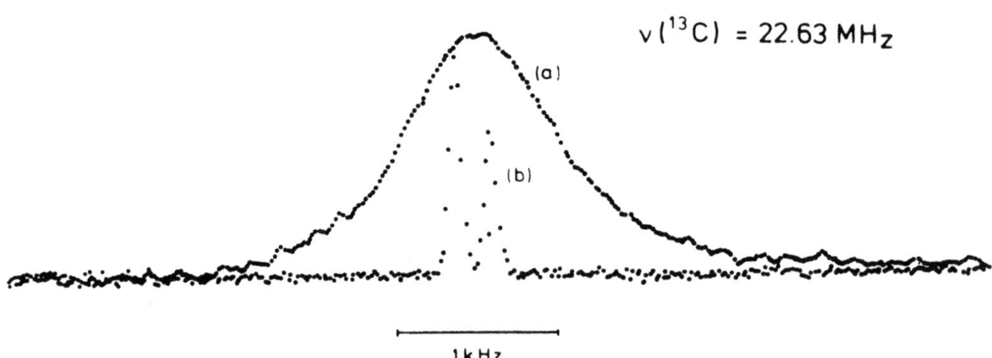

Fig. 1.5a, b. Application of the cross-polarization method of Pines, Gibby and Waugh [14] to ^{13}C in polycrystalline adamantane at room temperature. **a** Undecoupled ^{13}C spectrum governed by ^{13}C-^{1}H dipolar coupling. **b** Decoupled ^{13}C spectrum displaying two chemically inequivalent ^{13}C nuclei. Chemical shift anisotropy is not observed due to the rapid isotropic motion of the molecules. (Courtesy of G. Sinning)

referred to as "Proton Enhanced Nuclear Induction Spectroscopy" [14]. Patentholders: A. Pines, M. Gibby and J.S. Waugh, U.S. Pat. No. 3792,346 (1974). Figure 1.5 shows the ^{13}C spectrum of natural abundant (1.1%) ^{13}C in adamantane, obtained by this technique. There is considerable motion present in adamantane at room temperature to average out chemical shift anisotropies, leaving only the isotropic shift [15].

Very often, however, molecular motion is very slow in solids, leading to broad spectral features, known as powder pattern. If several nonequivalent ^{13}C nuclei are present in the sample, overlapping powder pattern may cause severe difficulties in spectrum analysis. Therefore magic angle spinning (MAS) of the sample is a useful technique to average all anisotropies, leaving only the trace, i.e. the isotropic part of the tensor interaction. An example of this technique is shown in Fig. 1.6.

There is, however, a different way of disentangling complicated spectra, called two-dimensional (2D) spectroscopy, originally proposed by Jeener [17] and extended and applied by Ernst and co-workers [18], Freeman and Morris [19] and others. Figure 1.7 shows an example due to Waugh and co-workers [20]. In these experiments the heteronuclear dipole-dipole interaction between ^{13}C and ^1H is used as a second frequency axis.

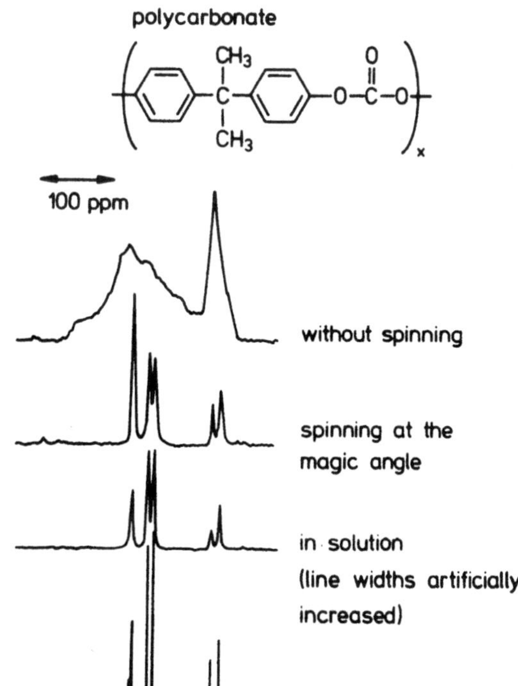

Fig. 1.6. Cross-polarization (CP) ^{13}C NMR spectra (22.6 MHz) of polycarbonate, with and without magic-angle spinning (MAS) (3 kHz) according to Schaefer, Stejskal and Buchdahl [16]

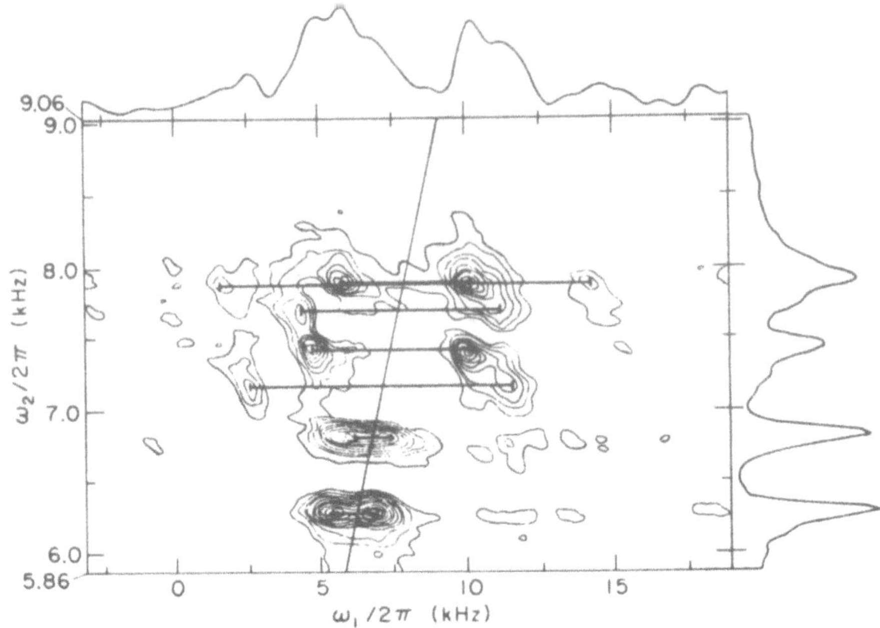

Fig. 1.7. Two-dimensional ^{13}C spectrum of calcium formate $Ca(HCO_2)_2$ according to Hester et al. [20b]. Dipolar splittings for each of the observed seven lines are indicated by the heavy lines. The ordinary one-dimensional spectrum is obtained by a projection onto the ω_2-axis as shown on the right hand side

Another new technique, recently initiated by Pines and co-workers [21] leads to multiple quantum coherence spectroscopy. Multiple quantum transitions are excited in order to resolve n-quantum spectra. An example is given in Fig. 1.8. This technique, so far mainly applied to liquids or liquid crystals may find interesting applications in solids too.

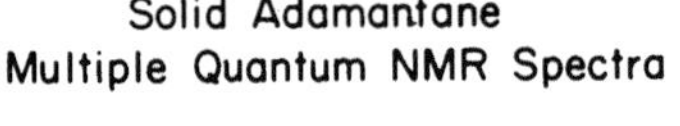

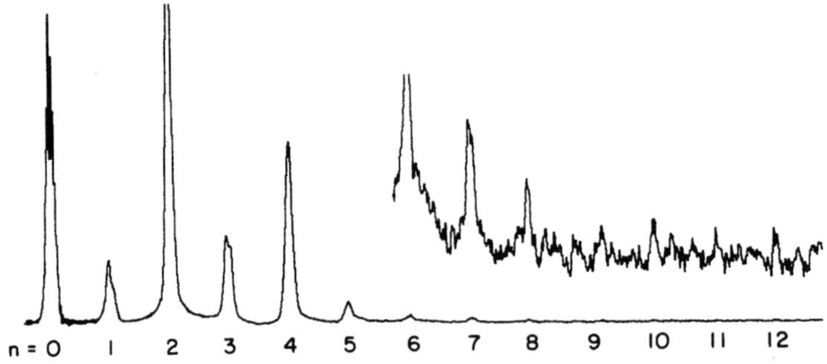

Fig. 1.8. Multiple-quantum NMR spectra of protons in adamantane $(C_{10}H_{16})$ according to A. Pines and Yu Sze Yen. Up to $n=12$ quanta are seen to be coherently absorbed. (Courtesy of A. Pines)

In Chap. 2 we remind the reader of the tensorial character of all spin interactions which manifest itself by second rank tensors in ordinary solids. There may be interactions of higher rank in principle, but no valid experimental verification has been given of this to my knowledge. There has been a discussion on the transformation properties of these interactions. We restrict ourselves, however, in this chapter to rotations in real space. In this context also magic angle spinning (MAS) is discussed as well as molecular reorientation.

Chapter 3 deals with multiple-pulse experiments, discussing the application of pulse cycles to the spin system, mainly serving the goal of *repressing* the dipolar interaction. Coherent averaging, second averaging and the influence of pulse imperfections are treated in detail.

Double resonance experiments on rare spins with the purpose of obtaining high resolution spectra in solids are analyzed in Chap. 4. After reviewing the principles of double resonance we turn to cross-polarization experiments, which recently have supplied a wealth of ^{13}C spectra in solids. Cross-polarization dynamics and spin decoupling dynamics are discussed also.

Two-dimensional (2D) spectroscopy in solids and the underlying principle is treated in Chap. 5. Although multiple-quantum spectroscopy has not yet been applied to solids extensively, we will nevertheless discuss its principles and potential usefulness in Chap. 6.

The main application of the techniques of *manipulation* and *dilution* has been the determination of magnetic shielding tensors, which we deal with in Chap. 7. After a brief introduction to the concepts we summarize the shielding tensors, thus far determined by these techniques.

Spin lattice relaxation has not played a dominant role in high resolution NMR in solids so far. However, as we see in Chap. 8 there are some useful applications of the techniques which have been described in the preceding chapters about the investigation of spin lattice relaxation processes. The *appendices* summarize for the convenience of the reader some useful aspects to which we refer in the text. Reviews on the subject have been written by P. Mansfield [22], R.W. Vaughan [23], R.G. Griffin [24] and E.R. Andrew [6]. The monograph by U. Haeberlen [25] covers part of this subject taking a slightly different approach and is highly recommended to the reader. This applies also to the review written more recently by H.W. Spiess [26], who deals with line shapes and relaxation of diluted spins.

2. Nuclear Spin Interactions in Solids

For the convenience of the reader we summarize in this section the basic nuclear spin interactions which occur in solids. The notation will be in tensorial form throughout to emphasize the anisotropy of these interactions. In order to keep the presentation as compact as possible we shall avoid detailed derivations and refer the reader to the outstanding book by A. Abragam [1]. Also Pool and Farach [2a] have given a detailed description of spin interactions in tensorial form. As a further reference the book by C.P. Slichter [2b] is recommended.

2.1 Basic Nuclear Spin Interactions in Solids

Figure 2.1 can serve as a guide to the different basic interactions occurring in solids. It represents the sevenfold way a nuclear spin can communicate with its surrounding. We shall refer to this figure later in the text. We first distinguish nuclear spin interactions between external fields (B_0, B_1) and internal fields:

$$\mathcal{H} = \mathcal{H}_{\text{ext}} + \mathcal{H}_{\text{int}} \tag{2.1}$$

where

$$\mathcal{H}_{\text{ext}} = \mathcal{H}_0 + \mathcal{H}_1 .$$

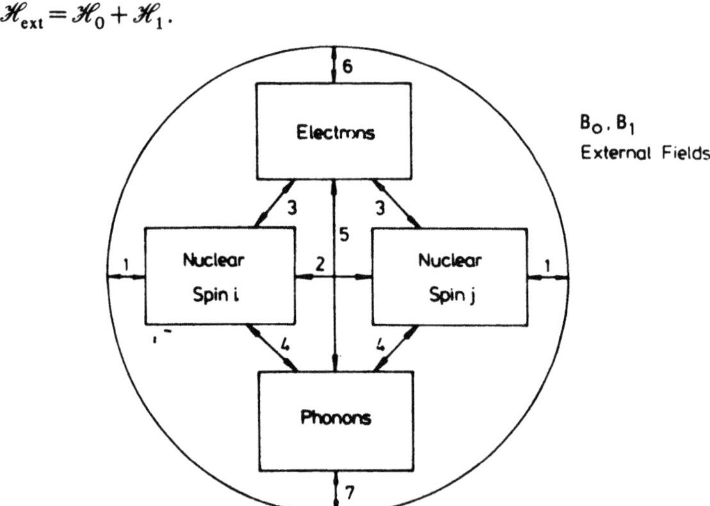

Fig. 2.1. The seven-fold way a nuclear spin system can interact with its surrounding. *1* Zeeman interaction of spins. *2* Direct spin interaction. *3* Nuclear spin-electron interaction and indirect spin interaction. *4* Direct spin-lattice interaction. *3–5* Indirect spin lattice interaction via electrons. *3–6* Shielding and polarization of nuclear spins by electrons. *4–7* Coupling of nuclear spins to sound fields etc.

$\mathcal{H}_0$ and $\mathcal{H}_1$ are the Zeeman interactions with the external fields B_0 and B_1, respectively. This distinction is appropriate in fact, because we are going to work in a regime where the "size" of $\mathcal{H}_0$ and $\mathcal{H}_1$ is much larger than the "size" of $\mathcal{H}_{int}$. By the "size" of an Hamiltonian we mean unless otherwise stated:

$$\|\mathcal{H}\| = [\mathrm{Tr}\{\mathcal{H}^2\}]^{1/2}. \tag{2.2}$$

The spin interactions of two different types of spins (I, S) (gyromagnetic ratios γ_I and γ_S) with internal fields may be written as

$$\mathcal{H}_{int} = \mathcal{H}_{II} + \mathcal{H}_{SS} + \mathcal{H}_{IS} + \mathcal{H}_Q + \mathcal{H}_S + \mathcal{H}_L \tag{2.3}$$

where $\mathcal{H}_{II}$ and $\mathcal{H}_{SS}$ represent the direct (dipolar) as well as the indirect interactions among I spins and S spins respectively (path 2 and 3 in Fig. 2.1). The same paths are involved in the I-S interaction as expressed by $\mathcal{H}_{IS}$, which covers the direct as well as the indirect interactions between I spins and S spins. $\mathcal{H}_Q$ is the quadrupole Hamiltonian of the I and S spins respectively. $\mathcal{H}_S$ contains all shielding Hamiltonians (chemical shift and Knight shift) of the I and S spins (path 3, 6 in Fig. 2.1).

$\mathcal{H}_L$ describes the spin lattice interaction (path 4 and 3, 5 in Fig. 2.1). We find it convenient to express all Hamiltonian in frequency units (ω-units) throughout this monograph. In the following we assume that the symmetry of the solid is such that all the spin interactions can be represented by second rank Cartesian tensors in the following way [1, 2] (see Table 2.1):

$$\mathcal{H} = \mathbf{I} \cdot \tilde{\mathbf{A}} \cdot \mathbf{S} = (I_x, I_y, I_z) \begin{pmatrix} A_{xx} & A_{xy} & A_{xz} \\ A_{yx} & A_{yy} & A_{zz} \\ A_{zx} & A_{zy} & A_{zz} \end{pmatrix} \begin{pmatrix} S_x \\ S_y \\ S_z \end{pmatrix} \tag{2.4}$$

Table 2.1. Interaction Hamiltonians

Interaction	Form of the Hamiltonian
Chemical shift	$\mathcal{H}_S = \gamma \mathbf{I} \cdot \tilde{\sigma} \cdot \mathbf{B}_0$
Dipole-dipole	$\mathcal{H}_D = \sum\limits_{i<j} \hbar \gamma_i \gamma_j r_{ij}^{-3} \left(\mathbf{I}_i \cdot \mathbf{I}_j - \dfrac{3(\mathbf{I}_i \cdot \mathbf{r}_{ij})(\mathbf{I}_j \mathbf{r}_{ij})}{r_{ij}^2} \right)$
	$\mathcal{H}_D = \sum\limits_{i=j} \mathbf{I}_i \cdot \tilde{\mathbf{D}} \cdot \mathbf{I}_j$
	$D_{\alpha\beta} = \hbar \gamma_i \gamma_j r_{ij}^{-3} (\delta_{\alpha\beta} - 3e_\alpha e_\beta)$
	$\alpha, \beta = x, y, z; \; e_\alpha$: α-component of the unit vector along r_{ij}
J-coupling	$\mathcal{H}_J = \sum\limits_{i \neq j} \mathbf{I}_i \cdot \tilde{\mathbf{J}} \cdot \mathbf{I}_j$
Spin-rotation	$\mathcal{H}_{CR} = \sum\limits_i \mathbf{I}_i \cdot \tilde{\mathbf{C}}_i \cdot \mathbf{J}$
Quadrupole	$\mathcal{H}_Q = \dfrac{eQ}{2I(2I-1)\hbar} \mathbf{I} \cdot \tilde{\mathbf{V}} \cdot \mathbf{I}$
	$\tilde{V} = \{V_{\alpha\beta}\}; \; \alpha, \beta = x, y, z$

where I and S are row and column vectors respectively and $\tilde{\mathbf{A}}$ is a second rank Cartesian tensor (3×3 matrix).

Coupling of the spin I to external field is now expressed as:

$$\mathcal{H}_{0I} = \mathbf{I} \cdot \tilde{\mathbf{Z}} \cdot \mathbf{B}_0$$
$$\mathcal{H}_{1I} = \mathbf{I} \cdot \tilde{\mathbf{Z}} \cdot \mathbf{B}_1$$

where $\mathbf{B}_0 = (B_x, B_y, B_z)$,

$$\mathbf{B}_1(t) = 2(B_{1x}(t), B_{1y}(t), B_{1z}(t)) \cos \omega t \quad \text{and} \quad \tilde{\mathbf{Z}} = -\gamma_I \tilde{\mathbf{1}}$$

with the unit matrix

$$\tilde{\mathbf{1}} = \begin{pmatrix} 1 & 0 & 0 \\ 0 & 1 & 0 \\ 0 & 0 & 1 \end{pmatrix}$$

being the coupling matrix between the spin I and the magnetic field B_0. Similarly we express the coupling between two homonuclear ($\omega_{0i} = \omega_{0j}$) spins $I^{(i)}$ and $I^{(j)}$ as

$$\mathcal{H}_{II} = \mathbf{I}^{(i)} \cdot \tilde{\mathbf{D}}_{II} \cdot \mathbf{I}^{(j)} \tag{2.6}$$

or between heteronuclear ($\omega_{0I} \neq \omega_{0S}$) I and S spins as

$$\mathcal{H}_{IS} = \mathbf{I} \cdot \tilde{\mathbf{D}}_{IS} \cdot \mathbf{S}. \tag{2.7}$$

Note that the tensor $\mathbf{D}$ in Eqs. (2.6) and (2.7) covers both, direct spin-spin interactions (dipolar interactions) as well as indirect spin-spin interactions (scalar coupling, pseudo dipolar interaction and its antisymmetric counterpart) i.e. in general

$$\text{Tr}\{\tilde{\mathbf{D}}\} \neq 0$$

and $\mathbf{D}$ is not necessarily symmetric i.e. it contains nine different elements. The symmetric part of such an electron-coupled spin-spin coupling tensor has been, recently determined by Tutunjian and Waugh [3] for ^{31}P in tetraethyldiphosphine disulfide. In the case of *dipolar interaction*, however, $\tilde{\mathbf{D}}$ is axially symmetric and additionally

$$\text{Tr}\{\tilde{\mathbf{D}}\} = 0.$$

In this case $\mathcal{H}_{II}$ and $\mathcal{H}_{IS}$ can be expanded into the dipolar alphabet (see A. Abragam [1]) and $D_{\alpha\beta}$ ($\alpha, \beta = x, y, z$) may be written as

$$D_{\alpha\beta} = \frac{\gamma_I \gamma_S \hbar}{r^3} (\delta_{\alpha\beta} - 3 e_\alpha e_\beta) \quad (\alpha, \beta) = x, y, z \tag{2.8}$$

where r is the distance between the spins, $\delta_{\alpha\beta}$ is the Kronecker delta and e_α ($\alpha = x, y, z$) are x, y and z-components of a unit vector pointing from one spin to the other.

The "dipolar alphabet" [1] can be expressed as

$$\mathcal{H}_{IS} = \gamma_I \gamma_S r_{IS}^{-3} \hbar [A + B + C + D + E + F] \tag{2.9}$$

where

$$A = (1 - 3\cos^2 \vartheta)\, I_z S_z$$
$$B = \tfrac{1}{2}(1 - 3\cos^2 \vartheta)\,(I_z S_z - \mathbf{I} \cdot \mathbf{S})$$
$$C = -\tfrac{3}{2}\sin \vartheta \cos \vartheta\, e^{-i\varphi}(I_z S_+ + I_+ S_z)$$
$$D = C^* = -\tfrac{3}{2}\sin \vartheta \cos \vartheta\, e^{i\varphi}(I_z S_- + I_- S_z)$$
$$E = -\tfrac{3}{4}\sin^2 \vartheta\, e^{-i2\varphi} I_+ S_+$$
$$F = E^* = -\tfrac{3}{4}\sin^2 \vartheta\, e^{i2\varphi} I_- S_- .$$

We consider the quadrupole Hamiltonian as being a degenerate case of Eq. (2.6) with

$$\mathcal{H}_{QI} = \mathbf{I} \cdot \tilde{\mathbf{Q}} \cdot \mathbf{I} \tag{2.10a}$$

where

$$\tilde{\mathbf{Q}} = \frac{eQ}{2I(2I-1)\hbar}\,\tilde{\mathbf{V}} \quad \text{with} \quad \tilde{\mathbf{V}} = \{V_{\alpha\beta}\} \tag{2.10b}$$

with the nuclear quadrupole moment Q and the electric field gradient tensor $\tilde{\mathbf{V}}$. The trace of $\tilde{\mathbf{V}}$ vanishes due to the Laplace equation.

The shielding Hamiltonian of the spin I can be visualized as the coupling of the spin I with the magnetic field B_0 via the shielding tensor $\tilde{\mathbf{S}}$:

$$\mathcal{H}_S = \mathbf{I} \cdot \tilde{\mathbf{S}} \cdot \mathbf{B}_0. \tag{2.11}$$

The tensor $\tilde{\mathbf{S}}$, however, is no longer the unit matrix as in the case of the Zeeman interaction [Eq. (2.5)], since the coupling is established via the electronic surrounding (path 3, 6 in Fig. 2.1) rather than being direct (path 1 in Fig. 2.1). This is why the investigation of shielding interactions can give valuable information about electronic states in solids and molecules. The main purpose of the techniques to be described in this book serves the goal of measuring just this quantity. Here we would like to point out that the corresponding tensor interaction in esr is the g-tensor. The shielding tensor $\tilde{\mathbf{S}}$ in Eq. (2.11) may be expressed in the case of the chemical shift tensor as

$$\tilde{\mathbf{S}} = \gamma_I \begin{pmatrix} \sigma_{xx} & \sigma_{xy} & \sigma_{xz} \\ \sigma_{yx} & \sigma_{yy} & \sigma_{yz} \\ \sigma_{zx} & \sigma_{zy} & \sigma_{zz} \end{pmatrix} \tag{2.12}$$

and similarly in the case of the Knight shift tensor as

$$\tilde{\mathbf{K}} = -\gamma_I \{K_{\alpha\beta}\}.$$

In the following we do not distinguish between $\tilde{\mathbf{S}}$ and $\tilde{\mathbf{K}}$ and assume all shielding interactions which are linear in the spin variable to be represented by $\tilde{\mathbf{S}}$. In general $\mathrm{Tr}\{\tilde{\mathbf{S}}\} \neq 0$ and $\tilde{\mathbf{S}}$ is not necessarily symmetric. However, as shown in Appendix C antisymmetric components of $\tilde{\mathbf{S}}$ contribute to the resonance shift only in second order and can usually be ignored.

The spin rotation interaction $\mathcal{H}_J$, which represents the coupling of the nuclear spin I with the magnetic moment produced by the angular momentum

J of the molecule can be expressed as

$$\mathscr{H}_J = \mathbf{J} \cdot \tilde{\mathbf{C}} \cdot \mathbf{I}. \qquad (2.13)$$

The spin rotation interaction tensor $\tilde{\mathbf{C}}$ has some common features with the shielding tensor $\mathbf{S}$, since both are associated with the "dequenching" of orbital angular momentum, whether by B_0 or by $\mathbf{J}$.

The corresponding Hamiltonians for the S spins may be easily obtained by exchanging I with S in Eqs. (2.5–2.7, 2.10, 2.11, 2.13). The spin lattice interaction $\mathscr{H}_L$ involves the scattering of a phonon by the spin. In NMR however, this interaction is usually treated semiclassically by rendering the spin interactions time dependent.

Under rapid isotropic molecular reorientation, which is the general case in liquids, but which can occur in solids also, the second rank tansor interaction is averaged and only the scalar invariant of the tensor $\{T_{\alpha\beta}\}$ i.e.

$$\tfrac{1}{3}\mathrm{Tr}\{T_{\alpha\beta}\} = \tfrac{1}{3}\sum_\alpha T_{\alpha\alpha}$$

is retained. Notice, that $\mathscr{H}_{\mathrm{ext}}$ is not changed under rotation of the sample. The other interaction Hamiltonians can be written in the isotropic average as follows:

Indirect spin spin interaction:

$$\mathscr{H}_{IS} = J\mathbf{I} \cdot \mathbf{S} \qquad (2.14)$$

where

$$J = \tfrac{1}{3}\mathrm{Tr}\{\tilde{\mathbf{D}}_{IS}\}. \qquad (2.15)$$

Chemical shift interaction:

$$\mathscr{H}_S = \gamma_I \sigma \mathbf{I} \cdot \mathbf{B}_0 \qquad (2.16)$$

where

$$\sigma = \tfrac{1}{3}\mathrm{Tr}\{\sigma_{\alpha\beta}\}$$

and similar in the case of Knight shift interaction.

In the case of spin rotation interaction:

$$\mathscr{H}_J = c\mathbf{I} \cdot \mathbf{J} \qquad (2.17)$$

where

$$c = \tfrac{1}{3}\mathrm{Tr}\{C_{\alpha\beta}\}.$$

With $\mathbf{B}_0 = (0, 0, B_0)$ the total Hamiltonian of the spin interactions may be expressed in the isotropic average as:

$$\mathscr{H}_{\mathrm{int}} = -\gamma_I B_0(1 - \sigma_I + K_I) I_z - \gamma_S B_0(1 - \sigma_S + K_S) S_z + J\mathbf{I} \cdot \mathbf{S}. \qquad (2.18)$$

This is the usual Hamiltonian in a liquid. "High Resolution NMR" is supplied by nature in this case without any extra manipulations on the spin system. We will not be concerned with this degenerate case in this book. Our ultimate goal, however, is to approach a related situation by *manipulation* and/or *dilution* in a solid sample, while retaining some tensorial character of the spin interactions.

The anisotropy of spin interactions can be expressed either in terms of Cartesian tensors $(3 \times 3$ matrix) as done above or by spherical tensors (see Appendix 2.A), leading to a Hamiltonian of the form [4]

$$\mathcal{H} = \sum_{k=0}^{2} \sum_{q=-k}^{+k} (-1)^q A_{kq} T_{k-q}. \tag{2.19}$$

Here A_{kq} is a spherical tensor of rank k due to the lattice variables (see Table 2.2). If rotations or other unitary transformations are involved it is convenient to work with spherical tensors.

In order to facilitate comparison between Cartesian and spherical notation, we treat as an example the general I-S interaction between a spin I and a spin S. In Cartesian form the Hamiltonian reads

$$\mathcal{H}_{IS} = (I_x, I_y, I_z) \begin{pmatrix} D_{xx} & D_{xy} & D_{xz} \\ D_{yx} & D_{yy} & D_{yz} \\ D_{zx} & D_{zy} & D_{zz} \end{pmatrix} \begin{pmatrix} S_x \\ S_y \\ S_z \end{pmatrix} \tag{2.20a}$$

or

$$\begin{aligned} \mathcal{H}_{IS} = \ & D_{xx} I_x S_x + D_{yy} I_y S_y + D_{zz} I_z S_z \\ & + D_{xy} I_x S_y + D_{yx} I_y S_x + D_{xz} I_x S_z \\ & + D_{zx} I_z S_x + D_{yz} I_y S_z + D_{zy} I_z S_y. \end{aligned} \tag{2.20b}$$

In spherical tensor form we write

$$\begin{aligned} \mathcal{H}_{IS} = \ & A_{00} T_{00} + A_{10} T_{10} - (A_{11} T_{1-1} + A_{1-1} T_{11}) \\ & + A_{20} T_{20} - (A_{21} T_{2-1} + A_{2-1} T_{21}) + A_{22} T_{2-2} + A_{2-2} T_{22} \end{aligned} \tag{2.21}$$

with nine terms in the general case as for the Cartesian notation. The A_{kq} may now be inserted according to Table 2.2, whereas the T_{kq} are given in Table 2.3. The same expression is obtained as given by Eq. (2.20) for the I-S interaction after some rearrangement of terms. Where the form of Eq. (2.21) has the advantage of having terms arranged according to their rotational symmetries.

In the case of dipolar interaction $\mathrm{Tr}\{D_{\alpha\beta}\} = 0$ and $D_{\alpha\beta} = D_{\beta\alpha}$, i.e. A_{00} and A_{1q} vanish.

In this case $D_{\alpha\beta}$ is given by Eq. (2.8) and only the five terms with $k=2$ survive and constitute the dipolar alphabet. Similarly in the case of quadrupolar interaction only the $k=2$ terms contribute and $D_{\alpha\beta}$ is replaced by $Q_{\alpha\beta}$ as defined in Eq. (2.10b) and (S_x, S_y, S_z) is replaced by (I_x, I_y, I_z). Chemical shifts and Knight shifts, however, do have non-vanishing trace and of course anti-symmetric parts in general, i.e. A_{00} and A_{1q} contribute.

In the case of chemical shifts $D_{\alpha\beta}$ is replaced by $\gamma\sigma_{\alpha\beta}$ and (S_x, S_y, S_z) by $(0, 0, B_0)$ if the z-axis is parallel to B_0. The T_{kq} in the case of chemical shift interaction can therefore be obtained from the I-S interaction by making the appropriate replacements (see line 1 in Table 2.3).

The principal axis system $(1, 2, 3)$ (of the chemical shift tensor) is the frame

Table 2.2. Connection between spherical tensors and Cartesian tensors for spin interactions

Interaction	$A_{\alpha\beta}(\alpha,\beta=x,y,z)$	A_{kq}
Chemical shift	$\gamma\sigma_{\alpha\beta}$	$A_{00} = -\dfrac{1}{\sqrt{3}}(A_{xx}+A_{yy}+A_{zz}) = -\dfrac{1}{\sqrt{3}}\mathrm{Tr}\{A_{\alpha\beta}\}$
Dipole-dipole ($k=2$)	$D_{\alpha\beta}$	$A_{10} = -\dfrac{i}{\sqrt{2}}(A_{xy}-A_{yx})$
J-coupling	$J_{\alpha\beta}$	$A_{1\pm1} = -\dfrac{1}{2}[A_{zx}-A_{xz}\pm i(A_{zy}-A_{yz})]$
Spin-rotation	$C_{\alpha\beta}$	$A_{20} = \dfrac{1}{\sqrt{6}}[3A_{zz}-(A_{xx}+A_{yy}+A_{zz})]$
Quadrupole ($k=2$)	$\dfrac{eQ}{2I(2I-1)\hbar}V_{\alpha\beta}$	$A_{2\pm1} = \mp\dfrac{1}{2}[A_{xz}+A_{zx}\pm i(A_{yz}+A_{zy})]$
		$A_{2\pm2} = \dfrac{1}{2}[A_{xx}-A_{yy}\pm i(A_{xy}+A_{yx})]$

Table 2.3. Spherical tensor representation of spin operators (with $I_{\pm}=I_x\pm iI_y$)

Interaction	T_{00}	T_{10}	$T_{1\pm1}$	T_{20}	$T_{2\pm1}$	$T_{2\pm2}$
Chemical shift	$-\dfrac{1}{\sqrt{3}}I_zB_0$	0	$-\dfrac{1}{2}I_{\pm}B_0$	$\sqrt{\dfrac{2}{3}}I_zB_0$	$\mp\dfrac{1}{2}I_{\pm}B_0$	0
Dipole-dipole	0	0	0	$\dfrac{1}{\sqrt{6}}(3I_zS_z-\mathbf{I}\cdot\mathbf{S})$	$\mp\dfrac{1}{2}(I_zS_{\pm}+I_{\pm}S_z)$	$\dfrac{1}{2}I_{\pm}S_{\pm}$
J-coupling	$-\dfrac{1}{\sqrt{3}}\mathbf{I}\cdot\mathbf{S}$	$-\dfrac{1}{2\sqrt{2}}(I_+S_--I_-S_+)$	$+\dfrac{1}{2}[I_zS_{\pm}-I_{\pm}S_z]$	$\dfrac{1}{\sqrt{6}}(3I_zS_z-\mathbf{I}\cdot\mathbf{S})$	$\mp\dfrac{1}{2}(I_zS_{\pm}+I_{\pm}S_z)$	$\dfrac{1}{2}I_{\pm}S_{\pm}$
Spin-rotation	$-\dfrac{1}{\sqrt{3}}\mathbf{I}\cdot\mathbf{J}$	$-\dfrac{1}{2\sqrt{2}}(I_+J_--I_-J_+)$	$+\dfrac{1}{2}[I_zJ_{\pm}-I_{\pm}J_z]$	$\dfrac{1}{\sqrt{6}}(3I_zJ_z-\mathbf{I}\cdot\mathbf{J})$	$\mp\dfrac{1}{2}(I_zJ_{\pm}+I_{\pm}J_z)$	$\dfrac{1}{2}I_{\pm}J_{\pm}$
Quadrupole	0	0	0	$\dfrac{1}{\sqrt{6}}(3I_z^2-I(I+1))$	$\mp\dfrac{1}{2}(I_zI_{\pm}+I_{\pm}I_z)$	$\dfrac{1}{2}I_{\pm}I_{\pm}$

in which the symmetric part (A_{2q}) is diagonal. The anti-symmetric part (A_{1q}) is not necessarily zero and the A_{kq} can be expressed in the principal axis frame $(1,2,3)$ as

$$A_{00} = -(1/\sqrt{3})\,\mathrm{Tr}\{\gamma\tilde{\sigma}\}$$
$$A_{10} = -i\sqrt{2}\,\gamma\sigma_{12}; \qquad A_{1\pm1}=\gamma(\sigma_{13}\pm i\sigma_{23})$$
$$A_{20} = \sqrt{3/2}\,\gamma(\sigma_{33}-\tfrac{1}{3}\mathrm{Tr}\{\tilde{\sigma}\})$$
$$A_{2\pm1}=0$$
$$A_{2\pm2}=\tfrac{1}{2}\gamma(\sigma_{11}-\sigma_{22}).$$

The advantage of spherical tensor notation will be fully exploited in the following chapters, when unitary transformations (rotations) have to be performed.

2.2 Spin Interactions in High Magnetic Fields

If we apply a static magnetic field B_0 to the spin system, forces (angular momentum) are exerted on the spin *via* the Zeeman interaction. As a consequence, the spin frame is accelerated with respect to the laboratory frame, i.e. the spin interactions become time dependent [5]. The Hamiltonian is no longer a scalar quantity, but becomes a "tensor operator". If B_0 is large in the sense $\|\mathscr{H}_0\| \gg \|\mathscr{H}_{int}\|$ only those parts of $\mathscr{H}_{int}$ which commute with $\mathscr{H}_0$ contribute in first order to the spectrum, i.e. we can split $\mathscr{H}_{int}$ into two parts

$$\mathscr{H}_{int} = \mathscr{H}'_{int} + \mathscr{H}''_{int} \tag{2.22}$$

with

$$[\mathscr{H}_0, \mathscr{H}'_{int}] = 0 \quad \text{and} \quad [\mathscr{H}_0, \mathscr{H}''_{int}] \neq 0$$

where $\mathscr{H}'_{int}$ is the so called secular part and $\mathscr{H}''_{int}$ is the non-secular part of $\mathscr{H}_{int}$. The NMR spectrum is "generated" by $\mathscr{H}'_{int}$ in first order. With

$$\mathscr{H}_0 = -\omega_{0S} S_z - \omega_{0I} I_z \tag{2.23}$$

where

$$\omega_{0I} = \gamma_I B_0 \quad \text{and} \quad \omega_{0S} = \gamma_S B_0$$

an interaction representation is imposed, which renders the interaction Hamiltonian $\mathscr{H}_{int}$ time dependent as follows:

$$\tilde{\mathscr{H}}_{int}(t) = e^{-i(\omega_{0I} I_z + \omega_{0S} S_z)t} \, \mathscr{H}_{int} \, e^{i(\omega_{0I} I_z + \omega_{0S} S_z)t}. \tag{2.24}$$

This comes about by the transformation into the "rotating frame". In the laboratory frame the equation of motion

$$\frac{d}{dt}\rho = -i[\mathscr{H}, \rho] \tag{2.25}$$

with

$$\mathscr{H} = \mathscr{H}_0 + \mathscr{H}_{int}$$

may be solved for $\rho(t)$ as

$$\rho(t) = e^{-i\mathscr{H}t} \rho(0) e^{i\mathscr{H}t}. \tag{2.26}$$

If $\|\mathscr{H}_0\| \gg \|\mathscr{H}_{int}\|$ it is convenient to express the spin density matrix in the interaction representation

$$\frac{d}{dt}\rho^* = -i[\mathscr{H}^*, \rho^*] \tag{2.27}$$

where

$$\rho^* = e^{i\mathscr{H}_0 t} \rho \, e^{-i\mathscr{H}_0 t}. \tag{2.28}$$

Comparing Eq. (2.25) with Eq. (2.27) leads to

$$\mathcal{H}^*(t) = e^{i\mathcal{H}_0 t}\,\mathcal{H}_{\text{int}}\,e^{-i\mathcal{H}_0 t}. \tag{2.29}$$

This is equivalent to Eq. (2.24). $\mathcal{H}^*(t)$ will contain in general time-independent (secular) and time-dependent (non-secular) terms. In first order the time-dependent terms may be neglected.

We note, that the application of the static magnetic field B_0 already imposes a manipulation of the spin system which separates certain parts of the interaction Hamiltonian. In fact, only the time average of $\tilde{\mathcal{H}}_{\text{int}}(t)$ survives in first order

$$\mathcal{H}'_{\text{int}} = \frac{1}{t_c} \int\limits_0^{t_c} dt\, \tilde{\mathcal{H}}_{\text{int}}(t) \tag{2.30}$$

where t_c is a suitable chosen cycle time, with the condition

$$\omega_{0I}\, t_c = m\, 2\pi \quad \text{and} \quad \omega_{0S}\, t_c = n\, 2\pi.$$

For the interaction between a spin I and a spin S we obtain in the case

(i) $\omega_{0I} \neq \omega_{0S}$ (heteronuclear)

$$\mathcal{H}'_{IS} = D_{zz}\, I_z S_z \tag{2.31}$$

where in the case of dipolar interaction

$$D_{zz} = \gamma_I \gamma_S \hbar\, r_{IS}^{-3}(1 - 3\cos^2 \vartheta_{IS})$$

according to Eqs. (2.8 and 2.9) and in the case

(ii) $\omega_{0I} = \omega_{0S}$ (homonuclear)

$$\mathcal{H}'_{IS} = D_{zz}\, I_z S_z + \tfrac{1}{2}(D_{xx} + D_{yy})(I_x S_x + I_y S_y) + \tfrac{1}{2}(D_{xy} - D_{yx})(I_x S_y - I_y S_x). \tag{2.32}$$

Equation (2.32) may be rewritten as:

$$\mathcal{H}'_{IS} = J\,\mathbf{I}\cdot\mathbf{S} + \tfrac{1}{2}(D_{zz} - J)(3 I_z S_z - \mathbf{I}\cdot\mathbf{S}) + \tfrac{1}{2}(D_{xy} - D_{yx})(I_x S_y - I_y S_x) \tag{2.33}$$

where $J = \tfrac{1}{3}\text{Tr}\{D_{\alpha\beta}\}$.

Note that antisymmetric components of the interaction appear in the secular Hamiltonian. However, these parts are not involved in line shifts in first order.

Often $\tilde{D}$ is a symmetric tensor i.e. $D_{\alpha\beta} = D_{\beta\alpha}$ and the last term in Eq. (2.33) vanishes. In the case of dipolar interaction in addition $\text{Tr}\{D_{\alpha\beta}\} = 0$ and the familiar truncated or secular dipolar Hamiltonian results ($\omega_{0I} = \omega_{0S}$):

$$\mathcal{H}'_D = \frac{\gamma_I \gamma_S \hbar}{r^3}\,\frac{1}{2}(1 - 3\cos^2 \vartheta_{IS})(3 I_z S_z - \mathbf{I}\cdot\mathbf{S}) \tag{2.34}$$

where ϑ_{IS} is the angle between the magnetic field B_0 and r is the vector connecting spin I and S.

The secular part of the shielding Hamiltonian is immediately obtained as

$$\mathcal{H}'_s = S_{zz} B_0 I_z. \tag{2.35}$$

It contains only components of the symmetric part of the tensor i.e. the eigenvalues of this symmetric part of the shielding tensor exist and can be determined by different orientations of the magnetic field B_0 with respect to the principal axes of the tensor. Antisymmetric and other non-secular constituents of the shielding tensor contribute only in second order (see Appendix H).

Suppose the interaction Hamiltonian is represented in the spherical tensor notation as in Eq. (2.21), the secular part of the interactions is given by [3, 4]

$$\mathcal{H}'_{\text{int}} = A_{00} T_{00} + A_{10} T_{10} + A_{20} T_{20}$$

since

$$[I_z, T_{kq}] = q T_{kq}$$

which vanishes besides trivial cases only for $q = 0$.

Inserting the Cartesian counterparts in A_{k0} and T_{k0} according to Tables 2.2 and 2.3 leads to the same expressions for the secular spin interactions as before.

As an example of non-secular contributions let us consider the chemical shift tensor $\tilde{\sigma}$.

The magnetic shielding Hamiltonian according to Eq. (2.11) may be also visualized as the interaction of the nuclear spin I with the magnetic field B_S as (see Appendix A)

$$\mathbf{B}_S = -\tilde{\sigma} \mathbf{B}_0.$$

The vector $\mathbf{B}_S$ is no longer parallel to the vector $\mathbf{B}_0$, because of the tensor $\tilde{\sigma}$ (see Fig. 2.2). The nucleus under consideration now "sees" an effective field

$$\mathbf{B}_{\text{eff}} = \mathbf{B}_0 + \mathbf{B}_S \tag{2.36}$$

where

$$\mathbf{B}_0 = (0, 0, B_0)$$

and

$$\mathbf{B}_S = -(\sigma_{xz}, \sigma_{yz}, \sigma_{zz}) B_0.$$

If we simply ignore the deviation of the effective field $\mathbf{B}_{\text{eff}}$ at the nuclear site from the magnetic field direction $\mathbf{B}_0$ we could use the projection of $\mathbf{B}_S$ onto $\mathbf{B}_0$, which results in

$$B_{\text{eff}} = (1 - \sigma_{zz}) B_0.$$

This was called the secular contribution before. If we want to calculate the size of B_{eff}, however exactly, we have to write

$$B_{\text{eff}}^2 = B_0^2 + B_S^2 + 2 \mathbf{B}_0 \cdot \mathbf{B}_S$$

with

$$B_{\text{eff}} = B_0 (1 + \sigma_{xz}^2 + \sigma_{yz}^2 + \sigma_{zz}^2 - 2\sigma_{zz})^{1/2}$$

or

$$B_{\text{eff}} = B_0 (1 - \sigma_{zz}) \left[1 + \frac{\sigma_{xz}^2 + \sigma_{yz}^2}{(1 - \sigma_{zz})^2} \right]^{1/2}. \tag{2.37}$$

Because the σ values are very small numbers we expand the square root in Eq. (2.37) to obtain

$$B_{\text{eff}} = B_0(1-\sigma_{zz})\left[1+\frac{1}{2}\frac{\sigma_{xz}^2+\sigma_{yz}^2}{(1-\sigma_{zz})^2}\right]. \tag{2.38}$$

Using Eq. (2.36) we may write

$$B_S = -B_0\sigma_{zz}\left(1-\frac{\sigma_{xz}^2+\sigma_{yz}^2}{2\sigma_{zz}(1-\sigma_{zz})}\right) = B_{\text{eff}} - B_0. \tag{2.39}$$

A similar expression is obtained in Appendix H [Eq. (H.14)], when the so called non-secular terms are taken into account. Since only the symmetric part of the shielding tensor contributes to σ_{zz} and the σ values are usually very small numbers, the contribution of antisymmetric and other non-secular parts can be neglected.

The symmetric part of the shielding tensor can always be diagonalized and σ_{zz} changes with the orientation of the static magnetic field B_0 with respect to the principal axis system (1, 2, 3) of this tensor.

Let us now turn to the pictorial representation of the shielding tensor. In Fig. 2.2 we have plotted the variation of $S = B_S/B_0$, which we refer to as the "shielding vector" with respect to the magnetic field. The shielding vector $S = (S_1, S_2, S_3)$ may be represented by

$$\frac{S_1^2}{\sigma_{11}^2} + \frac{S_2^2}{\sigma_{22}^2} + \frac{S_3^2}{\sigma_{33}^2} = 1. \tag{2.40}$$

This is the representation of an ellipsoid for the vector S which gives a direct measure of the σ_{ii} $(i=1,2,3)$ values, when the magnetic field B_0 points along

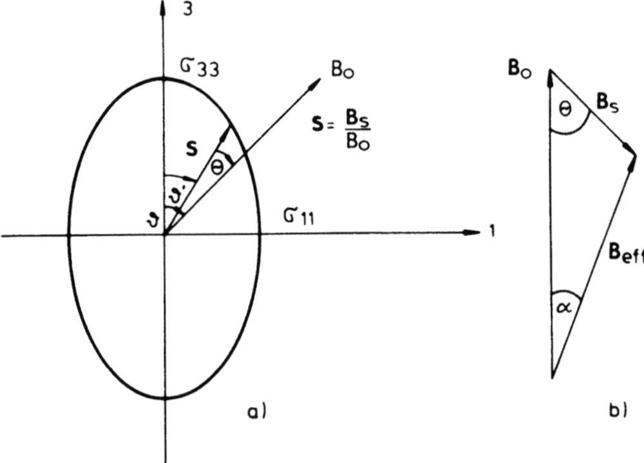

Fig. 2.2. a Graphical representation of the "shielding vector" $S=(S_1,S_2,S_3)$ according to Eq. (2.40). b Vectorial representation of the shielding field B_S due to the nuclear spin-electron coupling. Notice, that S and B_S are not parallel to B_0 due to the tensorial character of the interaction (see text)

the i-axis. In between of course, the $\mathbf{S}$ vector is no longer parallel to $\mathbf{B}_0$ as outlined above. We prefer this representation with respect to the so called "quartic" representation, where $1/\sqrt{\sigma_{ii}}$ appears on the i-axis.

There are still different possible choices for obtaining a pictorial representation of the shielding tensor, as far as the reference is concerned to which the measured σ_{ii} values are related. Since no absolute shielding data are available we are going to define a suitable reference to the pictorial representation as follows:

(i) we always use the convention: $\sigma_{11} \leqq \sigma_{22} \leqq \sigma_{33}$
(ii) choose reference (zero point of σ-scale) such that $\sigma_{11} = |\sigma_{33} - \sigma_{11}|/n$

where a suitable $n = 1, 2, 3, 4 \ldots$ may be chosen for aesthetical reasons. We find $n = 3$ most appealing. Condition (ii) has to be applied to each shielding tensor in a molecule individually. This assures that the anisotropy of the shielding is clearly visible. Most shielded, intermediate and least shielded axes are clearly distinguishable. Furthermore, the size of the anisotropy is directly related to the size of the ellipsiod. Examples of such an "aestheticized" tensor representation are given in Figs. 7.5 and 7.7. A circle of constant radius may be added to each nucleus, if this is found to be convenient for representing e.g. more or less isotropic tensors.

2.3 Transformation Properties of Spin Interactions in Real Space

Since we have expressed the spin interactions by second rank tensors, we can resort to the transformation properties of those tensors in order to express the interactions in different reference frames (see Appendix B). Suppose $\tilde{\mathbf{A}}$ is a second rank tensor expressed in the coordinate system (x, y, z) and $\tilde{\mathbf{A}}'$ is the same tensor expressed in the system (x', y', z') which is related to (x, y, z) by a unitary transformation $R(\alpha, \beta, \gamma)$ with the Euler angles (α, β, γ) i.e.

$$\mathbf{r}' = R\mathbf{r} \tag{2.41}$$

where $\mathbf{r}$ is a vector in (x, y, z) and $\mathbf{r}'$ the same vector in (x', y', z'). Now $\tilde{\mathbf{A}}'$ and $\tilde{\mathbf{A}}$ are related by the unitary transformation

$$\tilde{\mathbf{A}}' = R\tilde{\mathbf{A}}R^{-1}. \tag{2.42}$$

$\tilde{\mathbf{A}}$ represents here any of the spatial coupling tensors of the spin interactions as discussed in Sect. 2.1.

Instead of expressing the matrix R by the Euler angles, R may be represented also by the direction cosines [3, 4].

$$R = \{r_{ij}\} = \{\cos(\mathbf{r}', \mathbf{r})\} \tag{2.43}$$

with $\{r_{ij}\} = \{r_{ji}\}^{-1}$ because of the unitary nature of R.

The matrix elements r_{ij} are given in Appendix B Eq. (B.2).

Equation (2.43) can be readily expressed in Cartesian components as

$$A_{ij} = \sum_{k,l} A_{kl} r_{ik} r_{jl}. \tag{2.44a}$$

The tensor $\tilde{A}$ is expressed in the principal axis system as

$$A_{kl} = A_{kk}\delta_{kl} \quad \text{with the Kronecker} \quad \delta_{kl} = \begin{cases} 1 & \text{if } k=l \\ 0 & \text{if } k \neq l \end{cases}.$$

The spectral position of a line (i.e. the resonance frequency) is proportional to A_{zz} the z-component of the tensor $\tilde{A}$ in the laboratory frame (x, y, z). This component is readily calculated according to Eq. (2.44a) with $i = j = 3$

$$A'_{zz} = \sum_{k,l} A_{k3} r_{3k} r_{3l} \tag{2.44b}$$

where the A_{kl} are the Cartesian tensor elements in the goniometer frame (not necessarily the principal axis or even crystal frame). If only the z-component of the tensor is wanted, which is the case for the secular part of the shielding interaction for example, we obtain from Eq. (2.44b)

$$A'_{zz} = \sum_k A_{kk} r^2_{3k} = A_{11} r^2_{31} + A_{22} r^2_{32} + A_{33} r^2_{33} \tag{2.45}$$

or in terms of Euler angles by using the r_{3k} according to Appendix B as

$$A'_{zz} = A_{11} \cos^2 \alpha \sin^2 \beta + A_{22} \sin^2 \alpha \sin^2 \beta + A_{33} \cos^2 \beta \tag{2.46}$$

which is the well-known formula for the variation of the secular part of a second rank tensor with the Euler angles (α, β). Equation (2.46) may be applied to the shift tensor $\tilde{S}$ as well as to the field gradient tensor $\tilde{V}$, for example. A complete determination of the tensor elements A_{11}, A_{22}, and A_{33}, as well as the determination of the principal axis system with respect to some reference frame, can be achieved by measuring the secular part A'_{zz} of the spin interaction at different orientations of a single crystal.

In the following we always assume that dipolar couplings have been eliminated by the techniques of manipulation and dilution, mentioned in the introduction (Chap. 1).

Suppose a single crystal is mounted on a goniometer in a well defined fashion, i.e. the crystal axes have a known relation to the goniometer axes (x_g, y_g, z_g). The goniometer rotates the crystal by an angle φ about its rotation axis (z_g), which usually has an angle $\vartheta = \pi/2$ with respect to the magnetic field (see Fig. 2.3). How does the secular component of the spin interaction change with the angle φ?

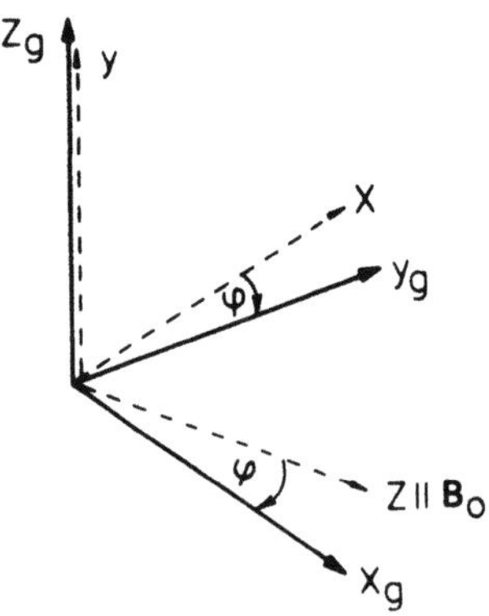

Fig. 2.3. Relation between the goniometer frame (x_g, y_g, z_g) and the laboratory frame (x, y, z). The angle φ represents the "rotation angle" by which the goniometer is rotated away from the starting position: $x \parallel y_g$; $y \parallel z_g$; $z \parallel x_g$

The transformation from the goniometer frame (x_g, y_g, z_g) to the laboratory frame (x, y, z) may be described by $R(\varphi, \vartheta, \psi)$:

$$
\begin{array}{ccc}
\mathbf{A} & & \mathbf{A'} \\
(x_g, y_g, z_g) & \xrightarrow{\;R(\varphi, \vartheta, \psi)\;} & (x, y, z) \\
z_g \parallel \text{rot. axis} & & z \parallel B_0
\end{array}
$$

where

$$R(\varphi, \vartheta, \psi) = R_{zg}(\psi)\, R_{yg}(\vartheta)\, R_{zg}(\varphi). \tag{2.47}$$

It immediately follows from Eq. (2.43) that the rotation in Fig. 2.3 can be represented by

$$R(\varphi, 90°, 90°) = \begin{pmatrix} -\sin\varphi & \cos\varphi & 0 \\ 0 & 0 & 1 \\ \cos\varphi & \sin\varphi & 0 \end{pmatrix}. \tag{2.48}$$

With

$$r_{31} = \cos\varphi; \quad r_{32} = \sin\varphi; \quad r_{33} = 0,$$

we obtain from Eq. (2.44b)

$$A'_{zz} = \tfrac{1}{2}(A_{11} + A_{22}) + \tfrac{1}{2}(A_{11} - A_{22})\cos 2\varphi + \tfrac{1}{2}(A_{12} + A_{21})\sin 2\varphi. \tag{2.49}$$

This means that the observed resonance line shift which is proportional to A'_{zz} varies harmonically with 2φ. The three distinct tensor elements (A_{11}, A_{22}, A_{12}) of the symmetric part $(A_{12} = A_{21})$ of the spin interaction tensor with respect to the goniometer axis system can be determined from the corresponding rotation pattern. Examples of this will be shown in later sections.

In general, the goniometer may have an arbitrary orientation with respect to the laboratory frame, describable by the three Euler angles $(\varphi', \vartheta, \psi)$.

If φ is still the angle by which B_0 is rotated with respect to the goniometer starting position, the corresponding transformation has to be performed in two steps:

$$R_{\text{total}} = R'(\varphi', \vartheta, \psi)\, R(\varphi).$$

The second rank Cartesian tensor representing the spin interactions in the lab-frame (x, y, z) can then be expressed by

$$\tilde{\mathbf{A}}' = R' R \tilde{\mathbf{A}} R^{-1} R'^{-1}$$

or by using the unitary nature of R and R' by

$$A'_{ij} = \sum_{k,l,m,n} A_{lm}\, r'_{ik}\, r'_{jn}\, r_{kl}\, r_{nm}$$

where r'_{ik} and r_{kl} are the corresponding matrix elements of R' and R respectively (see Appendix B).

As noted before, in most cases only the z-component $(i = j = 3)$ of the tensor $\tilde{\mathbf{A}}'$ is measured in an experiment leading to

$$A'_{zz} = C_0 + C_1 \cos\varphi + S_1 \sin\varphi + C_2 \cos 2\varphi + S_2 \sin 2\varphi \tag{2.50}$$

where it can be easily realized that

$$C_0 = \tfrac{1}{2}(A_{11} + A_{22})(r'^2_{31} + r'^2_{32}) + A_{33}\,r'^2_{33} \tag{2.50a}$$

$$C_1 = (A_{13} + A_{31})\,r'_{31}r'_{33} + (A_{23} + A_{32})\,r'_{32}r'_{33} \tag{2.50b}$$

$$S_1 = (A_{23} + A_{32})\,r'_{31}r'_{33} - (A_{13} + A_{31})\,r'_{32}r'_{33} \tag{2.50c}$$

$$C_2 = (A_{12} + A_{21})\,r'_{31}r'_{32} + \tfrac{1}{2}(A_{11} - A_{22})(r'^2_{31} - r'^2_{32}) \tag{2.50d}$$

$$S_2 = \tfrac{1}{2}(A_{12} + A_{21})(r'^2_{31} - r'^2_{32}) + (A_{22} - A_{11})\,r'_{31}\cdot r'_{32}. \tag{2.50e}$$

Notice that in this case $\tilde{\mathbf{A}}'_z$ is *not* invariant under a π rotation. If the goniometer axis does not coincide with a principal axis of the tensor, five of the six possible tensor elements can be determined from one orientation plot.

If the goniometer axis is however orthogonal to the magnetic field (i.e. $\vartheta = 90°$ or $r'_{33} = 0$) C_1 and S_1 vanish, which results in

$$A'_{zz} = C_0 + C_2 \cos 2\varphi + S_2 \sin 2\varphi \qquad \text{if } z_g \perp B_0. \tag{2.51}$$

The special case discussed earlier in this section (Eq. (2.49)) can be recovered with $r'_{31} = 1$ and $r'_{32} = r'_{33} = 0$.

It should be noted that the goniometer axis is indeed usually orthogonal to the magnetic field B_0 and Eq. (2.51) applies. However, Eq. (2.50) may be very useful for estimating the error if the orthogonality condition is not accurately fulfilled. On the other hand, it might be interesting to use a tilted goniometer axis, since in this case in principal already five of the six elements of the symmetric tensor can be determined with one orientation plot as mentioned before. If in addition the trace of the tensor is known, the full tensor can be determined in one orientation plot. This procedure has been applied by Haeberlen and co-workers in one case of a proton shielding tensor [6] (see Sect. 7.3).

As mentioned before, the machinary of unitrary transformations of spherical tensors is highly developed [3, 4]. Making use of the Wigner rotation matrices (see Appendix B)

$$D^{(j)}_{mn}(\alpha\beta\gamma) = e^{-im\alpha}\,d^{(j)}_{mn}(\beta)\,e^{-in\gamma}$$

the unitary transformation (rotation) of a spherical tensor is expressed as [3, 4]

$$A'_{kq} \equiv R(\alpha\beta\gamma)A_{kq}R^{-1}(\alpha\beta\gamma) \equiv \sum_{p=-k}^{+k} A_{kp}D^{(k)}_{pq}(\alpha\beta\gamma) \tag{2.52}$$

where A'_{kq} is the spherical tensor expressed in the frame (x', y', z') obtained by the transformation $R(\alpha\beta\gamma)$ from the frame (x, y, z) in which A_{kp} is defined. The transformation procedure can be pictured as follows

$$\begin{array}{ccc} A_{kp} & \xrightarrow{\;R(\alpha\beta\gamma)\;} & A'_{kq} \\ (x, y, z) & & (x', y', z'). \end{array}$$

As an example let us treat the preceding case of the tilted goniometer which is rotated by an angle φ and whoose goniometer axis is tilted by an angle ϑ with respect to the magnetic field B_0. The following transformation procedure applies

$$\underset{\substack{\text{principal}\\\text{axis system}}}{\overset{\tilde{\mathbf{A}}''}{(1,2,3)}} \xrightarrow{R_p(\alpha',\beta',\gamma')} \underset{\substack{\text{crystal}\\\text{frame}}}{\overset{\tilde{\mathbf{A}}^*}{(a,b,c)}} \xrightarrow{R_i(\alpha'',\beta'',\gamma'')} \underset{\substack{\text{goniometer}\\\text{frame}}}{\overset{\tilde{\mathbf{A}}}{(x_g,y_g,z_g)}} \xrightarrow{R_g(\varphi,\vartheta,\psi)} \underset{\text{Lab. frame}}{\overset{\tilde{\mathbf{A}}'}{(x,y,z)}}$$

where R_p is the transformation from the principal axis system to the crystal frame. R_i ($i = 1, 2, 3 \ldots$) are the three or more different transformations from the crystal frame to the goniometer frame, depending on how the crystal is mounted on the goniometer axis. Finally R_g is the transformation from the fixed goniometer to the laboratory frame where φ is the angle of rotation of the goniometer and ϑ is the angle between the goniometer axis and B_0.

The angle φ may be made timedependent ($\varphi = \omega_r t$) as in sample spinning experiments which will be discussed in Sect. 2.6. In order to keep the notation as compact as possible we combine the operations R_p and R_i as

$$R(\alpha\beta\gamma) \equiv R_i(\alpha''\beta''\gamma'') R_p(\alpha'\beta'\gamma')$$

leading to the total transformation

$$R_{\text{total}} = R_g(\varphi, \vartheta\psi) R(\alpha\beta\gamma).$$

The observed spectral frequency is determined in first order by the spherical tensor component A'_{20} as shown before. Following Eq. (2.52) we therefore write

$$A'_{20} = \sum_{q,p=-2}^{+2} A''_{2p} D^{(2)}_{pq}(\alpha\beta\gamma) D^{(2)}_{q0}(\varphi\vartheta\psi)$$

or

$$A'_{20} = \sum_{q,p=-2}^{+2} A''_{2p} e^{-ip\alpha} d^{(2)}_{pq}(\beta) e^{-iq\gamma} e^{-iq\varphi} d^{(2)}_{q0}(\theta) \tag{2.53}$$

where the $d^{(k)}_{pq}(\beta)$ are given in Table B.1. Note, that in the principal axis system $A_{2\pm1} = 0$. Equation (2.53) can immediately be cast into the same form as Eq. (2.50)

$$A'_{20} = C_0 + C_1 \cos\varphi + S_1 \sin\varphi + C_2 \cos 2\varphi + S_2 \sin 2\varphi \tag{2.54}$$

where

$$C_q; S_q = (-1)^q d^{(2)}_{q0}(\theta) \sum_{p=-2}^{+2} \tilde{A}''_{2p} e^{-ip\alpha} \Delta_{pq} \tag{2.55a}$$

with

$$\Delta_{p0} = d^{(2)}_{p0}(\beta) \quad \text{for } C_0 \tag{2.55b}$$

$$\Delta_{pq} = d^{(2)}_{p-q}(\beta) e^{iq\gamma} + (-1)^q d^{(2)}_{pq}(\beta) e^{-iq\gamma} \quad \text{for } C_1, C_2 \tag{2.55c}$$

$$\Delta_{pq} = i(d^{(2)}_{p-q}(\beta) e^{iq\gamma} - (-1)^q d^{(2)}_{pq}(\beta) e^{-iq\gamma}) \quad \text{for } S_1, S_2. \tag{2.55d}$$

As a relevant example let us consider the chemical shift tensor interaction. Using

$$A''_{20} = \sqrt{\tfrac{3}{2}}(\sigma_{33} - \sigma_i); \quad A''_{2\pm1} = 0; \quad A''_{2\pm2} = \tfrac{1}{2}(\sigma_{11} - \sigma_{22})$$

and the $d^{(2)}_{pq}(\beta)$ from Table B.1 one readily obtains

$$C_0 = \tfrac{1}{2}(3\cos^2\vartheta - 1)[\tfrac{1}{2}(3\cos^2\beta - 1)\sqrt{\tfrac{3}{2}}(\sigma_{33} - \sigma_i)$$
$$+ \tfrac{1}{2}\sqrt{\tfrac{3}{2}}\sin^2\beta\cos 2\alpha(\sigma_{11} - \sigma_{22})] \tag{2.56a}$$

$$C_1 = \sqrt{\tfrac{3}{2}}\sin\vartheta\cos\vartheta[-3\sin\beta\cos\beta\cos\gamma(\sigma_{33} - \sigma_i)$$
$$+ \sin\beta(\cos\beta\cos\gamma\cos 2\alpha - \sin\gamma\sin 2\alpha)(\sigma_{11} - \sigma_{22})] \tag{2.56b}$$

$$S_1 = \sqrt{\tfrac{3}{2}}\sin\vartheta\cos\vartheta[3\sin\beta\cos\beta\sin\gamma(\sigma_{33} - \sigma_i)$$
$$- \sin\beta(\cos\beta\sin\gamma\cos 2\alpha + \cos\gamma\sin 2\alpha)(\sigma_{11} - \sigma_{22})] \tag{2.56c}$$

$$C_2 = \tfrac{1}{4}\sqrt{\tfrac{3}{2}}\sin^2\vartheta[3\sin^2\beta\cos 2\gamma(\sigma_{33} - \sigma_i)$$
$$+ ((1 + \cos^2\beta)\cos 2\gamma\cos 2\alpha - 2\cos\beta\sin 2\gamma\sin 2\alpha)(\sigma_{11} - \sigma_{22})] \tag{2.56d}$$

$$S_2 = -\tfrac{1}{4}\sqrt{\tfrac{3}{4}}\sin^2\vartheta[3\sin^2\beta\sin 2\gamma(\sigma_{33} - \sigma_i)$$
$$+ ((1 + \cos^2\beta)\sin 2\gamma\cos 2\alpha + 2\cos\beta\cos 2\gamma\sin 2\alpha)(\sigma_{11} - \sigma_{22})]. \tag{2.56e}$$

These expressions will turn out to be useful when we discuss magic angle spinning (MAS) in Sect. 2.6. In fact Lippmaa et al. [7], Maricq and Waugh [8] as well as Herzfeld and Berger [9] have used similar expressions, when considering spinning sidebands.

In the following, however, we shall assume that the single crystal is rotated about an axis perpendicular to the magnetic field. In this case Eq. (2.51) applies and C_0, C_2 and S_2 are given by Eq. (2.56) with $\vartheta = \pi/2$. Three of the six tensor elements can be determined from one orientation plot. It would be sufficient to obtain one more orientationplot about an axis which gives non-degenerate results in order to determine the full tensor. However, in practice it is more convenient to rotate the crystal about three different orthogonal axes. By using different sets of transformations $R(\alpha''\beta''\gamma'')$ from the crystal frame (a, b, c) to the goniometer frame (x_g, y_g, z_g) the Cartesian tensor A_{ij} in the crystal frame can be determined from the orientation plots of the measured line position which is proportional to [10]

$$A'_{zz} = A + B\cos 2\varphi + C\sin 2\varphi$$

where

$$A = \tfrac{1}{2}\sum_{k,l} A^*_{kl}(r_{1k}r_{1l} + r_{2k}r_{2l}) \tag{2.57a}$$

$$B = \tfrac{1}{2}\sum_{k,l} A^*_{kl}(r_{1k}r_{1l} - r_{2k}r_{2l}) \tag{2.57b}$$

$$C = \tfrac{1}{2}\sum_{k,l} A^*_{kl}(r_{1k}r_{2l} + r_{2k}r_{1l}). \tag{2.57c}$$

The r_{ij} are the known matrix elements of the R_i rotation matrix, depending on how the crystal is mounted on the goniometer. Since the parameters A, B and C can be determined experimentally, a set of linear equations results which can be brought into the form

$$\mathbf{Y} = \tilde{\mathbf{a}} \cdot \mathbf{X} \tag{2.58}$$

where

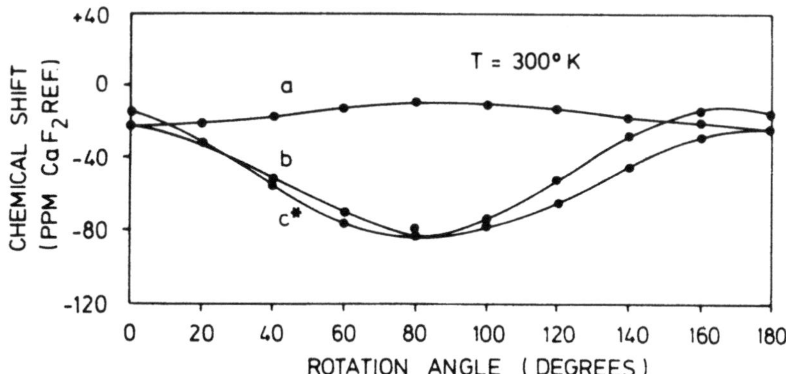

Fig. 2.4. Chemical shift of ^{19}F in solid CF$_3$COOAg at $T=300\,°$K versus the goniometer rotation angle, as obtained in a multiple-pulse experiment (WHH-4). The labels a, b, c* refer to the mutually orthogonal crystallografic axes (see Ref. [10])

$$\mathbf{Y} = (A_1, B_1, C_1, A_2, B_2, C_2 \ldots) \tag{2.59}$$

is a vector, defined by the parameters A, B, C from n different orientation plots and

$$\mathbf{X} = (A_{11}^*, A_{12}^*, A_{13}^*, A_{22}^*, A_{23}^*, A_{33}^*) \tag{2.60}$$

is a vector spanned by the six independent tensor elements A_{ij}^* in the crystal frame and $\tilde{\mathbf{a}}$ is a $n \times 6$ matrix which can be obtained from Eq. (2.57).

The vector $\mathbf{X}$ is determined by solving the set of linear equations or by a matrix inversion procedure

$$\mathbf{X} = \tilde{\mathbf{a}}^{-1}\mathbf{Y}. \tag{2.61}$$

These relations are used in high resolution NMR in solids, to obtain the chemical shift tensor in single crystals. As we have noted before only the symmetric part of the chemical shift tensor is determined in this way. A representative example of an orientation plot of the line shift in a single crystal in the case of ^{19}F in CF$_3$COOAg obtained in a four-pulse experiment is shown in Fig. 2.4. A single line is observed due to rapid motion of the CF$_3$-group at room temperature. Rotation of the single crystal was performed around three different axes closely related to the unit cell axes. The principal components and the principal axes of the shielding tensor were evaluated by Griffin et al. [10] following the procedure outlined above (see also Sect. 7.4).

2.4 Powder Spectrum Line Shape

Very often single crystals are not available or exceedingly difficult to grow. In this case a powder sample has to be investigated and the question arises which amount of information about the shift tensor can be gained.

Following Eq. (2.46), the observed frequency ω_{zz} of a spectral line can be expressed by the Euler angles (α, β) which relate the principal axes of the tensor $(\omega_{11}, \omega_{22}, \omega_{33})$ to the laboratory frame as

$$\omega_{zz} = \omega_{11} \cos^2\alpha \sin^2\beta + \omega_{22} \sin^2\alpha \sin^2\beta + \omega_{33} \cos^2\beta. \tag{2.62a}$$

For an axially symmetric tensor we write

$$\omega_{zz} = (\omega_{\parallel} - \omega_{\perp}) \cos^2\beta + \omega_{\perp} \tag{2.62b}$$

where β is the angle between the 3-axis and the direction of the magnetic field B_0 (z-axis) and where

$$\omega_{\parallel} = \omega_{33}; \quad \omega_{\perp} = \omega_{11} = \omega_{22}$$

leading to the total anisotropy

$$\Delta\omega = \omega_{\parallel} - \omega_{\perp}.$$

If $I(\omega)$ is the intensity of the NMR signal at the frequency ω and $p(\Omega) d\Omega$ is the probability of finding the tensor orientation in the range between the solid angle Ω and $\Omega + d\Omega$, we may write

$$p(\Omega) d\Omega = I(\omega) d\omega.$$

In a powder all angles Ω are equally probable i.e. $p(\Omega) = 1/4\pi$ leading to

$$I(\omega) = \frac{1}{4\pi} \left| \frac{d\Omega}{d\omega} \right|. \tag{2.63}$$

In the case of axial symmetry $d\Omega = \sin\beta \, d\beta$ we arrive at the line shape function, using Eqs. (2.62) and (2.63) as

$$I(\omega) = \tfrac{1}{2} [(\omega_{\parallel} - \omega_{\perp})(\omega - \omega_{\perp})]^{-1/2}. \tag{2.64}$$

The general case with $\omega_{33} > \omega_{22} > \omega_{11}$ (convention) is slightly more complicated and has been calculated by Bloembergen and Rowland [11] as

$$I(\omega) = \pi^{-1} (\omega - \omega_{11})^{-1/2} (\omega_{33} - \omega_{22})^{-1/2} \cdot K(m) \tag{2.65a}$$

with

$$m = (\omega_{22} - \omega_{11})(\omega_{33} - \omega)/(\omega_{33} - \omega_{22})(\omega - \omega_{11})$$

for

$$\omega_{33} \geqq \omega > \omega_{22}$$

and

$$I(\omega) = \pi^{-1} (\omega_{33} - \omega)^{-1/2} (\omega_{22} - \omega_{11})^{-1/2} K(m) \tag{2.65b}$$

with

$$m = (\omega - \omega_{11})(\omega_{33} - \omega_{22})/(\omega_{33} - \omega)(\omega_{22} - \omega_{11})$$

for

$$\omega_{22} > \omega \geqq \omega_{11}$$

$$I(\omega) = 0 \quad \text{in case } \omega > \omega_{33} \text{ and } \omega < \omega_{11}.$$

$K(m)$ is the complete elliptic integral of the first kind

$$K(m) = \int_0^{\pi/2} d\varphi \, [1 - m^2 \sin^2 \varphi]^{-1/2}. \tag{2.66}$$

The different line shapes according to Eqs. (2.64, 2.65) are plotted in Fig. 2.5 for two different cases. Usually a residual line broadening is taken into account by convolution of $I(\omega)$ with a broadening function (Lorentzian or Gaussian).

It is evident from Fig. 2.5 and the corresponding formulas for the powder spectrum how the principal shielding components ($\sigma_{11}, \sigma_{22}, \sigma_{33}$) can be obtained directly from the spectrum. However no information is obtained about the orientation of the principle axes of the tensor. One has to resort to theoretical arguments, symmetry considerations, consistency checks with related compounds or other molecular properties in order to assign the principal

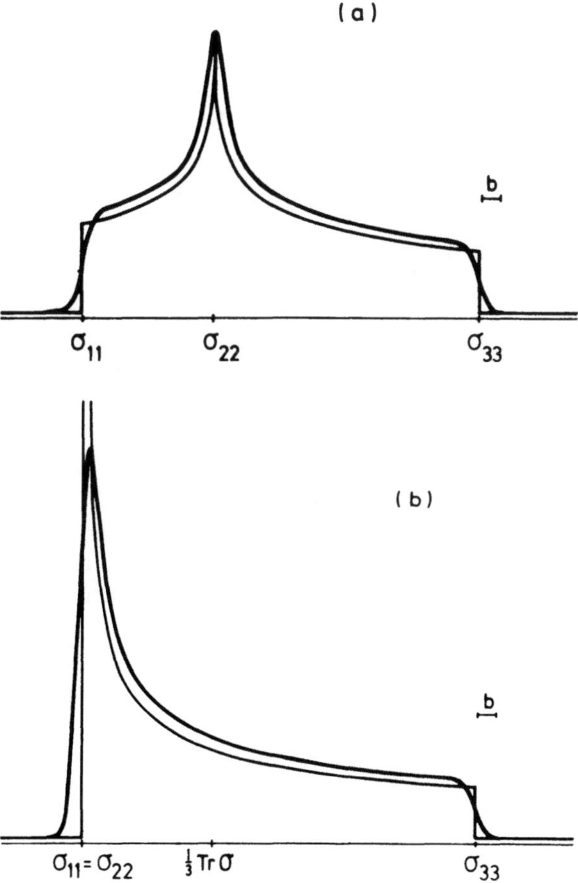

Fig. 2.5a, b. Schematic representation of theoretical powder line shapes for second rank tensor interactions (i.e., chemical shift tensor, Knightshift tensor, g-tensor etc.) according to Eqs. (2.64, 2.65). The theoretical curves are convoluted with Lorentzian broadening functions, whose width b is indicated. **a** Arbitrary second rank tensor, **b** axially symmetric second rank tensor

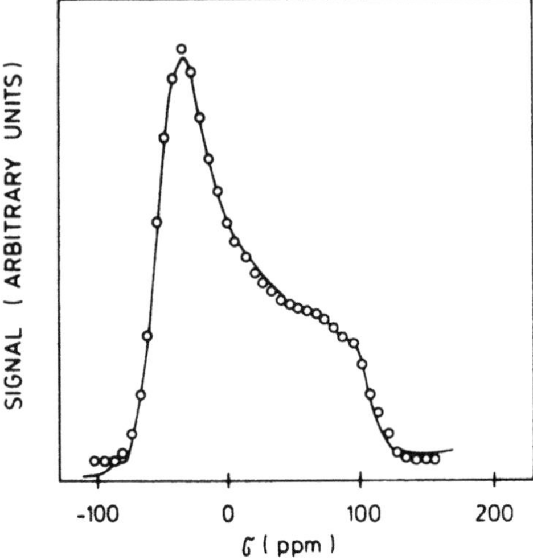

Fig. 2.6. ^{19}F spectrum of C_6F_6 obtained in a 4-pulse experiment (WHH-4) at 40 K [12]. The solid line is the experimental trace while the open circles are calculated points, obtained by convolution of the theoretical powder line shape Eq. (2.64) with a Gaussian broadening function [12]

axes tentatively. Such other properties can be e.g. molecular rotations which are known to proceed about certain symmetry axes. One case where such knowledge has been utilized to elucidate the orientation of the principal axes has been reported in Ref. [12]. Fig. 2.6 shows the powder spectrum of ^{19}F in C_6F_6 which displays an axially symmetric shielding tensor. Above $T = 100\,°K$ the molecule is known to reorient rapidly about its 6-fold axis and the shielding tensor is expected to be axially symmetric about this axis. At a temperature of $T = 50\,°K$ however this motion is slowed down to a rate very much below the shielding anisotropy and the "rigid" tensor should grow out of the spectrum. The big surprise at the time, however, was the fact that the spectrum was about the same at this low temperature. This leaves as the only possible conclusion, that the shielding tensor is in fact axially symmetric with the unique axis being the 6-fold axis of the molecule [12].

The same approach has been taken at several other occasions and is very helpful in assigning tentatively the principal axes to the molecular frame.

In solids or solid polymers, however, very often partial ordering occurs. In this case the spectrum is no longer represented by Eqs. (2.64) and (2.65). Instead the distribution function $p(\Omega)\,d\Omega$ of the orientational order modifies the spectra considerably. The lineshape $I(\omega)$ is given in this case by

$$I(\omega) = \int_0^{2\pi} \int_0^{\pi} \delta(\omega - \omega_{zz}) p(\Omega)\, d\alpha \sin\beta\, d\beta \tag{2.67}$$

where $\omega_{zz} = \omega_{zz}(\alpha\beta)$ as defined in Eq. (2.62a). NMR and ESR lineshapes resulting from such an orientational order have been treated by Sillescu, Spiess and co-workers [13]. Before discussing the general case, let us consider a fairly simple, but important example.

Suppose we have a sample where high order occurs in one direction, but complete disorder perpendicular to this direction. Such a situation occurs in a

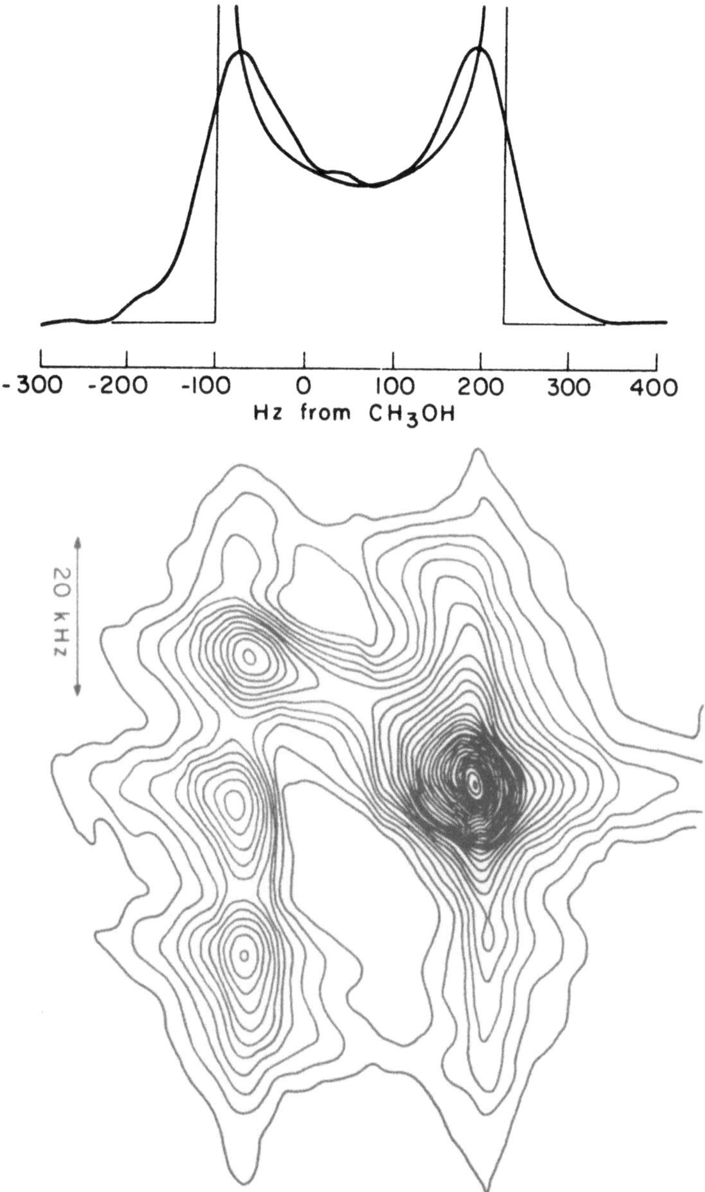

Fig. 2.7. ^{13}C spectrum of oriented polyethylene $(-CH_2-)_n$ where the magnetic field B_0 is perpendicular to the chain axis according to Opella and Waugh [15]. The theoretical lineshape due to a sinusoidal distribution of frequencies (see text) has been inserted. Bottom: 2D-spectrum of $^{13}C-^{1}H$ dipolar interaction (vertical) versus ^{13}C chemical shift (horizontal)

stretched or extruded polymer [13–15]. Opella and Waugh [15] and Spiess and co-workers [13b] have analyzed this situation. The typical spectrum which is obtained, when the unique direction is perpendicular to B_0 is shown in Fig. 2.7. The orientational disorder, corresponding to a "two-dimensional powder" when the magnetic field B_0 is perpendicular to the draw direction of polyethylene is clearly evident from Fig. 2.7. The situation can be described by a sinusoidal frequency dependence of the form

$$\omega = \omega_0 + \Delta\omega \cos(2\varphi + \Phi_0) \tag{2.68}$$

where φ is the orientational order parameter perpendicular to the unique direction and Φ_0 is some arbitrary initial value. It is interesting to note, that a similar situation arises in incommensurate systems as was pointed out by Blinc and co-workers [16]. The lineshape is readily obtained from

$$I(\omega)\,d\omega = p(\varphi)\,d\varphi$$

$$I(\omega) = p(\varphi)\left|\frac{d\omega}{d\varphi}\right|^{-1} \tag{2.69}$$

where $p(\varphi) = 1/\pi$ in our case of orientational disorder. Combining Eqs. (2.68) and (2.69) leads immediately to [13b, 15, 16]

$$I(\omega) = \frac{1}{2\pi}\,\frac{1}{\Delta\omega[1-x^2]^{1/2}} \quad \text{for} \ -1 \leqq x \leqq 1 \tag{2.70}$$

and

$$I(\omega) = 0 \quad \text{for} \ x^2 > 1$$

with

$$\int_{-\infty}^{+\infty} d\omega\, I(\omega) = 1$$

where

$$x = (\omega - \omega_0)/\Delta\omega.$$

Note the characteristic edge singularities at $x = \pm 1$, i.e. at frequencies $\omega = \omega_0 \pm \Delta\omega$.

2.5 The NMR Spectrum. Lineshapes and Moments

The NMR spectrum is completely determined by the eigenvalues of the interaction Hamiltonian and the matrix elements of the nuclear magnetic dipole operator I_+. In order to prepare for the later use of "time development operators" and the description of transient phenomena in multiple pulse experiments let us introduce the density matrix ρ and superoperator notation in Liouville space.

The spin system is assumed to be in thermal equilibrium initially, described by the Boltzman spin density matrix [1]

$$\rho_B = \exp(-\beta\mathcal{H})/\mathrm{Tr}\{\exp(-\beta\mathcal{H})\} \tag{2.71}$$

where $\beta = \hbar/kT$ and $\mathcal{H}$ is the stationary Hamiltonian of the system. At time $t = 0$ an rf δ pulse at the Larmor frequency may be applied in the y direction of the rotating frame preparing an initial superradiant state as

$$\rho(0) = \mathbf{P}_y(\vartheta)\rho_B\mathbf{P}_y^{-1}(\vartheta)$$

where

$$\mathbf{P}_y(\vartheta) = \exp(-i\vartheta \cdot I_y).$$

Here ϑ is the rotation angle of the δ pulse. In the case of the Bloch decay ϑ equals usually $\pi/2$.

With the Zeeman interaction $\mathscr{H}_z$ being the dominant contribution in high field we obtain

$$\rho(0) = \exp(-\beta\mathscr{H}_x)/\mathrm{Tr}\{\exp(-\beta\mathscr{H}_z)\}$$

where

$$\mathscr{H}_x \underset{\mathrm{def}}{=} \mathbf{P}_y(\pi/2)\,\mathscr{H}_z\mathbf{P}_y^{-1}(\pi/2) = -\omega_0 I_x \quad \text{and} \quad \omega_0 = \gamma B_0.$$

The initial state $\rho(0)$ is often expanded and truncated, according to the high temperature approximation as [17]

$$\rho(0) = (1 + \beta\omega_0 I_x)/\mathrm{Tr}\{1\} \tag{2.72}$$

where

$$\mathrm{Tr}\{1\} = (2I+1)^N$$

is the total number of states if only a single spin species I is involved.

This approximation is by no means straightforward and has to be justified by the corresponding expansion of observables. In NMR, where macroscopic observables like energy, magnetization etc. which are averages over a large number of states are considered, this approximation is always legitimate down to $T = 1$ K.

Since the trace of all spin operators vanishes we shall use the initial density matrix in the following truncated form

$$\rho_B = \frac{\beta\omega_0}{(2I+1)^N}\,I_z. \tag{2.73}$$

After application of a $\pi/2$ pulse in the y-direction of the rotating frame

$$\rho(0) = \frac{\beta\omega_0}{(2I+1)^N}\,I_x. \tag{2.74}$$

The superradiant state, prepared by the initiating pulse, now decays due to the spin interaction $\mathscr{H}_i(t)$, according to the Liouville-v. Neumann Equation

$$\frac{d}{dt}\rho(t) = -i[\mathscr{H}_i(t), \rho(t)] \tag{2.75}$$

which has the formal solution

$$\rho(t) = \mathbf{L}(t)\rho(0)\mathbf{L}^+(t) \tag{2.76}$$

with

$$\mathbf{L}(t) = T\exp\left[-i\int_0^t dt'\,\mathscr{H}_i(t')\right]. \tag{2.77}$$

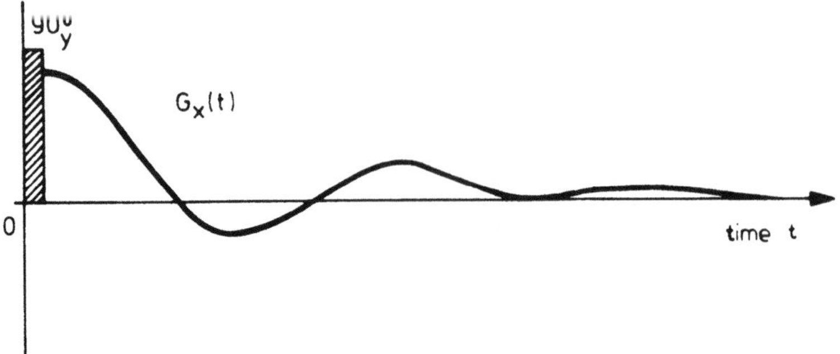

Fig. 2.8. Schematic representation of a Bloch-decay or free induction decay (FID) $G_x(t)$ of a spin system, excited by a 90° pulse in the y-direction of the rotating frame

T: Dyson time ordering operator and where $\mathbf{L}^+(t)$ is the adjoint of $\mathbf{L}(t)$ with $\mathbf{L}^+(t) = \mathbf{L}^{-1}(t)$ because $\mathscr{H}_i(t)$ is an Hermitian operator.

As a simple example let us consider a local field, which shifts the resonance of the nuclei away from the Larmor frequency by an amount ω_k. In this case $H_i = \omega_k I_z$ and

$$\mathbf{L}(t) = \exp[-it\,\omega_k I_z]$$

which describes a precession with frequency ω_k around the z direction.

The propagator $\mathbf{L}(t)$ which describes the time evolution of the system will be subject to detailed discussion in later sections. Actually this propagator $\mathbf{L}(t)$ is the important parameter, which governs the time evolution in multiple-pulse experiments as is discussed in the following chapter [18].

The transverse decay (Bloch decay) in the x direction of the rotating frame is now obtained as

$$G_x(t) = \mathrm{Tr}\{I_x\rho(t)\}/\mathrm{Tr}\{I_x\rho(0)\} \tag{2.78}$$

or by using Eq. (2.76)

$$G_x(t) = \mathrm{Tr}\{I_x\mathbf{L}(t)\rho(0)\mathbf{L}^{-1}(t)\}/\mathrm{Tr}\{I_x\rho(0)\} \tag{2.79}$$

which leads in the high temperature approximation to

$$G_x(t) = \mathrm{Tr}\{I_x\mathbf{L}(t)I_x\mathbf{L}^{-1}(t)\}/\mathrm{Tr}\{I_x^2\}. \tag{2.80}$$

Such a decay function is schematically drawn in Fig. 2.8. If dipolar interaction among many spins is involved, $G_x(t)$ cannot be calculated exactly and approximations as developed in many body theories have to be applied to Eq. (2.80) in order to describe $G_x(t)$ approximately.

As shown by van Vleck, a moment expansion of $G_x(t)$ can be performed exactly with [19]

$$G_x(t) = \sum_{n=0}^{\infty} \frac{(-it)^n}{n!} M_n. \tag{2.81}$$

Before proceeding with the discussion of approximation to $G_x(t)$ we introduce a shorthand notation for convenience, to be used later.

Introducing the Liouville operator [20] (see Appendix C)

$$\hat{\mathcal{H}} \underset{\text{def}}{=} [\mathcal{H}, \ldots] \tag{2.82}$$

which is a superoperator acting on other operators, which can be visualized as state vectors in Liouville space. We can rewrite the Liouville-v. Neumann Equation Eq. (2.75) as

$$\frac{d}{dt} |\rho(t)) = -i\hat{\mathcal{H}}_i(t) |\rho(t)) \tag{2.83}$$

with the formal solution

$$|\rho(t)) = \hat{L}(t) |\rho(0)) \tag{2.84}$$

where

$$\hat{L}(t) = T \exp\left[-i \int_0^t dt' \, \hat{\mathcal{H}}_i(t') \right].$$

Defining the scalar product of two vectors in Liouville space by [20]

$$(A|B) \underset{\text{def}}{=} \text{Tr}\{A^+ B\} \tag{2.85}$$

where A^+ is the adjoint of A, we can express the Bloch decay $G_x(t)$ as (see Appendix C)

$$G_x(t) = \frac{(I_x|\rho(t))}{(I_x|\rho(0))} \tag{2.86}$$

or

$$G_x(t) = \frac{(I_x|\hat{L}(t)|\rho(0))}{(I_x|\rho(0))}. \tag{2.87}$$

If $\mathcal{H}_i(t)$ is not explicitly time dependent we obtain with

$$\hat{L}(t) = \exp(-i\hat{\mathcal{H}}_i t) = \sum_{n=0}^{\infty} \frac{(-it)^n}{n!} \hat{\mathcal{H}}_i^n \tag{2.88}$$

$$G_x(t) = \sum_{n=0}^{\infty} \frac{(-it)^n}{n!} \frac{(I_x|\hat{\mathcal{H}}_i^n|\rho(0))}{(I_x|\rho(0))} \tag{2.89}$$

which defines the moments as

$$M_n = \frac{(I_x|\hat{\mathcal{H}}_i^n|\rho(0))}{(I_x|\rho(0))} \tag{2.90}$$

or in the high temperature approximation as

$$M_n = \frac{(I_x|\hat{\mathcal{H}}_i^n|I_x)}{(I_x|I_x)}. \tag{2.91}$$

Equation (2.91) together with Eqs. (2.82) and (2.85) leads to

$$M_2 = -\text{Tr}\{[\mathcal{H}_i, I_x]^2\}/\text{Tr}\{I_x^2\} \tag{2.92a}$$

and

$$M_4 = \mathrm{Tr}\{[\mathcal{H}_i,[\mathcal{H}_i,I_x]]^2\}/\mathrm{Tr}\{I_x^2\} \tag{2.92b}$$

which are the familiar second and fourth moment [1].

If the corresponding line shape is symmetric about the frequency $\omega_0 = 0$ (central moments vanish) all odd moments vanish and $G_x(t)$ may be expressed as

$$G_x(t) = 1 - \frac{M_2}{2!}t^2 + \frac{M_4}{4!}t^4 - + \dots. \tag{2.93}$$

Higher order moments are usually difficult to evaluate, although in the case of a simple cubic lattice all moments up to the eighth moment have been calculated. But in most cases just the second and fourth moment are known.

For homonuclear dipole-dipole interaction in high magnetic fields the interaction Hamiltonian $\mathcal{H}_i$ corresponds to the secular part

$$\mathcal{H}_i = \sum_{j<k} A_{jk}(3I_{zj}I_{zk} - \mathbf{I}_j \cdot \mathbf{I}_k) \tag{2.94}$$

where

$$A_{jk} = \gamma_I^2 \hbar\, r_{jk}^{-3}\, \tfrac{1}{2}(1 - 3\cos^2 \vartheta_{jk})$$

and the second and fourth moment are readily calculated as [1]

$$M_2 = 3I(I+1)\sum_k A_{jk}^2 \tag{2.95}$$

and

$$M_4 = [\tfrac{1}{3}I(I+1)]^2 \left\{ 9\sum_k A_{jk}^2 - \frac{27}{N}\sum_{jkl\,\ne} A_{jk}^2(A_{jl} - A_{kl})^2 \right.$$
$$\left. - \frac{81}{5}\sum_k A_{jk}^4 \left(8 + \frac{3}{2I(I+1)}\right) \right\}. \tag{2.96}$$

Since the moment expansion does not converge rapidly, the second and fourth moment describe only the very beginning of the decay function $G_x(t)$. That is why much effort has been put in approximative methods recently, to utilize the knowledge of low order moments most efficiently for predicting the full line shape of the spectrum (see Appendix C).

Let us now consider a few simple examples, which can be calculated rigorously. We always start from the expression for the FID $G_x(t)$ in the high temperature approximation as

$$G_x(t) = \frac{(I_x|\,\hat{\mathbf{L}}(t)\,|I_x)}{(I_x|I_x)} \tag{2.97}$$

where

$$\hat{\mathbf{L}}(t) = \exp[-i\hat{\mathcal{H}}_i \cdot t].$$

The lineshape function $g(\omega)$ is obtained by Laplace transform of $G_x(t)$ as

$$g(\omega) = \frac{(I_x|\,[i\omega + i\hat{\mathcal{H}}_i]^{-1}\,|I_x)}{(I_x|I_x)}$$

where its real part $I(\omega) = \mathrm{Re}(g(\omega))$ is usually called the NMR spectrum. In practice, however, one better calculates $G_x(t)$ first and then Fourier- or Laplace-transforms in order to obtain the spectrum.

(i) Distribution of local fields.

This class of spectrum includes nuclei with different chemical shifts as well as inhomogeneous broadening. Using the interaction Hamiltonian

$$\mathcal{H}_i = \sum_k \omega_k I_{zk}$$

the propagator $L(t)$ can be written as

$$\mathbf{L}(t) = \prod_k \exp[-i\omega_k t\, I_{zk}]$$

leading to a sum over different cosine functions for the FID according to Eq. (2.97)

$$G_x(t) = \sum_k \cos \omega_k t \tag{2.98}$$

with the corresponding spectrum $I(\omega)$ consisting of a sum over δ-functions

$$I(\omega) = \sum_k \delta(\omega - \omega_k).$$

(ii) Heteronuclear spin interaction.

We consider a single spin S with gyromagnetic ratio γ_S which is coupled to many I spins with γ_I. This is still an exactly solvable problem if interactions among the I spins are neglected. Experimentally this can be achieved by irradiation of the I spins at the "magic angle" a technique to be discussed later. The interaction Hamiltonian is given by

$$\mathcal{H}_i = \sum_k B_k I_{zk} S_z \tag{2.99}$$

where

$$B_k = \gamma_I \gamma_S \hbar\, r_k^{-3}(1 - 3\cos^2 \vartheta_k).$$

The FID may be now expressed as

$$G_{IS}(t) = \frac{(S_x|\, \hat{\mathbf{L}}(t)\, |S_x)}{(S_x|S_x)}$$

with

$$\mathbf{L}(t) = \prod_k \exp[-it\, B_k I_{zk} S_z]$$

and

$$(S_x|S_x) = \mathrm{Tr}\{S_x^2\} = \tfrac{1}{3} S(S+1)(2S+1)(2I+1)^{N_I}.$$

After some algebra one obtains [30]

$$G_{IS}(t) = \prod_j \frac{\mathrm{Tr}\{\cos(B_j t\, I_{zj})\}}{(2I+1)} \tag{2.100}$$

which can be expressed for some representative values of I as [30]

$$I = \tfrac{1}{2}: \quad G_{IS}(t) = \prod_j \cos(\tfrac{1}{2}B_j t) \tag{2.101a}$$

$$I = 1: \quad G_{IS}(t) = \prod_j \tfrac{1}{3}[1 + 2\cos(B_j t)] \tag{2.101b}$$

$$I = \tfrac{3}{2}: \quad G_{IS}(t) = \prod_j \tfrac{1}{2}[\cos(\tfrac{1}{2}B_j t) + \cos(\tfrac{3}{2}B_j t)]. \tag{2.101c}$$

Exactly solvable problems are rare and in general we will have to apply approximation schemes.

The simplest approximation would be to guess the line shape, or to determine it experimentally and to fit it by known moments. Two commonly used line shapes are the:

(i) Gaussian line shape

$$G(t) = \exp\left(-\frac{M_2}{2}t^2\right) \tag{2.102a}$$

with the cosine transform of $G(t)$ as

$$I(\omega) = \frac{1}{(2\pi M_2)^{1/2}} \exp(-\omega^2/2M_2) \tag{2.102b}$$

where the corresponding half width at half height equals

$$\delta_0 = [2\ln 2 \cdot M_2]^{1/2} \cong 1.18\, M_2^{1/2}. \tag{2.103}$$

Note, that the full width at half height (FWHH) in frequency units Δv is connected with δ_0 by

$$\Delta v = \delta_0/\pi \tag{2.104}$$

in general. In the following, however, we usually quote δ_0.

(ii) Truncated Lorentzian line shape [1, 17]

$$G(t) = \exp(-t/T_2) \quad \text{for } t > t^* \tag{2.105a}$$

and

$$I(\omega) = \frac{1}{\pi}\frac{\delta_0}{\omega^2 + \delta_0^2} \quad \text{for } \omega \leq \omega^* \tag{2.105b}$$

where

$$\delta_0 = 1/T_2 \tag{2.106}$$

is the half width at half height of the spectrum in rad s^{-1}.

Since the second and higher order moments diverge for a Lorentzian line shape, any experimentally observed line shape can be Lorentzian only in a limited frequency range. This justifies the limits in Eq. (2.105).

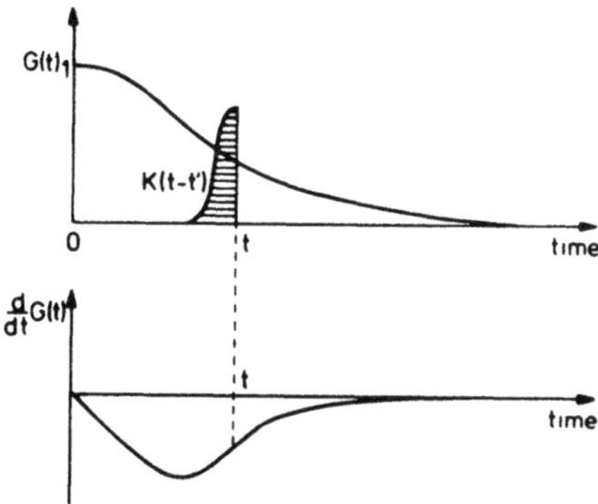

Fig. 2.9. Schematic drawing of the "memory function" approach to the calculation of FID's

If $I(\omega)$ vanishes outside $\pm\omega^*$ one arrives at the following moments for the truncated Lorentzian line shape [1, 17]:

$$M_2 = \frac{2\omega^* \delta_0}{\pi} \quad \text{and} \quad M_4 = \frac{2\omega^{*3} \delta_0}{3\pi}$$

which leads to the half width at half height as

$$\delta_0 = \frac{\pi}{2\sqrt{3}} \cdot \left[\frac{M_2}{\mu}\right]^{1/2} \tag{2.107}$$

where

$$\mu = M_4/M_2^2.$$

Since the direct approximation of $G(t)$ by assuming a functional shape which suits $G(t)$ cannot be made in general and on the other hand is very unsatisfactory from a theoretical point of view, more subtle methods have been designed which use an approximation at a less sensitive level. This is the "memory function approach" [21–26] which is extensively discussed in Appendix C.

In the case of a symmetric line shape function (vanishing central moments) the following exact integro-differential equation for the decay function $G(t)$ is obtained from the Liouville-v. Neumann Equation (see Appendix C)

$$\frac{d}{dt} G(t) = -\int_0^t dt' \, K(t-t') \, G(t') = -\int_0^t d\tau \, K(\tau) \, G(t-\tau) \tag{2.108}$$

where $K(t-t')$ is the so-called "memory function" which describes the memories $G(t')$ has of its early history in time development. A schematic drawing of this functional behavior is presented in Fig. 2.9. Since $G(t)$ is less sensitive to approximations on $K(\tau)$, rather than to approximations on $G(t)$ itself, we have reached a higher level of approximation. The next step further would be to express $K(\tau)$ again by an integro-differential equation like Eq. (2.108) whose memory function $K'(t-t')$ is approximated instead of $K(\tau)$ itself.

The simplest approximation to $K(t-t')$ would be a δ-function approximation i.e. no memory (correlation time $\tau_c = 0$)

$$K(t-t') = K_0 \delta(t-t') \tag{2.109}$$

leading to

$$G(t) = \exp(-t/T_2)$$

where

$$1/T_2 = K_0 = M_2^{1/2} \tag{2.110}$$

and M_2 is the second moment of $G(t)$. If $K(t-t')$ is much more rapidly decaying than $G(t)$ i.e. correlation time τ_c of $K(\tau)$ is much shorter than the decay time T_2, we may use a less stringent approximation than Eq. (2.109), namely

$$\frac{d}{dt} G(t) = -\int_0^\infty d\tau \, K(\tau) \cdot G(t) = -\frac{1}{T_2} G(t) \tag{2.111a}$$

where

$$\frac{1}{T_2} = \int_0^\infty d\tau \, K(\tau) \quad \text{for } \tau_c \ll T_2 \quad \text{``short correlation limit''.}$$

If we make a less severe approximation than Eq. (2.111a) by taking the upper limit of the integral to be t instead of infinity we arrive at the Andersson-Weiss [24c] Model

$$\frac{d}{dt} G(t) = -\int_0^t d\tau \, K(\tau) \, G(t). \tag{2.111b}$$

The exact line shape $I(\omega)$ can be obtained from Eq. (2.108) with

$$I(\omega) = \int_0^\infty dt \, G(t) \cos \omega t$$

by formal integration as [20]

$$I(\omega) = \frac{G(0) \, K'(\omega)}{[\omega - K''(\omega)]^2 + [K'(\omega)]^2} \tag{2.112}$$

where

$$K'(\omega) = \int_0^\infty dt \, K(t) \cos \omega t$$

and

$$K''(\omega) = \int_0^\infty dt \, K(t) \sin \omega t.$$

The half width δ_0 at half height of the lineshape $I(\omega)$ according to Eq. (2.112) can be obtained by iteration from

$$\delta = K''(\delta) + [2K'(0) \, K'(\delta) - K'^2(\delta)]^{1/2}. \tag{2.113}$$

If $G(t)$ and $K(t)$ are given by the following moment expansion [27]

$$G(t) = G(0) \sum_{n=0}^{\infty} (-1)^n \frac{t^{2n}}{(2n)!} M_{2n} \tag{2.114a}$$

and

$$K(t) = K(0) \sum_{n=0}^{\infty} (-1)^n \frac{t^{2n}}{(2n)!} N_{2n} \tag{2.114b}$$

it is straightforward to obtain the following relations by comparing both sides of Eq. (2.108): [20, 27, 28]

$$K(0) = M_2; \quad N_2 = M_2(M_4/M_2^2 - 1)$$
$$N_4 = M_2^2(M_6/M_2^3 - 2M_4/M_2^2 + 1) \quad \text{etc.} \tag{2.115}$$

See also Appendix C for a more rigorous discussion.

So far no approximation has been used in Eqs. (2.112–2.115). A convenient approximation which is still quite flexible in describing completely different line shapes including Gaussian and Lorentzian shapes, would be to use a Gaussian shape for $K(t)$ [27–29]

$$K(t) = K(0) \exp\left[-\frac{N_2}{2} t^2\right]. \tag{2.116}$$

Only the second moment of $K(t)$ i.e. the second and fourth moment of $G(t)$ is needed. The corresponding line-shape is discussed in more detail in Appendix C. As demonstrated in Appendix C a simple expression for the linewidth δ_0 of the corresponding spectral line is obtained by

$$\delta_0 = \sqrt{\frac{\pi}{2}} \left[\frac{M_2}{\mu - 1}\right]^{1/2}, \quad \mu \gg 1 \tag{2.117}$$

which is valid for large values of $\mu = M_4/M_2^2$. In general Eq. (2.117) will be multiplied by a function of the order of one, which will depend on the functional form used for the memory function. In order to cover a wide range of different line shapes it is shown in Appendix C that the following expression for the universal line width δ_0 is more appropriate

$$\delta_0 = \sqrt{\frac{\pi}{2}} \left[\frac{M_2}{\mu - 1.87}\right]^{1/2}, \quad \mu \geq 3. \tag{2.118}$$

In order to simplify Eq. (2.118) it is good enough considering the approximations involved to write

$$\delta_0 = \sqrt{\frac{\pi}{2}} \left[\frac{M_2}{\mu - 2}\right]^{1/2}, \quad \mu \geq 3 \tag{2.119}$$

where the FWHH $\Delta v = \delta_0/\pi$ is readily obtained from Eqs. (2.117–2.119).

With the help of these approximative methods it is thus possible to calculate lineshapes and the corresponding linewidth of "many body systems", i.e. where the time evolution of the system is governed by e.g. many body dipolar interactions. The above formulas may be used to calculate the residual linewidth in high resolution NMR in solids, as will be exploited in the next chapters. Note, that no adjustable parameters are employed in the theory.

The linewidth formulas have to be slightly modified, however, if coherent processes with frequency ω are involved. In this case the linewidth is obtained as

$$\Delta = \delta_0 \, J(\omega) + \Delta_0 \tag{2.120}$$

where $J(\omega)$ is the spectral density at frequency ω with $J(\omega=0)=1$ and Δ_0 is the linewidth when ω approaches infinity. Usually Δ_0 will be zero. An example of this short is the on-resonance decoupling case which is discussed in Sect. 4.4.

2.6 Magic Angle Spinning (MAS)

Manipulation of spin interactions can e.g. simply be achieved by rotating a sample about an axis which is tilted by an angle ϑ with respect to the magnetic field B_0 (see Fig. 2.10). As was first noted and exploited independently by Andrew [31] and by Lowe [32] that if the rotation axis is inclined at the "magic angle" second rank tensor interactions, which usually govern NMR anisotropies can be averaged out. This usually results in a narrow line displaying only the isotropic part of the spin interactions. When spinning the sample with a rotation rate ω_r all spin interactions become timedependent and sidebands with separation ω_r are produced in the spectrum (see Figs. 2.11 and 2.12). If the spinning rate ω_r is much larger than the anisotropic spin interaction the sidebands are well separated from the central line and become vanishingly small with increasing ω_r. Andrew and co-workers [31, 33–37] and others have exploited this technique extensively, to investigate isotropic shifts in solids. Such high spinning speeds are usually achieved with air bearing turbines of the Beams-Andrew type [38, 39, 40].

It is convenient to express the spin interaction Hamiltonian in terms of spherical tensor operators (see Appendix A)

$$\mathcal{H} = \sum_{k=0}^{2} \sum_{q=-k}^{+k} (-1)^q \, A_{kq}(t) \, T_{k-q}(t') \tag{2.121}$$

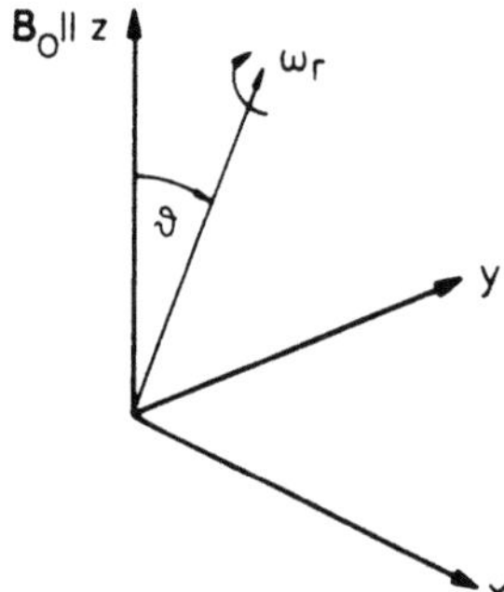

Fig. 2.10. Schematic drawing of the rotation axis (rotation rate ω_r) tilted with respect to the static magnetic field $\mathbf{B}_0$ by an angle ϑ

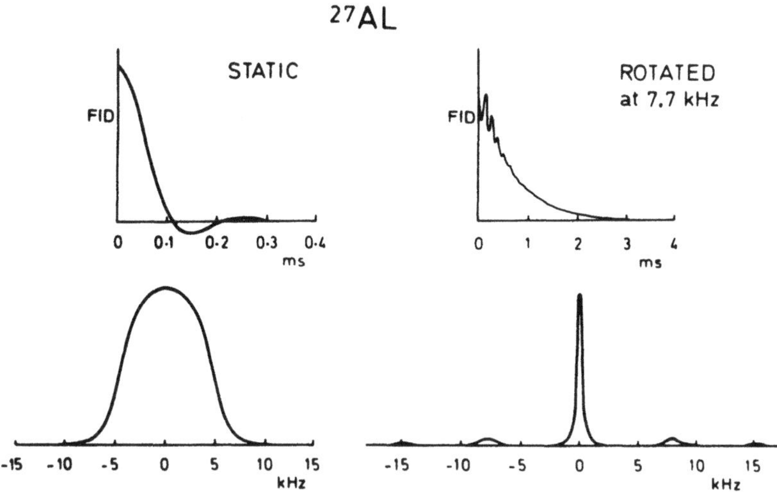

Fig. 2.11. Decay and spectrum of ^{27}Al in a powder sample of aluminium, with and without rotation at the "magic angle", according to Andrew *et al.* [36]. The generation of sidebands and the narrowing of the centerline is clearly demonstrated

where all spin parameters are contained in T_{kq}, whereas A_{kq} represents the lattice parameters. Two different types of time dependences are indicated in Eq. (2.121). The one imposed on $T_{kq}(t')$ stems from spin precession ω_0 in the static magnetic field B_0, whereas the time dependence of $A_{kq}(t)$ is imposed by the sample spinning. In the high field case usually encountered in NMR a first "averaging" [5] is achieved by observing stroboscopically at integer intervals of the Larmor period $n\,2\pi/\omega_0$, i.e. neglecting non-secular terms. In this case all terms with $q \neq 0$ in Eq. (2.121) are neglected as was discussed in Sect. 2.2. This is, however, only legitimate when $\omega_r \ll \omega_0$.

Under these conditions Eq. (2.121) reduces to

$$\mathcal{H}_0 = A_{00} T_{00} + A_{10}(t) T_{10} + A_{20}(t) T_{20}.$$

The antisymmetric part A_{10} of spin interactions does not contribute to the spectrum in first order and can safely be neglected. The isotropic part $A_{00} = -(1/\sqrt{3})\,\mathrm{Tr}\{A_{ij}\}$ is invariant under motion. We are therefore left with the calculation of

$$A_{20}(t) = [R(t)\,A_{2q}\,R^{-1}(t)]_{q'=0}. \tag{2.122}$$

The time dependence is introduced by rotating the sample by an angle $\varphi(t)$ about the spinner axis, with

$$\varphi(t) = \varphi_0 + \omega_r \cdot t.$$

If we classify the rotation transformation $R(\varphi\,\vartheta\,\psi)$ with the Euler angles φ, ϑ, ψ where ϑ is the angle between the spinner axis and the magnetic field B_0, we can express $A_{20}(t)$ according to Eq. (2.122) as

$$A_{20}(t) = \sum_{q=-2}^{+2} A_{2q}\, e^{-i\omega_r t q}\, e^{-iq\varphi_0}\, d_{q,0}^{(2)}(\vartheta). \tag{2.123}$$

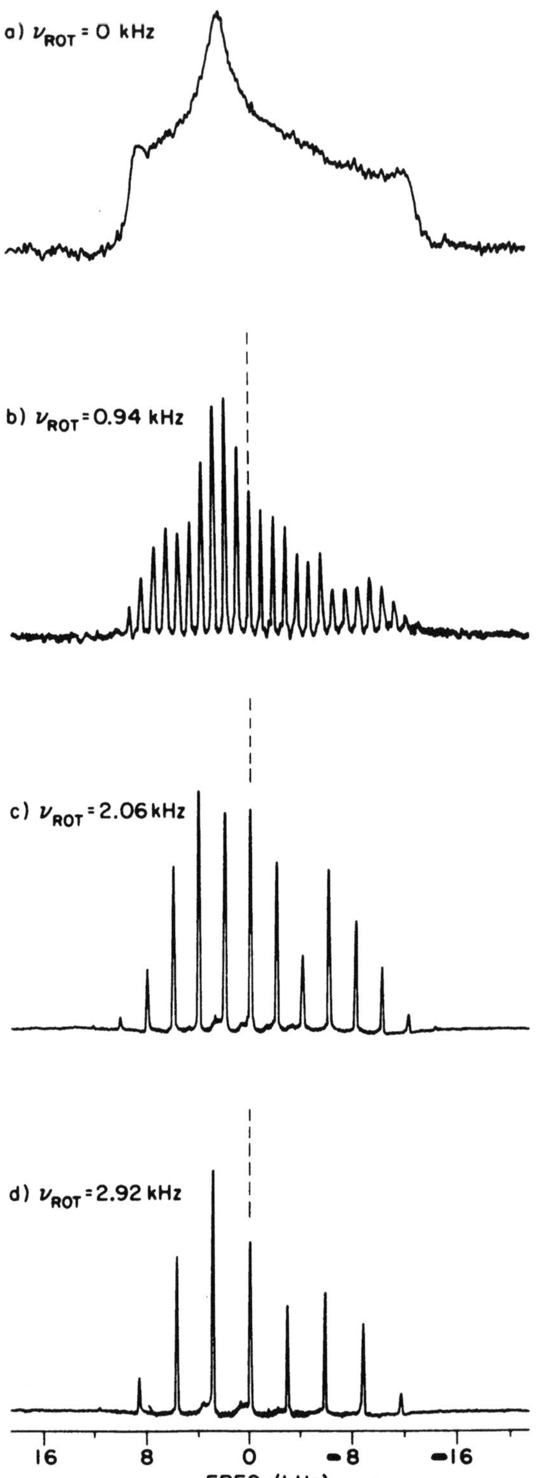

Fig. 2.12. Proton decoupled ^{31}P sideband spectra (119.05 MHz) of solid dipalmitoyl phosphatidylcholine spinning at the magic angle according to R.G. Griffin, published in [9]. ν_{Rot} indicates the spinning frequency

It is immediately evident from Eq. (2.123) that rotational echos [7–9, 41] and sidebands appear at multiples of the frequency ω_r and $2\omega_r$ away from the centerband. Figure 2.11 shows an early example of this technique by Andrew and co-workers [36], where this sideband structure is clearly visible. A more recent example obtained by Griffin [42] is shown in Fig. 2.12. Here the characteristic sideband pattern of a chemical shift tensor obtained at different spinning rates is displayed. Before discussing this aspect further, let us go back to the rapid spinning case. If stroboscopic observation at time intervals $\tau = n\,2\pi/\omega_r$ is performed or else ω_r is made very large, only the time independent part of $A_{20}(t)$ survives, i.e.

$$\bar{A}_{20}(t) = \tfrac{1}{2}(3\cos^2\vartheta - 1)\,A_{20}. \tag{2.124}$$

In this case, the time independent Hamiltonian

$$\bar{\mathscr{H}}_0 = A_{00}T_{00} + \tfrac{1}{2}(3\cos^2\vartheta - 1)\,A_{20}T_{20}$$

governs the spectrum. In the case of dipole-dipole and quadrupole interactions A_{00} vanishes and $A_{20}T_{20}$ is just the secular part of the corresponding Hamiltonian, where the z-axis of the geometric part now refers to the spinner axis. To be specific we quote different cases:

(i) Homonuclear coupling between two spins I_1 and I_2

$$\begin{aligned}\bar{\mathscr{H}}_{II} = {}& J\mathbf{I}_1\cdot\mathbf{I}_2 - \tfrac{1}{2}\cos\vartheta\,(D_{x'y'} - D_{y'x'})(I_{x_1}I_{y_2} - I_{y_1}I_{x_2}) \\ & + \tfrac{1}{2}(3\cos^2\vartheta - 1)\tfrac{1}{2}(D_{z'z'} - J)(3I_{z_1}I_{z_2} - \mathbf{I}_1\cdot\mathbf{I}_2)\end{aligned} \tag{2.125}$$

with

$$J = \tfrac{1}{3}\mathrm{Tr}\{D_{ij}\}$$

and where the primed axes refer to the sample rotation frame. Here we have also included for generality the antisymmetric part which occurs in this case in the secular Hamiltonian. Note, however, that it does not contribute to the spectrum in first order.

(ii) Heteronuclear coupling between two spins I and S

$$\bar{\mathscr{H}}_{IS} = J I_z S_z + \tfrac{3}{2}(3\cos^2\vartheta - 1)\tfrac{1}{2}(D_{z'z'} - J)I_z S_z. \tag{2.126}$$

(iii) Chemical shift tensor

$$\bar{\mathscr{H}}_S = \sigma_i\omega_0 I_z + \tfrac{1}{2}(3\cos^2\vartheta - 1)(\sigma_{z'z'} - \sigma_i)\omega_0 I_z \tag{2.127}$$

with the isotropic shift $\sigma_i = \tfrac{1}{3}\mathrm{Tr}\{\sigma_{ij}\}$ and where $\omega_0 = \gamma B_0$.

It is obvious, that a scaled anisotropy is observed if ϑ deviates slightly from the magic angle which was supposedly pointed out by A. Pines [43].

The most interesting case, however, is the one where the "magic angle condition" $3\cos^2\vartheta_m = 1$ is met. This was first exploited by Andrew and co-workers [31] and by Lowe [32].

In this case only the trace of the second rank tensor interaction survives, i.e. all tensor interactions which have zero trace like dipole-dipole and quadrupole-interaction vanish. This is true also for the heteronuclear dipolar interaction and no extra decoupling technique has to be applied. The measurable

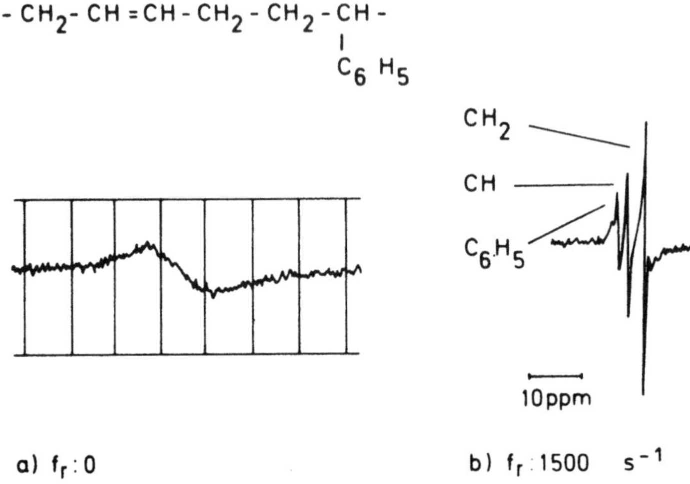

Fig. 2.13. Application of the "magic angle" rotation method to egg yolk lecithin [46b]. Different proton resonance lines due to CH_2, CH and C_6H_5 groups can be distinguished

quantities in the "magic angle" specimen rotation method are thus the isotropic chemical shift, Knight-shift and the scalar coupling. E.R. Andrew and co-workers have applied this method extensively to study these parameters in solids [33, 34]. One example of this technique has been presented in the introduction (Fig. 1.3) where the J-coupling between the ^{19}F and the ^{75}As in $KAsF_6$ reveals a four line spectrum for the ^{19}F nuclei [35].

Figure 2.11 shows the application of this technique to a metal powder [36], namely ^{27}Al. The development of sidebands with increasing rate of revolution combined with a narrowing of the center line is clearly visible. It was possible in this experiment to determine the isotropic Knight shift of ^{27}Al with respect to $AlCl_3$ solution with more than an order of magnitude improvement in precision. The result of $1,640 \pm 1$ ppm agrees within the combined errors with the value of $1,636 \pm 3$ ppm as obtained by multiple pulse experiments [44], to be discussed later (Sect. 7.6). The "magic angle" specimen rotation method has been successfully applied to polymers and compounds of biological interest also [45–47]. Babka et al. have resolved proton NMR spectra in lecithin and related membrane model systems [46b].

Müller and Zachmann [48] have investigated the proton spectrum in polyethylene and polyethyleneoxide using an axial spinner.

Figure 2.13 shows an example for egg yolk lecithin. The line narrowing techniques may fail however in these compounds, since the slow motions create a rapid spin lattice relaxation rate which may limit the resolution appreciably as is discussed in more detail in Chap. 8. For further reading on the application of " magic angle" rotation methods we refer the interested reader to the review papers on this subject by E.R. Andrew [33, 34] and the references cited therein.

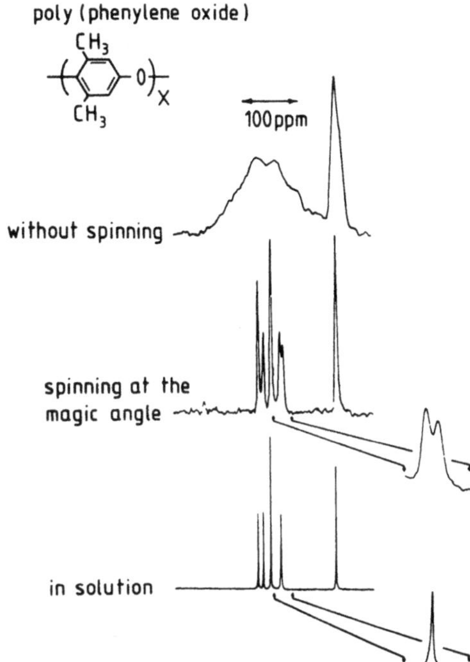

Fig. 2.14. Cross-polarization ^{13}C NMR spectra (22.6 MHz) of polyphenylene oxide with and without MAS (3 kHz) according to Schaefer et al. [47]

A recent revival of magic angle sample spinning has been initiated by Schäfer and Stejskal and co-workers [47, 49] by combining it with cross-polarization of ^{13}C nuclei. Highly resolved ^{13}C spectra are obtained even in complex systems like polymers. An example is shown in Fig. 2.14. The spectra are liquid like in the sense, that only the trace of the chemical shift tensor is observed. It were Maricq and Waugh [8] who pointed out, that utilizing slow spinning properly leads to resolved sideband structure from which information about the chemical shift tensor can be obtained by a moment analysis. Herzfeld and Berger [9] showed alternatively that the ratio of sideband intensities can be related to the chemical shift anisotropies. Other approaches to recover this information have been discussed by Lippmaa et al. [7a] and Waugh and co-workers [7b]. In the following we shall follow the lines of Maricq and Waugh [8] and of Herzfeld and Berger [9].

We readily obtain the timedependent resonance frequency $\omega(t)$ of a particular spin packet as

$$\omega(t)=\omega_0\{(-1/\sqrt{3})\,A_{00}+(2/\sqrt{6})\,A_{20}(t)\} \tag{2.128}$$

where $A_{20}(t)$ is given by Eq. (2.123) and with

$$(-1/\sqrt{3})\,A_{00}=(\sigma_{11}+\sigma_{22}+\sigma_{33})/3=\sigma_i \quad \text{and} \quad \omega_0=-\gamma B_0.$$

By using the transformation $R(\alpha\beta\gamma)$ from the principal axis frame $(1,2,3)$ to the spinner frame (x,y,z) we can express the spherical tensor components A_{2q} in the spinner frame as

$$A_{2q} = \sum_{q'=-2}^{+2} A''_{2q'} e^{-i\alpha q'} d^{(2)}_{q'q}(\beta) e^{-i\gamma q} \tag{2.129}$$

where

$$A''_{20} = \sqrt{\tfrac{3}{2}}(\sigma_{33} - \sigma_i); \quad A''_{2\pm1} = 0; \quad A''_{2\pm2} = \tfrac{1}{2}(\sigma_{11} - \sigma_{22}).$$

Combining this with Eqs. (2.123) and (2.128) leads to

$$\omega(t) = \omega_0 \left\{ \sigma_i + \sqrt{\tfrac{3}{2}} \sum_{q,q'=-2}^{+2} A''_{2q'} e^{-i\alpha q'} d^{(2)}_{q'q}(\beta) d^{(2)}_{q0}(\vartheta) e^{-iq(\gamma + \omega_r t)} \right\} \tag{2.130}$$

where ϑ is the angle between the static field B_0 and the spinner axis. The $d^{(2)}_{qq}$ are given in Table B.1. Note, that $\omega(t)$ can be separated into two parts

$$\omega(t) = \bar{\omega} + \omega^*(t)$$

where the static part

$$\bar{\omega} = \omega_0 \{ \sigma_i + \tfrac{1}{2}(3\cos^2\vartheta - 1)[\tfrac{1}{2}(3\cos^2\beta - 1)(\sigma_{33} - \sigma_i) \\ + \tfrac{1}{2}\sin^2\beta \cos2\alpha(\sigma_{11} - \sigma_{22})]\} \tag{2.131}$$

contains the terms with $q=0$ besides σ_i, whereas the $q \neq 0$ terms lead to [8, 9]

$$\omega^*(t) = C_1 \cos(\gamma + \omega_r \cdot t) + S_1 \sin(\gamma + \omega_r \cdot t) \\ + C_2 \cos(2\gamma + 2\omega_r t) + S_2 \sin(2\gamma + 2\omega_r t) \tag{2.132}$$

with

$$C_1 = \omega_0 \sin\vartheta \cos\vartheta \{ -3\sin\beta \cos\beta(\sigma_{33} - \sigma_i) + \sin\beta \cos\beta \cos2\alpha(\sigma_{11} - \sigma_{22})\}$$
$$S_1 = -\omega_0 \sin\vartheta \cos\vartheta \{\sin\beta \sin2\alpha(\sigma_{11} - \sigma_{22})\}$$
$$C_2 = \omega_0 \sin^2\vartheta \{\tfrac{3}{4}\sin^2\beta(\sigma_{33} - \sigma_i) + \tfrac{1}{4}(1 + \cos^2\beta)\cos2\alpha(\sigma_{11} - \sigma_{22})\}$$
$$S_2 = -\omega_0 \sin^2\vartheta \{\tfrac{1}{2}\cos\beta \sin2\alpha(\sigma_{11} - \sigma_{22})\}.$$

We are now equipped to calculate the free induction decay

$$g(t) = \frac{1}{8\pi^2} \int_0^{2\pi} \int_0^{\pi} \int_0^{2\pi} \exp[i\Phi(\alpha\beta\gamma, t)] \, d\alpha \sin\beta \, d\beta \, d\gamma \tag{2.133}$$

where

$$\Phi(\alpha\beta\gamma, t) = \int_0^t dt' \, \omega(\alpha\beta\gamma, t') \tag{2.134}$$

which can again be separated into two parts as

$$\Phi(\alpha\beta\gamma, t) = \Phi_0(t) + \Phi_a(t) \quad \text{with} \quad \Phi_0(t) = \bar{\omega}t$$

and

$$\Phi_a(t) = \frac{1}{\omega_r} [S_1 \cos\gamma - C_1 \sin\gamma + \tfrac{1}{2} S_2 \cos 2\gamma - \tfrac{1}{2} C_2 \sin 2\gamma]$$

$$+ \frac{C_1}{\omega_r} \sin(\gamma + \omega_r t) - \frac{S_1}{\omega_r} \cos(\gamma + \omega_r t)$$

$$+ \frac{C_2}{2\omega_r} \sin(2\gamma + 2\omega_r t) - \frac{S_2}{2\omega_r} \cos(2\gamma + 2\omega_r t). \tag{2.135}$$

Note that $\Phi_a(t) = \Phi_a(t - n\, 2\pi/\omega_r)$ with $n = 0, 1, 2, \ldots$ shows cyclic behaviour which leads to rotational spin echos at $t = n\, 2\pi/\omega_r$ as was pointed out by Waugh and co-workers [8, 41]. The echo envelope is governed by $\exp[i\Phi_0(t)]$ which reduces to the isotropic part under "magic angle condition" $\vartheta = \vartheta_m$. Maricq and Waugh [8] and later Herzfeld and Berger [9] have expressed $\exp[i\Phi_a(t)]$ by infinite sums over products of Bessel functions of the first kind. However, at the end this is not very useful and we prefer to use

$$e^{iz\sin\varphi} = \frac{1}{2\pi} \int\limits_0^{2\pi} d\Theta\, \delta(\Theta - \varphi)\, e^{iz\sin\Theta}$$

together with

$$\delta(\Theta - \varphi) = \sum_{N=-\infty}^{+\infty} e^{\pm iN(\Theta - \varphi)}.$$

A similar expression holds of course for $\exp[iz\cos\varphi]$. By using Eq. (2.135) one obtains after some algebra

$$e^{i\Phi_a(t)} = \sum_{N,N'=-\infty}^{+\infty} e^{i\gamma(N-N')} e^{iN\omega_r t} F_{N'}^* F_N \tag{2.136}$$

where

$$F_N = \frac{1}{2\pi} \int\limits_0^{2\pi} d\Theta\, e^{i\left[-N\Theta + \frac{C_1}{\omega_r}\sin\Theta - \frac{S_1}{\omega_r}\cos\Theta + \frac{C_2}{2\omega_r}\sin 2\Theta - \frac{S_2}{2\omega_r}\cos 2\Theta\right]}. \tag{2.137}$$

Integration over γ leads together with

$$\delta(N - N') = \frac{1}{2\pi} \int\limits_0^{2\pi} d\gamma\, e^{i\gamma(N-N')}$$

to

$$e^{i\Phi_a(t)} = \sum_{N=-\infty}^{+\infty} I_N\, e^{iN\omega_r t} \tag{2.138}$$

where the intensity of the N-th sideband is defined as

$$I_N = F_N^* F_N = |F_N|^2. \tag{2.139}$$

The following properties of I_N are immediately evident from Eqs. (2.136–2.139)

$$\sum_{N=-\infty}^{+\infty} I_N = 1; \qquad \lim_{N\to\pm\infty} I_N = 0$$

$$\lim_{\omega_r\to\infty} = \begin{cases} 1 & \text{for } N = 0 \\ 0 & \text{for } N \neq 0. \end{cases}$$

From the definition

$$J_k(z) = \frac{1}{2\pi} \int\limits_0^{2\pi} d\Theta \, \exp[-i(k\Theta - z\sin\Theta)]$$

for the Besselfunction $J_k(z)$ it follows that F_N can be expressed as [8, 9]

$$F_N = \sum_{k,l,m=-\infty}^{+\infty} J_{N-k-2l-2m}(C_1/\omega_r) \, J_k(S_1/\omega_r)$$
$$\cdot J_l(C_2/2\omega_r) \, J_m(S_2/2\omega_r) \, e^{-i\frac{\pi}{2}(k+m)}. \tag{2.140}$$

Maricq and Waugh [8] have proposed a moment analysis to obtain the chemical shift anisotropy parameters $\delta = (\sigma_{33} - \sigma_i)$ and $\eta = (\sigma_{11} - \sigma_{22})/(\sigma_{33} - \sigma_i)$.

From the definition of the k-th moment

$$M_k = \int\limits_{-\infty}^{+\infty} d\omega \, \omega^k \, g(\omega) = \omega_r^k \sum_{N=-\infty}^{+\infty} N^k I_N \tag{2.141}$$

expressions for the first eight moments ($k = 0, 1, \ldots 7$) were obtained by Maricq and Waugh [8]. After powder average of the form

$$I_N \equiv \frac{1}{4\pi} \int\limits_0^{2\pi} \int\limits_0^{\pi} I_N(\alpha\beta) \, d\alpha \sin\beta \, d\beta$$

they obtained for the first few moments [8]

$$M_0 = 1, \quad M_1 = 0$$
$$M_2 = (\delta^2/15)(3 + \eta^2); \quad M_3 = (2\delta^3/35)(1 - \eta^2)$$
$$M_4 = (15/7) M_2^2 + 2\omega_r^2 M_2.$$

The fourth and higher moments do not contain any new information which is not already available from M_2 and M_3. Note, that M_4 increases with ω_r whereas M_2 and M_3 are independent of the motion. In the spirit of lineshape theory (see Sect. 2.5) the parameter $\mu = M_4/M_2^2$ increases with ω_r^2 leading to quasi-Lorentzian lines of decreasing halfwidth.

The moment analysis may be problematic in practical cases since in principal an infinite number of sidebands has to be taken into account or else the short time behaviour of the rotational spin echos has to be analyzed very accurately. Herzfeld and Berger [9] therefore proposed as an alternative to use the sideband intensity ratio I_N/I_M instead. Conveniently the sidebands are related to the centerline with intensity I_0. Herzfeld and Berger [9] present contour plots of $I_{\pm N}/I_0$ for $N = 1$-5 calculated numerically from expressions like Eqs. (2.139) and (2.140). In order make contact with Herzfeld and Berger our parameters C_1, S_1, C_2 and S_2 have to be replaced by

$$C_1/\omega_r = \bar{A}_1; \quad -S_1/\omega_r = \bar{B}_1; \quad C_2/(2\omega_r) = \bar{A}_2; \quad -S_2/(2\omega_r) = \bar{B}_2.$$

Furthermore the principal elements σ_{xx}, σ_{yy} and σ_{zz} of reference [9] are not necessarily ordered and parameters ρ and μ are defined based on the principal elements $\sigma_{33} > \sigma_{22} > \sigma_{11}$

$$\mu = \gamma B_0(\sigma_{33} - \sigma_{11})/\omega_r$$
$$\rho = (\sigma_{11} + \sigma_{33} - 2\sigma_{22})/(\sigma_{33} - \sigma_{11})$$

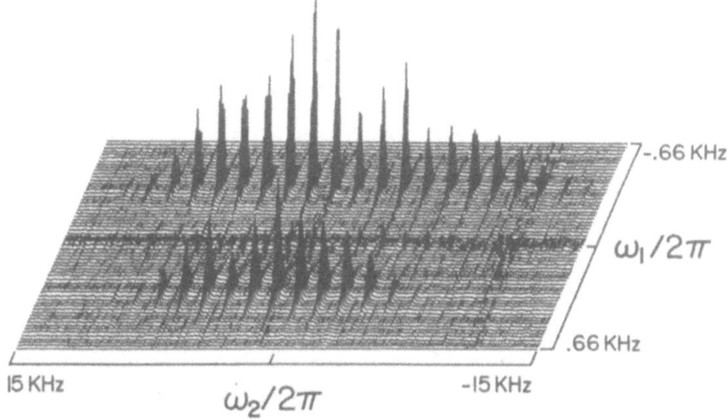

Fig. 2.15. Two-dimensional resolution of isotropic and anisotropic chemical shifts for the ^{31}P MAS sideband spectrum of the mixture of barium diethylphosphate (BDEP; upper half spectrum) and brushite (CaHPO$_4$ 2HPO; lower half spectrum) according to Aue, Ruben and Griffin [51]

where

$$-1 \leqq \rho \leqq 1 \quad \text{and} \quad \mu > 0.$$

It was demonstrated by Herzfeld and Berger [9], that reliable chemical shift tensor elements σ_{11}, σ_{22} and σ_{33} can be determined from powder samples using this method. In practice, usually several non-equivalent ^{13}C nuclei complicate the spectra, however, and a method to separate the different sideband spectra with respect to the isotropic shift σ_i is desirable. Proposals of different sorts using among others rotation synchromized pulses [7] have been made to solve this problem. However, only very recently Griffin and co-workers [51] found a very simple, but elegant solution in terms of two-dimensional spectroscopy. An example is shown in Fig. 2.15. Whereas one frequency axis spans the full anisotropic sideband pattern, the other frequency axis corresponds only to the isotropic shift.

The free induction decay can be expressed under the "magic angle condition" $\vartheta = \vartheta_m$ according to Eqs. (2.131) and (2.135) as

$$g(t) = e^{i\omega_0\sigma_i t}\, e^{i\Phi_a(t)}$$

where $\Phi_a(t)$ is periodic with $t - t_n$. If the FID is sampled at times $t_n = n\, 2\pi/\omega_r$, $\Phi_a(t_n) = 0$ and the FID represents the isotropic spectrum. Suppose now, that FID's $g(t_2)$ are observed, beginning at different times $t_1 = t_n$. After Fourier-transformation a two-dimensional spectrum as shown in Fig. 2.15 results, where the ω_1-axis corresponds to the isotropic spectrum, whereas the ω_2-axis represents the anisotropic part. Spectral resolution is enhanced, if a π-pulse is applied at the beginning of the FID at $t_1 = t_n$ [51].

Dixon [52] has made an interesting proposal recently to eliminate all sidebands in the spectrum by destructive interference caused by the proper timing of four π-pulses. This method can be extended to lead to 2D-spectroscopy where one frequency axis represents the isotropic shift.

Consider now, that we just observe the zeroth order sideband (isotropic line). The question arises, what is the linewidth caused by? Garroway et al. [53]

and Van der Hart et al. [54] have investigated this problem very thoroughly. In essence there are static and dynamic broadening mechanisms, namely (i) homonuclear dipole-dipole interaction of the diluted spins, (ii) molecular and bulk susceptibility effects, (iii) decoupling inefficiency (frequency off-set and insufficient field strength) and (iv) relaxation. The last effect was carefully investigated by Suwelack et al. [55] who demonstrated the importance of the spectral density of the molecular motion at the spinning frequency in the slow motion limit.

So far we have considered MAS experiments mainly in the context of chemical shift anisotropy. Although already Andrew and co-workers [36] have shown that quadrupole interaction can be eliminated by MAS, only recently this aspect has attracted renewed interest [56]. With highly stabilized spinners it was possible to obtain sideband spectra of ^{2}D in denterated samples by Pines and co-workers [56].

Another recent development is the combination of MAS with multiple-pulse experiments introduced by Gerstein and co-workers [57] and Schnabel and co-workers [58]. Application of this technique at 270 MHz by Scheler et al. [59] has demonstrated reasonable resolution which allowed e.g. to distinguish between intra- and inter-hydrogen bounded protons in maleic acid. Even higher resolution may be obtainable if the Burum-Rhim BR-24 multiple-pulse sequence [60] in combination with MAS is used as was demonstrated by Ryan et al. [57b].

Coming to the technical aspects of MAS it is worth noting, that two different types of spinners are currently used, namely the Andrew-Beams type [31, 38-40] and the axial spinner [32, 61-63]. Both have air bearings and air drive. Whereas the Andrew-Beams spinner can be driven beyond 10 KHz [33, 41] by using helium gas, the axial spinner usually runs in the 3-5 KHz range. The axial spinner, however, has the advantage that the tilt angle with respect to the magnetic field can be kept very stable. It also allows the rotating frequency to be varied over a wide range which is useful when analysing sideband pattern. Technical details of the Andrew-Beams type [31, 38-41] and the axial spinner [32, 61-63] can be found in the literature.

2.7 Rapid Anisotropic Molecular Rotation

It has been realized for a long time that even in solids, especially "organic solids", rapid molecular reorientation can occur at high rates (e.g. up to $10^{13}\,\text{s}^{-1}$) at intermediate temperatures (e.g. 100 K to 500 K). The best known examples are benzene and methyl groups. The variation of relaxation rates and the change in the second moment of the resonance line have been extensively studied in the past. We concentrate here on the spectrum or line shape under condition of rapid anisotropic molecular reorientation.

As in the preceding sections, the interaction Hamiltonian $\mathcal{H}_{\text{int}}$ is devided into a time averaged Hamiltonian $\mathcal{H}'_{\text{int}}$ (secular) and a time dependent Hamiltonian (non-secular) $\mathcal{H}''_{\text{int}}(t)$

$$\mathcal{H}_{\text{int}} = \mathcal{H}'_{\text{int}} + \mathcal{H}''_{\text{int}}(t).$$

In this section we assume the random process governing the motion with the correlation time τ_c, to be rapid, i.e. $1/\tau_c \gg \|\mathcal{H}_{int}\|$. In this sense only secular contributions have to be taken into account and it is sufficient to calculate the average Hamiltonian only. We can readily use the equations of the preceding section. If we consider the rotation of two homonuclear spins I about an axis orthogonal to their internuclear vector ($\vartheta_{ij} = 90°$) as an example, we find from Eq. (2.125) for their dipolar interaction

$$\mathcal{H}'_{II} = \tfrac{1}{4}(3\cos^2\vartheta - 1)\cdot\frac{\gamma_I^2\hbar}{r^3}(3I_{z1}I_{z2} - \mathbf{I}_1\cdot\mathbf{I}_2)$$

where ϑ is the angle between the rotation axis and the magnetic field B_0. This equation again reflects the well known fact that the line splitting vanishes at the "magic angle" ($\vartheta_m = 54°\,44'$).

We turn now to the impact rapid anisotropic molecular reorientation has on the chemical shift anisotropy, since this is one of the quantities usually observed in high resolution NMR in solids. It is obvious that the component of the shielding tensor in the direction of the rotation axis is unchanged, whereas the components perpendicular to the rotation axis are averaged, leading to an axially symmetric tensor with the rotation axis as the symmetry axis [12]. The trace of the tensor is, of course, unchanged. The relevant equation describing this fact is Eq. (2.127) of the preceding section which we repeat here for the convenience of the reader in a slightly different notation

$$\overline{S^*_{zz}} = \tfrac{1}{2}(3\cos^2\vartheta - 1)\,S^*_{ZZ}$$

where

$$S^*_{ii} = S_{ii} - \tfrac{1}{3}\mathrm{Tr}\{\tilde{\mathbf{S}}\}$$

and the Z-axis corresponds to the rotation axis. In this expression all values are referred to the trace of the tensor. The axial symmetry is quite evident, leaving as the motionally narrowed anisotropy

$$\overline{\Delta S} = S_{\parallel} - S_{\perp} = \tfrac{3}{2}S^*_{ZZ} \tag{2.142a}$$

where

$$S^*_{ZZ} = S^*_{11}\cos^2\alpha\sin^2\beta + S^*_{22}\sin^2\alpha\sin^2\beta + S^*_{33}\cos^2\beta$$

with the tensor elements S^*_{11}, S^*_{22}, and S^*_{33} and the Euler angles (α,β) defining the transformation from the principal axis system of the tensor $(1,2,3)$ to the molecular rotating frame (Z-axis).

Equation (2.142a) may be rewritten in the following way [12]:

$$\overline{\Delta S} = \tfrac{1}{2}(3\cos^2\beta - 1)\,[S_{33} - \tfrac{1}{2}(S_{11} + S_{22})] + \tfrac{3}{4}(S_{11} - S_{22})\sin^2\beta\cos 2\alpha. \tag{2.142b}$$

This equation is valid for the "starred" and "unstarred" values of S. If the shift tensor is axially symmetric in the "rigid" case, as is the case of ^{19}F in C_6F_6, Eq. (2.142b) reduces to

$$\overline{\Delta S} = \tfrac{1}{2}(3\cos^2\beta - 1)\,\Delta S$$

leaving the shift tensor unchanged if $\beta=0$ (e.g. C_6F_6). If the rotation occurs about an axis perpendicular ($\beta=90°$) to the symmetry axis of an axially symmetric shielding tensor, we obtain for example $\overline{\Delta S}=-\frac{1}{2}\Delta S$, with the old $S_\perp$ component being the new unique component of the averaged tensor.

Those motionally averaged powder spectra have been observed in multiple pulse experiments [12] and it was possible from the powder spectra of ^{19}F in CF_3COOAg at different temperatures to assign the direction of the principle axes to the molecular frame tentatively. This assignment was confirmed in a later single crystal investigation [10].

Let us adopt a simple model for the shielding tensor of the ^{19}F spins in a fluoromethyl group as $S_{11}=-1$, $S_{22}=0$, $S_{33}=+1$ with the 3-axis parallel to the CF bond and the 1-axis perpendicular to the CF bond in the CCF plane. This is realistic besides a factor of about 70 ppm in the S_{ii} values [10]. If we apply Eq. (2.142a) with the angles $\beta=70.5°$ and $\alpha=0°$ we obtain $\overline{\Delta S}=-1.166$,

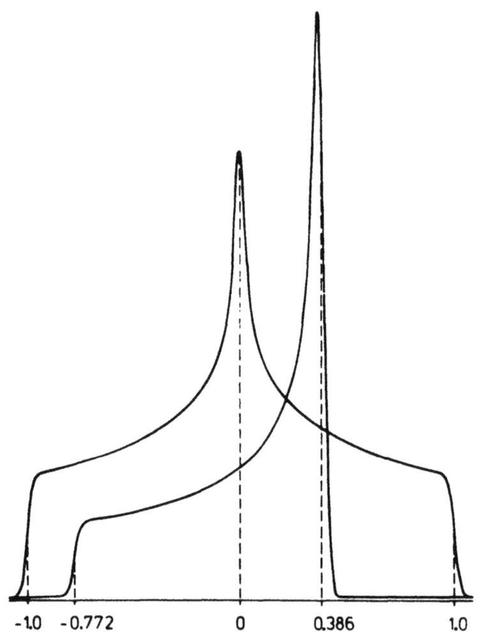

Fig. 2.16. Powder line shape for a model shift tensor with $S_{11}=-1$, $S_{22}=0$, $S_{32}=+1$, corresponding to a fluoromethyl group, with the 3-axis parallel to the CF-bond and the 1-axis perpendicular to the CF-bond in the CCF plane. Rapid rotation about the threefold axis of the methyl group leads to an axially symmetric tensor according to Eq. (2.142a) as indicated by the corresponding powder pattern

with the trace unchanged. Figure 2.16 shows the corresponding powder spectrum. Similar spectra were obtained experimentally in the case of ^{19}F in CF_3COOAg [12].

2.8 Line Shapes in the Presence of Molecular Reorientation

In the preceding section we have treated the most simple case of how the line shape changes under rapid molecular rotations. Now we are going to deal with the more general case, i.e. molecular reorientation at an arbitrary rate. This problem has been dealt with in a number of publications [14, 64–74]. Especially interesting is the region where the time scale of the motion is comparable with the size of the interaction to be studied. One can hope to differentiate between different types of molecular reorientation in this case.

We shall concentrate here on NMR spectra governed by shielding tensors, although it is evident that the same arguments can be applied to other types of interaction. There is particularly a close relationship to ESR line shapes in liquids governed by g-tensors. The usefulness of the lineshape analysis in this case applied to spectra displaying molecular motion has been demonstrated recently by Sillescu and co-workers [68, 69]. On the other hand Spiess et al. [71, 72] have applied this analysis, to NMR powder spectra for the first time (see Fig. 2.17).

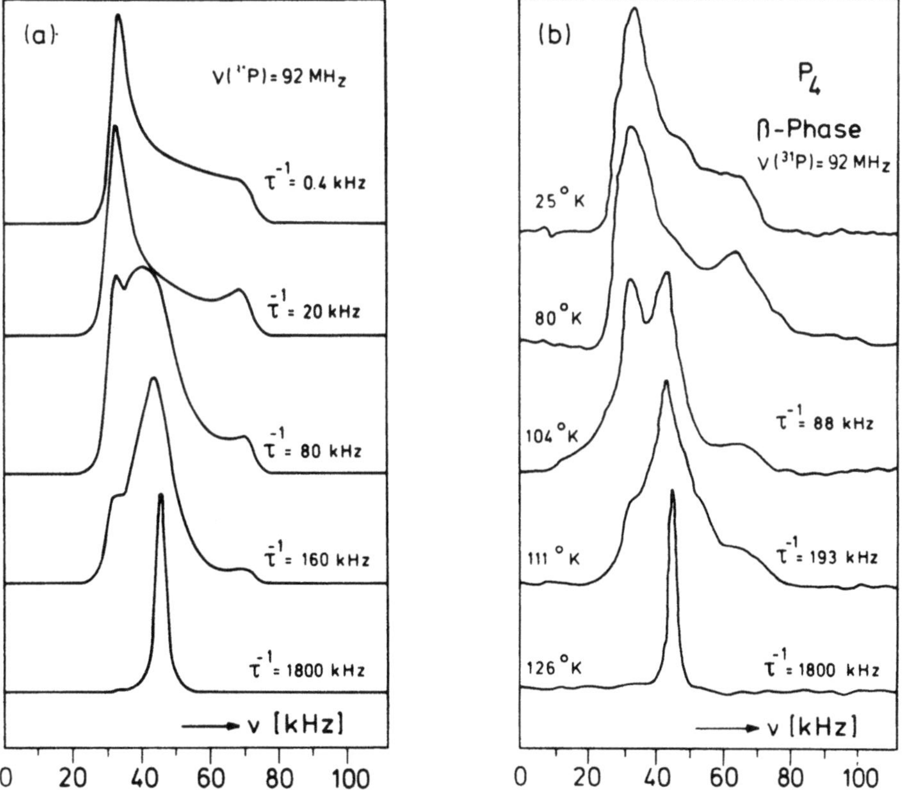

Fig. 2.17a, b. Calculated (**a**) and measured (**b**) powder line shapes for a random jumping tetrahedron with jump rates τ^{-1} according to Spiess et al. [71, 72]. The ^{31}P spectra are obtained from solid white phosphorus in the β-phase at various temperatures. The values of the jump rates τ^{-1} for the experimental spectra were obtained from T_1 data

Let us investigate the spectral changes which occur, when a molecule is jumping between N different sites j in a single crystal. In the "rigid case", i.e. when no hopping occurs j different spectral lines at frequencies ω_j, $j = 1 \ldots, N$ are observable. At elevated temperatures, however, the molecule begins to jump and the nuclear spins connected with it are exchanging local fields, i.e. frequencies ω_j depending on the orientation of the external magnetic field. Let the site exchange rate be $\kappa_{jk} = \tau_{jk}^{-1}$ and W_j the a priori probability for occupation of site j, then detailed balance requires

$$W_j \kappa_{jk} = W_k \kappa_{kj}.$$

In the following we always assume this condition to hold. Following Abragam [1] the equation of motion for the magnetization $M_j(t)$ at site j follows from the equation of motion for the density matrix and may be expressed as

$$\frac{d}{dt} M_j(t) = i\omega_j M_j(t) - \frac{1}{T_{2j}} M_j(t) + \sum_k \pi_{jk} M_k(t) \tag{2.143}$$

where non-secular terms have been neglected. It is assumed, that the transverse relaxation time T_{2j} is caused by other interactions than the exchange process. The exchange matrix $\tilde{\pi}$ describes the jumping process and depends on the specific model used to describe the molecular motion. Before considering this process further we want to derive expressions for the free induction decay $G(t)$ and the lineshape $I(\omega)$ for arbitrary $\tilde{\pi}$.

Let $M(t)$ be a column vector with N elements $M_j(t)$ Eq. (2.143) can be expressed as

$$\frac{d}{dt} \mathbf{M}(t) = (i\tilde{\omega} + \tilde{\pi}) \mathbf{M}(t) \tag{2.144}$$

where $\tilde{\pi}$ is the exchange matrix and $\tilde{\omega}$ is a diagonal matrix, with matrix elements $\omega_j + i/T_{2j}$. In most cases $1/T_{2j}$ will be ignored. Equation (2.144) is formally solved as

$$\mathbf{M}(t) = \exp[(i\tilde{\omega} + \tilde{\pi}) t] \, \mathbf{M}(t=0)$$

where the initial magnetization vector $\mathbf{M}(t=0)$ equals the a priori probability vector $\mathbf{W} = (W_1, \ldots W_N)$. The FID $G(t) = \sum_j M_j(t)$ is now conveniently expressed as [1]

$$G(t) = \mathbf{1} \cdot \exp[i(\tilde{\omega} + \tilde{\pi}) t] \cdot \mathbf{W} \tag{2.145}$$

where $\mathbf{1} = (1, 1, \ldots, 1)$ is a row "one-vector". Fouriertransformation leads to

$$\mathbf{M}(\omega) = \tilde{\mathbf{A}}(\omega)^{-1} \mathbf{M}(t=0)$$

with

$$\tilde{\mathbf{A}}(\omega) = i(\omega\tilde{\mathbf{1}} - \tilde{\omega}) - \tilde{\pi}$$

where $\tilde{\mathbf{1}}$ is the $N \times N$ unit matrix. The lineshape is readily expressed as

$$g(\omega) = \mathbf{1} \cdot \tilde{\mathbf{A}}(\omega)^{-1} \cdot \mathbf{W}. \tag{2.146}$$

In NMR the real part of $g(\omega)$ is usually termed the lineshape $I(\omega)$, i.e.

$$I(\omega) = \mathrm{Re}\{\mathbf{1} \cdot \tilde{\mathbf{A}}(\omega)^{-1} \cdot \mathbf{W}\}. \tag{2.147}$$

The calculation of spectra according to Eq. (2.147) may become very time consuming even on a fast digital computer. Gordon and McGinnis [75] proposed therefore a procedure, which uses the so-called QR transformation [76] to diagonalize the non-Hermitian matrix $i\tilde{\omega} + \tilde{\pi}$ as

$$\tilde{\mathbf{S}}^{-1}(i\tilde{\omega} + \tilde{\pi})\tilde{\mathbf{S}} = \tilde{\lambda}$$

where $\lambda_{ij} = \lambda_j \delta_{ij}$ with the eigenvalues λ_j.

The FID can now be expressed as

$$G(t) = \mathbf{1} \cdot \tilde{\mathbf{S}} \cdot \exp[\tilde{\lambda}t]\, \tilde{\mathbf{S}}^{-1} \cdot \mathbf{W} \tag{2.148a}$$

or

$$G(t) = \sum_j (\mathbf{1} \cdot \tilde{\mathbf{S}})_j\, e^{\lambda_j t}(\tilde{\mathbf{S}}^{-1}\mathbf{W})_j \tag{2.148b}$$

and the lineshape correspondingly

$$I(\omega) = \mathrm{Re}\{\mathbf{1} \cdot \tilde{\mathbf{S}}[i\omega\tilde{\mathbf{1}} - \tilde{\lambda}]^{-1}\tilde{\mathbf{S}}^{-1} \cdot \mathbf{W}\}$$

$$I(\omega) = \mathrm{Re}\sum_j \frac{(\mathbf{1} \cdot \tilde{\mathbf{S}})_j\,(\tilde{\mathbf{S}}^{-1} \cdot \mathbf{W})}{i\omega - \lambda_j}. \tag{2.149}$$

Thus the whole line shape problem of N lines is reduced to a single sum over the N lines once the diagonalization has been carried out. However, not only the diagonalization procedure has to be performed, but also the transformation matrix $\tilde{\mathbf{S}}$ and its inverse must be determined.

So far, we have considered only one specific orientation of the magnetic field B_0 with respect to the crystal frame. In a powder sample, however, all orientations of crystallites usually appear in the sample with equal probability and we will have to average $G(t, \Omega)$ and $I(\omega, \Omega)$ over all solid angles Ω. If the sample contains already some anisotropic orientational distribution $P(\Omega)$ we have to calculate in general

$$G(t) = \int d\Omega\, P(\Omega)\, G(t, \Omega) \tag{2.150a}$$

and

$$I(\omega) = \int d\Omega\, P(\Omega)\, I(\omega, \Omega) \tag{2.150b}$$

where the site frequencies $\omega_j(\Omega)$ are usually known from rigid lattice spectra. In the case of an isotropic powder $P(\Omega) = 1/4\pi$ when $d\Omega = \sin\vartheta\, d\vartheta\, d\varphi$ is used, whereas $P(\Omega) = 1/8\pi^2$ when $d\Omega = \sin\vartheta\, d\vartheta\, d\varphi\, d\psi$.

Because of singularities in $I(\omega, \Omega)$ the proper integration steps have to be chosen and numerical instabilities leading to fake spikes and kinks can easily occur. It is therefore adventageous to calculate $G(t)$ and obtain $I(\omega)$ by Fouriertransform. This procedure is always numerically stable and can be performed, using a fast algorithm.

As an illustrative example let us consider a two site exchange problem with the frequencies $\omega_1 = \omega(\Omega_1)$ and $\omega_2 = \omega(\Omega_2)$.

Using the exchange matrix

$$\tilde{\pi} = \kappa \begin{pmatrix} -1 & 1 \\ 1 & -1 \end{pmatrix} \tag{2.151}$$

with the residence time $\tau_r = \kappa^{-1}$ of a particular site we arrive at

$$\tilde{\mathbf{A}} = \begin{pmatrix} i(\omega - \omega_1) + \kappa & -\kappa \\ -\kappa & i(\omega - \omega_2) + \kappa \end{pmatrix}. \tag{2.152}$$

With $\mathbf{W} = \frac{1}{2}(1, 1)$ and $\mathbf{1} = (1, 1)$ the following spectrum according to Eq. (2.147) results

$$I(\omega) = \frac{1}{2} \frac{\kappa(\omega_1 - \omega_2)^2}{[(\omega - \omega_1)(\omega - \omega_2)]^2 + \kappa^2[2\omega - (\omega_1 + \omega_2)]^2} \tag{2.153}$$

which has been discussed in detail by Abragam [1]. As mentioned above it might be numerically more stable to sum up free induction decays when additional integration has to be performed over different orientations.

$G(t)$ is readility obtained by Fouriertransform and can be expressed as [77]

$$G(t, \Omega) = \exp(-\kappa t)[(\kappa/R) \sinh Rt + \cosh Rt] \tag{2.154}$$

with $R = [\kappa^2 + \omega_1 \omega_2]^{1/2}$ and where the average frequency $\langle \omega \rangle = (\omega_1 + \omega_2)/2$ has been set to zero.

Suppose there is a flipping of the molecule or a molecular group about an angle δ i.e. the interaction tensor connected with this molecule is flipped by an angle δ about an arbitrary axis. To determine the secular part of the interaction connected with this flipping, we write

$$\mathcal{H}'_{\text{int}} = A_{00} T_{00} + A_{20} T_{20}$$

which for interactions linear in the spin variable may be rewritten as

$$\mathcal{H}'_{\text{int}} = \left[\frac{1}{3} \text{Tr}\{A_{ij}\} + 2\sqrt{6}\, A_{20}(\delta) \right] B_0 I_z. \tag{2.155}$$

The two characteristic resonance frequencies ω_1 and ω_2 can be expressed as

$$\begin{aligned} \omega_1 &= \omega_0 + \omega(\delta = 0) \\ \omega_2 &= \omega_0 + \omega(\delta) \end{aligned} \tag{2.156}$$

where

$$\omega_0 = \frac{1}{3} \text{Tr}\{A_{ij}\} \quad \text{and} \quad \omega(\delta) = \frac{2}{\sqrt{6}} A_{20}(\delta).$$

Suppose $R''(\alpha, \beta, \gamma)$ is the transformation from the principal axis system $(1, 2, 3)$ to the molecular flipping frame (x_M, y_M, z_M) with the z_M-axis being the flipping axis and where $R'_z(\delta)$ represents the flippling about the z_M-axis by an angle δ. A further transformation $R(\varphi, \vartheta, \psi)$ has to be applied to transform from the molecular frame (x_M, y_M, z_M) to the laboratory frame (x, y, z). The complete transformation strategy is outlined as follows

$$\tilde{\mathbf{A}}'' \xrightarrow{R''(\alpha, \beta, \gamma)} \tilde{\mathbf{A}}_{M'} \xrightarrow{R'_z(\delta)} \tilde{\mathbf{A}}_M \xrightarrow{R(\varphi, \vartheta, \psi)} \tilde{\mathbf{A}}'$$
$$(1, 2, 3) \qquad\qquad\qquad (x_M, y_M, z_M) \qquad (x, y, z)$$

and the total transformation may be written as

$$R_{\text{total}} = R'(\varphi, \vartheta, \psi) R'_z(\delta) R''(\alpha, \beta, \gamma). \tag{2.157}$$

The powder average has to be performed over the angles $(\varphi, \vartheta, \psi)$ whereas (α, β, γ) represent the relation of the principle axis system to the flipping axis. Using the transformation properties of irreducible spherical tensors as outlined in Appendix A we obtain

$$A_{20}(\delta) = \sum_{q,p} A''_{2q} D^{(2)}_{qp}(\alpha, \beta, \gamma) e^{-ip\delta} D^{(2)}_{p0}(\varphi, \vartheta, \psi) \tag{2.158}$$

or

$$A_{20}(\delta) = \sum_{q,p} A''_{2q} d^{(2)}_{qp}(\beta) d^{(2)}_{p0}(\vartheta) e^{-ip(\gamma + \varphi + \delta)}. \tag{2.159}$$

Using (see Appendix A)

$$A''_{20} = \frac{3}{\sqrt{6}}(S_{33} - \tfrac{1}{3}\mathrm{Tr}\{S_{ij}\}) = \frac{3}{\sqrt{6}} S^*_{33}$$

$$A''_{2\pm 1} = 0 \quad \text{and} \quad A''_{2\pm 2} = \tfrac{1}{2}(S_{11} - S_{22})$$

we arrive at

$$\begin{aligned}
\omega(\delta) = \frac{2}{\sqrt{6}} A_{20}(\delta) = {}& S^*_{33} \sum_{p=-2}^{+2} d^{(2)}_{0p}(\beta) d^{(2)}_{p0}(\vartheta) e^{-ip(\gamma + \varphi + \delta)} \\
& + \tfrac{1}{2}(S_{11} - S_{22}) \sum_{p=-2}^{+2} (d^{(2)}_{2p}(\beta) e^{-i2\alpha} + d^{(2)}_{-2p}(\beta) e^{i2\alpha}) \\
& \cdot d^{(2)}_{p0}(\vartheta) e^{-i(\gamma + \varphi + \delta)}.
\end{aligned} \tag{2.160}$$

Further reduction of Eq. (2.160) is straightforwardly possible using the Wigner rotation matrices as listed in Appendix B and results in

$$\begin{aligned}
\omega(\delta) = {}& S^*_{33}[P_2(\cos\beta) P_2(\cos\vartheta) - \tfrac{3}{4}(\sin 2\beta \sin 2\vartheta \cos(\gamma + \varphi + \delta) \\
& - \sin^2\beta \sin^2\vartheta \cos 2(\gamma + \varphi + \delta))] \\
& + \sqrt{\tfrac{3}{8}}(S_{11} - S_{22})[P_2(\cos\vartheta)\sin^2\beta \cos 2\alpha + \sin\beta \sin 2\vartheta \\
& \cdot (\cos\beta \cos 2\alpha \cos(\gamma + \varphi + \delta) - \sin 2\alpha \sin(\gamma + \varphi + \delta)) \\
& + \sin^2\vartheta(\tfrac{1}{2}(1 + \cos^2\beta) \cos 2\alpha \cos 2(\gamma + \varphi + \delta) \\
& - \cos\beta \sin 2\alpha \sin 2(\gamma + \varphi + \delta))]
\end{aligned} \tag{2.161}$$

where P_2 is the Legendre Polynomial. (α, β, γ) are the Euler angles of the flipping axis with respect to the principal axis system of the tensor, with δ being the flipping angle. (ϑ, φ) are the Euler angles of the magnetic field B_0 with respect to the molecular frame containing the flipping axis as the z-axis.

Equation (2.161) becomes particularly simple in the case of axial symmetry $(S_{11} = S_{22})$. As a simple example let us consider a water molecule (H_2O or DHO) in a solid, like in gypsum. The proton shielding tensor is supposed to be axially symmetric about the bond direction and the water molecule is assumed to perform 180° jumps about an axis, bisecting the HOH-angle ($\beta = 54°44'$).

Under these conditions Eq. (2.161) reduces to

$$\omega(\delta) = S_{33}^* \left[\tfrac{1}{2}\sin^2\vartheta \cos 2(\varphi+\delta) - \frac{1}{\sqrt{2}}\sin 2\vartheta \cos(\varphi+\delta) \right]. \qquad (2.162\,\text{a})$$

With the two conditions $\delta = 0$, $180°$ we obtain

$$\omega_1 - \omega_2 = S_{33}^* \sqrt{2}\sin 2\vartheta \cos\varphi \qquad (2.162\,\text{b})$$

$$\tfrac{1}{2}(\omega_1 + \omega_2) = S_{33}^* \sin^2\vartheta \cos 2\varphi. \qquad (2.162\,\text{c})$$

If these relations are inserted into Eqs. (2.154) and (2.153) the free induction decay and the corresponding spectrum can be calculated for different orientations of the magnetic field and for different flipping rates κ of the molecule. In the case of a powder sample in addition a powder average over (ϑ, φ) has to be performed. It may be convenient to define a normalized flipping rate

$$\eta = \frac{2\kappa}{\Delta S} = \frac{4\kappa}{3 S_{33}^*}.$$

Figure 2.18 represents a computer calculation of powder line shapes for different values of η for a water molecule, flipping by $180°$ about its two-fold axis [77].

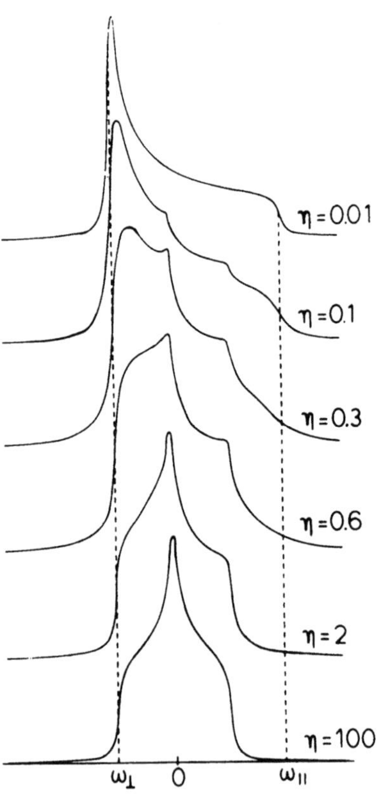

Fig. 2.18. Powder lineshapes of an axially symmetric shielding tensor, performing $180°$ jumps about an axis, which makes an angle of $54°44''$ with respect to the symmetry axis of the tensor (Becker [77]). Here η is the relative jump rate with respect to the anisotropy $\omega_\parallel - \omega_\perp$. This is considered to be a realistic model for "crystal water", like in $CaSO_4 \cdot 2H_2O$

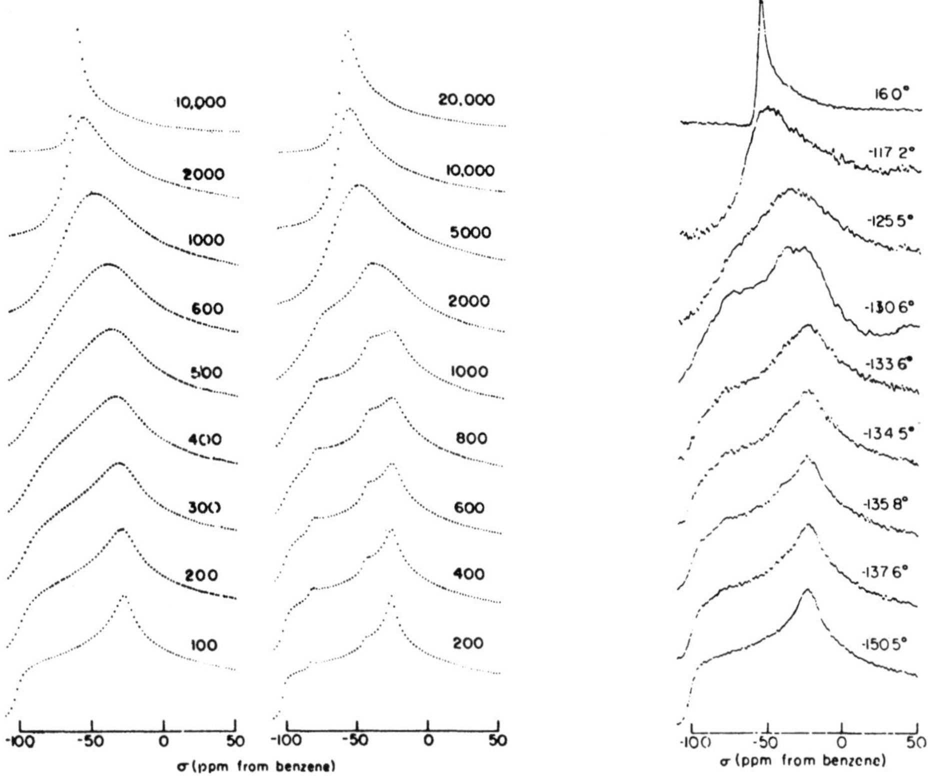

Fig. 2.19. Left part of the powder pattern of the ^{13}C spectrum of hexamethylbenzene at different jumping (or rotation) rates, according to Pines et al. [74]. Left: Calculated spectra assuming rotational diffusion with different rates. Middle: Calculated spectra assuming sixfold jumps about the C_6 axis with different jumping rates in Hertz, $v(^{13}C) = 45$ MHz. Right: Experimental spectra at various temperatures in degrees Celsius. (Courtesy of A. Pines)

Notice, that a completely non-axially symmetric pattern evolves at a very high flipping rate η. This can be rationalized easily by considering the fact that the tensor component orthogonal to the molecular plane (which is $\omega_\perp$) and the one along the flipping axis ($\omega = 0$) are not affected by the flipping motion. Since the trace of the tensor must be unchanged the third component should appear at $\omega_\parallel/2$.

Alexander et al. [73] and Pines et al. [74] have investigated powder line-shapes due to planar rotational jumps, both theoretically and experimentally. Figure 2.19 shows those spectra for hexamethylbenzene at various jumping rates about the sixfold axis [74]. There is a characteristic difference between sixfold jumps and rotational diffusion which is borne out by the spectra. Clearly, the experimental spectra seem to favour the jump model.

Similar spectra as the one shown in Figs. 2.18 and 2.19 have been obtained in ESR [67–70] and ^{2}D-NMR [78–83].

Let us now turn to an N-site exchange problem. Two different models may be considered:

(i) Nearest neighbour exchange $(\kappa_{jj+1}=\kappa_{jj-1}=\kappa=\tau^{-1})$ with the residence time in each position $\tau_r=(2\kappa)^{-1}=\tau/2$. The exchange matrix $\tilde{\pi}$ can be expressed as

$$\tilde{\pi}=\kappa\begin{pmatrix} -2 & 1 & 0 & \cdots & & 1 \\ 1 & -2 & 1 & 0 & \cdots & 0 \\ 0 & & & & & \vdots \\ \vdots & & & & & 1 \\ 1 & 0 & & \cdots & 1 & -2 \end{pmatrix} \qquad (2.163)$$

$$\text{or} \quad \pi_{jk}=\kappa[\delta_{jj+1}+\delta_{jj-1}-2\delta_{jk}] \quad \text{with } \delta_{jk}=\begin{cases}1 & \text{for } j=k \\ 0 & \text{for } j\neq k.\end{cases}$$

(ii) All sites exchange $(\kappa_{jk}=\kappa_{kj}=\kappa=\tau^{-1})$ with residence time $\tau_r=[(N-1)\kappa]^{-1}=\tau/(N-1)$. The exchange matrix $\tilde{\pi}$ can be expressed as

$$\tilde{\pi}=\kappa\begin{pmatrix} -(N-1) & 1 & 1 & \cdots & 1 \\ 1 & -(N-1) & 1 & \cdots & 1 \\ \vdots & & & & \vdots \\ 1 & & \cdots & \cdots & -(N-1) \end{pmatrix} \qquad (2.164)$$

or

$$\pi_{jk}=\kappa(1-N\delta_{jk}).$$

If the initial probabilities W_j are not equivalent, we may express $\tilde{\pi}$ as

$$\pi_{jk}=N\kappa[W_j-\delta_{jk}]. \qquad (2.165)$$

Notice that $\sum_k \pi_{jk}=0$ in cases (i) and (ii).

Wemmer et al. [84, 85] have shown that group theory, based on an approach of Rigny [86] can be used advantageously to simplify lineshape calculations. In the case of the all site exchange model (ii), Wemmer [84] has applied matrix manipulation to invert the matrix $\tilde{A}$. We shall follow this reasoning here in a slightly different may. By noting, that the exchange matrix $\tilde{\pi}$ of the all site exchange model (ii) can be written as

$$\tilde{\pi}=-N\kappa\tilde{1}+\kappa\tilde{I} \qquad (2.166)$$

with the unit matrix $\tilde{1}$ and the "one-matrix" $\tilde{I}$ which consists of only ones, we can express matrix

$$\tilde{A}=i(\omega\tilde{1}-\tilde{\omega})-\tilde{\pi} \qquad (2.167)$$

as

$$\tilde{A}=\tilde{B}+\tilde{C}$$

where

$$\tilde{B}=(i\omega+N\kappa)\tilde{1}-i\tilde{\omega} \quad \text{(diagonal)}$$

and

$$\tilde{C}=-\kappa\tilde{I}.$$

The matrix $\tilde{\mathbf{B}}$ is diagonal and its eigenvalues λ_j are given by

$$\lambda_j = i(\omega - \omega_j) + N\kappa \tag{2.168}$$

where the site frequency ω_j may still contain a local T_{2j} in the form $\omega_j + i/T_{2j}$ as discussed before. In order to calculate the lineshape $g(\omega)$ according to Eq. (2.147) we need to invert $\tilde{\mathbf{A}}$. We therefore write

$$\tilde{\mathbf{A}}^{-1} = \tilde{\mathbf{B}}^{-1}[\tilde{\mathbf{1}} + \tilde{\mathbf{C}}\tilde{\mathbf{B}}^{-1}]^{-1}$$

and expand

$$[\tilde{\mathbf{1}} + \tilde{\mathbf{C}}\tilde{\mathbf{B}}^{-1}] = \tilde{\mathbf{1}} - \tilde{\mathbf{C}} \cdot \tilde{\mathbf{B}}^{-1} + [\tilde{\mathbf{C}}\tilde{\mathbf{B}}^{-1}]^2 - [\tilde{\mathbf{C}}\tilde{\mathbf{B}}^{-1}]^3 + \dots$$

which can be expressed after some algebra as

$$\tilde{\mathbf{A}}^{-1} = \tilde{\mathbf{B}}^{-1}\left[\tilde{\mathbf{1}} - \frac{\kappa}{1 - \kappa L}\tilde{\mathbf{I}}\tilde{\mathbf{B}}^{-1}\right] \tag{2.169}$$

where

$$L = \sum_{j=1}^{N} \lambda_j^{-1}.$$

Assuming equal probability for all sites, i.e.

$$W_j = W_k = 1/N$$

the lineshape $g(\omega) = (1/N)\tilde{\mathbf{1}} \cdot \tilde{\mathbf{A}}^{-1} \cdot \tilde{\mathbf{1}}$ is obtained in the simple and compact form

$$g(\omega) = \frac{1}{N}\frac{L}{1 - \kappa L} \tag{2.170}$$

where

$$L = \sum_{j=1}^{N} [i(\omega - \omega_j) - 1/T_{2j} + N\kappa]^{-1}. \tag{2.171}$$

Let us discuss some special cases. In order to simplify notation we assume $1/T_{2j} = 0$, i.e. other than exchange motion effects are ignored. If we set $N = 2$ we immediately obtain $I(\omega) = \mathrm{Re}\, g(\omega)$ according to Eq. (2.153).

The case $N = 3$ is not only applicable to a methyl group or other three-site exchange problems like e.g. diffusion in tellurium and selenium [87], but is relevant to octahedral motion also as was pointed out by Spiess [14]. The reason is, that only three distinct frequencies occur due to the 180°-symmetry of the symmetric part of the shielding tensor. Inserting $N = 3$ in Eqs. (2.158–2.163) we obtain after some algebra

$$g(\omega) = \frac{1}{3}\frac{(\Omega_2 - 27\kappa^2) - i18\kappa(\omega - \langle\omega\rangle)}{2\kappa\Omega_2 + i[\Omega_3 - 9\kappa^2(\omega - \langle\omega\rangle)]} \tag{2.172}$$

where

$$\Omega_2 = (\omega - \omega_1)(\omega - \omega_2) + (\omega - \omega_1)(\omega - \omega_3) + (\omega - \omega_2)(\omega - \omega_3)$$

$$\Omega_3 = (\omega - \omega_1)(\omega - \omega_2)(\omega - \omega_3)$$

and

$$\langle\omega\rangle = (\omega_1 + \omega_2 + \omega_3)/3.$$

The spectrum $I(\omega) = \mathrm{Re}\, g(\omega)$ is readily obtained as [14]

$$I(\omega) = \tfrac{1}{3} \frac{2\kappa\Omega_2(\Omega_2 - 27\kappa^2) - 18\kappa(\omega - \langle\omega\rangle)[\Omega_3 - 9\kappa^2(\omega - \langle\omega\rangle)]}{4\kappa^2\Omega_2^2 + [\Omega_3 - 9\kappa^2(\omega - \langle\omega\rangle)]^2}. \tag{2.173}$$

The case $N = 4$ corresponds to the exchange motion of a tetrahedral molecule. By using Eqs. (2.158–2.163) one obtains [78a]

$$g(\omega) = \tfrac{1}{4} \frac{-i\Omega_3 - 8\kappa\Omega_2 + i3\cdot 4^3\kappa^2(\omega - \langle\omega\rangle)}{\Omega_4 - i3\kappa\Omega_3 - 8\kappa^2\Omega_2 + i4^3\kappa^3(\omega - \langle\omega\rangle)} \tag{2.174}$$

where

$$\Omega_4 = (\omega - \omega_1)(\omega - \omega_2)(\omega - \omega_3)(\omega - \omega_4)$$

$$\Omega_3 = (\omega - \omega_2)(\omega - \omega_3)(\omega - \omega_4) + (\omega - \omega_1)(\omega - \omega_2)(\omega - \omega_3)$$
$$\quad + (\omega - \omega_1)(\omega - \omega_2)(\omega - \omega_4) + (\omega - \omega_1)(\omega - \omega_3)(\omega - \omega_4)$$

$$\Omega_2 = (\omega - \omega_1)(\omega - \omega_2) + (\omega - \omega_1)(\omega - \omega_3) + (\omega - \omega_1)(\omega - \omega_4)$$
$$\quad + (\omega - \omega_2)(\omega - \omega_3) + (\omega - \omega_2)(\omega - \omega_4) + (\omega - \omega_3)(\omega - \omega_4)$$

and

$$\langle\omega\rangle = (\omega_1 + \omega_2 + \omega_3 + \omega_4)/4.$$

It is convenient to set $\langle\omega\rangle = 0$. Under this condition the spectrum $I(\omega) = \mathrm{Re}\, g(\omega)$ can be expressed as

$$I(\omega) = \tfrac{1}{4}\kappa \frac{\Omega_2[8\kappa^2\Omega_2 - \Omega_4][\Omega_3 - 3\cdot 4^3\kappa^2\omega][3\Omega_3 - 4^3\kappa^2\omega]}{[8\kappa^2\Omega_2 - \Omega_4]^2 + \kappa^2[3\Omega_3 - 4^3\kappa^2\omega]^2}. \tag{2.175}$$

Other exchange models may be calculated in a similar way.

3 Multiple-Pulse NMR Experiments

Multiple-pulse experiments have been applied to nuclear spins since the early days of NMR, beginning with the two pulse Hahn-echo [1], followed by three pulse echoes and the Carr-Purcell sequence [2], being the first cyclic multiple-pulse experiment at all. Later on modifications like the Gill-Meiboom [3] modification have been designed in order to reduce the influence of pulse errors which often have an accumulative effect in multiple-pulse sequences. Although these pulse sequences were first applied to liquids, it was soon realized, that similar pulse sequences could be applied to solids as well [4-8]. Even though only partial refocusing of the initial state was achieved, substantial insight into the spin dynamics of solids was gained in these experiments. Modified Carr-Purcell sequences applied to solids by Ostroff and Waugh [4] and by Mansfield and Ware [5] were unexpectedly successful in achieving a considerably lengthened decay of transverse magnetization. This effect contradicts the spin temperature hypothesis and was unexpected at first sight. Much effort was applied in designing more and more efficient multiple-pulse sequences. Since these sequences were promising in achieving a long transverse decay, but leaving resonance offset and chemical shift interactions effective, high resolution spectra in solids were expected. This development culminated in the four-pulse sequence by Waugh, Huber and Haeberlen [9], which is the most efficient basic multiple-pulse sequence up to date. Many modifications of this sequence have been proposed recently for correcting pulse errors.

Prior to any NMR experiment the spin system is supposed to be in thermal equilibrium described by the Boltzman spin density matrix [10, 11]

$$\rho_B = \exp(-\beta\mathcal{H})/\mathrm{Tr}\{\exp(-\beta\mathcal{H})\} \tag{3.1}$$

where $\beta = \hbar/k_B T$ and $\mathcal{H}$ is the stationary Hamiltonian of the system.

The observable in NMR is usually the magnetic dipole moment of the spin which induces a voltage in the sample coil. Any observable Q can be expressed by [10, 11]

$$\langle Q \rangle = \mathrm{Tr}\{Q\rho\} \tag{3.2}$$

which may be expanded as

$$\langle Q \rangle = \sum_{h=0}^{\infty} \frac{(-\beta)^n}{n!} \mathrm{Tr}\{Q\mathcal{H}^n\} \Big/ \sum_{k=0}^{\infty} \frac{(-\beta)^k}{k!} \mathrm{Tr}\{\mathcal{H}^k\}. \tag{3.3}$$

For bounded operators like spin operators the trace over all odd powers of $\mathcal{H}$ vanishes as does the trace over Q. We therefore define the "size" of a Hamiltonian as

$$\|\mathcal{H}\| \equiv [\mathrm{Tr}\{\mathcal{H}^2\}]^{1/2}. \tag{3.4}$$

At temperatures $T \geq 1 \, \text{mK}$ the quantity $\beta \|\mathscr{H}\| \ll 1$ and the expansion Eq. (3.3) converges rapidly. In fact only the leading term

$$\langle Q \rangle = -\beta \, \text{Tr}\{Q\mathscr{H}\}/\text{Tr}\{\tilde{1}\} \tag{3.5}$$

where

$$\text{Tr}\{\tilde{1}\} = (2I+1)^{N_I}$$

has to be taken into account. Referring to these arguments we shall use loosely speaking a "truncated density matrix"

$$\rho = -\beta\mathscr{H}/\text{Tr}\{\tilde{1}\} \tag{3.6}$$

in the following.

In NMR magnetic fields B_0 are applied to the sample which are much larger than any local field. Therefore the Zeeman interaction

$$\mathscr{H}_z = -\omega_{0I} I_z \tag{3.7}$$

where

$$\omega_{0I} = \gamma_I B_0$$

is always the leading term. The Boltzmann density matrix in truncated form

$$\rho_B = \frac{\beta \omega_{0I}}{(2I+1)^{N_I}} I_z \tag{3.8}$$

may be further abbreviated simply by $\rho_z = I_z$.

However, the principle of time resolved spectroscopy as applied here relies on the fact that a non-equilibrium state is created initially which then evolves under the influence of the systems Hamiltonian. The equation of motion of the density matrix $\rho(t)$ at time t is governed by the Liouville-v. Neumann equation

$$\frac{d}{dt}\rho(t) = -i[\mathscr{H}(t), \rho(t)] \tag{3.9}$$

which has the formal solution [12, 13]

$$\rho(t) = \mathbf{L}(t)\,\rho(0)\,\mathbf{L}^+(t) \tag{3.10}$$

where

$$\mathbf{L}(t) = T \exp\left[-i \int_0^t dt'\, \mathscr{H}(t')\right] \tag{3.11}$$

is a unitary operator, T being the Dyson time ordering operator and where $\mathbf{L}^+(t)$ is the adjoint of the "propagator" $\mathbf{L}(t)$ with $\mathbf{L}^+(t) = \mathbf{L}^{-1}(t)$ because $\mathscr{H}(t)$ is an Hermitian operator. If $\mathscr{H}$ does not explicitly depend on time $\mathbf{L}(t)$ assumes the simple form

$$\mathbf{L}(t) = \exp[-i\mathscr{H}t]. \tag{3.12}$$

In order to simplify notations we shall use in the following Liouville space notation as introduced in Appendix C. Operators which are matrices in Hilbert space are considered vectors in Liouville space by using the notation $|Q)$

for a "ket" and $(\mathbf{Q}|$ for a "bra" Liouville space vector [14-16]. The scalar product between two Liouville space vectors is then defined as

$$(\mathbf{B}|\mathbf{A})^* = (\mathbf{A}|\mathbf{B}) \equiv \text{Tr}\{\mathbf{A}^+\mathbf{B}\} \tag{3.13}$$

where $\mathbf{A}^+$ is the adjoint of $\mathbf{A}$. Liouville operators $\hat{\mathbf{C}}$ labelled by a hat are defined as superoperators [14-16]

$$\hat{\mathbf{C}}|\mathbf{B}) \equiv |[\mathbf{C},\mathbf{B}]) \tag{3.14a}$$

$$(\mathbf{A}|\hat{\mathbf{C}} \equiv ([\mathbf{C}^+,\mathbf{A}]| \tag{3.14b}$$

equivalently

$$\exp[-it\hat{\mathscr{H}}]|\mathbf{B}) \equiv |e^{-it\mathscr{H}}\mathbf{B}e^{+it\mathscr{H}}) \tag{3.15a}$$

as can be immediately realized by a series expansion of the exponential operator and by using Eq. (3.14). Equivalently

$$(\mathbf{A}|e^{-it\hat{\mathscr{H}}} \equiv (e^{it\mathscr{H}}\mathbf{A}e^{-it\mathscr{H}}| \tag{3.15b}$$

or

$$(\mathbf{A}|e^{-it\hat{\mathscr{H}}}|\mathbf{B}) \equiv \text{Tr}\{\mathbf{A}^+e^{-it\mathscr{H}}\mathbf{B}e^{it\mathscr{H}}\} = \text{Tr}\{e^{it\mathscr{H}}\mathbf{A}^+e^{-it\mathscr{H}}\mathbf{B}\}. \tag{3.16}$$

Other technicalities may be obtained from Appendix C.

Note, that we do not restrict ourselves to spin 1/2 nor do we need to know the wavefunctions. On the contrary we do not care about eigenfunctions or which basis set is the most appropriate one. All these subtleties are built into the scalar product i.e. into the trace operation which is independent of the representation. In fact the picture we are using is more physical in the sense that we visualize the spin operators as vectors which are rotated under the action of a unitrary operator which is determined by the interaction Hamiltonian. These arguments will be more obvious in the following.

As a simple example let us consider the application of a linearly polarized radio-frequency field perpendicular to B_0 beginning at $t=0$.

$$\mathscr{H}_1 = -2\omega_1 I_y \cos\omega t \tag{3.17}$$

where $\omega_1 = \gamma B_1$. Equation (3.17) may be written as

$$\mathscr{H}_1 = -\omega_1 I_y(e^{i\omega t} + e^{-i\omega t}).$$

The equation of motion for the density matrix reads now

$$\frac{d}{dt}|\rho(t)) = -i(\hat{\mathscr{H}}_z + \hat{\mathscr{H}}_1(t))|\rho(t)). \tag{3.18}$$

It turns out to be convenient to transform into a frame which rotates with the frequency ω of the radio-frequency field about the z-direction [10]. The density matrix $\rho^*(t)$ in this "rotating frame" may be defined as

$$|\rho^*(t)) = e^{-i\omega t \hat{I}_z}|\rho(t)). \tag{3.19}$$

The equation of motion for $\rho^*(t)$ is readily obtained by using Eqs. (3.18, 3.19) and $\mathscr{H}_z = -\omega_0 I_z$ as

$$\frac{d}{dt}|\rho^*(t)) = -i\hat{\mathscr{H}}^*(t)|\rho^*(t)) \tag{3.20}$$

where the Hamiltonian in the "rotating frame" is given by

$$\mathscr{H}^*(t) = (\omega - \omega_0)I_z + \mathscr{H}_1^*(t) \tag{3.21a}$$

with

$$\mathscr{H}_*(t) = e^{-i\omega t I_z}\,\mathscr{H}_1(t)\,e^{i\omega t I_z} \tag{3.21b}$$

or

$$\mathscr{H}_1^*(t) = -\omega_1 I_y[1 + \cos 2\omega t] - \omega_1 I_x \sin 2\omega t. \tag{3.22}$$

The time dependent or "non-secular" part does not contribute in first order (see however Appendix H) and may be dropped. The truncated Hamiltonian in the rotating frame reads now

$$\mathscr{H}^*(t) = -(\Delta\omega I_z + \omega_1 I_y) \tag{3.23}$$

where the resonance offset frequency is defined as $\Delta\omega = \omega_0 - \omega$. The solution of the equation of motion Eq. (3.20) of the density matrix $\rho^*(t)$ in the rotating frame is

$$|\rho^*(t)) = e^{it(\Delta\omega \hat{I}_z + \omega_1 \hat{I}_y)}|\rho(0)). \tag{3.24}$$

With the initial density matrix $\rho(0) = I_z$ we obtain

$$|\rho^*(t)) = e^{i\vartheta \hat{I}_x} e^{it\omega_e \hat{I}_z} e^{-i\vartheta \hat{I}_x}|I_z) \tag{3.25}$$

where we have used the property

$$e^{i\vartheta I_x} I_z e^{-i\vartheta I_x} = I_z \cos\vartheta + I_y \sin\vartheta$$

of rotational operators as discussed in Appendix B.
Here

$$\omega_e^2 = \Delta\omega^2 + \omega_1^2 \tag{3.26a}$$

and

$$\tan\vartheta = \omega_1/\Delta\omega. \tag{3.26b}$$

Equation (3.25) can be readily evaluated with the help of Appendix B as

$$|\rho^*(t)) = -\sin\vartheta \sin\omega_e t|I_x) + \sin\vartheta \cos\vartheta(1 - \cos\omega_e t)|I_y)$$
$$+ (\cos^2\vartheta + \sin^2\vartheta \cos\omega_e t)|I_z). \tag{3.27}$$

The situation is much simpler, if irradiation is performed on-resonance, i.e. $\Delta\omega = \omega_0 - \omega = 0$. In this case

$$e^{it\omega_1 \hat{I}_y}|I_z) = \cos\omega_1 t|I_z) - \sin\omega_1 t|I_x). \tag{3.28}$$

If the radio-frequency field is applied for a time t_p such that corresponding to a $\pi/2$-pulse the initial z-magnetization is completely transferred into x-magnetization $|\rho^*(\pi/2)) = -|I_x)$.

In the following we shall always work in the rotating frame and the star in ρ^* and $\mathscr{H}^*$ will be dropped to simplify notation.

The Bloch decay or free induction decay (FID) which is governed by the interaction Hamiltonian $\mathscr{H}_{int}$ in the rotating frame is in our notation expressed as

$$g_x(t) = (I_x|\rho(t))/(I_x|I_x) \tag{3.29a}$$

$$g_y(t) = (I_y|\rho(t))/(I_y|I_y) \tag{3.29b}$$

where

$$(I_x|I_x) = (I_y|I_y) = (I_z|I_z) = \mathrm{Tr}\{I_z^2\}$$

and

$$|\rho(t)) = e^{-it\hat{\mathscr{H}}_{\mathrm{int}}}\mathbf{P}(\beta)|I_z) \tag{3.30}$$

and where $P(\beta)$ stands for the unitary transformation (rotating by an angle β) due to the initial pulse. If the rf-field is sufficiently strong, i.e. $\omega_1 \gg \|\mathscr{H}_{\mathrm{int}}\|$ the interaction $\mathscr{H}_{\mathrm{int}}$ may be neglected during pulsed excitation and in case of a $\pi/2$-pulse we can write

$$\mathbf{P}(\pi/2)|I_z) \equiv e^{-i\frac{\pi}{2}\hat{I}_y}|I_z) = |I_x) \tag{3.31}$$

leading to

$$g_x(t) = (I_x|e^{-it\hat{\mathscr{H}}_{\mathrm{int}}}|I_x)/(I_z|I_z) \tag{3.32a}$$

$$g_y(t) = (I_y|e^{-it\hat{\mathscr{H}}_{\mathrm{int}}}|I_x)/(I_z|I_z). \tag{3.32b}$$

In the case of inhomogeneous line broadening

$$\mathscr{H}_{\mathrm{int}} = \sum_j \omega_j I_{zj} \tag{3.33}$$

the FID is readily expressed as

$$g_x(t) = \sum_j \cos \omega_j t$$

$$g_y(t) = \sum_j \sin \omega_j t.$$

Before discussing more sophisticated multiple-pulse sequences let us take a brief look at the two-pulse Hahn echo [1], where a second pulse P_2 follows the first pulse P_1 a time τ later. The time evolution after the second pulse

$$|\rho(t > \tau)) = e^{-i\hat{\mathscr{H}}_{\mathrm{int}}(t-\tau)}\mathbf{P}_2 e^{-i\hat{\mathscr{H}}_{\mathrm{int}}\tau}\mathbf{P}_1|I_z) \tag{3.34}$$

may be written as

$$|\rho(t > \tau)) = e^{-i\hat{\mathscr{H}}_{\mathrm{int}}(t-\tau)}e^{-i\hat{\mathscr{H}}'_{\mathrm{int}}\tau}\mathbf{P}_2\mathbf{P}_1|I_z)$$

where

$$\hat{\mathscr{H}}'_{\mathrm{int}} = \mathbf{P}_2\hat{\mathscr{H}}_{\mathrm{int}}\mathbf{P}_2^{-1}.$$

If P_2 is a π-pulse in the sense $\mathbf{P}_2\mathscr{H}_{\mathrm{int}}\mathbf{P}_2^{-1} = -\mathscr{H}_{\mathrm{int}}$ we obtain

$$|\rho(t > \tau)) = e^{-i\hat{\mathscr{H}}_{\mathrm{int}}(t-2\tau)}\mathbf{P}_2\mathbf{P}_1|I_z) \tag{3.35}$$

which describes a symmetric time evolution around $t = 2\tau$ where the echo appears.

3.1 Idealized Multiple Pulse Sequences

In this and some of the following sections we make the assumptions: (i) spin
lattice relaxation is neglected (ii) the applied rf pulses are assumed to be δ
pulses.

Let us first discuss the well-known

a) Carr-Purcell Sequence [2]

As shown in Fig. 3.1 a preparation pulse P_0 (usually a $\pi/2$ pulse) is applied to
initiate the first coherent state, which decays during time τ to some intermediate state due to the static secular spin interactions. At time τ a π pulse
P_1 refocusses all the spins (only if the interactions are linear in the spin
variable) to produce an echo at 2τ. The state at 2τ can now be considered as a
new initial coherent state whose decay can be inverted by another π pulse a
time τ later and so fourth (see Fig. 3.1).

With

$$\mathbf{P}_0 = e^{-i\frac{\pi}{2}\hat{I}_y}; \quad \mathbf{P}_k = e^{-i\pi\hat{I}_y} \quad k = 1, 2 \ldots n$$

and

$$\mathbf{D}(\tau) = e^{-i\mathcal{H}_i\tau}$$

where $\mathcal{H}_i$ is the interaction Hamiltonian we obtain for the spin density matrix
at $t = 2\tau$

$$|\rho(2\tau)) = \mathbf{D}(\tau)\mathbf{P}_1\mathbf{D}(\tau)|\rho(0)) \tag{3.36}$$

or

$$|\rho(2\tau)) = \mathbf{L}(2\tau)|\rho(0)) \tag{3.37}$$

where

$$\mathbf{L}(2\tau) = \mathbf{D}(\tau)\mathbf{P}_1\mathbf{D}(\tau) \tag{3.38}$$

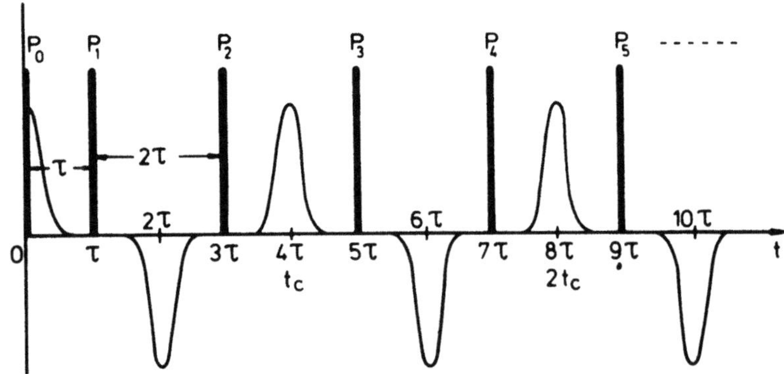

Fig. 3.1. Schematic drawing of a Carr-Purcell sequence, generated by P_0 being a $\pi/2$ pulse and the
π pulse P_i, $i \geqq 1$. Notice, that the phase of the spin echoes is alternating

and

$$|\rho(0)) = \mathbf{P}_0 |\rho_B).$$

The time evolution operator $\mathbf{L}(2\tau)$ in Eq. (3.38) may be rewritten as

$$\mathbf{L}(2\tau) = \mathbf{D}(\tau)\tilde{\mathbf{D}}(\tau)\mathbf{P}_1 \tag{3.39}$$

where

$$\tilde{\mathbf{D}}(\tau) = \mathbf{P}_1\mathbf{D}(\tau)\mathbf{P}_1^{-1} = e^{-i\tilde{\mathscr{H}}_i\tau} \tag{3.40}$$

with

$$\tilde{\mathscr{H}}_i = \mathbf{P}_1\mathscr{H}_i\mathbf{P}_1^{-1}. \tag{3.41}$$

Let us assume $\tilde{\mathscr{H}}_i = -\mathscr{H}_i$ i.e. the pulses $\mathbf{P}_k$ invert the interaction Hamiltonian in the case of δ pulses which is consistent with interactions, which depend linearly on the spin variable (e.g. shielding Hamiltonian); we obtain:

$$\mathbf{D}(\tau)\tilde{\mathbf{D}}(\tau) = \tilde{\mathbf{1}} \quad \text{and} \quad \mathbf{L}(2\tau) = \mathbf{P}_1$$

where $\tilde{\mathbf{1}}$ is the unit operator.

In general

$$|\rho(n2\tau)) = \mathbf{L}(n2\tau)|\rho(0)) \tag{3.42}$$

where

$$\mathbf{L}(n2\tau) = \prod_{k=1}^{n} \mathbf{P}_k = \mathbf{P}_1^n$$

which results in

$$\mathbf{L}(n2\tau) = [\exp(-i\pi\hat{I}_y)]^n = \exp(-in\pi\hat{I}_y) \tag{3.43}$$

or with the initial state $|\rho(0)) = |I_x)$ in

$$|\rho(n2\tau)) = \cos n\pi |I_x) = (-1)^n |I_x) \tag{3.44}$$

displaying the alternating sign of the echo observed in Carr-Purcell experiments, when phase sensitive detection is used. Notice that identical states of the density matrix occur after each "cycle" containing two pulses. This leads us to defining a "cycle" by

$$\prod_{k=1}^{n} \mathbf{P}_k = \tilde{\mathbf{1}} \tag{3.45}$$

if the lowest possible n is chosen. We realize that $n=2$ for the Carr-Purcell sequence with a "cycle time" $t_c = 4\tau$.

An accumulation of pulse length errors may spoil the beauty of the classical Carr-Purcell experiment. Assuming

$$\mathbf{P}_k = e^{-i(\pi+\varepsilon)\hat{I}_y}$$

we see, that because of

$$\prod_{k=1}^{n} \mathbf{P}_k = e^{-in\varepsilon\hat{I}_y}e^{-in\pi\hat{I}_y} \tag{3.46}$$

the pulse length error ε accumulates, leading to a destruction of the echo envelope.

Gill and Meiboom [3] suggested a modification which cures this defect by a 90° phase shift of the initializing pulse with respect to the cyclic pulses as follows

$$\mathbf{P}_0 = e^{-i\frac{\pi}{2}\hat{I}_y}; \qquad \mathbf{P}_k = e^{-i\pi\hat{I}_x}. \tag{3.47}$$

Since $\mathbf{P}_k$ commutes with the initial state, we now have

$$\prod_{k=1}^{n} \mathbf{P}_k |\rho(0)\rangle = |\rho(0)\rangle$$

and only the error due to the violation of the condition $\tilde{\mathscr{H}}_i = -\mathscr{H}_i$ is still effective. If the rotation angle is close to π, this is far less dramatic than without the 90° phase shift. Notice that the echo phase is no longer alternating.

We turn next to the

b) Phase Alternated Sequence (PAS) [4–7]

which is merely a series of equally spaced δ pulses alternating in phase (see Fig. 3.2). An initializing pulse $\mathbf{P}_0$, as in the Carr-Purcell sequence, may be applied but is not a necessity.

With

$$\mathbf{P}_1 = e^{-i\beta\hat{I}_y}; \qquad \mathbf{P}_2 = \mathbf{P}_1^{-1} \quad \text{and} \quad \mathbf{P}_k\mathbf{P}_{k+1} = \tilde{\mathbf{1}}$$

we obtain for the propagator $\mathbf{L}(t_c)$ at the cycle time $t_c = 4\tau$

$$\mathbf{L}(t_c) = \mathbf{D}(2\tau)\mathbf{P}_2\mathbf{D}(2\tau)\mathbf{P}_1 \tag{3.48}$$

or

$$\mathbf{L}(t_c) = e^{-i\tilde{\mathscr{H}}_i 2\tau} e^{-i\tilde{\mathscr{H}}_i 2\tau} \tag{3.49}$$

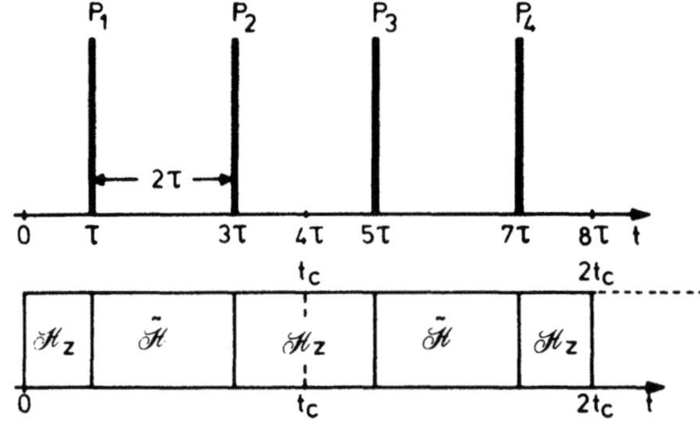

Fig. 3.2. Phase alternating sequence with two pulses per cycle alternating in phase. The interaction Hamiltonian in the "toggling rotating frame" at different times during the cycle is schematically drawn

where

$$\hat{\tilde{\mathcal{H}}}_i = \mathbf{P}_1^{-1} \hat{\mathcal{H}}_i \mathbf{P}_1.$$

$\mathbf{L}(t_c)$ may be formally rewritten as

$$\mathbf{L}(t_c) = e^{-i\hat{\mathcal{H}}_i^* t_c} \tag{3.50}$$

where $\hat{\mathcal{H}}_i^*$ is defined by equating Eqs. (3.49) and (3.50).

The spin density at any integer multiple of the cycle time can be expressed as

$$|\rho(Nt_c)) = \mathbf{L}(Nt_c)|\rho(0)) \tag{3.51}$$

where

$$\mathbf{L}(Nt_c) = \mathbf{L}^N(t_c) = \exp(-i\hat{\mathcal{H}}_i^* Nt_c). \tag{3.52}$$

As stated before, we are mainly concerned with the time evolution of the magnetization, the issue is to find expressions for $\mathbf{L}(t_c)$. The mathematical effort is considerably reduced by exploiting the cyclic property of the multiple-pulse sequences, i.e. the time evolution operator of the system has to be evaluated over one cycle only, in order to describe the total response of the system at any integer multiple of the cycle time.

Using Eq. (3.49), we obtain in the case of a shift Hamiltonian

$$\mathcal{H}_i = \sum_k \omega_k I_{zk}$$

$$\mathbf{L}(t_c) = \exp(-i\sum_k 2\omega_k \tau \hat{I}_{zk}) \cdot \exp[-i\sum_k 2\omega_k \tau(\hat{I}_{zk}\cos\beta - \hat{I}_{xk}\sin\beta)] \tag{3.53}$$

where β is the rotation angle of the pulses.

The general system response can be calculated exactly, using this propagator [Eq. (3.53)], but it is more elucidating to discuss the response in the limit of small pulse spacing 2τ.

In this limit, i.e.

$$\|\mathcal{H}_1\|\tau; \quad \|\mathcal{H}_2\|\tau \ll 1$$

using Eq. (3.53) and

$$\exp(-i\mathcal{H}_1\tau)\exp(-i\mathcal{H}_2\tau) \simeq \exp(-i(\mathcal{H}_1 + \mathcal{H}_2)\tau)$$

with

$$\mathbf{L}(t_c) = \exp(-i\hat{\mathcal{H}}^* t_c) \quad t_c = 4\tau$$

we find for the "average" Hamiltonian $\mathcal{H}^*$

$$\mathcal{H}^* = \sum_k \frac{\omega_k}{2}[I_{zk}(1+\cos\beta) - I_{xk}\sin\beta]. \tag{3.54}$$

An effective quantization axis in the x, z plane is defined by Eq. (3.54), which makes an angle ϑ with the z axis and where

$$-\tan\vartheta = \sin\beta/(1+\cos\beta).$$

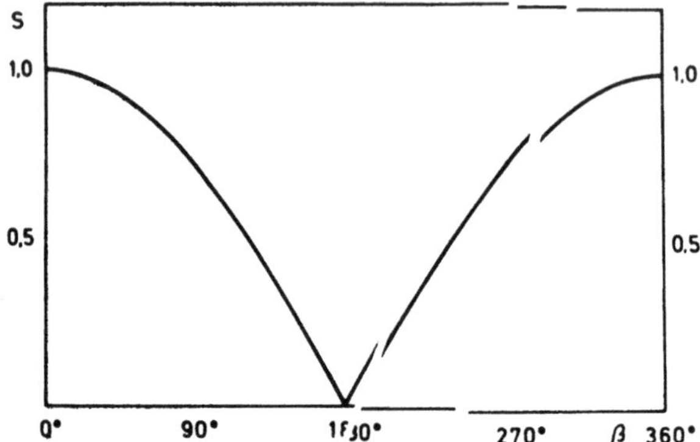

Fig. 3.3. Theoretical scaling factor S of interactions, which are linear in the spin variable (e.g. chemical shift, Knight-Shift etc.) versus the rotation angle of the rf pulse in the phase alternating sequence according to Eq. (3.55)

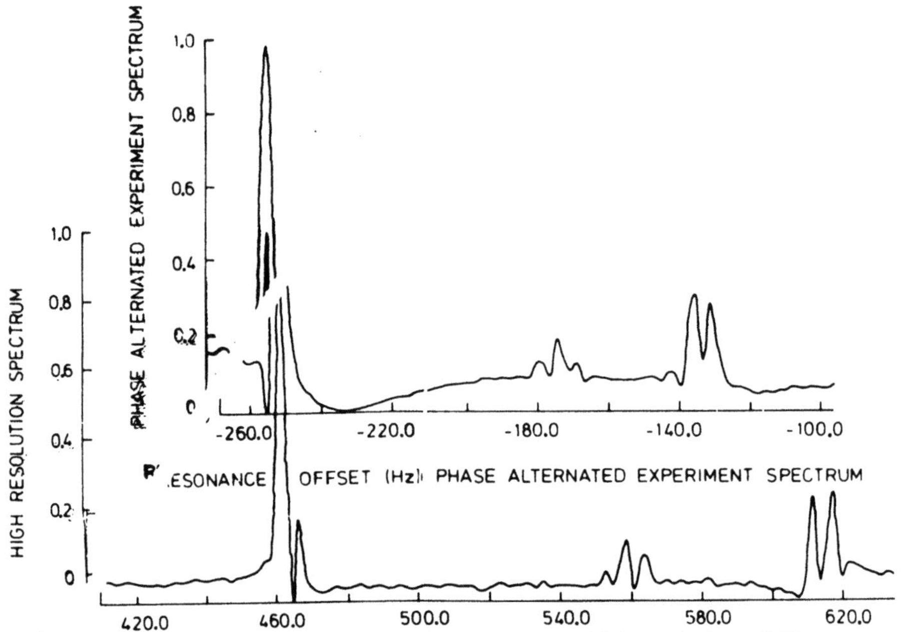

Fig. 3.4. Application of the phase-alternated sequence to a high resolution NMR spectrum by J.D. Ellett and J.S. Waugh [17] to demonstrate the scaling effect

The effective frequency ω_{ke} which describes the oscillations of the decay can be straightforwardly derived from Eq. (3.54) as

$$\omega_{ke} = S\omega_k$$

where

$$S = [(1 + \cos \beta)/2]^{1/2} = |\cos(\beta/2)| \tag{3.55}$$

is the "scaling factor" by which all the spectral line shifts are scaled. This effect has been named "chemical shift concertina" and was discussed in detail by Ellett and Waugh [17].

By varying the rotation angle β of the rf pulses different scaling factors can be obtained, showing up as a simulation of different smaller static magnetic fields although the spectrometer is working at a high field. These features are demonstrated in Figs. 3.3 and 3.4.

We have only treated the simple shift Hamiltonian so far, since the theoretical results are especially easy to visualize. If we turn to secular dipolar interactions with

$$\mathcal{H}_{Dz} = \sum_{i<j} A_{ij}(3I_{zi}I_{zj} - \mathbf{I}_i \cdot \mathbf{I}_j) \quad \text{where} \quad A_{ij} = \tfrac{1}{2}\gamma_I^2 \hbar(1 - 3\cos^2 \vartheta_{ij})/r_{ij}^3$$

we immediately realize that $\mathcal{H}_{Dz}$ is invariant under a π rotation, i.e. it cannot be inverted by a π pulse. As a consequence we expect a completely different behavior as compared with the shift Hamiltonian.

For a phase alternated 90° pulse sequence, we obtain according to Eq. (3.49)

$$\mathbf{L}(t_c) = \exp(-i\hat{\mathcal{H}}_{Dz} 2\tau) \cdot \exp(-i\hat{\mathcal{H}}_{Dx} 2\tau)$$

where

$$\hat{\mathcal{H}}_{Dx} = \exp\left(-i\frac{\pi}{2}\hat{I}_y\right) \hat{\mathcal{H}}_{Dz} \exp\left(i\frac{\pi}{2}\hat{I}_y\right)$$

or

$$\mathcal{H}_{Dx} = \sum_{i<j} A_{ij}(3I_{xi}I_{xj} - \mathbf{I}_i \cdot \mathbf{I}_j).$$

In the limit $\|\mathcal{H}_D\| \tau \ll 1$, the one cycle propagator $\mathbf{L}(t_c)$ may be written as

$$\mathbf{L}(t_c) = \exp[-it_c(\hat{\mathcal{H}}_{Dz} + \hat{\mathcal{H}}_{Dx})/2]$$

or

$$\mathbf{L}(t_c) = \exp\left[it_c \tfrac{1}{2}\hat{\mathcal{H}}_{Dy}\right] \tag{3.56}$$

where

$$\mathcal{H}_{Dx} + \mathcal{H}_{Dy} + \mathcal{H}_{Dz} = 0$$

has been used.

The dipolar interaction is consequently reduced by a factor of two, leading to twice the free induction decay time. If the initial condition is prepared in the y direction by a $\pi/2$ x-pulse no decay results due to $\mathbf{L}(t_c)$, since $|I_y)$ is invariant under the operation of $\mathbf{L}(t_c)$ as given by Eq. (3.56). This is the case of spin-locking. Ostroff and Waugh [4] and also Mansfield and Ware [5] have analyzed this sequence in more detail. If shift interaction and dipolar interaction are present at the same time, shift interactions are resolved only up to the point where the largest shift is just resolved [4].

Since both groups Mansfield and co-workers and Waugh and co-workers have extensively investigated the behavior of dipolar interaction in this two pulse sequence, we shall refer to it in later sectons as MW-2 sequence. The corresponding 90° pulse sequence without phase alternation which is actually a four pulse sequence will be referred to as MW-4.

3.2 The Four-Pulse Sequence [9] (WHH-4)

The basic multiple-pulse experiment with vanishing average dipolar and quadrupolar Hamiltonian, retaining shift interactions is the four-pulse experiment, proposed by J.S. Waugh, L.M. Huber and U. Haeberlen [9] (see Fig. 3.5). The rf pulses P_i are assumed to be δ pulses with a rotation angle of 90°. If P_x represents the rotation operator about the x direction of the rotating frame, this sequence may be symbolized in short form by $(P_{\bar{y}} - 2\tau - P_y - \tau - P_x - 2\tau - P_{\bar{x}} - \tau -)_n$, where the sequence with the cycle time t_c is repeated n times. But it is also legitimate to visualize the cycle from $0^* \to t_c^*$ in short form $(-\tau - P_y - \tau - P_x - 2\tau - P_{\bar{x}} - \tau - P_{\bar{y}} - \tau)_n$ as a four-pulse cycle which is repeated n times. In both cases $t_c = t_c^* = 6\tau$, where τ is the closest pulse spacing. The magnetization is usually sampled once every cycle during one of the large "windows" following the $\bar{y}$ or x pulse (see Fig. 3.5).

The one cycle propagator $\mathbf{L}(t_c)$ of the four-pulse experiment can be written down immediately, following Fig. 3.5 as:

$$\mathbf{L}(t_c) = \mathbf{D}_z(\tau)\,\mathbf{P}_{\bar{x}}\,\mathbf{D}_z(2\tau)\,\mathbf{P}_x\,\mathbf{D}_z(\tau)\,\mathbf{P}_y\,\mathbf{D}_z(2\tau)\,\mathbf{P}_{\bar{y}} \tag{3.57}$$

where

$$\mathbf{P}_\alpha = \exp\left(-i\frac{\pi}{2}\hat{I}_\alpha\right); \quad \mathbf{P}_{\bar{\alpha}} = \exp\left(i\frac{\pi}{2}\hat{I}_\alpha\right)$$

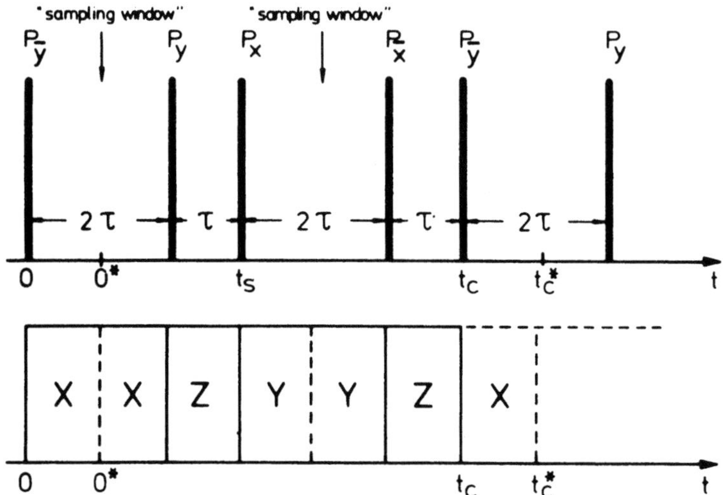

Fig. 3.5. Schematical drawing of the four-pulse sequence (WHH-4). The pulses are assumed to be 90° pulses applied to the spins in different directions of the rotating frame as indicated. Here $\bar{x}$ means $-x$ etc. The interaction Hamiltonian in the "toggling rotating frame" is labelled in short form by capital letters as explained in the text. The magnetization is usually sampled during one of the large windows

and

$$\mathbf{D}_\alpha(\tau) = \exp(-i\hat{\mathscr{H}}_\alpha\tau); \quad \alpha = \pm x; \ \pm y; \ \pm z.$$

After applying the transformation operation produced by the rf pulses P_α which is usually referred to as going into the "toggling rotating frame" [18], we obtain

$$\mathbf{L}(t_c) = \mathbf{D}_z(\tau)\,\mathbf{D}_y(2\tau)\,\mathbf{D}_z(\tau)\,\mathbf{D}_x(2\tau). \tag{3.58}$$

By applying a short form notation for the "switching Hamiltonian" in the "toggling rotating frame" as

$$\mathbf{X} \underset{\text{def}}{=} \hat{\mathscr{H}}_x = \mathbf{P}_y\,\hat{\mathscr{H}}_z\,\mathbf{P}_y^{-1}$$

$$\mathbf{Y} \underset{\text{def}}{=} \hat{\mathscr{H}}_y = \mathbf{P}_x^{-1}\,\hat{\mathscr{H}}_z\,\mathbf{P}_x \tag{3.59}$$

$$\mathbf{Z} \underset{\text{def}}{=} \hat{\mathscr{H}}_z$$

we can characterize the Hamiltonian, switching through the different states $\mathbf{X}$, $\mathbf{Y}$, $\mathbf{Z}$ as shown in Fig. 3.5. The one cycle propagator $\mathbf{L}(t_c)$ may be conveniently expressed in the Mansfield notation [19] as

$$\mathbf{L}(t_c) = [2\mathbf{X}, \mathbf{Z}, 2\mathbf{Y}, \mathbf{Z}] \tag{3.60}$$

which is selfexplanatory from Fig. 3.5.

Figure 3.5 depicts also another cycle with the cycle time t_c^*. This cycle is recognized immediately as being symmetrical by writing down its one cycle propagator in the Mansfield notation [19] as

$$\mathbf{L}(t_c^*) = [\mathbf{X}, \mathbf{Z}, \mathbf{Y}; \ \mathbf{Y}, \mathbf{Z}, \mathbf{X}] = [\mathbf{X}, \mathbf{Z}, \mathbf{Y}]\,[\mathbf{Y}, \mathbf{Z}, \mathbf{X}] \tag{3.61a}$$

or

$$\mathbf{L}(t_c^*) = [\![\mathbf{X}, \mathbf{Z}, \mathbf{Y}]\!] \quad \text{reflection symmetry [19].} \tag{3.61b}$$

Aspects of symmetry [19] in the one cycle propagator will be discussed later in connection with higher order correction terms in the Magnus expansion (see Sect. 3.4). In this section we are mainly interested in the average Hamiltonian [12]. With

$$\bar{\mathscr{H}} = \frac{1}{t_c}\int_0^{t_c} dt'\,\hat{\mathscr{H}}(t')$$

we obtain from Eqs. (3.60) and (3.61)

$$\bar{\mathscr{H}} = \tfrac{1}{3}(\mathscr{H}_x + \mathscr{H}_y + \mathscr{H}_z). \tag{3.62}$$

The following cases are discussed:

(i) dipolar interaction (and quadrupolar interaction)

$$\bar{\mathscr{H}}_D = 0 \quad \text{since} \quad \mathscr{H}_{Dx} + \mathscr{H}_{Dy} + \mathscr{H}_{Dz} = 0$$

(ii) shift interaction

$$\bar{\mathscr{H}}_S = \tfrac{1}{3}\sum_k (\delta_k + \Delta\omega)(I_{xk} + I_{yk} + I_{zk}) \tag{3.63}$$

where $\Delta\omega$ is the resonance offset.

Equation (3.63) may be written in short form as

$$\bar{\mathscr{H}}_S = \frac{1}{\sqrt{3}} \sum_k (\delta_k + \Delta\omega)\, I_{k\,111} \tag{3.64}$$

where

$$I_{111} = (I_x + I_y + I_z)/\sqrt{3}$$

and the "scaling factor" equals $1/\sqrt{3}$ by which all shift interactions, including the resonance offset are scaled.

Some representative examples of the application of the four-pulse experiment (WHH-4) are given in Figs. 3.6 to 3.13. A spectrum is obtained in a WHH-4 experiment in the following way:

(i) The magnetization is sampled once every cycle, conveniently in one of the large "windows". The resulting decay is stored in some sort of memory.

(ii) The decay is Fourier-transformed by means of a computer and phase shifted to obtain the proper "absorption signal".

The first single crystal spectrum of ^{19}F in a non-cubic solid using the WHH-4 sequence was obtained by L.M. Stacey, R.W. Vaughan, and D.D. Elleman [20]. Figure 3.6 represents a ^{19}F multi-line spectrum, obtained from a single crystal of CF_3COOAg at low temperature (40 K) [21]. The rotation of the fluoro-methyl group is essentially frozen at this temperature (jumping rate is small compared to the line width), resulting in six different lines from the two non-equivalent CF_3-groups in the CF_3COOAg dimer [21]. Upon crystal rotation all these lines shift according to the different ^{19}F shielding tensors which have been analyzed in Ref. [21]. The rotation plot for a rotation about

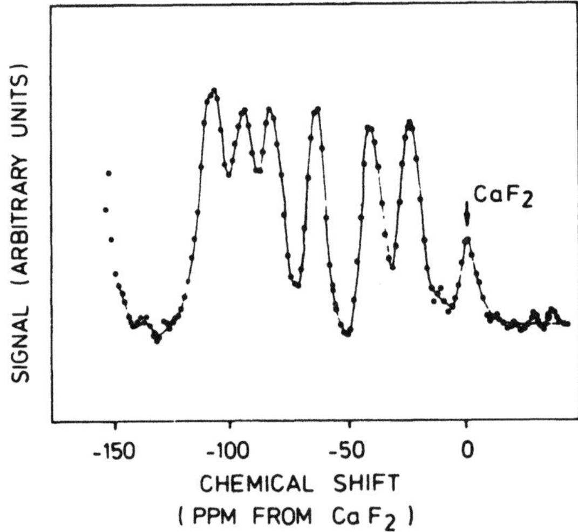

Fig. 3.6. Four-pulse spectrum of ^{19}F in a single crystal of CF_3COOAg at 40 K (methyl group "rotation rate" much less than line width) obtained by Griffin *et al.* [21]. The six lines from the molecular dimer are clearly distinguished. This spectrum corresponds to a rotation angle of 100° in Fig. 3.7

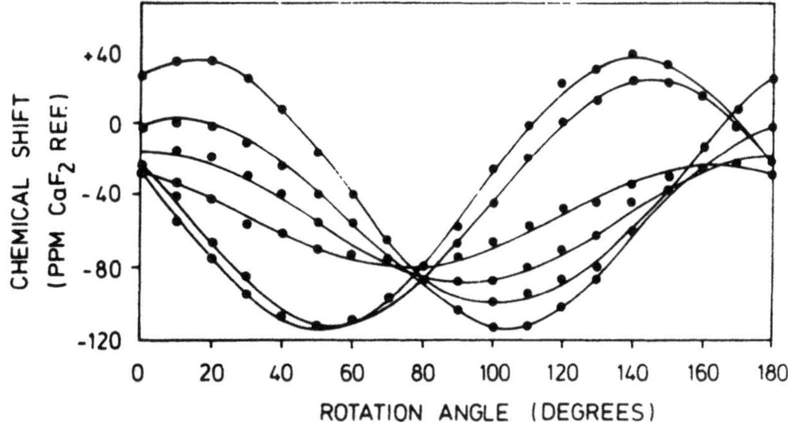

Fig. 3.7. Variation of the spectral lines in Fig. 3.8. with rotation about approximately the c^* axis of the CF₃COOAg unit cell. $T=40$ K (Griffin *et al.* [21])

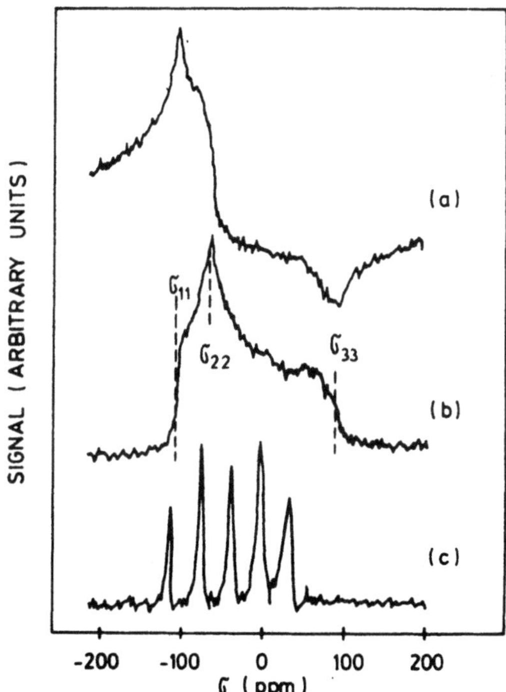

Fig. 3.8a–c. ¹⁹F powder spectra of fluoranil, $C_6F_4O_2$ at 300 K. **a** and **b** are the dispersion and adsorption signal respectively, with the tensor elements σ_{11}, σ_{22}, and σ_{33} as indicated, while **c** gives the calibration markers obtained with liquid benzene [23]

the c^* crystalline axis is shown in Fig. 3.7. By rotation about the two other crystalline axes a and b the ¹⁹F shielding tensors for the three different fluorine of the methyl group have been determined [21]. See Sect. 7.4 for a detailed discussion. The first chemical shielding tensor powder spectrum [22] as obtained by the four-pulse sequence is shown in Fig. 3.8. The shift tensor of ¹⁹F in fluoranil has three distinct tensor elements which are readily obtained from the powder pattern. The principal axes of the tensor are, however, undetermined from such an experiment. These can be assigned unambiguously only by a single crystal study. However, a tentative assignment may be possible by checking the consistency with related compounds, such as substituted fluoro-benzenes (see Sect. 7.4).

^{19}F shielding tensors were investigated in the pioneer era of multiple-pulse experiments because of their large values [23–27]. However, recently also other nuclei have been studied, including protons, which have a very small shift anisotropy [28–37]. Besides a few exceptions, most of these detailed investigations of proton shielding tensors we done by Haeberlen and co-workers [28–31].

A representative example of this work is demonstrated in Figs. 3.9 and 3.10. It turned out that protons in carboxyl groups displayed the largest values of shielding anisotropies among the protons. The shielding tensor is mostly axially symmetric with the unique axis being the "hydrogen bond" direction [18]. Its size is about 20 ppm and extends over the entire range of isotropic shift values, as demonstrated in the case of "squaric acid" [32] in Fig. 3.11. For a further discussion of proton shielding tensors see Sect. 7.3. Also metals have been investigated using the WHH-4 experiments [38] (see Figs. 3.12 and 3.13). A special obstacle in these cases is the skin effect which attenuates and phase shifts the rf field, when penetrating into the metal. This can be circumvented by using powder samples with a grain size less than the skin depth. Since aluminium is a cubic metal, second rank tensor shift interactions vanish

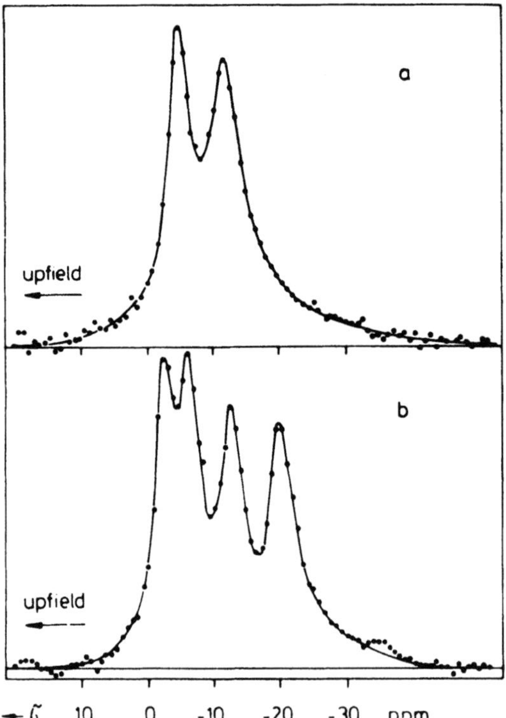

Fig. 3.9a, b. Multiple pulse spectra of protons in a single crystal of malonic acid according to Haeberlen et al. [29]. **a** Degenerate spectrum, where the two lines of the CH$_2$ group coalesce into a single line (left) and the same holds for the two carboxyl (OH) protons (right). **b** Arbitrary orientation, where all four lines are visible

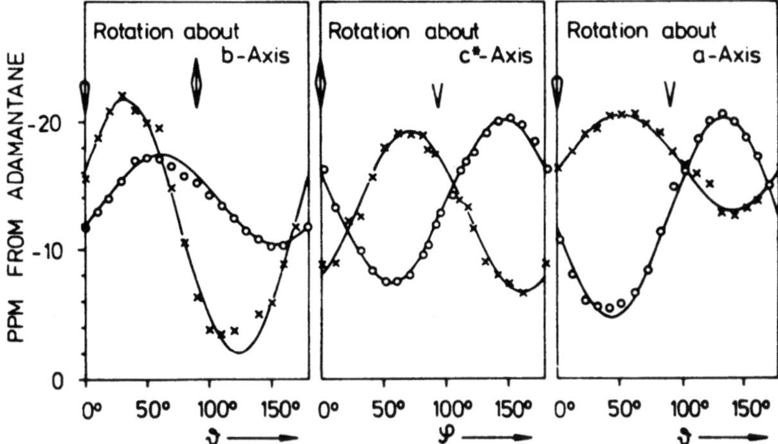

Fig. 3.10. Rotation patterns for the carboxyl proton lines in malonic acid (see Fig. 3.11) for three different crystal orientations. The markers indicate equivalent orientations of B_0 (Haeberlen et al. [29])

and the four-pulse experiment may be effectively used to reduce the dipolar as well as the strain produced quadrupolar line width [39] as shown in Fig. 3.12. Similarly in Fig. 3.13, the large quadrupole splitting in ^9Be has been reduced considerably [38]. Although Knight shift anisotropy could arise in principal in beryllium, none has been observed. The isotropic Knight shifts, however, have been measured with respect to aqueous solutions of $AlCl_3$ and $BeCl_2$ respectively as [38]: $1,636 \pm 3$ ppm for ^{27}Al and -10 ± 3 ppm for ^9Be. The Knight shift of aluminium is in accord with the high precision values of $1,640 \pm 1$ ppm as obtained by the sample rotation method (see Sect. 2.5).

Since dipolar interactions and quadrupolar interactions vanish in the multiple-pulse experiment, broadenings of the line due to shift interactions as produced by e.g. paramagnetic impurities can be effectively studied using this method.

One difficulty which one encounters in determining shifts by multiple-pulse experiments in metals is due to the skin effect, which decreases the finite rotation angle of the pulses for those nuclei which are further away from the surface [40]. A distribution of rotation angles results over the sample which leads to a different scaling factor (see Sect. 3.5) for the metal and e.g. a liquid sample, even if an "internal standard" is used. It is a necessity therefore to determine the line shift versus resonance offset for both the metal and the reference at the same time. The correct resonance frequency is then determined from the intersection of the two lines (which may have a different slope!) with the off-resonance shift axis. The following puzzling situation can occur, which has been observed in ^9Be, that the distance between the liquid and the metal resonance line changes with resonance offset and even changes sign i.e. there are cases where the reference line is observed at the same side of the metal line, no matter what the spectrometer frequency is, above or below resonance. For a further discussion of the influence of the skin effect on NMR pulse experi-

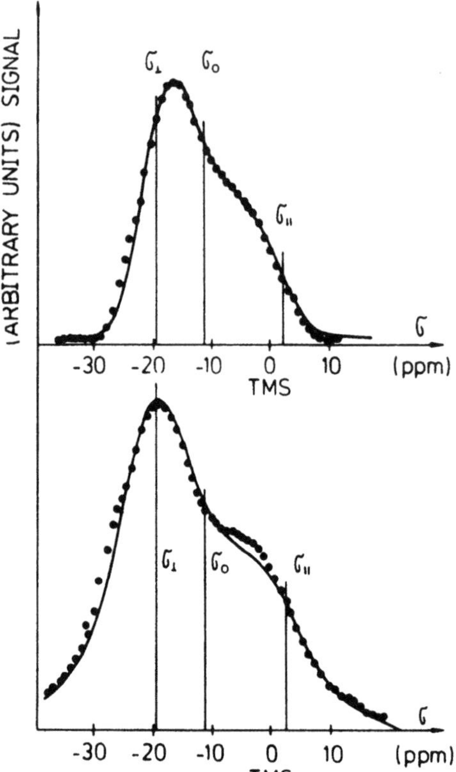

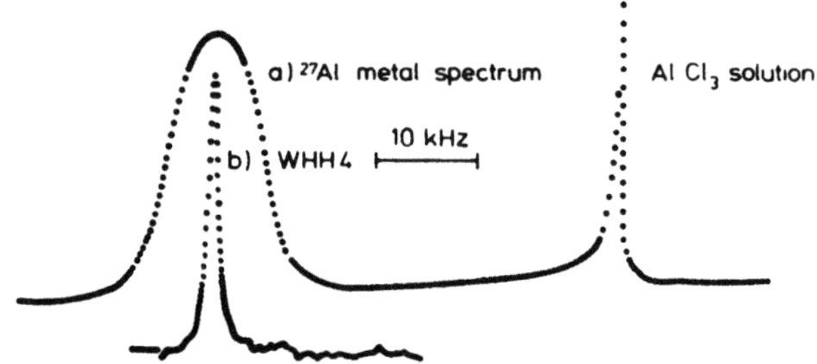

Fig. 3.11. Powder spectra of protons in squaric acid obtained in a four-pulse experiment by Raber et al. [32]. Notice, that the spectrum covers more than the isotropic shift range of protons. Absorption spectrum (top) and power spectrum (bottom) are plotted for comparison

Fig. 3.12a, b. Ordinary (**a**) and four-pulse spectrum (**b**) of ^{27}Al in aluminium powder, containing an aqueous solution of AlCl$_3$. Dipolar interaction between ^{27}Al nuclei and quadrupole interactions due to lattice defects are effectively reduced in the four-pulse experiment (see Ref. [38])

ments, see Ref. [40]. So far only homonuclear spin systems with only I spins present have been discussed.

This is, however, not the general case and the line narrowing efficiency may deteriorate substantially if a second spin species S is present. In a high magnetic field only the secular part of the I and S spin interaction survives, as was shown in Sect. 2.2

$$\mathcal{H}'_{IS} = \sum_{i,j} B_{ij} I_{zi} S_{zj} \quad \text{with} \quad B_{ij} = \gamma_I \gamma_S \hbar (1 - 3 \cos^2 \vartheta_{ij})/r_{ij}^3.$$

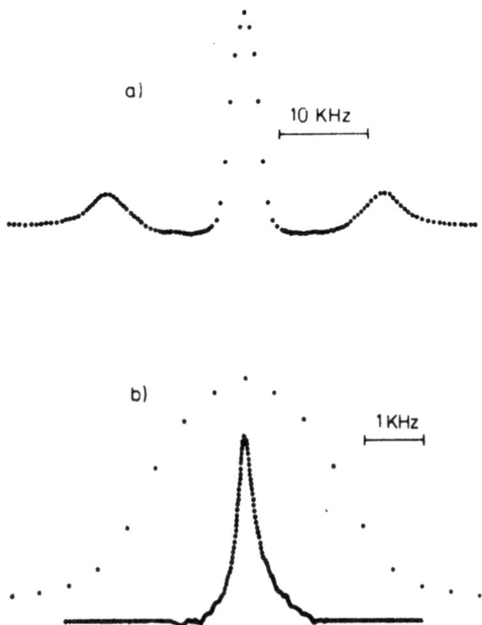

Fig. 3.13a, b. Spectrum of ^{9}Be in beryllium powder (a) displaying the characteristic quadrupole powder spectrum. **b** Blow up of the central line with the corresponding four-pulse spectrum inserted (Ref. [38])

This I-S interaction is of course altered when a multiple-pulse experiment is applied to the I spins. It is e.g. in a WHH-4 experiment scaled by a factor $1/\sqrt{3}$ which is not a bargain in terms of resolution. Thinking in terms of an abverage Hamiltonian, we would have to quench the average of $S_z(t)$ by some means of irradiation at the Larmor frequency of the S spins. A continous irradiation of the S spins at their Larmor frequency with an rf field B_1 in say the x direction would result in

$$S_z(t) = S_z \cos \omega_1 t - S_y \sin \omega_1 t$$

with

$$\omega_1 = \gamma B_1$$

which gives $\overline{S_z(t)} = 0$. This method is extensively applied in double resonance experiments, as discussed in Sect. 4.2 and 4.4. However, the reader may convince himself that his does not in general lead to a vanishing average I-S Hamiltonian in a WHH-4 experiment i.e. the condition for vanishing I-S interaction has to be stated more correctly as

$$\overline{\mathscr{H}'_{IS}}(t) = \sum_{i,j} B_{ij} \overline{I_{zi}(t) S_{zj}(t)} = 0.$$

One way of achieving this is to apply a π pulse every WHH-4 cycle to the S spins as outlined in Fig. 3.14 [41]. Other means have been discussed by Haeberlen [18]. The first experiment of this type was performed on ^{23}Na^{19}F,

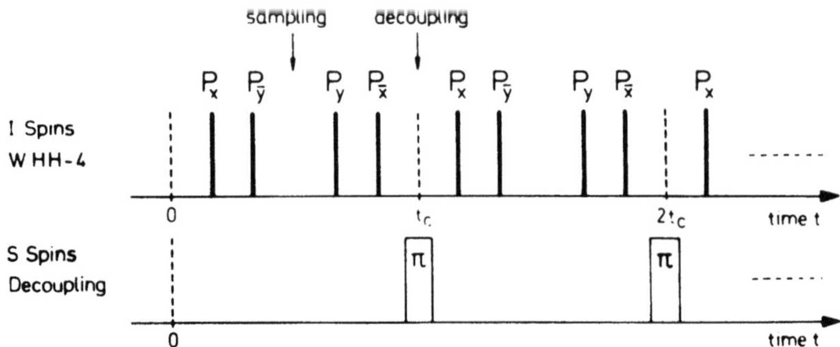

Fig. 3.14. Pulse timing of a WHH-4 sequence applied to the *I* spins, while applying decoupling pulses to the *S* spins once every four-pulse cycle

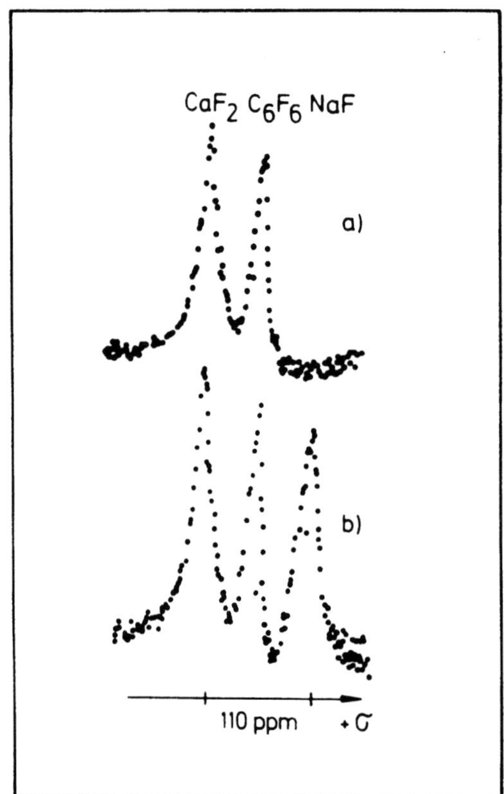

Fig. 3.15a, b. Four-pulse spectra of ^{19}F in a sample containing a mixture of C_6F_6, CaF_2 and NaF. **a** Without decoupling pulses applied to the ^{23}Na spins. **b** Decoupling pulses applied to the ^{23}Na spins as outlined in Fig. 3.14 (Ref. [41])

where a WHH-4 experiment was applied to the ^{19}F spins while the ^{23}Na spins were irradiated by π pulses once every cycle [41] (see Fig. 3.15). Without the decoupling pulses the ^{19}F resonance was still too broad to be observed on the scale of Fig. 3.15. This technique has been extensively used in investigations of other heteronuclear spin systems by multiple pulse NMR [37, 42]. Especially the determination of ^{19}F shielding tensors in partially fluorinated benzenes as discussed in Sect. 7.4 could not have been done without the decoupling of the

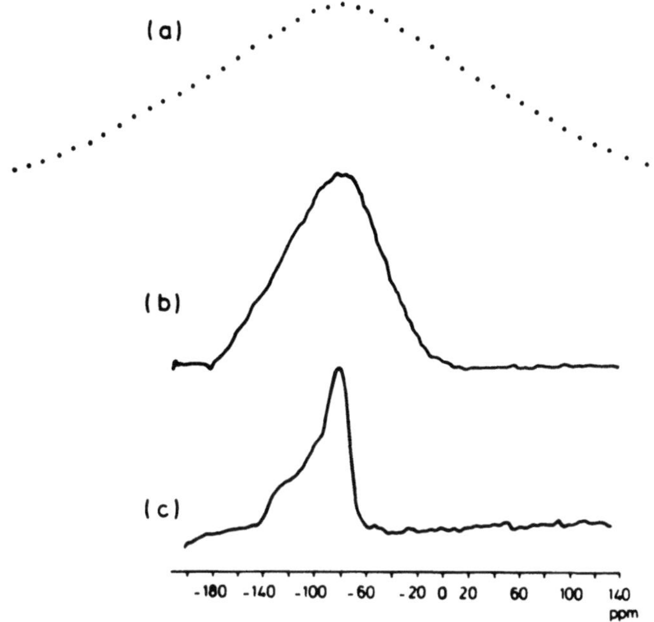

Fig. 3.16a–c. Different stages of resolution enhancement, demonstrated in the case of ^{19}F resonance of a CF$_3$-group in CF$_3$C$_6$H$_4$COOH at room temperature. **a** Fourier transform of FID. **b** Four-pulse spectrum. **c** Same as **b** but decoupling pulses are applied to the proton spins. Curve **c** shows the characteristic powder spectrum of a rapidly rotating fluoromethyl group. (Courtesy of D. Suwelack)

protons [42]. Figure 3.16 represents such a case, where the different stages of line narrowing are clearly visible.

A shift of the resonance frequency occurs in all these decoupling experiments due to the non-secular contribution of the decoupling field. This effect is also termed "Bloch Siegert shift" [43], which originally accounted for the non-secular contribution of the counter-rotating component of a linear polarized rf field. This effect will be discussed in more detail in Appendix H. Because of the second order character this shift is usually very small and amounts to about 10 ppm in the case of ^{19}F–^{1}H with ordinary rf fields (see Appendix H). This may seem to be appreciable. However, in NMR never absolute frequencies, but rather differences with respect to some reference line are determined, where the reference line is shifted by about the same amount, if it is recorded in the course of the same experiment. This of course is mandatory. The remaining Bloch Siegert shift between different lines in the spectrum is far below detectability, even if highest resolution is achieved [44].

The efficiency of spin decoupling and some more spin dynamical aspects are discussed in greater detail in Sect. 4.4.

3.3 Coherent Averaging Theory [12, 18]

In Sects. 3.1 and 3.2 we have dealt with multiple pulse experiments simply by taking the average Hamiltonian over the cycle time. This certainly represents the correct Hamiltonian of the propagator in the limit of $\|\mathcal{H}_{int}\| t_c \rightarrow 0$. The cycle time, however, cannot be made arbitrarily small because of technical reasons. On the contrary, usually

$$\|\mathcal{H}_{int}\| t_c \cong 1$$

and any further reduction of t_c demands considerable effort. Coherent averaging theory [12, 25] now describes the coherent evolution of nuclear states, taking into account higher order terms i.e. interference terms of non-commuting operators in the cycle. Here we follow closely the paper by U. Haeberlen and J.S. Waugh [12], who have introduced this concept to multiple-pulse NMR. The general Hamiltonian in the rotating frame is split into two parts

$$\mathcal{H} = \mathcal{H}_{ext}(t) + \mathcal{H}_{int}$$

leading to the propagator

$$\mathbf{L}(t) = T\exp\left[-i\int_0^t dt'(\hat{\mathcal{H}}_{ext}(t') + \hat{\mathcal{H}}_{int})\right]. \tag{3.65}$$

T: Dyson time ordering operator.

This propagator may be split into a product

$$\mathbf{L}(t) = \mathbf{L}_1(t)\,\mathbf{L}_0(t) \tag{3.66}$$

where

$$\mathbf{L}_1(t) = T\exp\left[-i\int_0^t dt'\,\hat{\mathcal{H}}_{ext}(t')\right] \tag{3.66a}$$

and

$$\mathbf{L}_0(t) = T\exp\left[-i\int_0^t dt'\,\tilde{\hat{\mathcal{H}}}(t')\right] \tag{3.66b}$$

and with

$$\tilde{\hat{\mathcal{H}}}(t) = \mathbf{L}_1^{-1}(t)\,\hat{\mathcal{H}}_{int}\,\mathbf{L}_1(t). \tag{3.66c}$$

This product corresponds to the Heisenberg picture i.e. to an interaction representation or speaking colloquially to going into the "toggling frame". $\mathcal{H}_{ext}$ describes the spin interactions with external fields like

$$\mathcal{H}_{ext} = \Delta\omega I_z + \mathcal{H}_1(t)$$

where $\mathcal{H}_1(t)$ may result from the pulsed rf field and

$$\mathcal{H}_{int} = \mathcal{H}_D + \mathcal{H}_Q + \mathcal{H}_S + \dots$$

represents the spin interactions with "internal fields". In general $\|\mathcal{H}_{ext}\| \gg \|\mathcal{H}_{int}\|$, although this is not necessary for Eq. (3.66) to be valid. It is, however, assuring rapid convergence, when approximating $\mathbf{L}_0(t)$ by a truncated cumulant expansion.

We now further assume, that the external field may be periodic [12, 18, 45] i.e.

$$\mathcal{H}_{ext}(t + nt_c) = \mathcal{H}_{ext}(t); \quad n = 0, 1, 2, \dots \tag{3.67}$$

from which follows

$$\mathbf{L}_1(nt_c) = \mathbf{L}_1^n(t_c) \tag{3.68}$$

and which also leads to a periodicity of $\tilde{\mathscr{H}}(t)$ with

$$\tilde{\mathscr{H}}(t) = \tilde{\mathscr{H}}(t + nt_c) \tag{3.69}$$

and

$$\mathbf{L}_0(nt_c) = \mathbf{L}_0^n(t_c). \tag{3.70}$$

The general propagator for periodical external fields may now be expressed as:

$$\mathbf{L}(nt_c) = \mathbf{L}_1^n(t_c) \cdot \mathbf{L}_0^n(t_c).$$

If in addition the external field is cyclic, in the sense

$$\mathbf{L}_1(t_c) = \tilde{\mathbf{1}} \tag{3.71}$$

the complete time evolution is described by the one cycle propagator $\mathbf{L}_0(t_c)$. Equation (3.71) is e.g. fulfilled in all phase-alternated multiple-pulse experiments, i.e. in all cases where $\mathscr{H}_{ext}(t) = -\mathscr{H}_{ext}(t_c - t)$. Equation (3.71), however, is a too stringent condition, i.e. it is a sufficient, but not a necessary condition for the external field to be cyclic. A less stringent condition would be

$$\mathbf{L}_1^{-1}(t_c) I_i \mathbf{L}_1(t_c) = I_i \quad \text{with} \quad i = x, y \text{ or } z \tag{3.72}$$

i.e. $\mathbf{L}_1(t_c)$ leaves the spin operator I_i invariant. This is for example the case in a series of not phase alternated pulses, where the rotation angle β of the rf pulses is an integer multiple of 2π. Condition Eq. (3.72), however, is not fulfilled if a resonance offset $\Delta\omega$ is included in $\mathscr{H}_{ext}$. Even in a phase alternated experiment the resonance offset part $\Delta\omega I_z$ has to be included into $\mathscr{H}_{int}$ for the external field $\mathscr{H}_{ext}(t)$ to be cyclic.

The one-cycle propagator $\mathbf{L}_0(t_c)$ was expressed by a *product* of unitary operators in Eqs. (3.49) and (3.58). Our goal, however, is to represent $\mathbf{L}_0(t_c)$ as

$$\mathbf{L}_0(t_c) = \exp[-it_c \,\hat{\mathscr{H}}_e(t_c)] \tag{3.73}$$

where $\mathscr{H}_e(t_c)$ is the effective Hamiltonian which may be expanded as follows [45]

$$\mathscr{H}_e(t_c) = \sum_{k=0}^{\infty} \bar{\mathscr{H}}^{(k)}(t_c) = \bar{\mathscr{H}}^{(0)} + \sum_{k=1}^{\infty} \bar{\mathscr{H}}^{(k)}(t_c). \tag{3.74}$$

Here $\bar{\mathscr{H}}^{(0)}$ is the average Hamiltonian which has been used in Sects. 3.1 and 3.2 exclusively. Although $\bar{\mathscr{H}}^{(0)}$ is independent of the cycle time t_c, all higher order terms depend on the cycle time explicitly and may be ordered with increasing exponent of the cycle time as

$$\bar{\mathscr{H}}^{(k)} = t_c^k \, \mathbf{F}_k \quad k = 0, 1, 2, \ldots \tag{3.75}$$

where $\mathbf{F}_k$ is a Hermitian operator and is independent of the cycle time. The effective Hamiltonian $\mathscr{H}_e(t_c)$ may be obtained by equating Eqs. (3.73) and (3.66b) and comparing with increasing exponent of t_c. This results in the so-called Magnus-expansion [12, 13, 45–49]

$$\mathbf{L}_0(t_c) = \exp[-it_c(\bar{\mathscr{H}}^{(0)} + \bar{\mathscr{H}}^{(1)} + \bar{\mathscr{H}}^{(2)} + \ldots)] \tag{3.76}$$

where

$$\bar{\mathscr{H}}^{(0)} = \frac{1}{t_c} \int_0^{t_c} dt\, \tilde{\mathscr{H}}(t) \tag{3.76a}$$

$$\bar{\mathscr{H}}^{(1)} = \frac{-i}{2t_c} \int_0^{t_c} dt_2 \int_0^{t_2} dt_1\, [\tilde{\mathscr{H}}(t_2), \tilde{\mathscr{H}}(t_1)] \tag{3.76b}$$

$$\bar{\mathscr{H}}^{(2)} = -\frac{1}{6t_c} \int_0^{t_c} dt_3 \int_0^{t_3} dt_2 \int_0^{t_2} dt_1\, \{[\tilde{\mathscr{H}}(t_3), [\tilde{\mathscr{H}}(t_2), \tilde{\mathscr{H}}(t_1)]]$$
$$+ [\tilde{\mathscr{H}}(t_1), [\tilde{\mathscr{H}}(t_2), \tilde{\mathscr{H}}(t_3)]] \tag{3.76c}$$

etc.

Notice, that the effective Hamiltonian in the Magnus expansion stays Hermitian at every stage of approximation, thus leaving $L_0(t_c)$ always unitary, independent of the degree of approximation used.

For the demonstrative purposes it is usually sufficient to consider the average Hamiltonian Eq. (3.76a) only, as we have done in the preceding sections. In practice it turns out that no higher correction terms than the second term $\bar{\mathscr{H}}^{(2)}$ have to be taken into account. Otherwise one has to question the interaction representation chosen. Exploiting the symmetry of the "switching Hamiltonian" $\tilde{\mathscr{H}}(t)$ during one cycle it is possible to reduce the computational effort considerably, as has been first shown by Mansfield [19] and later by several other authors [45, 50-52]. We follow now the arguments of Wang and Ramshaw [45]. The one cycle propagator $L_0(t_c)$ according to Eqs. (3.73) and (3.74) is a unitary operator with

$$L_0^+(t_c) = L_0^{-1}(t_c) = \exp[it_c\, \hat{\mathscr{H}}_e(t_c)]. \tag{3.77}$$

For a
  (i) symmetric cycle, i.e. $\tilde{\mathscr{H}}(t) = \tilde{\mathscr{H}}(t_c - t)$ it follows

$$L_0^+(t_c) = L_0(-t_c) \tag{3.78}$$

which leads to

$$\exp[it_c\, \mathscr{H}_e(t_c)] = \exp[it_c\, \mathscr{H}_e(-t_c)]$$

or

$$\mathscr{H}_e(t_c) = \mathscr{H}_e(-t_c). \tag{3.79}$$

Expanding $\mathscr{H}_e(t_c)$ according to Eq. (3.74) leads to

$$\sum_{k=0}^{\infty} t_c^k\, F_k = \sum_{k=0}^{\infty} (-t_c)^k\, F_k$$

which results in $F_k = 0$ if k is odd, i.e. all correction terms $\bar{\mathscr{H}}^{(k)}$ vanish with $k = 1, 3, 5 \ldots$ odd.

Because of this fact, multiple pulse cycles are conveniently designed symmetrically, leaving only the average Hamiltonian and even order correction terms in the Magnus expansion to be looked at more closely. We shall take advantage of this fact in the subsequent sections. Furthermore it can be shown for a symmetrical cycle that [50]

$$\bar{\mathscr{H}}^{(2)}(t_c) = \bar{\mathscr{H}}^{(2)}(\tfrac{1}{2}t_c) \quad \text{if} \quad \bar{\mathscr{H}}^{(0)} = 0 \tag{3.80}$$

which states that $\bar{\mathscr{H}}^{(2)}$ has to be computed over half the cycle only in this case, which may save some labour.

For an

(ii) antisymmetric cycle, i.e. $\mathscr{H}(t) = -\mathscr{H}(t_c - t)$ it follows

$$\mathbf{L}_0^+(t_c) = \mathbf{L}_0(t_c) \tag{3.81}$$

which leads to

$$\mathscr{H}_e(t_c) = -\mathscr{H}_e(t_c)$$

or

$$\mathscr{H}_e(t_c) = 0 \quad \text{and} \quad \mathbf{L}_0(t_c) = \tilde{\mathbf{1}}. \tag{3.82}$$

There is no change of the initial state whatsoever in this case. This merely amounts to stating, that the initial state can always be recovered if the full Hamiltonia n of the system can be inverted. This phenomenon is called the Loschmid demon and is the essential part of every spin echo experiment [1, 51].

There is, however, a different approach to obtain an averaged Hamiltonian, called "se cular averaging" to be discussed in detail in Appendix G.

3.4 Af plication of Coherent Averaging Theory to Mu ltiple-Pulse Sequences

In this section we want to discuss the correction terms $\bar{\mathscr{H}}^{(1)}$ and $\bar{\mathscr{H}}^{(2)}$ to the averal ze Hamiltonian $\bar{\mathscr{H}}^{(0)}$ for a) the phase-alternated sequence (MW-2), b) the spin-l ocking sequence (MW-4) and c) the four-pulse experiment (WHH-4).

a) The Phase-Alternated Sequence (MW-2) [4-7]

The phase-alternated sequence may be expressed as $(\tau - P_{\bar{y}} - 2\tau - P_y - \tau)_n$ or by the "switched Hamiltonian" $\mathscr{H}(t)$ as represented in Fig. 3.17.

The one cycle propagator $\mathbf{L}_0(t_c)$ can be expressed by

$$\mathbf{L}_0(t_c) = [\mathbf{Z}, 2\mathbf{X}, \mathbf{Z}] = [\![\mathbf{Z}, \mathbf{X}]\!]. \tag{3.83}$$

We obtain immediately

$$\bar{\mathscr{H}}^{(0)} = \tfrac{1}{2}(\mathscr{H}_x + \mathscr{H}_z)$$

and in the special case with $\mathscr{H}$ being the dipolar Hamiltonian

$$\bar{\mathscr{H}}_D^{(0)} = -\tfrac{1}{2}\mathscr{H}_{Dy}.$$

In case the interaction is represented by a shift Hamiltonian

$$\bar{\mathscr{H}}_S^{(0)} = \tfrac{1}{2}\sum_k (\delta_k + \Delta\omega)(I_{xk} + I_{zk}) \tag{3.84}$$

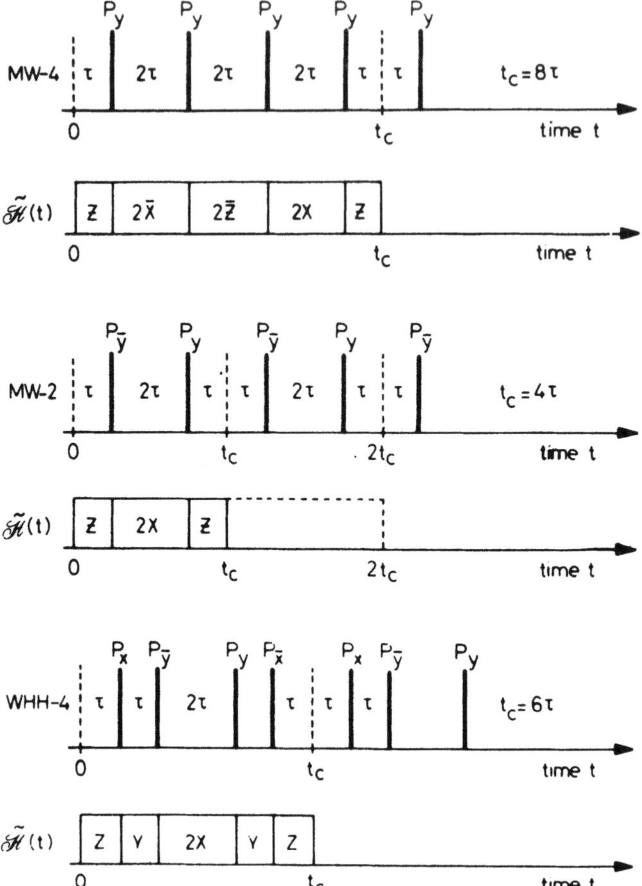

Fig. 3.17. Pulse timing and interaction Hamiltonian in the "toggling rotating frame" for the MW-4, MW-2, and the WHH-4 pulse sequences

which may be rewritten as

$$\bar{\mathscr{H}}_S^{(0)} = \sum_k \frac{1}{\sqrt{2}} (\delta_k + \Delta\omega)\, I_{101}^{(k)}$$

to emphasize the scaling factor $S = 1/\sqrt{2}$, and where

$$I_{101}^{(k)} = (I_{xk} + I_{zk})/\sqrt{2}.$$

Because of the symmetry of the cycle ($\mathscr{H}(t) = \mathscr{H}(t_c - t)$) all odd order correction terms vanish and the leading correction term equals $\mathscr{H}^{(2)}$, which may be obtained by applying Eq. (3.76c) as

$$\bar{\mathscr{H}}^{(2)} = -\frac{\tau^2}{12} [(2\mathbf{X} + \mathbf{Z}), [\mathbf{X}, \mathbf{Z}]]; \quad t_c = 4\tau \tag{3.85}$$

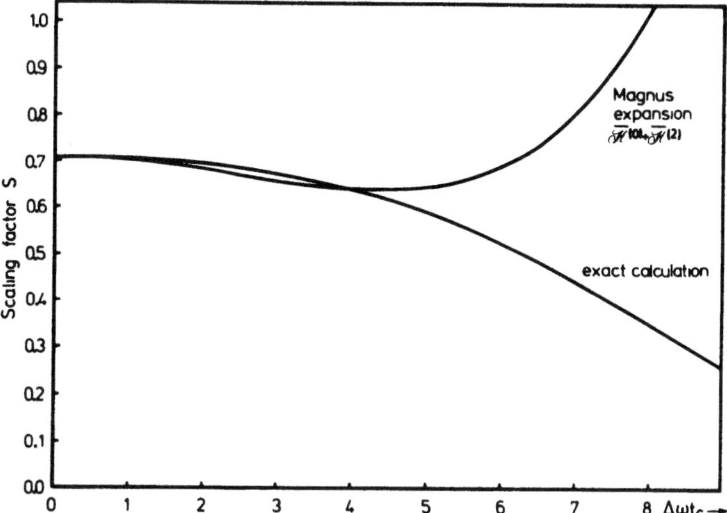

Fig. 3.18. Scaling factor S for shift interactions in a phase-alternated sequence applying 90° pulses (MW-2) versus resonance offset $\Delta\omega t_c$. The exact calculation by applying Eq. (3.53) is compared with the Magnus expansion for a symmetric cycle up to second order [Eq. (3.88)]. Up to $\Delta\omega t_c = 1$ the difference is negligible. (Courtesy of D. Suwelack)

which results in the case of shift interactions in

$$\overline{\mathcal{H}}_S^{(2)} = - \sum_k K(\delta_k + \Delta\omega)(2I_{zk} - I_{xk}) \tag{3.86}$$

where $K = \tau^2(\delta_k + \Delta\omega)^2/12$.

The total average shift Hamiltonian $\overline{\mathcal{H}}_S = \overline{\mathcal{H}}_S^{(0)} + \overline{\mathcal{H}}_S^{(2)}$ is obtained by combining Eqs. (3.84) and (3.86) to

$$\overline{\mathcal{H}}_S = \sum_k S(\delta_k + \Delta\omega) I_{\mu k} \tag{3.87}$$

with the scaling factor

$$S = \frac{1}{\sqrt{2}}(1 - 2K + 10K^2)^{1/2} \tag{3.88}$$

and

$$I_\mu = e^{-i\beta I_y} I_z e^{i\beta I_y}$$

where

$$\tan\beta = \frac{1 + 2K}{1 - 4K}. \tag{3.89}$$

In Fig. 3.18 the exact scaling factor obtained from an exact solution of the equation of motion using Eq. (3.53), is compared with this approximative method, including correction terms up to second order. A similar calculation has been made by Ellet and Waugh [17], there however the symmetry of the cycle was not realized.

b) Spin-Locking Sequence (MW-4) [4, 7]

The spin-locking sequence as shown in Fig. 3.17 and represented by $(\tau - P_y - 2\tau - P_y - 2\tau - P_y - 2\tau - P_y - \tau)_n$ is similar to the phase-alternated sequence, besides the phase alternation. This leads to a cycle time of $t_c = 8\tau$ and a slightly changed "switched Hamiltonian" with the propagator

$$\mathbf{L}_0(t_c) = [\mathbf{Z}, 2\bar{\mathbf{X}}, 2\bar{\mathbf{Z}}, 2\mathbf{X}, \mathbf{Z}] \tag{3.90}$$

where $\bar{\mathbf{X}}$ for example, means a minus sign in front of each I_x occuring in the Hamiltonian. Since the dipolar Hamiltonian is invariant under such a change of sign, we have exactly the same expression for $\bar{\mathscr{H}}_D^{(k)}$ as in the phase alternated experiment where especially all odd order correction terms vanish, because of the symmetry of the cycle. The spin-locking cycle, however, is not a symmetrical cycle in the above defined sense for shift interactions, since $\bar{\mathbf{X}}$ and $\bar{\mathbf{Z}}$ are different from $\mathbf{X}$ and $\mathbf{Z}$ respectively. It can be shown in a straightforward manner, that

$$\bar{\mathscr{H}}^{(1)} = -\frac{i\tau}{4} [\bar{\mathbf{Z}}, (\bar{\mathbf{X}} - \mathbf{X})] \tag{3.91}$$

and we leave it as an exercise for the interested reader, to derive the correction terms of the average shift and dipolar Hamiltonian and the corresponding cross correction terms, using Eq. (3.91)

c) The Four-Pulse Sequence (WHH-4) [9, 12]

Here we use a slightly different cycle than in Sect. 3.2, which is shown in Fig. 3.17. This cycle possesses reflection symmetry as expressed by the one cycle propagator

$$\mathbf{L}_0(t_c) = [\mathbf{Z}, \mathbf{Y}, \mathbf{X}; \mathbf{X}, \mathbf{Y}, \mathbf{Z}] = [\![\mathbf{Z}, \mathbf{Y}, \mathbf{X}]\!] \tag{3.92}$$

and it follows, that

$$\bar{\mathscr{H}}^{(0)} = \tfrac{1}{3}(\mathbf{X} + \mathbf{Y} + \mathbf{Z})$$

with

$$\bar{\mathscr{H}}_D^{(0)} = 0$$

and

$$\bar{\mathscr{H}}_S^{(0)} = \sum_k \frac{1}{\sqrt{3}} (\delta_k + \Delta\omega) I_{111}^{(k)}$$

as obtained in Sect. 3.2. Because of the reflection symmetry $\bar{\mathscr{H}}^{(1)}$ and all other odd order correction terms vanish, leaving as the leading correction term

$$\bar{\mathscr{H}}^{(2)} = \frac{\tau^2}{18} \{[(\mathbf{Z} + 2\mathbf{X} + 2\mathbf{Y}), [(\mathbf{X} + \mathbf{Y}), \mathbf{Z}]] - [(\mathbf{Y} + 2\mathbf{X}), [\mathbf{Y}, \mathbf{X}]]\} \tag{3.93}$$

which in the case of the dipolar interaction reduces to [12]

$$\bar{\mathscr{H}}_D^{(2)} = \frac{\tau^2}{18} [(\mathscr{H}_{Dx} - \mathscr{H}_{Dz}), [\mathscr{H}_{Dy}, \mathscr{H}_{Dx}]] \tag{3.94}$$

and in the case of shift interactions to

$$\bar{\mathcal{H}}_S^{(2)} = -\frac{\tau^2}{18} \sum_k (\delta_k + \Delta\omega)^3 (I_{yk} - 2I_{xk} + 4I_{zk}). \tag{3.95}$$

The interested reader may want to calculate cross terms between shift and dipolar interactions, which can be done straightforwardly by using Eq. (3.93). It should be noted, that $\bar{\mathcal{H}}_D^{(2)}$ is responsible for the ultimate resolution of the WHH-4 experiment *on-resonance*, if no pulse imperfections are present, whereas, cross terms between the shift Hamiltonian and the dipolar Hamiltonian may become important *off-resonance* [18].

d) Eight-Pulse Sequences (HW-8 [12] and MREV-8 [19, 26, 52, 53])

Haeberlen and Waugh [12] realized very soon, that using their four-pulse sequence (WHH-4) as a mother cycle, a wealth of daughter cycles could be designed to suit different purposes. Some of them are "a beauty" others are not. The basic WHH-4 cycle is used as a subcycle and repeated with different pulse settings (phase, width, timing etc.) to form a 2^n pulse cycle, where n went up to as high a value as $n=6$ (theoretically) [44, 54]. The idea behind this procedure is to design compensation schemes for the different diseases real pulse sequences suffer from, such as rf inhomogeneity, pulse width errors, phase errors etc. It was Mansfield [19] who realized that symmetrization of pulse sequences and cascading cycles with different symmetries could lead to cancellation of errors.

We are going to discuss here merely eight-pulse cycles, namely the Haeberlen-Waugh [12] (HW-8) and another cycle, which was cryptically proposed by Mansfield [19] and later independently discovered by Rhim, Elleman and Vaughan [26, 53] to which we will therefore refer as MREV-8 sequence. Figure 3.21 represents the pulse timing and the "switching Hamiltonian" for these two pulse sequences. Of course, other sequences are possible by changing the initial condition, however, we do not regard those to be *different*.

The one cycle propagator $\mathbf{L}(t_c)$ is immediately written down according to Fig. 3.19 as

$$\text{HW-8:} \quad \mathbf{L}_0(t_c) = [\mathbf{Z}, \mathbf{Y}, 2\mathbf{X}, \mathbf{Y}, \bar{\mathbf{Z}}][\bar{\mathbf{Z}}, \mathbf{Y}, 2\bar{\mathbf{X}}, \mathbf{Y}, \mathbf{Z}] \tag{3.96}$$

$$\text{MREV-8:} \quad \mathbf{L}_0(t_c) = [\mathbf{Z}, \mathbf{Y}, 2\mathbf{X}, \mathbf{Y}, \mathbf{Z}][\mathbf{Z}, \bar{\mathbf{Y}}, 2\mathbf{X}, \bar{\mathbf{Y}}, \mathbf{Z}]$$

$$\text{or} \quad \mathbf{L}_0(t_c) = [[\mathbf{Z}, \mathbf{Y}, \mathbf{X}]][\mathbf{Z}, \bar{\mathbf{Y}}, \mathbf{X}]] \tag{3.97}$$

where $t_c = 12\tau$.

It is straightforward to show that $\bar{\mathcal{H}}_D^{(0)} = 0$, in both cases and

$$\text{HW-8:} \quad \bar{\mathcal{H}}_S^{(0)} = \tfrac{1}{3} \sum_k (\delta_k + \Delta\omega) I_{010}^{(k)} \tag{3.98}$$

where $I_{010} = I_{yk}$, scaling factor $S = 1/3$ and

$$\text{MREV-8:} \quad \bar{\mathcal{H}}_S^{(0)} = \frac{\sqrt{2}}{3} \sum_k (\delta_k + \Delta\omega) I_{101}^{(k)} \tag{3.99}$$

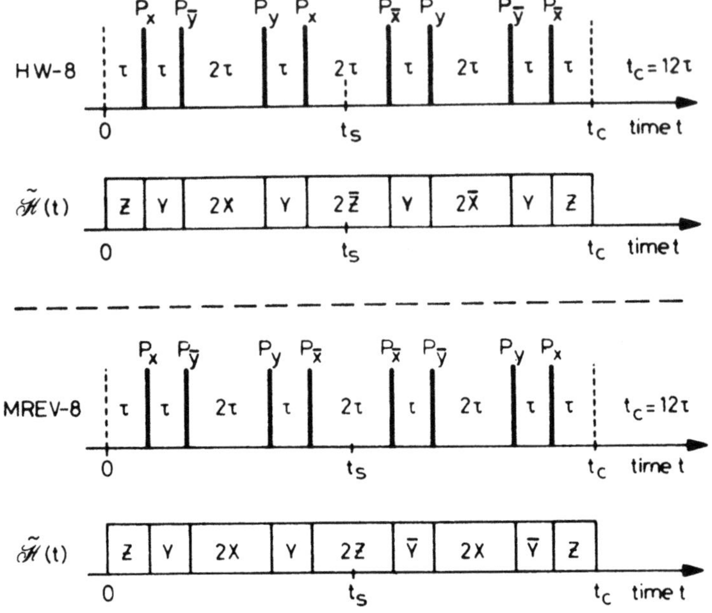

Fig. 3.19. Pulse timing and interaction Hamiltonian in the "toggling rotating frame" for the HW-8 and the MREV-8 pulse sequences

where

$$I_{101}^{(k)} = (I_{xk} + I_{zk})/\sqrt{2}, \quad \text{scaling factor} \quad S = \sqrt{2}/3.$$

For the dipolar interaction the bars in Eqs. (3.96, 3.97) are of no significance since $\mathcal{H}_{Dz} = \mathcal{H}_{D-z}$ and both cycles are symmetric leading to $\mathcal{H}_D^{(n)} = 0$ for all n which are odd, especially $\bar{\mathcal{H}}_D^{(1)} = 0$. This is not so for the shift Hamiltonian, since the cycle symmetry is broken, according to Eqs. (3.96, 3.97) for Hamiltonians which are not invariant under π rotation:

This first correction term is readily obtained as

$$\text{HW-8:} \quad \bar{\mathcal{H}}^{(1)} = \frac{-i\tau}{12}\,[(\mathbf{X} - \bar{\mathbf{X}}), (\mathbf{Z} - \bar{\mathbf{Z}}) - 2\mathbf{Y}]; \quad t_c = 12\tau \tag{3.100}$$

$$\text{MREV-8:} \quad \bar{\mathcal{H}}^{(1)} = \frac{-i\tau}{6}\,[(\mathbf{X} + \mathbf{Z}), (\mathbf{Y} - \bar{\mathbf{Y}})]. \tag{3.101}$$

The corresponding corrections to the shift interaction

$$\text{HW-8:} \quad \bar{\mathcal{H}}_S^{(1)} = \frac{-\tau}{6} \sum_k (\delta_k + \Delta\omega)^2 (I_{yk} + I_{zk}) \tag{3.102}$$

$$\text{MREV-8:} \quad \bar{\mathcal{H}}_S^{(1)} = \frac{\tau}{3} \sum_k (\delta_k + \Delta\omega)^2 (I_{zk} - I_{xk}) \tag{3.103}$$

and crossterms between shift interactions and dipolar interactions are readily obtained. The second correction term $\bar{\mathcal{H}}_D^{(2)}$ for dipolar interactions need to be calculated over the subcycle $[\mathbf{Z}, \mathbf{Y}, \mathbf{X}]$ only and is thus identical for the WHH-4, HW-8 and the MREV-8 sequence, namely

$$\bar{\mathcal{H}}_D^{(2)} = -\frac{\tau^2}{18}\left[(\mathcal{H}_{Dx} - \mathcal{H}_{Dz}), [\mathcal{H}_{Dx}, \mathcal{H}_{Dy}]\right]. \tag{3.104}$$

In summarizing we note, that nothing seems to be gained by using the eight-pulse cycles compared with the WHH-4 cycle, on the contrary, shift Hamiltonians are even scaled further than in the four-pulse experiment. It will turn out, however, later, when we treat rf inhomogeneity (see Sect. 3.7) that the eight-pulse experiments are in some respects superior to the WHH-4 sequence.

Rhim and co-workers [53, 55] have demonstrated in a series of publications this superiority in resolution of the MREV-8 sequence. A chemical shift spectrum for ^{1}H in ice using MREV-8 was reported by Gerstein and co-workers [56].

e) The 24-Pulse (BR-24) and 52-Pulse (BR-52) Sequences [57–60]

In order to improve the resolution of multiple pulse cycles even higher order multiple-pulse sequences were developed. Some of them were not so effective in reducing the residual linewidth because higher order admixture of pulse imperfections come into play and may ruin what one has calculated on the basis of ideal pulses. It is therefore essential to find a simple procedure to analyse complex multiple-pulse sequences and to find ways to design sequences which have the desired property. Burum and Rhim [57] introduced a scheme which they termed "pulse cycle decoupling" which allowed them to simplify the calculation of crossterms in the Magnus expansion for complex multiple pulse cycles. This finally led to the proposal of the 52-pulse and 24-pulse cycles [57–60] which seem to be the multiple-pulse sequences with the best resolution known today.

The detailed analysis of these sequences is rather complicated and the reader is referred to the interesting papers by Burum and Rhim [57–60]. Writing down the complete pulse sequence and the effective Hamiltonian in the "toggling frame" for the BR-52 sequence already takes a lot of space. We restrict ourselves here therefore to a brief discussion of the BR-24 sequence, which anyway seems to have a similar resolution as the BR-52 sequence.

Following the notation of Burum and Rhim [57] we use

$$(XY) \equiv \tau - (\pi/2)_x - \tau - (\pi/2)_y - \tau$$

as a shorthand notation for the $x - y$ solid echo sequence. In the same spirit

$$(XY)(\bar{Y}\bar{X}) \equiv \tau - (\pi/2)_x - \tau - (\pi/2)_y - 2\tau - (\pi/2)_{-y} - \tau - (\pi/2)_{-x} - \tau$$

is one of the familiar WHH-4 sequences. In order to compress the notation further, we introduce

$$W_1 \equiv (XY)(\bar{Y}X)$$
$$W_2 \equiv (\bar{X}Y)(\bar{Y}X)$$
$$W_3 \equiv (YX)(\bar{X}Y)$$
$$W_4 \equiv (\bar{Y}X)(\bar{X}Y)$$

as a compact notation for the different WHH-4 sequences. We are now well equipped to write down the BR-24 pulse-sequence in the form

$$W_1 W_2 W_3(\bar{Y}X) W_3 W_4(\bar{X}Y)$$

where the cycle time $t_c = 36\tau$. The one-cycle propagator $L_0(t_c)$ which governs the time-evolution is readily derived as [57]

$$\mathbf{L_0}(t_c) = [\![\mathbf{XYZ}]\!][\![\mathbf{X\bar{Y}Z}]\!][\![\mathbf{XZY}]\!][\![\mathbf{X\bar{Z}Y}]\!][\![\mathbf{YXZ}]\!][\![\mathbf{Y\bar{X}Z}]\!][\![\mathbf{Y\bar{Z}X}]\!] \tag{3.105}$$

where we have used the same compact notation as defined before in Eqs. (3.59–3.61) for the Hamiltonian state. The zero order chemical shift Hamiltonian is straightforwardly derived as

$$\bar{\mathcal{H}}_S^{(0)} = \frac{2}{3\sqrt{3}}(\delta + \Delta\omega)I_{111} \tag{3.106}$$

with $I_{111} = (I_x + I_y + I_z)/3$. The scaling factor of the BR-24 sequence is therefore $2/(3\sqrt{3}) = 0.385$.

When finite pulsewidth t_w is taken into account, the scaling factor is changed to [61]

$$S = \frac{2}{3\sqrt{3}}(1 + 2a)$$

where

$$a = \frac{t_w}{4\tau}\left(\frac{4}{\pi} - 1\right).$$

Because of its superior resolution, the BR-24 sequence will most likely become to be the "working horse" of high resolution multiple-pulse NMR.

3.5 Arbitrary Rotations and Finite Pulse Width in Multiple-Pulse Experiments

So far we always have considered idealized pulse sequences by assuming the rf pulses to be δ pulses, usually with a rotation angle $\beta = \pi/2$ or π.

Although we have treated the phase alternated sequence in Sect. 3.1 for arbitrary β, we still assumed the pulses to be δ pulses there. Moreover, we have not taken into account their effect on the dipolar interactions. In this section we are going to lift all these constraints. Arbitrary rotations are conveniently treated with the interaction Hamiltonians expressed by irreducible tensor operators (see Appendix A) as

$$\mathcal{H}_{\mathrm{int}} = \sum_k \sum_{q=-k}^{k} (-1)^q A_{k,-q} T_{kq} \tag{3.107}$$

which in the special case of
   (i) shift interactions (see Appendix A) is written as

$$\mathcal{H}_S = \sum_{q=-1}^{1} (-1)^1 A_{1,-q} T_{1q} \tag{3.108}$$

where

$$A_{10} = B_{Sz}; \qquad A_{1\pm1} = \mp\frac{1}{\sqrt{2}}(B_{Sx} \pm iB_{Sy})$$

and

$$T_{10} = I_z; \qquad T_{1\pm1} = \mp\frac{1}{\sqrt{2}}(I_x \pm iI_y).$$

With $\mathbf{B}_0 = (0, 0, B_0)$ we can write

$$B_{Sx} = S_{xz}B_0; \qquad B_{Sy} = S_{yz}B_0; \qquad B_{Sz} = S_{zz}B_0.$$

In the case of

(ii) dipolar interaction (see Appendix A) we obtain

$$\mathcal{H}_D = \sum_{i<j} \sum_{q=-2}^{2} (-1)^q A_{2,-q}^{(i,j)} T_{2q}^{(i,j)} \tag{3.109}$$

where

$$A_{20}^{ij} = -dC_{20}^{(i,j)}; \qquad A_{2\pm1}^{(i,j)} = \mp dC_{2\pm1}^{(i,j)}; \qquad A_{2\pm2}^{(i,j)} = -dC_{2\pm2}^{(i,j)} \tag{3.110}$$

with $d = \sqrt{6}\gamma_I^2\hbar/r_{ij}^3$ and where the modified spherical harmonics C_{kq} are defined by the spherical harmonics Y_{kq} as

$$C_{kq} = \left[\frac{4\pi}{2k+1}\right]^{1/2} Y_{kq} \tag{3.111}$$

and

$$T_{20}^{(i,j)} = \frac{1}{\sqrt{6}}(3I_{zi}I_{zj} - \mathbf{I}_i\cdot\mathbf{I}_j)$$

$$T_{2\pm1}^{(i,j)} = \mp\tfrac{1}{2}(I_{zi}I_{\pm j} + I_{\pm i}I_{zj}) \tag{3.112}$$

$$T_{2\pm2}^{(i,j)} = \tfrac{1}{2}I_{\pm i}I_{\pm j}.$$

In the following we drop the indices i, j on the tensor operators for convenience. The external Hamiltonian performing the rotational operation in the rotating frame may be expressed as:

$$\mathcal{H}_{\text{ext}} = -\boldsymbol{\omega}(t)\cdot\mathbf{I} = -\omega_1(t)\mathbf{n}\cdot\mathbf{I} \tag{3.113}$$

where $\mathbf{n}$ is a unit vector in the direction of the applied field. The rotation is now described by the unitary operator

$$\mathbf{L}_1(t) = \exp[-i\beta(t)\mathbf{n}\cdot\mathbf{I}] \tag{3.114}$$

where

$$\beta(t) = \int_0^t dt'\,\omega_1(t').$$

In multiple-pulse experiments the time evolution operator $\mathbf{L}_0(t)$ which governs the nuclear response is expressed by the "modulated" or "switched" Hamiltonian

$$\mathscr{\tilde{H}}(t) = \mathbf{L}_1^{-1}(t)\mathscr{H}_{int}\mathbf{L}_1(t)$$

as shown in Sect. 3.3. Since we operate in spin space, the rotation operation is applied to the T_{kq} only, and if only secular contributions are taken into account we may write

$$\mathscr{\tilde{H}}(t) \sim \tilde{T}_{k0}(t)$$

where

$$\tilde{T}_{kq}(t) = \mathbf{L}_1^{-1}(t)\, T_{kq}\mathbf{L}_1(t). \tag{3.115}$$

In the following we discuss the secular part $\tilde{T}_{k0}(t)$ when rf irradiation is applied in the x, y plane of the rotating frame. In this case (see Appendix B) [62-64]

$$\tilde{T}_{k0}(t) = \sum_{q=-k}^{k} T_{kq}D_{q0}^{(k)}(\alpha, \beta(t), -\alpha) \tag{3.116}$$

with the Wigner matrices [62-64]

$$D_{q0}^{(k)}(\alpha, \beta(t), -\alpha) = e^{-iq\alpha}d_{q0}^{(k)}(\beta) \tag{3.117}$$

where α is the angle between the rf field direction and the y axis and $\beta(t)$ is the rotating angle during rf irradiation

(i) Irradiation in y direction ($\alpha=0$) [64]

$$\tilde{T}_{k0}(t) = \sum_{q=-k}^{k} T_{kq}d_{q0}^{(k)}(\beta)$$

with

$$\tilde{T}_{10}(t) = T_{10}\cos\beta(t) + \frac{1}{\sqrt{2}}(T_{1-1} - T_{1+1})\sin\beta(t) \tag{3.118}$$

and

$$\tilde{T}_{20}(t) = T_{20}\tfrac{1}{2}(3\cos^2\beta(t) - 1) + \sqrt{\tfrac{3}{2}}(T_{2-1} - T_{21})\sin\beta(t)\cos\beta(t)$$
$$+ \sqrt{\tfrac{3}{8}}(T_{22} + T_{2-2})\sin^2\beta(t) \tag{3.119}$$

(ii) irradiation in x direction ($\alpha = -\pi/2$)

$$\tilde{T}_{k0}(t) = \sum_{q=-k}^{k} T_{kq}e^{iq\pi/2}d_{q0}^{(k)}(\beta(t)) \tag{3.120}$$

from where it follows, that the terms with $q = \pm 1$ in (i) have to be multiplied by $\pm i$, whereas the terms with $q = \pm 2$ have to be multiplied by -1.

The average Hamiltonian during rf irradiation is defined as [64]

$$\bar{T}_{k0}(\beta_1) = \frac{1}{\beta_1}\int_0^{\beta_1} d\beta\,\tilde{T}_{k0}(\beta) \tag{3.121}$$

where

$$\beta_1 = \int_0^{t_w} dt\,\omega_1(t) \tag{3.122}$$

with the pulse width t_w.

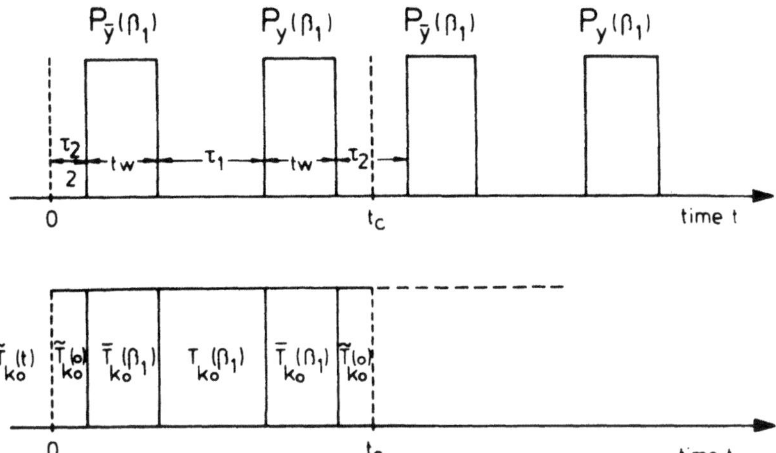

Fig. 3.20. Pulse timing and tensor representation of the interaction Hamiltonian in a phase alternated sequence with arbitrary pulse width

We are now well equipped to analyse the phase-alternated sequence (MW-2) and the four-pulse (WHH-4) cycle with arbitrary pulse width. According to Fig. 3.20 we may use the following parameters:

$$t_c = \tau_1 + \tau_2 + 2t_w \qquad \text{cycle time} \tag{3.123a}$$

$$\delta = 2t_w/t_c \qquad \text{duty factor} \tag{3.123b}$$

$$\kappa = \frac{\tau_1 + t_w}{\tau_2 + t_w} \qquad \text{pulse timing} \tag{3.123c}$$

The cycle has reflection symmetry which leads to $\bar{\mathscr{H}}^{(1)} = 0$. In the following we discuss the average Hamiltonian only, which is straightforwardly obtained for an arbitrary direction of the irradiation, as [64])

$$\bar{T}_{k0} = \frac{1}{t_c} \{ \tau_2 \tilde{T}_{k0}(0) + \tau_1 \tilde{T}_{k0}(\beta_1) + 2t_w \bar{T}_{k0}(\beta_1) \}. \tag{3.124}$$

Using the above parameters this may be expressed as (there is a slight change in parameters compared with Ref. [64])

$$\bar{T}_{k0} = \frac{2 - (1 + \kappa)\delta}{2(1 + \kappa)} \tilde{T}_{k0}(0) + \frac{2\kappa - (1 + \kappa)\delta}{2(1 + \kappa)} \tilde{T}_{k0}(\beta_1) + \delta\bar{T}_{k0}(\beta_1) \tag{3.125}$$

or by writing this in compact form as

$$\bar{T}_{k0} = \sum_{q = -k}^{k} T_{kq} \overline{D_{q0}^{(k)}(\alpha, \beta_1, -\alpha)}. \tag{3.126}$$

The following abbreviations are used in the above equation for $\alpha = 0$

$$D_0 = \overline{D_{00}^{(2)}(\beta_1)}; \qquad D_1 = \mp\overline{D_{\pm10}^{(2)}(\beta_1)}; \qquad D_2 = \overline{D_{\pm20}^{(2)}(\beta_1)} \tag{3.127}$$

$$C_0 = \overline{D_{00}^{(1)}(\beta_1)}; \qquad C_1 = \mp\overline{D_{\pm10}^{(1)}(\beta_1)} \tag{3.128}$$

with

$$D_0 = \tfrac{3}{4}\cos\beta_1 \left[\delta\,\frac{\sin\beta_1}{\beta_1} + \left(\frac{2\kappa}{1+\kappa}-\delta\right)\cos\beta_1\right] + \frac{2-\kappa}{2(1+\kappa)} \tag{3.129a}$$

$$D_1 = \sqrt{\tfrac{3}{8}}\,\sin\beta_1 \left[\delta\,\frac{\sin\beta_1}{\beta_1} + \left(\frac{2\kappa}{1+\kappa}-\delta\right)\cos\beta_1\right] \tag{3.129b}$$

$$D_2 = \tfrac{1}{2}\sqrt{\tfrac{3}{8}}\left[\sin\beta_1\left(\frac{2\kappa}{1+\kappa}-\delta\right) - \delta\,\frac{\cos\beta_1}{\beta_1}\right]\cdot\sin\beta_1 + \tfrac{1}{2}\sqrt{\tfrac{3}{8}}\,\delta \tag{3.129c}$$

$$C_0 = \delta\,\frac{\sin\beta_1}{\beta_1} + \tfrac{1}{2}\left(\frac{2\kappa}{1+\kappa}-\delta\right)\cos\beta_1 + \tfrac{1}{2}\left(\frac{2}{1+\kappa}-\delta\right) \tag{3.130a}$$

$$C_1 = \frac{\delta}{\sqrt{2}}(1-\cos\beta_1)/\beta_1 + \frac{1}{2\sqrt{2}}\left(\frac{2\kappa}{1+\kappa}-\delta\right)\sin\beta_1. \tag{3.130b}$$

Equations (3.129, 3.130), were misprinted in Ref. [64] and are expressed here for convenience using a slightly different definition of the parameter κ.

The average dipolar Hamiltonian in the WHH-4 experiment ($\kappa=2$) can now be expressed as

$$\bar{T}_{10} = C_0 T_{10} + C_1\left[\tfrac{1}{2}(1-i)T_{1-1} - \tfrac{1}{2}(1+i)T_{11}\right] \tag{3.131a}$$

$$\bar{T}_{20} = D_0 T_{20} + D_1\left[\tfrac{1}{2}(1-i)T_{2-1} - \tfrac{1}{2}(1+i)T_{21}\right]. \tag{3.131b}$$

In order to digest these formulas, we take as a simple example a case which we have treated already in Sect. 3.2, i.e. $\delta=0$, $\kappa=1$, $\beta=\pi/2$.

We obtain from Eqs. (3.129) and (3.130) in this case

$$D_0=\tfrac{1}{4}; \quad D_1=0; \quad D_2=\tfrac{1}{2}\sqrt{\tfrac{3}{8}}$$

and

$$C_0=\tfrac{1}{2}; \quad C_1=\frac{1}{2\sqrt{2}}.$$

$\bar{T}_{k0}$ may now be calculated by using Eqs. (3.126–3.128).

In the case of dipolar interaction this results in

$$\bar{T}_{20} = -\tfrac{1}{2}\cdot\frac{1}{\sqrt{6}}(3I_{yi}I_{yj}-\mathbf{I}_i\cdot\mathbf{I}_j)$$

or

$$\mathcal{H}_D^{(0)} = -\tfrac{1}{2}\mathcal{H}_{Dy}$$

and in the case of shift interactions in

$$\bar{T}_{10} = \tfrac{1}{2}(I_z+I_x)$$

or

$$\mathcal{H}_S^{(0)} = \sum_k \tfrac{1}{2}(\delta_k+\Delta\omega)(I_{zk}+I_{xk}).$$

This is identical with the results obtained in Sect. 3.1 under the δ pulse assumption. Equations (3.129) and (3.130), however, are much more powerful,

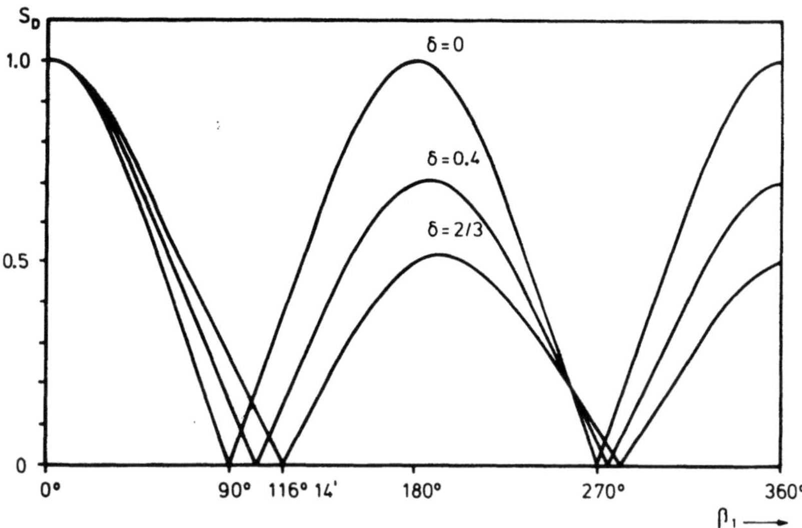

Fig. 3.21. Dipolar scaling factor $S_D = \|\overline{\mathcal{H}}_D^{(0)}\| / \|\mathcal{H}_D'\|$ in a four-pulse experiment versus the rotation angle β_1 of the rf pulses for different values of the duty factor δ. For each value $0 \leq \delta \leq 2/3$ an angle β_1 can be found, where $S_D = 0$. Notice, that rf inhomogeneity (variation of β_1 over the sample) becomes less important with increasing δ. We remark further, that S_D is invariant to a π or 2π pulse only if $\delta = 0$

since they describe the general case and can be readily applied to different multiple-pulse sequences.

The considered cycle in Fig. 3.20, is a subcycle of the WHH-4 sequence, if the subsequent cycle is performed under irradiation in the x direction. As noted before, Eqs. (3.124), (3.129), and (3.130) are still valid in this case, the terms with $q = \pm 1$ have to be multiplied by $\pm i$, however, whereas the terms with $q = \pm 2$ have to be multiplied by -1. This cancels the D_2 term in the average over the full cycle and the D_1 and D_0 term have to be considered only. For high resolution purposes the dipolar terms D_0 and D_1 must vanish. This may be achieved by making the square bracket in Eqs. (3.129a, b) zero. This is possible under the condition $\kappa = 2$, which appears to be the characteristic timing of the four-pulse cycle which we shall call the "ideal timing". A "dipolar scaling factor" $S_D = \|\overline{\mathcal{H}}_D^{(0)}\| / \|\mathcal{H}_D'\| = (D_0^2 + D_1^2)^{1/2}$ may be defined, which is plotted versus the total rotation angle β_1 in Fig. 3.21 for different duty factors δ [64]. As follows from Eq. (3.129) D_0 and D_1 always vanish for the same value of β_1. The corresponding condition for β_1 can be easily derived from Eqs. (3.129a, b) for $\kappa = 2$ to be [64]

$$\delta(1 - \tan\beta_0/\beta_0) = \tfrac{4}{3}; \qquad \beta_0 \geq 90°. \tag{3.131c}$$

This condition is plotted in Fig. 3.22, demonstrating that for every duty factor δ up to $\delta = 2/3$ a rotation angle β_0 can be found which leads to a vanishing average dipolar Hamiltonian. Expanding $\tan\beta_0$ for $\beta_0 = \pi/2 + \varepsilon$ leads to

$$\delta = \tfrac{2}{3}\pi\varepsilon$$

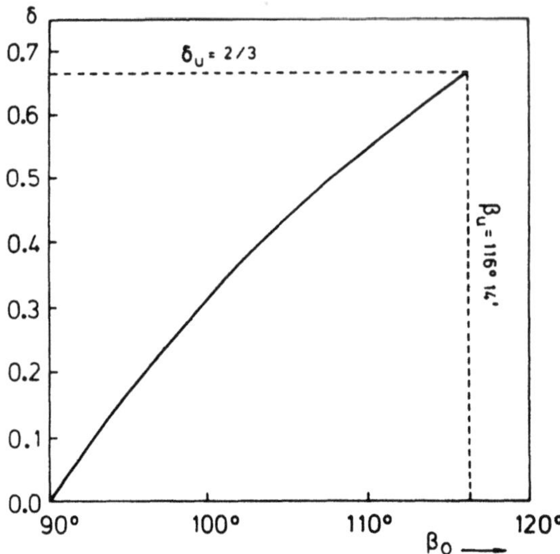

Fig. 3.22. Relation between the duty factor of a four-pulse cycle and the rotation angle β_0 for vanishing S_D, according to Eq. (3.131c) (see Ref. [64])

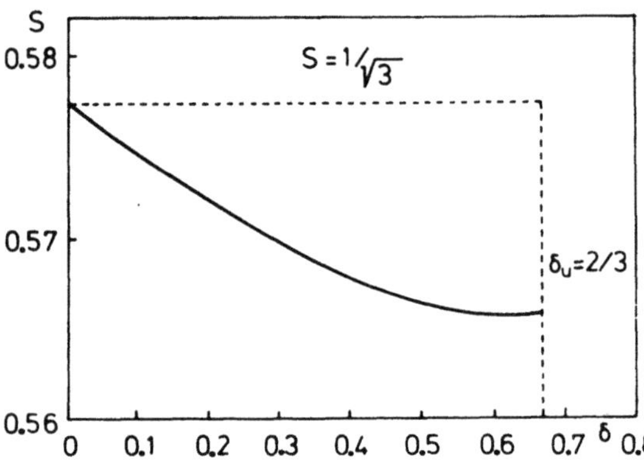

Fig. 3.23. Variation of the scaling factor S for shift interactions with the duty factor δ of a four-pulse sequence according to Eq. (3.132) (see Ref. [64])

which has been used before to account for the influence of finite pulse width [12]. The upper pulse width to be used in the four-pulse experiment is according to Eq. (3.131a) given by ($\delta = 2/3$)

$$\tan\beta_u = -\beta_u$$

which leads to [64]

$$\beta_u = 116°14'21''.$$

The scaling factor of the shift interaction is changed correspondingly, and may be obtained from Eq. (3.130a, b) by

$$S = [C_0^2 + C_1^2]^{1/2} = [\mathrm{Tr}\{\bar{T}_{10}^2\}/\mathrm{Tr}\{T_{10}^2\}]^{1/2}. \tag{3.132}$$

Treating the familiar case $\kappa = 2$, $\delta = 0$, $\beta_1 = \pi/2$ we obtain immediately by using Eqs. (3.131) and (3.132) $S = 1/\sqrt{3}$. The general case for arbitrary δ up to $\delta = 2/3$ is plotted in Fig. 3.23 under the "high resolution" condition $\kappa = 2$ and obeying

Eq. (3.131c). The scaling factor drops monotonically by about 2% when δ approaches $\delta_u = 2/3$, i.e. its effect on the scaling factor is negligible. We have demonstrated that coherent averaging takes place also during the rf pulse in multiple-pulse experiments, and conditions can be found for the average dipolar Hamiltonian to vanish. Thus finite pulse width is not to be considered as a pulse imperfection, as long as all pulses have the same width and the same rotation angle β_0.

A similar treatment has been given by Haeberlen [18] and H. Ernst et al. [65] using a slightly different approach.

More recently Burum, Linder and Ernst [61] have compared the resolution of different multiple-pulse experiments when using finite pulse width. They even propose the "windowless" multiple pulse cycles BLEW-12 and BLEW-48 [61].

3.6 Second Averaging

In the preceding sections we have shown that the time evolution operator governing the nuclear response during a multiple-pulse sequence can be expressed by a single average Hamiltonian $\bar{\mathcal{H}}$, which may include all orders of correction. This was achieved by imposing a cyclic property on the rf pulses, where the cycle time t_c has to fulfil the condition

(i) $\quad \dfrac{1}{t_c} \gg \| \mathcal{H}_{\mathrm{int}} \|$

in order to ensure a rapid convergence of the Magnus expansion. So far we always have included the offset Hamiltonian $\Delta\omega I_z$ in the interaction Hamiltonian.

We are now going to separate the offset Hamiltonian (modified by the multiple-pulse sequence) from the other terms of the interaction Hamiltonian

$$\bar{\mathcal{H}} = \overline{\Delta\omega}\, I_{\bar{\mu}} + \bar{\mathcal{H}}_{\mathrm{int}} \tag{3.133}$$

since it may constitute a new "Zeeman term" with the new quantization axis $\bar{\mu}$, under the condition

(ii) $\quad \overline{\Delta\omega} \gg \| \bar{\mathcal{H}}_{\mathrm{int}} \|.$

In this case now the modified offset Hamiltonian acts as an external field and the spins are precessing around this "average offset field" i.e., around the μ axis with the cycle time $t_\Delta = 2\pi/\overline{\Delta\omega}$.

The offset term may be factored out of $\bar{\mathcal{H}}$ [Eq. (3.124)] and coherent averaging theory (see Sect. 3.3) can be readily applied to

$$\tilde{\bar{\mathcal{H}}}_{\mathrm{int}} = e^{-i\overline{\Delta\omega}\, I_{\bar{\mu}} \cdot t}\, \bar{\mathcal{H}}_{\mathrm{int}}\, e^{i\overline{\Delta\omega}\, I_{\bar{\mu}} \cdot t}. \tag{3.134}$$

Again averaging over the cycle time t_Δ results in the "second averaged" Hamiltonian $\bar{\mathcal{H}}_{\mathrm{int}}^{(0)}$ as characterized by the two bars. However, this procedure is legitimate only under the constraints of the conditions (i) and (ii), summarized into the following condition for second averaging:

$$\frac{1}{t_c} \gg \overline{\Delta\omega} \gg \|\bar{\mathcal{H}}_{\text{int}}\|. \tag{3.135}$$

The main consequence of "second averaging" is that it provides another means of *manipulation* in spin space which is capable of averaging all interactions, which are orthogonal to the direction of the "effective off-resonance field". With the corresponding cycle time t_Δ we write

$$\bar{\mathcal{H}}_{\text{int}}^{(0)} = \frac{1}{t_\Delta} \int\limits_0^{t_\Delta} dt' \, \tilde{\mathcal{H}}_{\text{int}}(t'). \tag{3.136}$$

Many of the residual interaction Hamiltonian in multiple-pulse experiments, such as $\bar{\mathcal{H}}_D^{(2)}$ and some Hamiltonian due to pulse imperfections etc., are averaged to zero according to Eq. (3.136). Thus enhanced resolution is observed in multiple pulse experiments, when the spectrometer frequency is shifted off-resonance.

Second averaging has been dealt with by several authors [66, 67]. We find, however, the approach of Pines and Waugh [67] especially appealing and shall follow along those lines.

We define a unit spin vector μ along the z axis and apply to it the time evolution operator $L_1(t)$, see Eq. (3.66), to obtain the motion of the vector μ as

$$\mu(t) = \mathbf{L}_1^{-1}(t) \, \mu \mathbf{L}_1(t).$$

The average unit spin vector μ may now be expressed as

$$\mu = \int\limits_0^{t_c} dt \, \mu(t) \Bigg/ \left| \int\limits_0^{t_c} dt \, \mu(t) \right|. \tag{3.137}$$

Since the offset Hamiltonian transforms like μ it is evident, that

$$\overline{\Delta\omega} = S \cdot \Delta\omega \tag{3.138}$$

where the scaling factor is

$$S = \frac{1}{t_c} \int\limits_0^{t_c} dt \, \mu(t). \tag{3.139}$$

On the other hand μ may be obtained by a rotation

$$\mu = R(\varphi, \vartheta, \psi) \, \mu R^{-1}(\varphi, \vartheta, \psi) \tag{3.140}$$

where R may be derived from Eq. (3.137).

Using the rotation R we can define any Hamiltonian with respect to the μ axis by

$$\mathcal{H}_{\bar{\mu}} = \dot{R} \, \mathcal{H}_z R^{-1}. \tag{3.141}$$

The average interaction Hamiltonian $\bar{\mathcal{H}}_{\text{int}}$ usually is not "parallel" to $\mathcal{H}_{\bar{\mu}}$ as is demonstrated in Fig. 3.24. However, $\bar{\mathcal{H}}_{\text{int}}$ may be separated into a "parallel" and an "orthogonal" component (see Fig. 3.25)

$$\bar{\mathcal{H}}_{\text{int}} = \mathcal{H}_{\parallel} + \mathcal{H}_{\perp} \tag{3.142}$$

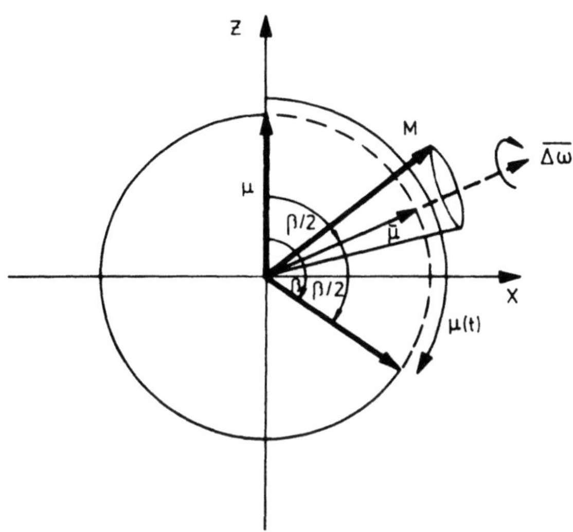

Fig. 3.24. Pictorial description of resonance offset averaging (second averaging) in the phase-alternated sequence, employing rf pulses in the y direction with a rotation angle β [67]. The unit magnetization vector μ is flipped by an angle β and $-\beta$ alternatively, leading to an average unit vector $\bar{\mu}$. The average offset frequency $\overline{\Delta\omega} = \cos(\beta/2)\,\Delta\omega$ points along $\bar{\mu}$ and any magnetization **M** processes on the average about $\bar{\mu}$ with frequency $\overline{\Delta\omega}$

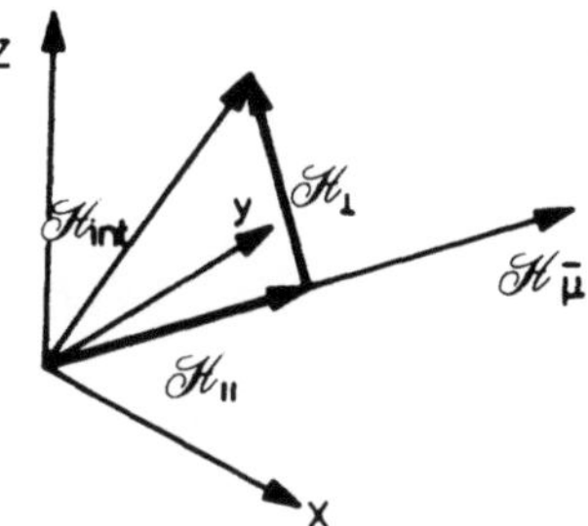

Fig. 3.25. Mnemonic vectorial representation of the projection procedure of an interaction Hamiltonian $\mathscr{H}_{int}$ onto a Hamiltonian $\mathscr{H}_{\bar{\mu}}$. $\mathscr{H}_{int}$ is separated into a parallel and an orthogonal part with respect to $\mathscr{H}_{\bar{\mu}}$ by means of a projection operator technique as defined in the text

or by using the projector $P_{\bar{\mu}}$

$$\bar{\mathscr{H}}_{int} = P_{\bar{\mu}}\,\bar{\mathscr{H}}_{int} + (1 - P_{\bar{\mu}})\,\bar{\mathscr{H}}_{int} \tag{3.143}$$

where P is idempotent ($P^2 = P$) and

$$P_{\bar{\mu}} = \frac{|\mathscr{H}_{\bar{\mu}})\,(\mathscr{H}_{\bar{\mu}}|}{(\mathscr{H}_{\bar{\mu}}|\mathscr{H}_{\bar{\mu}})}. \tag{3.144}$$

The second averaged Hamiltonian may now be expressed as the "parallel" component of $\bar{\mathscr{H}}_{int}$, whereas, the orthogonal part is averaged out

$$\bar{\mathscr{H}}^{(0)}_{int} = \mathscr{H}_{\parallel} = P_{\bar{\mu}}\,\bar{\mathscr{H}}_{int}. \tag{3.145}$$

Using Eqs. (3.142–3.145) we can write

$$\bar{\mathscr{H}}^{(0)}_{int} = p\,\mathscr{H}_{\bar{\mu}} \tag{3.146}$$

where p is a scalar with

$$p = \frac{(\mathscr{H}_{\bar{\mu}}|\bar{\mathscr{H}}_{int})}{(\mathscr{H}_{\bar{\mu}}|\mathscr{H}_{\bar{\mu}})}. \tag{3.147}$$

If we restrict ourselves to the zeroth order average Hamiltonian $\bar{\mathcal{H}}_{\text{int}}^{(0)}$ and remember, that

$$|\bar{\mathcal{H}}_{\text{int}}^{(0)}) = \frac{1}{t_c} \int\limits_0^{t_c} dt\, \mathbf{L}_1^{-1}(t)\, |\mathcal{H}_z)$$

we may write, using Eq. (3.141)

$$p = \frac{1}{t_c} \int\limits_0^{t_c} dt\, \frac{(\mathcal{H}_z|\, R^{-1}\, \mathbf{L}_1^{-1}(t)\, |\mathcal{H}_z)}{(\mathcal{H}_z|\mathcal{H}_z)}. \tag{3.148}$$

A very convenient expression for p can be obtained as shown by Pines and Waugh [67], with

$$\frac{(\mathcal{H}_z|\, R^{-1}\, \mathbf{L}_1^{-1}(t)\, |\mathcal{H}_z)}{(\mathcal{H}_z|\mathcal{H}_z)} = P_k\, \bar{\mu} \cdot \mu(t)) \tag{3.149}$$

where P_k is the Legendre polynomial of order k, and $\bar{\mu}$ and $\mu(t)$ are given by Eqs. (3.136) and (3.137). The order k depends on the rank of $\mathcal{H}_z$ e.g. $k=2$ for dipolar interaction and $k=1$ for shift interaction.

Following Eqs. (3.148) and (3.149) p may be expressed as

$$p = \overline{P_k(\bar{\mu} \cdot \mu(t))} \tag{3.150a}$$

where

$$\overline{P_k(\bar{\mu} \cdot \mu(t))} = \frac{1}{t_c} \int\limits_0^{t_c} dt\, P_k(\bar{\mu} \cdot \mu(t)). \tag{3.150b}$$

Combining Eqs. (3.146) and (3.150) leads to

$$\bar{\mathcal{H}}_{\text{int}}^{(0)} = \overline{P_k(\bar{\mu} \cdot \mu(t))} \cdot \mathcal{H}_{\bar{\mu}} \tag{3.151}$$

where $\mathcal{H}_{\bar{\mu}}$ is according to Eq. (3.141) just the truncated or secular interaction Hamiltonian with respect to the axis $\bar{\mu}$. The simple yet remarkable result obtained by Pines and Waugh [67] is, that no matter how complicated the multiple-pulse cycle is, as long as condition Eq. (3.135) is fulfilled, the average interaction Hamiltonian is just the secular interaction Hamiltonian itself with respect to some average axis in the rotating frame, scaled by the factor p. This factor p can be much more easily evaluated, than dealing with spin operators in a complicated multiple pulse sequence.

We shall treat now the phase-alternated sequence MW-2 as a simple example, to illustrate the usefulness of this approach. The phase-alternated sequence (MW-2) as shown in Fig. 3.17 consists of β pulses in the y and $\bar{y}$ direction alternatively. If we assume equal pulse spacing for all pulses ($\kappa = 1$) we find using Eq. (3.151) for the average dipolar Hamiltonian

$$\bar{\mathcal{H}}_D^{(0)} = \mathcal{H}_{D\bar{\mu}}(3 p_\delta(\beta) + 1)/4 \tag{3.152}$$

where

$$p_\delta(\beta) = (1 - \delta)\cos + \delta\, \frac{\sin \beta}{\beta}. \tag{3.153}$$

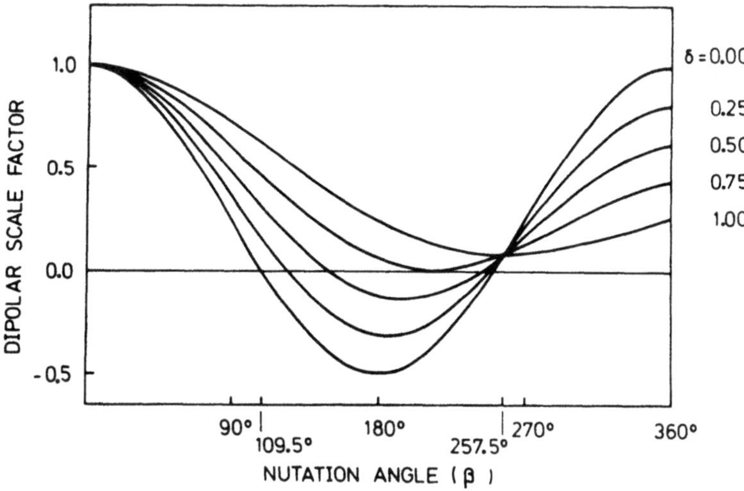

Fig. 3.26. Dipolar scaling factor $\overline{P_2(\bar{\mu} \cdot \mu(t))} = (3p_\delta(\beta)+1)/4$ [see Eq. (3.152)] for a phase-alternated sequence with a duty factor δ employing β pulses. (A. Pines and J.S. Waugh [67].) Line narrowing in solids is achieved with this sequence when $\bar{P}_2 = 0$, e.g. at 109.5° for $\delta = 0$ (phase-alternated tetrahedral experiment)

The $\bar{\mu}$ axis lies in the x-z plane. If there is a shift Hamiltonian $\mathcal{H}_S^{(0)}$ present in the rotating frame Hamiltonian, the second average shift Hamiltonian is expressed as

$$\bar{\bar{\mathcal{H}}}_S^{(0)} = \mathcal{H}_{S\bar{\mu}} \cdot p_\delta(\beta/2) \tag{3.154}$$

where $p_\delta(\beta)$ is given by Eq. (3.153).

Figure 3.26 shows a plot of $(3p_\delta(\beta)+1)/4$ as a function of β for different values of the duty factor δ. We see that $\bar{\bar{\mathcal{H}}}_D^{(0)}$ can be made to vanish for $\delta < 0.75$ opening the possibility for line narrowing experiments. If $\delta = 0$ it is evident from Fig. 3.26 and Eq. (3.152) that $\bar{\bar{\mathcal{H}}}_D^{(0)} = 0$ for $\beta = \beta_t$, where β_t is the tetrahedral angle (109° 28′). This sequence was named "phase-alternated tetrahedral" PAT sequence [66]. Line narrowing is still achieved if $\delta > 0$, however, the rotation angle β has to be larger than β_t in this case. The negative value of the dipolar scaling factor indicates, that "magic echoes" may be obtained in this region. For further details we refer the interested reader to Refs. [66, 67].

3.7 The Influence of Pulse Imperfection on Multiple-Pulse Experiments

One of the basic problems in multiple-pulse NMR concerns the question which resolution or line narrowing efficiency can be obtained. A number of derivates of the four-pulse cycle (WHH-4) have been proposed with the promise of better resolution. First, the second order correction term $\bar{\mathcal{H}}_D^{(2)}$ of the average dipolar Hamiltonian was felt to be the prime source for the limit in resolution, however, it was soon realized that pulse imperfections play a major role. The remedy usually prescribed is, to design cascaded cycles with different symmetry properties in order to cancel the effect of specific pulse imperfections [19, 53, 55].

Let us first start with the effect of magnetic field inhomogeneity. It is evident that inhomogeneity of the static magnetic field B_0 results in the same line broadening as in conventional NMR, however, scaled by the shift scaling factor S.

Inhomogeneity of the static magnetic field may be produced by the field itself or by the static magnetic susceptibility of the sample, if a non-spherical sample is used. The last effect, however, can be eliminated if a spherical sample is used. This is highly recommended, as is readily seen, when comparing the results in Table 3.1. The influence of the B_1 inhomogeneity, however, is more subtle and needs further consideration. A detailed discussion can be found in Refs. [19, 52, 53, 55, 65]. If a train of "inhomogenous" rf pulses is applied to a spin system, each individual spin I_i will experience a different rotation angle β_i, resulting in a distribution of rotation angles over the sample. An average rotation angle β_0 may be defined by

$$\beta_0 = \overline{\beta_i} \tag{3.155}$$

and the B_1 inhomogeneity by

$$\varepsilon_i = (\beta_i - \beta_0)/\beta_0 \tag{3.156}$$

with

$$\sum_i \varepsilon_i = 0$$

and the standard deviation

$$\varepsilon = [\overline{\varepsilon_i^2}]^{1/2}. \tag{3.157}$$

Table 3.1. Influence of the static magnetic susceptibility of the sample on the linewidth (W.K. Rhim, D.D. Elleman and R.W. Vaughan [26])

Sample shape	Linewidth[a]
Sphere[b]	0.4 ppm
Cylinder[c]	1.1 ppm
Rectangular parallelepiped[d]	1.4 ppm

[a] The chemical shift scaling factor has been taken into account for these values.
[b] 4 mm in diameter and spherical within 0.1 %.
[c] 4 mm in diameter and 4.5 mm in length; cylinder axis was perpendicular to the external field.
[d] 4 mm × 4 mm × 5 mm with cubic face perpendicular to field.

If a Gaussian distribution of the rf inhomogeneity is assumed, we obtain [68]

$$p(\varepsilon_i) = \frac{1}{(2\pi M_\varepsilon)^{1/2}} \exp(-\varepsilon_i^2/2M_\varepsilon) \tag{3.158a}$$

$$p(\beta_i) = \frac{N_0}{2\sqrt{\pi}} \exp\left[-\left(\frac{N_0}{2}\right)^2 \left(\frac{\beta_i - \beta_0}{\beta_0}\right)^2\right] \tag{3.158b}$$

where the second moment of the rf inhomogeneity distribution is given by

$$M_\varepsilon = \left(\frac{4}{\pi}\right)^2 \varepsilon^2$$

and where

$$N_0 = \frac{\pi}{2\sqrt{2}\varepsilon}. \tag{3.159b}$$

There is a direct loss of coherence among the spins due to this rf inhomogeneity with a coherence time proportional to $M_\varepsilon^{-1/2}$. This direct effect can be observed in a continuous rf field or a train of pulses with equal phase. A train of equally spaced $\pi/2$ pulses applied to a liquid sample is a suitable experiment for observing this direct effect and measuring the value M_ε, ε^2 or N_0 directly. The nuclear signal after the N-th $\pi/2$ pulse follows [68].

$$S(N) = \sin(N\pi/2) \exp[-(N/N_0)^2]. \tag{3.159c}$$

In phase-alternated cycles, however, this disorder is remedied due to the fact, that the initial state is recovered after each pair of phase-alternated pulses.

One should be aware, however, that the inhomogeneity effect will be transmitted to $\mathcal{H}_{\text{int}}(t)$, the "switched interaction Hamiltonian" under the condition $[\mathcal{H}_1(t), \mathcal{H}_{\text{int}}] \neq 0$. This indirect B_1 inhomogeneity effect is essential in all line narrowing sequences utilizing phase alternation. Since the average interaction Hamiltonian $\bar{\mathcal{H}}_0(\beta)$ depends on the rotation angle β of the rf pulses, the contribution of B_1 inhomogeneity to the average Hamiltonian may be conveniently expressed by

$$\bar{\mathcal{H}}_\varepsilon = \left[\frac{d\bar{\mathcal{H}}_0(\beta)}{d\beta}\right]_{\beta_0} \cdot \varepsilon_i \tag{3.160a}$$

if we limit ourselves to effects linear in ε_i.

Here β_0 denotes the value of β, for which maximum line narrowing efficiency is obtained.

Equation (3.160a) is applicable to any pulse sequence, employing arbitrary pulse width. The total average Hamiltonian can be expressed as

$$\bar{\mathcal{H}}_{\text{tot}} = \bar{\mathcal{H}}_0 + \bar{\mathcal{H}}_\varepsilon \tag{3.160b}$$

where $\bar{\mathcal{H}}_0$ is the average Hamiltonian according to the average pulse rotation angle β_0 of the B_1 inhomogeneity distribution. In order to introduce a measure for the influence the B_1 inhomogeneity has on the NMR signal, we consider the total scaling factor

$$S_{\text{tot}} = [\text{Tr}\{\bar{\mathcal{H}}_{\text{tot}}^2\}/\text{Tr}\{\mathcal{H}_{\text{int}}^2\}]^{1/2} \tag{3.160c}$$

where $\mathcal{H}_{\mathrm{int}}$ is the interaction Hamiltonian in the rotating frame i.e. without application of any pulse sequence. Let us first discuss the effect of indirect

(i) off-resonance B_1 inhomogeneity coupling

The magnitude of this effect will be directly proportional to the B_1 inhomogeneity and the resonance offset and can become very large in multiple-pulse experiments, as can be easily demonstrated in a liquid sample.

Applying Eqs. (3.160a–c), we obtain ($\varepsilon_i \ll 1$).

$$S_{\mathrm{tot}} = S_0(1 + 2\Delta S \varepsilon_i + \Delta S' \varepsilon_i^2)^{1/2} \cong S_0(1 + \Delta S \varepsilon_i) \tag{3.161a}$$

where

$$S_0 = [\mathrm{Tr}\{\bar{\mathcal{H}}_0^2\}/\mathrm{Tr}\{\mathcal{H}_S^2\}]^{1/2}$$

is the usual scaling factor of the multiple pulse experiment and

$$\Delta S = \mathrm{Tr}\left\{\bar{\mathcal{H}}_0 \left(\frac{d}{d_\beta} \bar{\mathcal{H}}_0\right)_{\beta_0}\right\} \bigg/ \mathrm{Tr}\{\bar{\mathcal{H}}_0^2\} \tag{3.161b}$$

and

$$\Delta S' = \mathrm{Tr}\left\{\left(\frac{d}{d_\beta} \bar{\mathcal{H}}_0\right)^2_{\beta_0}\right\} \bigg/ \mathrm{Tr}\{\bar{\mathcal{H}}_0^2\}$$

are the scaling factors of the B_1 inhomogeneity.

Since ε_i is a small quantity the contribution of $\Delta S' \cdot \varepsilon_i^2$ is negligible and the important parameter which describes the spectral distribution of the NMR line for a given B_1 inhomogeneity ε_i is ΔS, as far as shift interactions are concerned. In order to compare different pulse sequences we summarize in Table 3.2 the different expressions for $\bar{\mathcal{H}}_\varepsilon$ and $\bar{\mathcal{H}}_0$ in the limit $\delta = 0$; $\beta_0 = \pi/2$ (see Tables 3.3 and 3.4) and include the parameter ΔS.

Notice, that the MW-2 and MREV-8 sequence have the smallest value of ΔS, whereas WHH-4 and especially HW-8 have larger ones. This means, e.g. that a HW-8 sequence shows four times stronger broadening due to B_1 inhomogeneity to off-resonance coupling than the MREV-8 sequence. Let us consider a simple example. Suppose the mean variation of the B_1 field in the sample coil is 2%, the MW-2 and MREV-8 sequences would show a line broadened by 1% with respect to the resonance offset, i.e. line width 100 Hz for $\Delta v = 10$ kHz. Figure 3.27 demonstrates this effect in the case of a MREV-8 sequence applied to a liquid sample. The same procedure as outlined by Eqs. (3.160, 3.161) can be applied when arbitrary pulse width is assumed in the

Table 3.2. Influence of B_1 inhomogeneity (ε_i) on shift interactions (resonance offset). ΔS is a measure of the spectral distribution due to B_1 inhomogeneity (see text)

	MW-2	WHH-4	HW-8	MREV-8
$\bar{\mathcal{H}}_0$	$\frac{1}{2}\Delta\omega\cdot(I_z - I_y)$	$\frac{1}{3}\Delta\omega\cdot(I_x + I_y + I_z)$	$\frac{1}{3}\Delta\omega\cdot I_z$	$\frac{1}{3}\Delta\omega\cdot(I_x + I_z)$
$\bar{\mathcal{H}}_\varepsilon$	$-\frac{1}{2}\Delta\omega\varepsilon I_z$	$-\frac{1}{3}\Delta\omega\varepsilon(I_y + I_z)$	$-\frac{2}{3}\Delta\omega\varepsilon I_z$	$-\frac{1}{3}\Delta\omega\varepsilon I_z$
$|\Delta S|$	$\frac{1}{2}$	$\frac{2}{3}$	2	$\frac{1}{2}$

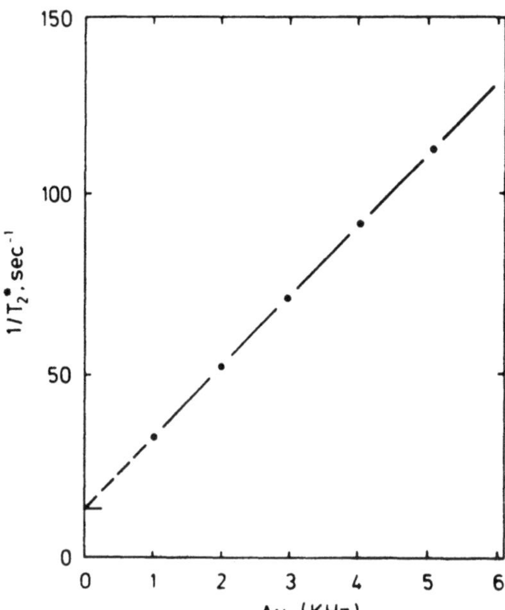

Fig. 3.27. Decay rate $1/T_2^*$ for a liquid sample (^{19}F in C_6F_6) in a MREV-8 pulse experiment versus resonance offset Δv, demonstrating the off-resonance B_1 inhomogeneity effect (Rhim et al. [26])

different multiple pulse sequences. As an example we shall treat the WHH-4 sequence, however, the extension to other pulse sequences is straightforward. Using Eqs. (3.161b, 3.130, 3.131a, c) we may write

$$\Delta S = \frac{C_0 \left(\frac{d}{d_\beta} C_0\right)_{\beta_0} + C_1 \left(\frac{d}{d_\beta} C_1\right)_{\beta_0}}{C_0^2 + C_1^2}.$$

Using the expressions for C_0 and C_1 as given by Eq. (3.130) for $\kappa = 2$, ΔS can be calculated for different duty factors δ or optimum pulse rotation angle β_0 in the WHH-4 experiment. Figure 3.28 represents such a calculation of $|\Delta S|$ versus δ or β_0 respectively. There is a slight change for ΔS although not a very drastic one with an increasing duty factor.

The B_1 inhomogeneity effect discussed so far vanishes at resonance ($\Delta \omega = 0$). In order to discuss

(ii) on-resonance B_1 inhomogeneity effects

we shall investigate how the B_1 inhomogeneity is transmitted to the average dipolar interactions.

This effect has been analyzed by Pfeifer and coworkers [68] and others [52, 53, 55]. The procedure is similar as before and Eq. (3.160) has to be applied. Let us again consider the WHH-4 sequence as an example. Since $\mathcal{H}_D^{(0)} = 0$, if the pulse rotation angle $\beta = \beta_0$, we obtain for the total dipolar scaling factor according to Eq. (3.160)

$$S_{\text{tot}} = \left[\text{Tr}\left\{\left(\frac{d}{d\beta} \bar{\mathcal{H}}_D\right)^2\right\} \Big/ \text{Tr}\{\mathcal{H}_D^2\}\right]^{1/2} \varepsilon_i = \Delta D \varepsilon_i$$

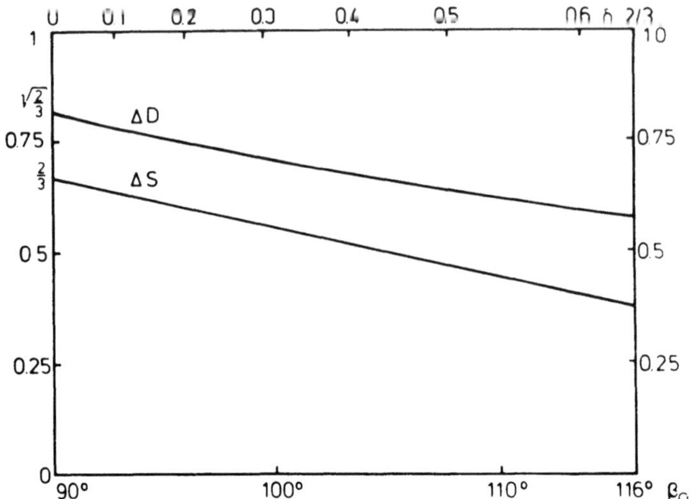

Fig. 3.28. Influence of the B_1 inhomogeneity coupling in a WHH-4 experiment versus duty factor δ or pulse rotation angle β_0 respectively. The coupling of B_1 inhomogeneity of the (a) dipolar interaction is represented by ΔD and (b) shift interaction and resonance offset is represented by ΔS (see text)

where

$$\Delta D = \left[\left(\frac{d}{d\beta} D_0 \right)^2_{\beta_0} + \left(\frac{d}{d\beta} D_1 \right)^2_{\beta_0} \right]^{1/2}. \tag{3.162}$$

The coupling parameter ΔD of the dipolar interaction to the B_1 inhomogeneity has been calculated for the WHH-4 experiment ($\kappa = 2$) using Eqs. (3.129, 3.162) and is plotted versus β_0 in Fig. 3.28. Notice, that ΔD is of about the same size as ΔS, the coupling parameter for resonance-offset to B_1 inhomogeneity. Both effects have to be considered in the WHH-4 experiment and can severely limit the resolution.

It can be readily shown, that there is no B_1 inhomogeneity effect on the average dipolar Hamiltonian in the case of the HW-8 and the MREV-8 sequence [66, 68, 70] in the limit $\delta = 0$, $\beta = \pi/2$ (see Table 3.4). This fact combined with the small offset coupling is what makes the MREV-8 sequence superior to the others as claimed by its designers. However, if $\delta \neq 0$ B_1 inhomogeneity becomes effective.

The WHH-4 sequence shows the following contributions of B_1 inhomogeneity coupling to the dipolar interaction in the limit $\delta = 0$, $\beta_0 = \pi/2$:

$$\bar{\mathcal{H}}_D^{(0)}(\varepsilon_i) = -i \sum_i \tfrac{1}{3} \varepsilon_i [\mathcal{H}_{Dz}, (I_{xi} - I_{yi})] \tag{3.163a}$$

$$\bar{\mathcal{H}}_D^{(1)}(\varepsilon_i) = \tau \sum_i \tfrac{1}{3} \varepsilon_i [\mathcal{H}_{Dx}[(I_{xi} - I_{yi}), \mathcal{H}_{Dz}]]. \tag{3.163b}$$

We neglect the contribution of ε_i to $\bar{\mathcal{H}}_D^{(2)}$ and still use

$$\bar{\mathcal{H}}_D^{(2)} = \frac{\tau^2}{18} [(\mathcal{H}_{Dx} - \mathcal{H}_{Dy}), [\mathcal{H}_{Dy}, \mathcal{H}_{Dx}]] \tag{3.163c}$$

for the WHH-4, HW-8, and MREV-8 sequences.

Using the Gaussian B_1 field distribution [Eq. (3.158)] with the second moment M_ε due to the B_1 inhomogeneity, Pfeifer et al. [68] obtained for the average second moments of the different multiple-pulse experiments the following expressions

$$\bar{M}_2^{\text{WHH-4}} = \frac{2}{9} M_\varepsilon \left[M_2 + \frac{(1+C_0)}{2} M_4 \tau^2 \right] + \frac{C_1}{324} M_6 \tau^4 \tag{3.164}$$

and

$$\bar{M}_2^{\text{MREV-8}} = \bar{M}_2^{\text{HW-8}} = \frac{C_1}{324} M_6 \tau^4 \tag{3.165}$$

where M_2, M_4, and M_6 are the corresponding moments of the dipolar interaction and where C_0 and C_1 are numbers defined by

$$C_0 = \frac{(I_z | \hat{\mathscr{H}}_{Dx} \hat{\mathscr{H}}_{Dz}^2 \hat{\mathscr{H}}_{Dx} | I_z)}{(I_x | \hat{\mathscr{H}}_{Dz}^4 | I_x)} \tag{3.166a}$$

and

$$C_1 = \frac{(I_z | \hat{\bar{\mathscr{H}}}_D^{(2)^2} | I_z)}{(I_x | \hat{\mathscr{H}}_{Dz}^2 | I_x)}. \tag{3.166b}$$

Using the following values for CaF_2 with H_0 parallel to the (111)-direction [68]

$$M_2 = 1.23 \, K^2; \quad M_4 = 3.56 \, K^4; \quad M_6 = 15.7 \, K^6$$
$$C_0 = 0.668; \quad C_1 = 2.523; \quad K = \gamma^2 \hbar \, d^{-3}$$

where d is the lattice constant, we obtain [68]

$$\bar{M}_2^{\text{WHH-4}} = M_\varepsilon \cdot (0.22 \, M_2 + 0.217 \, M_2^2 \tau^2) + 0.065 \, M_2^3 \tau^4 \tag{3.167a}$$
$$\bar{M}_2^{\text{MREV-8}} = 0.065 \, M_2^3 \tau^4. \tag{3.167b}$$

Figure 3.29 confirms this dependence of the WHH-4 sequence on the on-resonance effect of the B_1 inhomogeneity in the case of CaF_2 in the (111)-direction.

The overall dependence of the MREV-8 sequence on the B_1 inhomogeneity is demonstrated in Fig. 3.30. On resonances as well as off-resonance B_1 inhomogeneity coupling is clearly visible. A detailed discussion of these points has been given by Garroway et al. [52].

Other pulse imperfections may be investigated in a similar fashion. Indeed, this has been done by several authors [52, 53, 55, 69]. We shall follow here the meticulous work of Rhim, Elleman, Schreiber, and Vaughan [55]. Although B_1 inhomogeneity preserves the cyclic condition in all phase-alternated experiments, this may not be true for other pulse imperfections, such as phase transients, phase misadjustment, and errors in pulse length. In the following it is assumed that the violation of the cyclic condition is weak in the sense that the rf Hamiltonian $\mathscr{H}_1(t)$ can be split into a major part $\mathscr{H}_1^0(t)$, which still satisfies the cyclic condition and represents the ideal part of the rf pulse and a small part $\mathscr{H}_1^1(t)$, which represents the non-ideal part of $\mathscr{H}_1(t)$

$$\mathscr{H}_1(t) = \mathscr{H}_1^0(t) + \mathscr{H}_1^1(t). \tag{3.168}$$

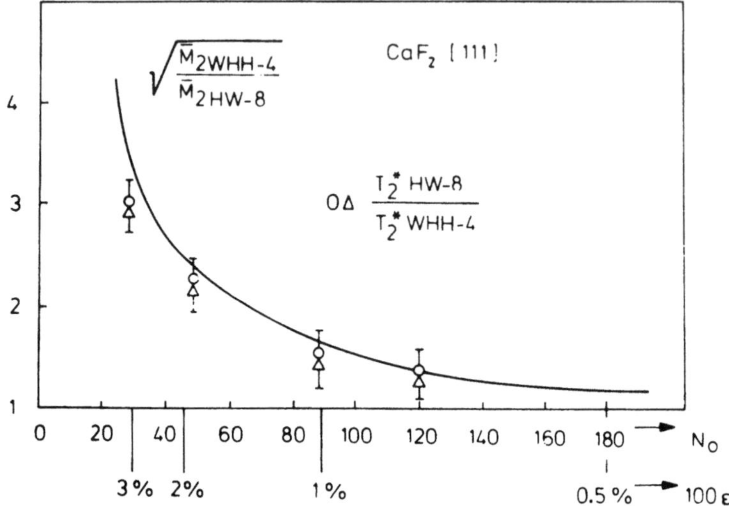

Fig. 3.29. Influence of the relative B_1 inhomogeneity (ε) on the decay time T_2^* in a HW-8 and WHH-4 experiment (Pfeifer and coworkers [68]). The theoretical curve (full line) corresponds to a modified second moment ratio as discussed in the text. N_0 is a measure of the B_1 inhomogeneity as defined in Eq. (3.159b)

We assume that

$$\|\mathcal{H}_1^0(t)\| \gg \|\mathcal{H}_1^1(t)\|$$

and combine $\mathcal{H}_1^1(t)$ with the interaction Hamiltonian $\mathcal{H}_{int}$. The ideal pulses now operate on the combined Hamiltonian $\mathcal{H}_{int}+\mathcal{H}_1^1(t)$ and coherent averaging theory may be applied to $\tilde{\mathcal{H}}_{int}(t)+\mathcal{H}_1^1(t)$. In general there are different imperfections present in multiple-pulse experiments and $\mathcal{H}_1^1(t)$ may be expressed by a sum as

$$\mathcal{H}_1^1(t)=\sum \mathcal{H}_k(t)$$

where k represents in the following:

P: phase misadjustment,

T: phase transients,

δ: pulse length misadjustments,

ε: B_1 inhomogeneity,

O: resonance offset.

Then the average Hamiltonian takes the following form:

$$\bar{\mathcal{H}}_{int}^{(0)} = \bar{\mathcal{H}}_O^{(0)} + \bar{\mathcal{H}}_D^{(0)} + \sum_k \bar{\mathcal{H}}_k^{(0)}, \tag{3.169a}$$

$$\bar{\mathcal{H}}_{int}^{(1)} = \bar{\mathcal{H}}_O^{(1)} + \bar{\mathcal{H}}_D^{(1)} + \bar{\mathcal{H}}_{OD}^{(1)} + \sum_k (\bar{\mathcal{H}}_{Ok}^{(1)} + \bar{\mathcal{H}}_{Dk}^{(1)}), \tag{3.169b}$$

$$\bar{\mathcal{H}}_{int}^{(2)} = \bar{\mathcal{H}}_O^{(2)} + \bar{\mathcal{H}}_D^{(2)} + \bar{\mathcal{H}}_{OD}^{(2)} + \sum_k (\bar{\mathcal{H}}_{Ok}^{(2)} + \bar{\mathcal{H}}_{Dk}^{(2)}). \tag{3.169c}$$

The notation here is self-explaining with, for instance, $\bar{\mathcal{H}}_{Ok}^{(1)}$ representing the first-order coupling between resonance offset $\mathcal{H}_O$ and the k-th imperfection

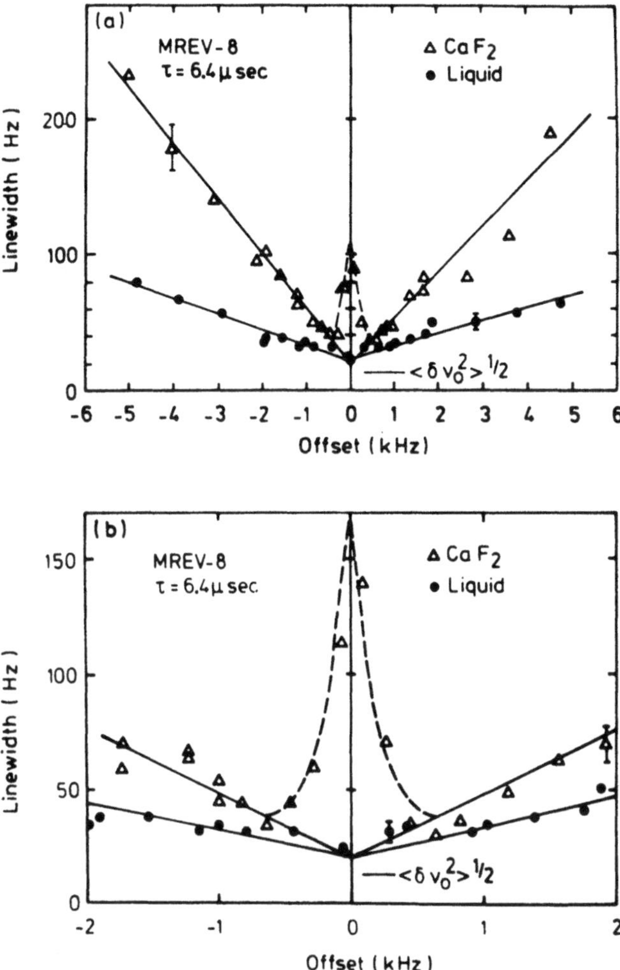

Fig. 3.30. a ^{19}F linewidth in CaF$_2$ $B_0\|\langle 111\rangle$ and in CF$_3$COOH (liquid) versus resonance offset for the MREV-8 sequence (Mansfield and co-workers [52]). **b** Behavior near resonance. The linewidth is defined here as the full width at half height as observed in the multiple-pulse experiment. $\langle \delta v_0^2 \rangle^{1/2}$ is the linewidth due to the static magnetic field inhomogeneity

$\mathcal{H}_k(t)$. Since the pulse imperfections $\mathcal{H}_1^1(t)$ will be in practice much smaller than the spin interactions $\mathcal{H}_{\text{int}}$, we have neglected cross terms between two imperfections. However, this can be incorporated if it is felt to be significant. Table 3.3 and Table 3.4, according to Rhim et al. [55] summarize the average interaction Hamiltonian due to different pulse imperfections for the following multiple-pulse sequences:

$$(-\tau - P_x - 2\tau - P_x - 2\tau - P_x - 2\tau - P_x - \tau)_n \qquad \text{MW-4}$$

$$(-\tau - P_x - 2\tau - P_{\bar{x}} - \tau -)_n \qquad \text{MW-2}$$

$$(-\tau - P_{\bar{x}} - \tau - P_y - 2\tau - P_{\bar{y}} - \tau - P_x - \tau)_n \qquad \text{WHH-4}$$

$$(-\tau - P_x - \tau - P_y - 2\tau - P_{\bar{y}} - \tau - P_{\bar{x}} - 2\tau - P_{\bar{x}} - \tau - P_y - 2\tau$$
$$- P_{\bar{y}} - \tau - P_x - \tau)_n. \qquad \text{MREV-8}$$

Table 3.3. Average interaction Hamiltonians for tune-up cycles, assuming the pulse rotation angle β_0 close to 90° (W.K. Rhim et al. [55]). (O, offset; P, phase error; T, phase transient; δ, pulse width error; ε, rf inhomogeneity; d, powder droop)

Average Hamiltonian	$\left(\frac{\pi}{2}\right)_x - \left(\frac{\pi}{2}\right)_x - \left(\frac{\pi}{2}\right)_x - \left(\frac{\pi}{2}\right)_x$ (MW-4)	Flip-flop cycle (MW-2)
$\bar{\mathcal{H}}_O^{(0)\,a}$	0	$\frac{1}{2}\Delta\omega(1+\frac{2}{3}a)(I_z-I_y)$
$\bar{\mathcal{H}}_O^{(1)}$	$-\frac{1}{16}t_c(\Delta\omega)^2 I_x$	0
$\bar{\mathcal{H}}_P^{(0)}$	0	$(1/t_c)(-\Phi_x+\Phi_{-x})(I_y+I_z)$
$\bar{\mathcal{H}}_{PO}^{(1)}$	0	$\frac{1}{4}\Delta\omega(\Phi_x+\Phi_{-x})I_x$
$\bar{\mathcal{H}}_T^{(0)\,b}$	0	$(1/t_c)J_1(I_z-I_y)$
$\bar{\mathcal{H}}_{TO}^{(1)}$	$-\frac{1}{2}\Delta\omega J_1 I_x$	$-\frac{1}{4}\Delta\omega J_2 I_x$
$\bar{\mathcal{H}}_\delta^{(0)}$	$-(4/t_c)\delta_x I_x$	$(1/t_c)(-\delta_x+\delta_{-x})I_x$
$\bar{\mathcal{H}}_{\delta O}^{(1)}$	$\frac{1}{2}\Delta\omega\,\delta_x I_z$	$-\frac{1}{4}\Delta\omega(\delta_x+\delta_{-x})I_z$
$\bar{\mathcal{H}}_\varepsilon^{(0)}$	$-(4/t_c)\sum_i \varepsilon_i I_{xi}$	0
$\bar{\mathcal{H}}_{\varepsilon O}^{(1)}$	$\frac{1}{2}\Delta\omega\sum_i \varepsilon_i I_{zi}$	$-\frac{1}{2}\Delta\omega\sum_i \varepsilon_i I_{zi}$
$\bar{\mathcal{H}}_d^{(0)\,c}$ (n-th cycle)	$-\frac{\omega_s t_w}{t_c}4(1-e^{-8n/b})I_x$	$-\frac{\omega_s t_w}{t_c}\frac{2}{b}e^{-4n/b}I_x$

[a] $a=3t_w/t_c(4/\pi-1)$.

[b] $J_1=\int_0^{t_w}\omega_T(t)(\sin\omega_1 t-\cos\omega_1 t)\,dt$; $J_2=\int_0^{t_w}\omega_T(t)(\sin\omega_1 t+\cos\omega_1 t)\,dt$.

[c] $\bar{\mathcal{H}}_d^{(0)}(n\text{-th cycle})=\bar{\mathcal{H}}_d^{(0)}[nt_c\to(n+1)t_c]$. Exponential decay of $\omega_{rf}(t)$ with time constant $b\tau$; $b\gg1$ is assumed.

These four sequences are the most important ones in multiple-pulse NMR. The MW-4 sequence is usually applied to a liquid sample in order to adjust the pulse length precisely and to measure the B_1 inhomogeneity, whereas, the MW-2 sequence allows the precise adjustment of the 180° phase difference between two pulses. The WHH-4 and the MREV-8 sequences are the best known and most efficient line narrowing sequences in solids which suppress dipolar and quadrupolar interaction.

Due to the different pulse imperfections such as B_1 inhomogeneity (ε), pulse shape and pulse size (δ), phase errors (Φ), and effects of phase transients ($\omega_I(t),\omega_T(t)$) the x pulse Hamiltonian for example becomes [55]

$$\mathcal{H}_1(t)=-\omega_1 I_x-\sum_i \frac{\varepsilon_i}{t_w}I_{xi}-\frac{\delta_x}{t_w}I_x$$
$$-\omega_1\Phi_x I_y-\omega_I(t)I_x+\omega_T(t)I_y \tag{3.170}$$

where the following conditions are assumed to be satisfied:

$$\left|\frac{\varepsilon_i}{t_w}\right|;\quad \left|\frac{\delta_x}{t_w}\right|;\quad |\Phi_x\omega_1|;\quad |\omega_I(t)|;\quad |\omega_T(t)|\ll|\omega_1|. \tag{3.171}$$

Table 3.4. Average interaction Hamiltonians for line narrowing multiple-pulse experiments, assuming the pulse rotation angle β_0 close to 90° (W.K. Rhim et al. [55]). (O, offset; D, dipolar; P, phase error; T, phase transient; δ, pulse width error; ε, rf inhomogeneity; d, powder droop)

	Four pulse cycle (WHH-4); $t_c = 6\tau$	Eight pulse cycle (MREV-8); $t_c = 12\tau$
$\bar{\mathcal{H}}_O^{(0)a}$	$\frac{1}{3}\sum_i (\Delta\omega + \omega_0\sigma_{zzi})(1+a)(I_{xi}+I_{yi}+I_{zi})$	$\frac{1}{3}\sum_i (\Delta\omega + \omega_0\sigma_{zzi})(1+2a)(I_{xi}+I_{zi})$
$\bar{\mathcal{H}}_O^{(1)}$	0	$\frac{\tau}{3}\sum_i (\Delta\omega + \omega_0\sigma_{zzi})^2(I_{xi}-I_{zi})$
$\bar{\mathcal{H}}_D^{(0)}$	0^b	0
$\bar{\mathcal{H}}_D^{(2)}$	$(\tau^2/18)[\mathcal{H}_D^{(x)} - \mathcal{H}_D^{(z)}, [\mathcal{H}_D^{(x)}, \mathcal{H}_D^{(y)}]]$	$(\tau^2/18)[\mathcal{H}_D^{(x)} - \mathcal{H}_D^{(z)}, [\mathcal{H}_D^{(x)}, \mathcal{H}_D^{(y)}]]$
$\bar{\mathcal{H}}_{DO}^{(1)}$	0	0
$\bar{\mathcal{H}}_{DO}^{(2)c}$	d	$(\tau^2/6)(\Delta\omega)[I_x, [\mathcal{H}_D^{(x)}, \mathcal{H}_D^{(y)}]] + (\tau^2/3)(\Delta\omega)^2(\mathcal{H}_D^{(z)} - \mathcal{H}_D^{(x)})$
$\bar{\mathcal{H}}_P^{(0)}$	$\frac{1}{6\tau}[(\Phi_y - \Phi_{-y})I_x + (-\Phi_x + \Phi_{-x} - \Phi_y + \Phi_{-y})I_y + (\Phi_x - \Phi_{-x})I_z]$	$\frac{1}{6\tau}[(\Phi_y - \Phi_{-y})I_x + (-\Phi_x + \Phi_{-x})I_y]$
$\bar{\mathcal{H}}_{PO}^{(1)}$	$\frac{1}{6}\sum_i (\Delta\omega + \omega_0\sigma_{zzi})[-(\Phi_x + \Phi_{-x})(I_{xi}-I_{yi}) + (\Phi_x + \Phi_{-x} - \Phi_y - \Phi_{-y})I_{zi}]$	$\frac{1}{6}\sum_i (\Delta\omega + \omega_0\sigma_{zzi})[(-\Phi_y + \Phi_{-y})I_{xi} + 2\Phi_x I_{yi}$
$\bar{\mathcal{H}}_{PD}^{(1)}$	$\frac{1}{2}(-\Phi_x - \Phi_{-x} + \Phi_y + \Phi_{-y})\sum_{i<j} A_{ij}(I_{xi}I_{zj} + I_{zi}I_{xj})$	0
$\bar{\mathcal{H}}_T^{(0)e}$	$\frac{1}{6\tau}J_1(I_x + 2I_y + I_z)$	$\frac{1}{6\tau}J_1(I_x + I_z)$
$\bar{\mathcal{H}}_{TO}^{(1)e}$	$\frac{1}{6}J_2\sum_i (\Delta\omega + \omega_0\sigma_{zzi})(I_{xi}-I_{yi})$	$\frac{1}{6}\sum_i (\Delta\omega + \omega_0\sigma_{zzi})[3J_1(I_{xi}-I_{zi}) - J_2 I_{yi}]$
$\bar{\mathcal{H}}_{TD}^{(1)}$	0	0
$\bar{\mathcal{H}}_\delta^{(0)}$	$\frac{1}{6\tau}[(-\delta_x + \delta_{-x})I_x + (\delta_y - \delta_{-y})I_z$	$\frac{1}{6\tau}(-\delta_x + \delta_{-x})I_x$
$\bar{\mathcal{H}}_{\delta D}^{(1)}$	$\frac{1}{2}\sum_{i<j} A_{ij}[(\delta_x + \delta_{-x})(I_{yi}I_{zj} + I_{zi}I_{yj}) + (\delta_y + \delta_{-y})(I_{xi}I_{yj} + I_{yi}I_{xj})]$	0

Table 3.4 (continued)

	Four pulse cycle (WHH-4); $t_c = 6\tau$	Eight pulse cycle (MREV-8); $t_c = 12\tau$
$\bar{\mathcal{H}}_{\delta O}^{(1)}$	$-\frac{1}{6}\sum_i(\varDelta\omega + \omega_0\sigma_{zzi})[(\delta_x + \delta_{-x})I_{zi} + (\delta_y + \delta_{-y})I_{yi}]$	$\frac{1}{6}\sum_i(\varDelta\omega + \omega_0\sigma_{zzi})[-2\delta_{-x}I_{zi} + (\delta_y - \delta_{-y})I_{yi}]$
$\bar{\mathcal{H}}_{\varepsilon}^{(0)}$	0	0
$\bar{\mathcal{H}}_{\varepsilon O}^{(1)}$	$-\frac{1}{3}\sum_i\varepsilon_i(\varDelta\omega + \omega_0\sigma_{zzi})(I_{yi} + I_{zi})$	$-\frac{1}{3}\sum_i\varepsilon_i(\varDelta\omega + \omega_0\sigma_{zzi})I_{zi}$
$\bar{\mathcal{H}}_{\varepsilon D}^{(1)f}$	$\sum_{i<j}\varepsilon_i A_{ij}[I_{yi}(I_{xj} + I_{zj}) + (I_{xi} + I_{zi})I_{yj}]$	0
$\bar{\mathcal{H}}_{d}^{(0)}$ (n-th cycle)^g	$-\dfrac{\omega_s t_w}{3\tau}\dfrac{1}{b}e^{-6n/b}(2I_x + I_z)$	$\dfrac{\omega_s t_w}{\tau}\dfrac{1}{b^2}e^{-12n/b}(2I_x + I_z)$
$\bar{\mathcal{H}}_{dD}^{(1)}$ (n-th cycle)^g	$\omega_s t_w[1 + (-1 + 3/b)e^{-6n/b}\cdot\sum_{i<j}A_{ij}(I_{yi}(I_{xj} + I_{zj}) + (I_{xi} + I_{zi})I_{yj}]$	$(3\omega_s t_w/b)e^{-12n/b}\cdot\sum_{i<j}A_{ij}[I_{yi}(I_{xj} + I_{zj}) + (I_{xi} + I_{zi})I_{yj}]$

[a] $a = (3t_w/t_c)(4/\pi - 1)$, assuming 90° pulses (see however Sect. 3.5).

[b] Pulse rotation angle β_0 has to be adjusted according to duty factor (see Sect. 3.5).

[c] This calculation assumes all nuclei are chemically and geometrically equivalent.

[d] $\bar{\mathcal{H}}_{DO}^{(2)} = (t_c^2/648)\{-3\varDelta\omega[\mathcal{H}_D^{(x)}, [\mathcal{H}_D^{(y)}, I_x]] - 3\varDelta\omega[\mathcal{H}_D^{(z)}, [\mathcal{H}_D^{(x)}, I_x]] + (\varDelta\omega)^2[I_y + I_z, [\mathcal{H}_D^{(y)}, I_x]] + (\varDelta\omega)^2[4I_x + 3I_y + I_z, [\mathcal{H}_D^{(x)}, I_y]]$
$+ i(\varDelta\omega)^2[\mathcal{H}_D^{(y)}, I_x] - i(\varDelta\omega)^2[\mathcal{H}_D^{(x)}, I_y] + i(\varDelta\omega)^2[\mathcal{H}_D^{(x)}, I_z]\}$.

[e] $J_1 = \int_0^{t_w}\omega_T(t)(\sin\omega_1 t - \cos\omega_1 t)\,dt;\ J_2 = \int_0^{t_w}\omega_T(t)(\sin\omega_1 + \cos\omega_1 t)\,dt.$

[f] It is assumed that ω_1 is constant over a scale of molecular dimensions.

[g] $\bar{\mathcal{H}}_d^{(0)}$ (n-th cycle $= \bar{\mathcal{H}}_d^{(0)}[nt_c \to (n+1)t_c]$ and similarly for $\bar{\mathcal{H}}_{dD}^{(1)}$. Exponential decay of $\omega_{rf}(t)$ with time constant $b\tau$; $b \gg 1$ is assumed.

We note again, that we are dealing here with rf pulses which are close to δ pulses. The whole procedure, however, may be readily applied to pulses of arbitrary width. In Eq. (3.170) the ideal x pulse is represented by $\omega_1 I_x$, whereas ε_i takes into account the deviation of the rotation angle at the i-th nucleus caused by B_1 inhomogeneity, with

$$\sum_i \varepsilon_i = 0.$$

δ_x represents the pulse size misadjustment; Φ_x is a phase angle misadjustment; $\omega_I(t)$ is the difference between the in-phase component of the real rf pulse and the idealized pulse with the condition

$$\int\limits_0^{t_w} dt\, \omega_I(t) = 0.$$

Rhim et al. [55] have investigated the effect of these different pulse imperfections on the above mentioned multiple-pulse sequences in great detail, and the interested reader is referred to Ref. [55]. A detailed treatment of different pulse imperfections has also been given by Haeberlen [18].

3.8 Resolution of Multiple-Pulse Experiments

One of the most important questions is concerned with the resolution obtainable in a given multiple-pulse experiment [45, 52, 53, 70, 71]. As shown in the preceding section, the resolution of line narrowing sequences may be considerably reduced by pulse imperfections. Although all line narrowing type multiple-pulse sequences are designed to achieve vanishing average dipolar (and quadrupolar) interaction, leftover higher order correction terms, such as $\mathcal{H}_D^{(2)}$ and other higher order terms may limit the resolution even in the case of ideal rf pulses. Which of the many possible Hamiltonian, listed in Tables 3.3 and 3.4 of the preceding section gives the major contribution, has to be investigated separately in each case. We shall present here a general approach to the problem of resolution and discuss some representative examples. The average interaction Hamiltonian in a multiple-pulse experiment can be expressed in general as [45]

$$\bar{\mathcal{H}}_{\text{int}} = \mathbf{F}_0 + t_c \mathbf{F}_1 + t_c^2 \mathbf{F}_2 + \ldots \tag{3.172}$$

where $\mathbf{F}_0, \mathbf{F}_1, \mathbf{F}_2$ etc. are the different contributions due to zeroth, first and second order correction terms according to the Magnus expansion.

The nuclear response under the action of the average interaction Hamiltonian $\bar{\mathcal{H}}_{\text{int}}$ will be a decay of magnetization $G(t)$ and an oscillation. Shift Hamiltonian do not contribute to the decay time constant T_2^*, whereas dipolar interaction and cross coupling of dipolar interactions with shift interactions including B_1 inhomogeneity are directly responsible for a decay in magnetization. $\bar{\mathcal{H}}_{\text{int}}$ in Eq. (3.172) may represent those interactions. In an ideal line narrowing experiment we usually have $\mathbf{F}_0 = \mathbf{F}_1 = 0$, leaving

$$\bar{\mathcal{H}}_{\text{int}} = t_c^2 \mathbf{F}_2 \tag{3.173}$$

as the leading correction term.

According to Eq. (3.173) we could argue, that simply the time scale of the decay function is scaled by t_c^2, leading to a decay time $T_2^* \sim t_c^{-2}$. The last conclusion is wrong, although the beginning of the decay may depend on t_c^2.

In order to describe the total decay we apply the line shape theory as discussed in Sect. 2.5 and Appendix C. There we have shown that an exact derivation starting from the Liouville-v. Neumann Equation leads under certain conditions to

$$\frac{d}{dt} G(t) = -\int_0^t dt'\, K(t-t')\, G(t')$$

where the "memory function" $K(t-t')$ may be replaced by an approximate expression. Since the decay time constant T_2^* in multiple-pulse experiments is usually long compared with the cycle time t_c, we invoke the "short correlation" limit, i.e. we expect the correlation time of $K(t-t')$ to be on the order of $t_c \ll T_2^*$. In this case an exponential decay results or a Lorentzian line shape with [see Eq. (2.111a)]

$$\frac{1}{T_2^*} = \int_0^\infty d\tau\, K(\tau).$$

This is confirmed by experiments, i.e. the line shapes observed in multiple-pulse experiments are usually close to a Lorentzian. As shown in Sect. 2.5 Eq. (2.117) this leads to a line width δ of

$$\delta = \frac{1}{T_2^*} = \sqrt{\frac{\pi}{2}} \left[\frac{M_2}{\mu - 1}\right]^{1/2} \tag{3.174}$$

where in multiple-pulse experiments, we usually find

$$\mu = M_4/M_2^2 \gg 1.$$

Using the average interaction Hamiltonian as given by Eq. (3.172) we define the following "multiple-pulse moments"

$$M_2^* = \frac{(I_\mu | \bar{\mathcal{H}}_{\text{int}}^2 | I_\mu)}{(I_z | I_z)} \tag{3.175}$$

$$M_4^* = \frac{(I_\mu | \bar{\mathcal{H}}_{\text{int}}^{*4} | I_\mu)}{(I_z | I_z)}$$

where the appropriate component I_μ has to be chosen, accordingly. Now we obtain the general expression for the multiple-pulse decay time T_2^* by using Eqs. (3.174) and (3.173) as

$$\frac{1}{T_2^*} = \sqrt{\frac{\pi}{2}} \left[\frac{M_2^{*\,3}}{M_4^*}\right]^{1/2} \tag{3.176}$$

where $\mu - 1$ has been replaced by μ because of $\mu \gg 1$.

Let us now discuss several cases:

(i) $\quad \bar{\mathcal{H}}_D = t_c^2 \mathbf{F}_2$

which is the case for example in the WHH-4, HW-8 and MREV-8 sequence on resonance.

It follows immediately that

$$M_2^* = a t_c^4 \quad \text{and} \quad M_4^* = b t_c^8 \tag{3.177}$$

where

$$a = \frac{(I_\mu | \hat{\mathbf{F}}_2^2 | I_\mu)}{(I_z | I_z)}$$

and

$$b = \frac{(I_\mu | \hat{\mathbf{F}}_2^4 | I_\mu)}{(I_z | I_z)}.$$

The decay time T_2^* according to Eq. (3.176) equals in this case

$$\frac{1}{T_2^*} = \sqrt{\frac{\pi}{2}} \left[\frac{a^3}{b} \right]^{1/2} \cdot t_c^2. \tag{3.178}$$

It is, however, well known that the t_c dependence of $1/T_2^*$ in multiple-pulse experiments ranges from t_c^2 to t_c^5 as observed experimentally (see Fig. 3.31).

It will be shown in the following that this is to be expected, if different correction terms of the average Hamiltonian are governing the decay.

The next case we are going to discuss is represented by a strong zeroth order contribution as is the case in the MW-2 and MW-4 experiment

(ii) $\bar{\mathcal{H}}_D = \mathbf{F}_0 + t_c^2 \mathbf{F}_2$

with $\mathbf{F}_0 = -\frac{1}{2} \mathcal{H}_{Dx}$ in the case of MW-2 and MW-4.

The $\mathbf{F}_2$ term will be usually represented by the $\bar{\mathcal{H}}_D^{(2)}$ part of the average dipolar Hamiltonian. In multiple-pulse line narrowing experiments, however, the above condition is also applicable, if cross terms between dipolar interaction and B_1 inhomogeneity, resonance-offset etc. are taken into account. Schmiedel [70], Ernst et al. [71] and others [52, 53], have discussed these effects in detail.

In general, we obtain in this case

$$M_2^* = a_0 + a_1 t_c^2 + a_2 t_c^4 \tag{3.179}$$

where

$$a_0 = (I_\mu | \hat{\mathbf{F}}_0^2 | I_\mu)/(I_z | I_z)$$
$$a_1 = (I_\mu | \hat{\mathbf{F}}_0 \hat{\mathbf{F}}_2 | I_\mu)/(I_z | I_z)$$
$$a_2 = (I_\mu | \hat{\mathbf{F}}_2^2 | I_\mu)/(I_z | I_z)$$

and

$$M_4^* = C_0 + C_1 t_c^2 + C_2 t_c^4 + C_3 t_c^6 + C_4 t_c^8 \tag{3.180}$$

where

$$C_0 = \text{Tr}\{k_0^2\}/\text{Tr}\{I_z^2\}; \quad C_1 = 2\,\text{Tr}\{k_0 k_1\}/\text{Tr}\{I_z^2\}$$
$$C_2 = \text{Tr}\{(2 k_0 k_2 + k_1^2)\}/\text{Tr}\{I_z^2\}$$

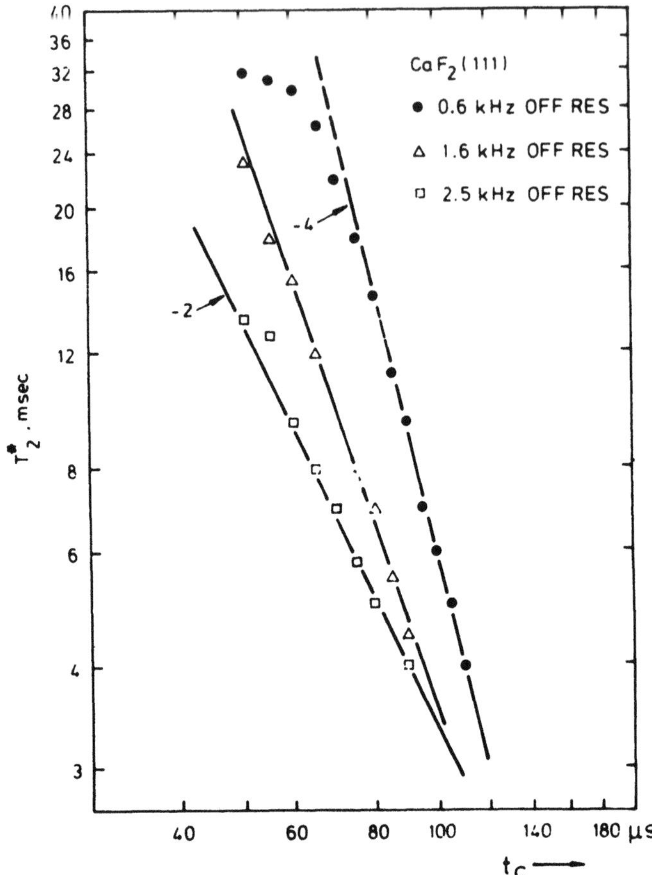

Fig. 3.31. Decay time constant T_2^* in CaF$_2$ observed in a MREV-8 pulse experiment versus cycle time t_c for three different resonance offsets according to W.K. Rhim, D.D. Elleman and R.W. Vaughan [26]

$$C_3 = 2\,\mathrm{Tr}\{k_1 k_2\}/\mathrm{Tr}\{I_z^2\}$$

$$C_4 = \mathrm{Tr}\{k_2^2\}/\mathrm{Tr}\{I_z^2\}$$

with the commutators

$$k_0 = \hat{\mathbf{F}}_0^2 I_\mu$$

$$k_1 = (\hat{\mathbf{F}}_0 \hat{\mathbf{F}}_2 + \hat{\mathbf{F}}_2 \hat{\mathbf{F}}_0) I_\mu$$

$$k_2 = \hat{\mathbf{F}}_2^2 I_\mu.$$

Inserting M_2^* and M_4^* into Eq. (3.176) yields any t_c dependence of T_2^* up to t_c^6, depending on the relative size of the coefficients a_i and c_i.

Let us analyse in more detail the phase-alternated sequence (MW-2) for which we have

$$\mathbf{F}_0 = -1/2\,\mathcal{H}_{Dx}; \qquad \mathbf{F}_2 = \tfrac{1}{192}[(\mathcal{H}_{Dy} - \mathcal{H}_{Dx}), [\mathcal{H}_{Dy}, \mathcal{H}_{Dz}]]. \tag{3.181}$$

Because of $[\mathcal{H}_{Dx}, I_x] = 0$ we obtain $a_0 = a_1 = 0$ resulting in $M_2^* = a_2 t_c^4$.

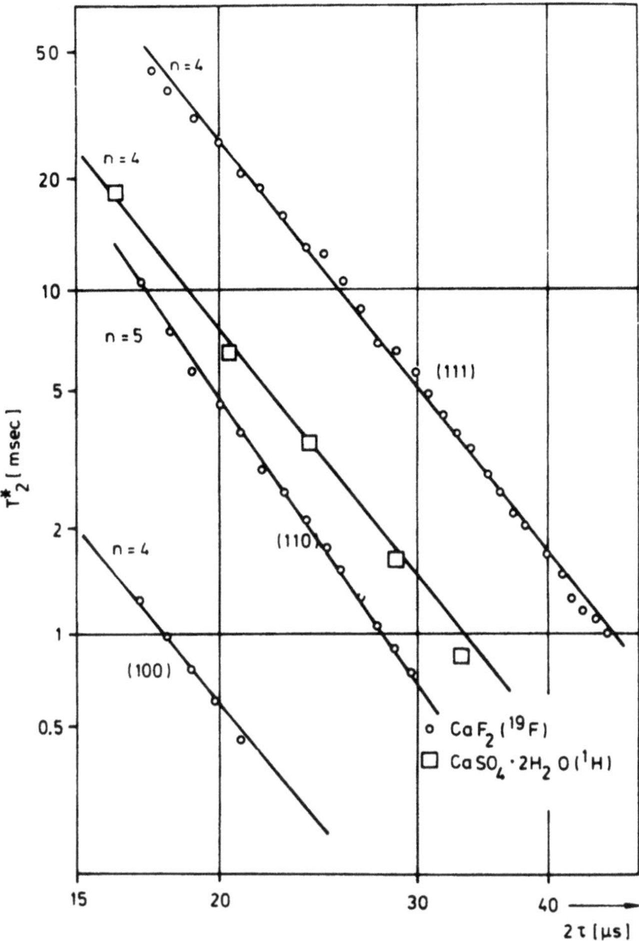

Fig. 3.32. Decay time T_2^* versus pulse spacing 2τ in different MW-4 multiple pulse experiments according to H. Ernst et al. [65]

For the same reason we have $k_0 = 0$ and $C_0 = C_1 = 0$, leaving as the leading term

$$C_2 = \mathrm{Tr}\{[\mathbf{F}_0,[\mathbf{F}_2,I_x]]^2\}/\mathrm{Tr}\{I_z^2\}$$

and

$$M_4^* = C_2 t_c^4. \tag{3.182}$$

Inserting M_2^* and M_4^* into Eqs. (3.176) result in

$$\frac{1}{T_2^*} = \sqrt{\frac{\pi}{2}}\left[\frac{a_2^3}{c_2}\right]^{1/2}\cdot t_c^4. \tag{3.183}$$

This is verified experimentally as can be seen in Figs. 3.32 and 3.33. In line narrowing multiple-pulse experiments, however, it is difficult to disentangle the different contributions of the pulse imperfections to line broadening in order to

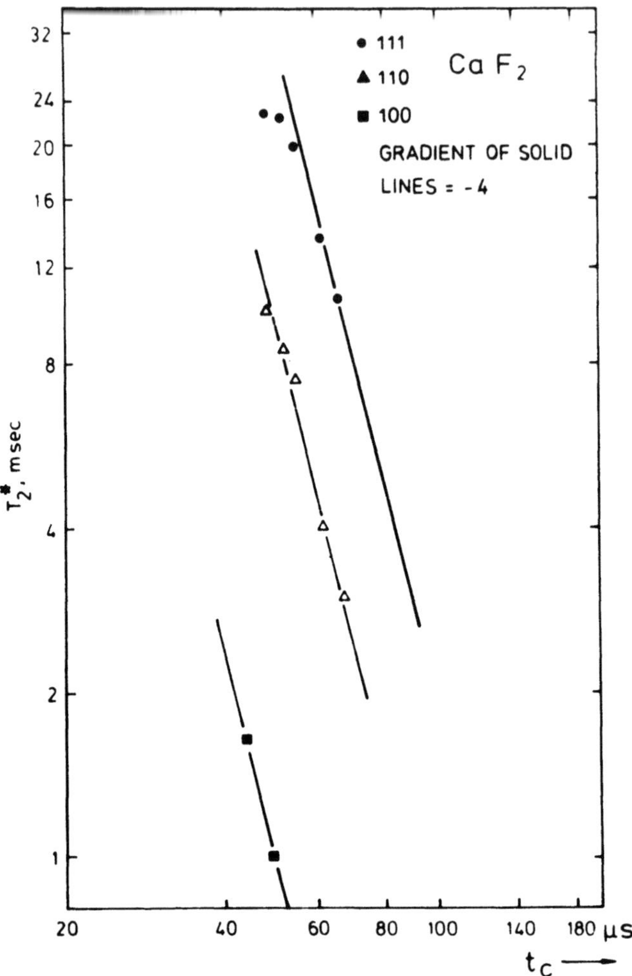

Fig. 3.33. Decay time T_2^* versus cycle time t_c in CaF$_2$ for different orientations of the magnetic field, observed in a MREV-8 pulse experiment (Rhim et al. [26])

make theoretical predictions of the t_c dependence of T_2^*. The t_c dependence of T_2^* on the other hand may give some hints as to which terms contribute significantly.

3.9 Magic Angle Rotating Frame Line Narrowing Experiments

For completeness we shall briefly touch on a second class of line narrowing experiments. These experiments are equivalent to the "magic angle" specimen rotation method discussed in Sect. 2.6, besides that, rotation is performed in spin space. This has the advantage of leaving all ansisotropies in the shift Hamiltonian unchanged and allowing, on the other hand, a very rapid "rotation" without any moving part simply by applying magnetic fields. Fig. 3.34 shows how this may be achieved. The external field Hamiltonian $\mathcal{H}_1(t)$ in the

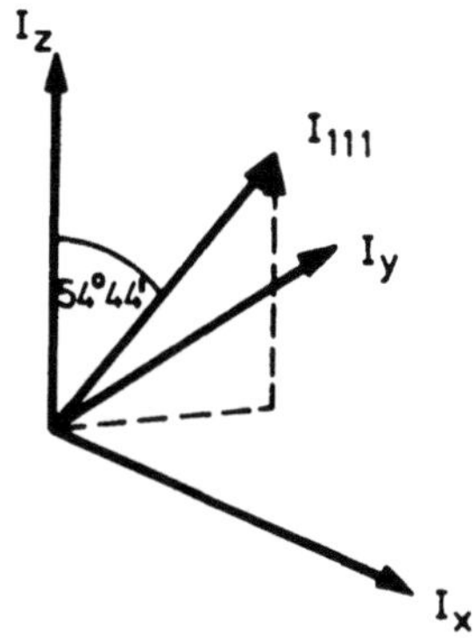

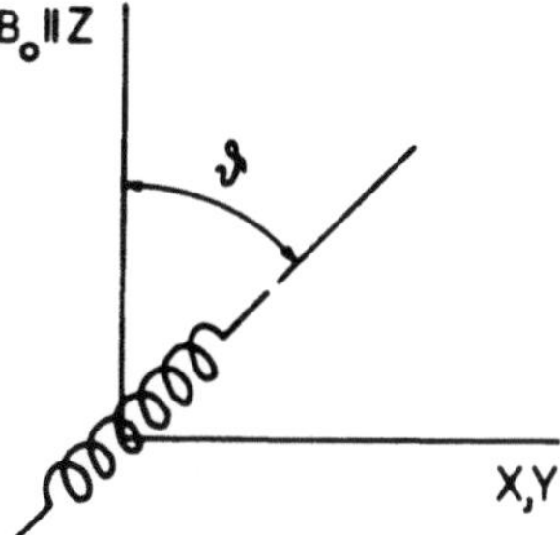

Fig. 3.34. Pictorial representation of the operator I_{111} in spin space

Fig. 3.35. Tilted coil arrangement for producing rf field and video field in the same coil in order to perform magic angle experiments in the rotating frame

rotating frame must be proportional to $I_{111}=(I_x+I_y+I_z)/\sqrt{3}$. One way of producing this field, would be to apply the B_1 field off-resonance by $\Delta\omega$ i.e.

$$\mathcal{H}_1=\Delta\omega I_z+\omega_1 I_x=\omega_e(I_z\cos\beta+I_x\sin\beta)$$

where

$$\omega_e^2=\Delta\omega^2+\omega_1^2$$

and

$$\tan\beta=\frac{\omega_1}{\Delta\omega}.$$

The angle β can be adjusted easily to fulfill the magic angle condition $\tan\beta_m=\sqrt{2}$.

This technique has been applied by Lee and Goldburg (LG) [72] to obtain a lengthened decay in a solid. Notice, however, that no phase alternation can be performed in this case, since a phase reversal of ω_1 does not reverse $\mathcal{H}_1$. Even application of the B_1 field below resonance $(-\Delta\omega)$ does not correspond to a phase reversal, since the two different reference frames involved are in general not coherent. One way around this is a dc "video" field in the z direction which may be switched to $+\Delta\omega$. A convenient means of achieving this is the tilted coil arrangement as shown in Fig. 3.35. The coil makes an angle ϑ with the direction of B_0 and is excited at the same time by a "video" field B_0 and a rf field B_1.

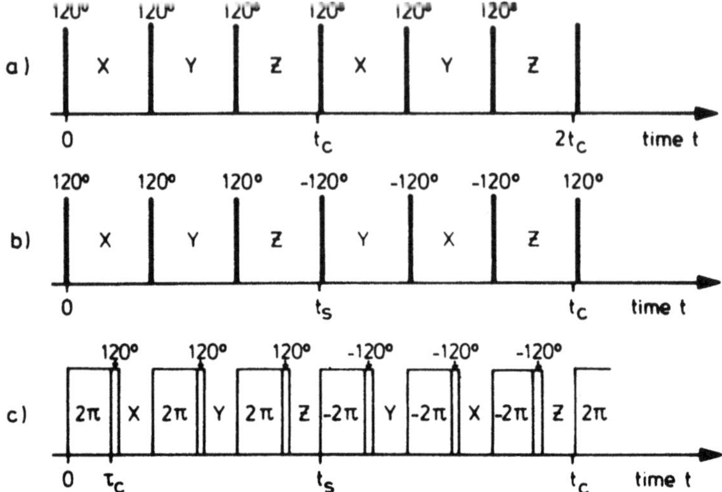

Fig. 3.36a–c. Pulse timing for three different magic angle experiments in spin space. a Three pulse cycle. b Six pulse cycle. c Nested cycles (see Ref. [50])

The corresponding fields in the z and x, y direction are

$$B_z = B_0 \cos\vartheta; \quad B_{x,y} = B_1 \sin\vartheta.$$

The magic angle condition can now easily be satisfied by

$$\tan\beta_m = \frac{B_{x,y}}{B_z} = \frac{B_1}{B_v} \tan\vartheta = \sqrt{2}.$$

In practice ϑ is conveniently chosen to be $\vartheta = 45°$. Haeberlen and Waugh [12] have carried out a modification of the LG experiment by applying a train of 120° pulses under the magic angle to CaF_2, as shown in the schematic diagram in Fig. 3.36a. The one cycle propagator can be written down readily as

$$L(t_c) = [\mathbf{X}, \mathbf{Y}, \mathbf{Z}]$$

with

$$\bar{\mathcal{H}}^{(0)} = \tfrac{1}{3}(\mathbf{X} + \mathbf{Y} + \mathbf{Z})$$

as in any line narrowing multiple-pulse experiment, i.e. $\bar{\mathcal{H}}_D^{(0)} = 0$. Thus line narrowing in solids is obtained in such a sequence and is demonstrated in Fig. 3.37. It is evident, however, that this cycle possesses no reflection symmetry and indeed the first correction term $\bar{\mathcal{H}}^{(1)} \neq 0$ and especially $\bar{\mathcal{H}}_D^{(1)} \neq 0$, explaining the destruction of the decay in Fig. 3.37. Therefore, it has been proposed by HW [12] to apply phase reversed LG cycles alternatively in order to reduce the first order correction term $\bar{\mathcal{H}}^{(1)}$. This is achieved in the second sequence in Fig. 3.36b. The leading correction term to the average dipolar interaction is then $\bar{\mathcal{H}}_D^{(2)}$, which may be readily obtained like in the other multiple-pulse experiments as

$$\bar{\mathcal{H}}_D^{(2)} = \frac{\tau^2}{18}\left[(\mathcal{H}_{Dy} - \mathcal{H}_{Dx}), [\mathcal{H}_{Dx}, \mathcal{H}_{Dy}]\right].$$

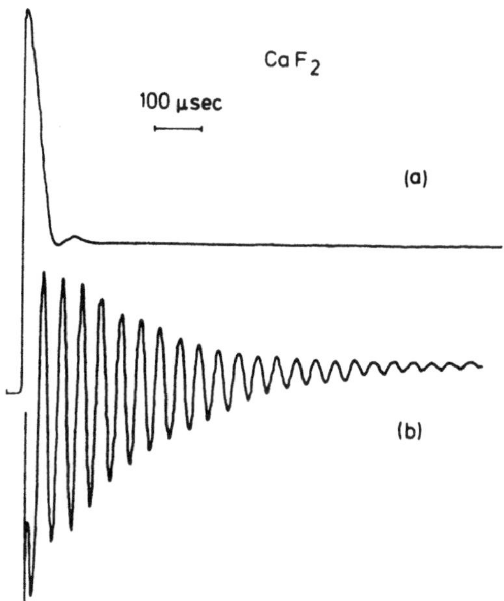

Fig. 3.37. a Bloch decay and **b** three pulse magic angle decay (see Fig. 3.36a) of ^{19}F in CaF$_2$ (U. Haeberlen and J.S. Waugh [12])

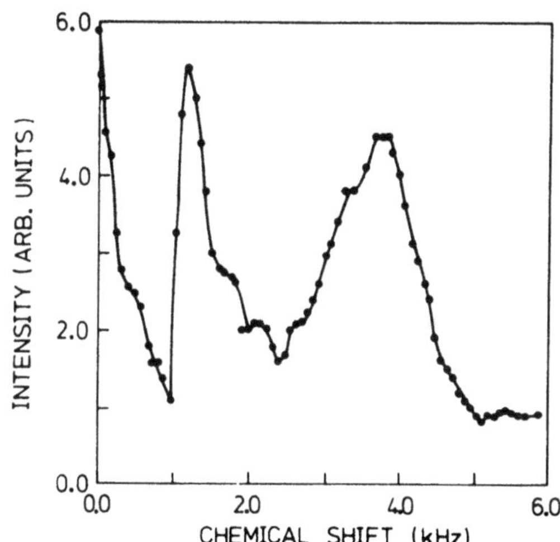

Fig. 3.38. ^{19}F spectrum observed in solid TFE/PFMVE copolymer at 54 MHz by applying a nested cycle magic angle experiment according to Fig. 3.36c (see Ref. [50]). The sharp peak corresponds to the rapidly rotating OCF$_3$-group

The scaling factor in these experiments is evidently $S = 1/\sqrt{3}$. Several modifications of these pulsed LG experiments have been applied to solids [50]. One of these sequences is schematically drawn in Fig. 3.36c. The idea behind these sequences is to implement a fast LG cycle, which corresponds to a 2π rotation around the (111) direction in the rotating frame, in front of each 120° pulse in order to reduce second order correction terms more effectively. The observation windows (X, Y, Z) are "precessing" in the sense that averaging of the dipolar interaction is obtained in the windows alone. Nesting full LG cycles between "precessing windows" in this way has been successfully applied to solids and considerable line narrowing has been achieved. A representative example is given in Fig. 3.38. For further discussion consult Ref. [12, 50].

3.10 Modulation Induced Line Narrowing

Yannoni and Vieth [73] proposed a modulation technique to remove dipolar interaction in solids. The combined action of a strong rf field $\omega_1 = \gamma B_1$ and a field- or frequency-modulation with frequency ω_m "dresses" the nuclear states in the rotating frame such that under certain conditions the "average" dipole-dipole interaction between two spins vanishes. The total Hamiltonian in the rotating frame is expressed as

$$\mathcal{H} = -\omega_1(I_{x_1} + I_{x_2}) - \kappa\omega_m \cos\omega_m t(I_{z_1} + I_{z_2}) - A(3I_{z_1}I_{z_2} - \mathbf{I}_1 \cdot \mathbf{I}_2) \tag{3.184}$$

where A is the coupling constant of the two spins considered and κ is the modulation index. After applying a suitable set of transformations, Yannoni and Vieth [73] arrive at a first order average Hamiltonian of the form

$$\bar{\mathcal{H}}_{II} = \tfrac{1}{4}A(3\cos^2\beta - 1)(3I_{z_1}I_{z_2} - \mathbf{I}_1 \cdot \mathbf{I}_2) \tag{3.185}$$

where β is one of the tilt angles and is given by

$$\tan\beta = \kappa(1-x)/[(\omega_1/\omega_m)J_0(\kappa x) - 1] \tag{3.186}$$

and x is a solution of the equation

$$\kappa(1-x) = 2J_1(\kappa x)\omega_1/\omega_m \tag{3.187}$$

$J_0(\kappa x)$ and $J_1(\kappa x)$ are Bessel functions. For $\omega_m > \omega_1$ and $\omega_m/\omega_1 \lesssim 1.4$, κ can be adjusted so that $3\cos^2\beta = 1$.

This opens the possibility for eliminating dipole-dipole interaction of homonuclear spins. Figure 3.39 demonstrates this behaviour. The experimental

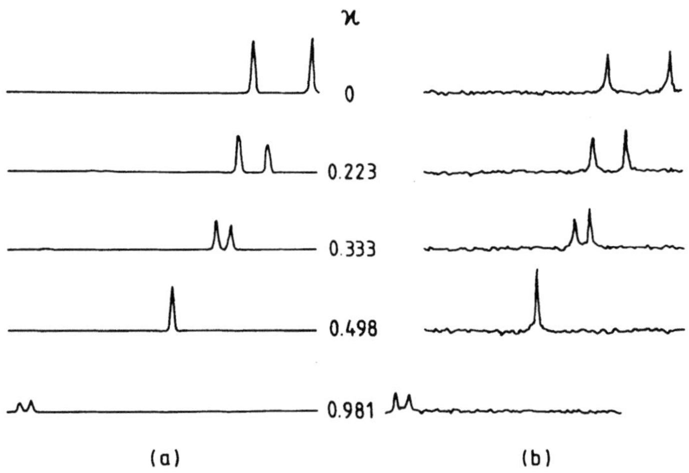

Fig. 3.39a, b. Fourier-transform spectra of modulated transient nutations of the two coupled ^{19}F nuclei in 1,1-difluorotetrachloroethane oriented in a nematic liquid crystal solvent, according to Yannoni and Vieth [73]. The rf (23 MHz) is centered on the doublet (splitting 710 Hz, $A = 237$ Hz). **a** Theory with $\omega_1 = 17.58\,A$ and $\omega_m/\omega_1 = 1.168$. **b** Experiment with $\omega_1 = 3,982$ Hz and $\omega_m = 4,661$ Hz. The value of the modulation index, κ, is shown with each pair of plots

spectra (b) of two coupled ^{19}F spins in 1,1-difluorotetrachloroethane oriented in a nematic liquid crystal solvent are compared with calculated spectra (a). The experiment was performed in a quasicontinous time sharing mode at 23 MHz. The transient nutation of the magnetization was observed after suddenly turning on the rf irradiation at the center of the doublet. In order to observe the magnetization rf pulses were used instead of continous rf and the transient magnetization was observed during the "windows". The following parameters hold in the experiment shown in Fig. 3.39.

$$\omega_1 = 17.58\ A; \qquad \omega_m/\omega_1 = 1.168 \quad \text{with} \quad \omega_1 = 3,982\ \text{Hz}$$

$$\omega_m = 4,661\ \text{Hz} \quad \text{and} \quad A = 237\ \text{Hz}.$$

This technique has not found much use in linenarrowing experiments yet, but certainly has some interesting features.

3.11 Applications of Multiple-Pulse Experiments

As was shown in the previous sections multiple-pulse experiments are designed to eliminate the homonuclear dipole-dipole interaction "on the average" leaving the linear spin interactions like chemical shift and Knight shift and the scalar interaction. Some representative data of chemical shift tensors are collected in Chap. 7. Here we want to review briefly some examples where the chemical shift tensor has been used to monitor molecular motion.

One of the earliest applications of multiple-pulse NMR was already concerned with the influence of molecular motion on the chemical shift tensor of ^{19}F in some fluorinated solid compounds [21–23]. Among the molecular solids polymers are particular interesting, since they show a wealth of structure (cristalline and amorphous) and a complexity of molecular motions. It is therefore not surprising, that polytetrafluoroethylene (PTFE) was the first polymer investigated by multiple pulse experiments [23, 74].

Later more detailed investigations by Garroway et al. [75] and English and Vega [76a] have demonstrated that information about orientational ordering and molecular motion in the different phases of PTFE can be obtained from ^{19}F multiple-pulse NMR. Vega and English [76b] have combined chemical shift information with relaxation measurements to unravel fine details of the motion in PTFE. Dybowski and co-workers [77] determined the orientation parameter of stretched PTFE from ^{19}F multiple-pulse spectra.

In fact ^{19}F has proved to be a versatile label in biological applications. Sykes and co-workers [78] used ^{19}F spin labelled tyrosine attached to biological molecules for the investigation of molecular motion. Later Ho and co-workers [79] were able to draw some conclusions on molecular motions and orientation in biological compounds by using known chemical shift tensors in ^{19}F labelled compounds.

Dybowsky and Vaughan [80] noticed that anisotropic lattice motion can be determined with the help of multiple pulse NMR. They investigated the lineshape of protons in natural rubber and cis-polyisoprene under a modified

MREV-8 sequence. Relaxation and chemical shift contributions could be distinguished.

Application of multiple-pulse techniques to surface studies will become very important in the near future. Duncan and Dybowski [81] have written a review article on this subject recently.

Multiple-pulse experiments with the feature of eliminating homonuclear dipole-dipole interaction are used also in two-dimensional spectroscopy (Chap. 5) multiple-quantum spectroscopy (Chap. 6) and relaxation experiments (Chap. 8).

Blinc and co-workers [82] have developed multiple-pulse sequences in combination with gradient pulses to study diffusion. By this technique slow diffusion is measurable in solids in a direct fashion.

Rhim and co-workers [83, 84] used multiple-pulse sequences in spin-temperature experiments like spin-locking and adiabatic demagnetization in the rotating frame (ADRF).

The combination of multiple-pulse experiments with magic angle spinning (MAS) was first applied by Pembleton et al. [85] and by Schnabel and co-workers [86]. Considerable increase in resolution was obtained leading to the accurate determination of isotropic shifts in OH groups of silicates [87]. Further increase in resolution was demonstrated by Gerstein and co-workers [88] by employing the high resolution Burum-Rhim sequence BR-24 together with MAS. Increasing the resonance frequency of the protons to 270 MHz with the help of a superconducting magnet increases the resolution in principle further as shown by Scheler et al. [89].

Further technicalities like phase compensation [90], extraction of quadrature phase information from multiple pulse NMR signals [91] and pulse error and average Hamiltonian corrections [55] have been discussed in detail.

4 Double Resonance Experiments

There are many nuclei, namely, ^{13}C, ^{15}N etc. which can give valuable information about the electronic structure of molecules. Because of their low natural abundance their homonuclear dipolar coupling is very weak and the coupling to abundant spins such as protons can be eliminated by decoupling techniques, resulting in a high resolution spectrum.

The NMR signals to be observed, however, are extremely weak, because of (i) the low natural abundance, (ii) the small gyromagnetic ratio and (iii) usually long spin lattice relaxation times. To make an estimate on the detectability of those spins let us make the following assumptions, which are appropriate in NMR pulse experiments:

a) Signals are to be compared in a constant static magnetic field B_0.

b) The quality factor Q of the probe coil and the filling factor are constant at different frequencies.

c) The detector bandwidth is constant and less than the bandwidth of the resonance circuit.

d) The measuring time is constant.

Under these conditions the observed signal to noise ratio S/N depends on the spin quantum number I, the gyromagnetic ratio γ and the number of spins as [4a]

$$S/N \sim I(I+1)N\gamma^{5/2}.$$

We can increase S/N by relaxing assumption d) and by accumulating more and more spectra i.e. by increasing the measuring time. Let us define a normalized measuring time T_N, which is the time a measurement has to be performed to give the same signal to noise ratio as an equivalent proton sample. If we assume in addition that e) the spin lattice relaxation time T_1 is constant for all our samples (which would be a favourable case, if we refer to the T_1 of protons), we may write

$$T_N \sim (S/N)^{-2}.$$

Some representative values of S/N and T_N are given in Table 4.1. There is clear evidence that it is a formidable task to detect those rare spins.

There is another class of rare spins which are neighbours to impurities, such as: (i) point defects in alkali halides (ii) paramagnetic impurities in metals (Kondo effect) (iii) paramagnetic impurities in diamagnetic crystals etc. The number of these spins depends on the number of the impurities which may be very small.

In order to detect these different types of rare spins, a wealth of double resonance techniques have been developed. The basic ideas were proposed by Hartmann and Hahn [1] from which different schemes are derived. Since the

Table 4.1. Normalized signal to noise ratio S/N and normalized measuring time T_N with respect to protons for different natural abundant nuclei in the same static magnetic field

Spin		Natural abundance %	S/N	Measuring time T_N
^{1}H	$(I=1/2)$	100	1	1
^{19}F	$(I=1/2)$	100	0.858	1.36
^{31}P	$(I=1/2)$	100	0.104	92
^{39}K	$(I=3/2)$	93.08	$2.2 \cdot 10^{-3}$	$2.1 \cdot 10^5$
^{13}C	$(I=1/2)$	1.108	$3.5 \cdot 10^{-4}$	$8.1 \cdot 10^6$
^{109}Ag	$(I=1/2)$	48.65	$2.27 \cdot 10^{-4}$	$2.0 \cdot 10^7$
^{43}Ca	$(I=7/2)$	0.13	$3.2 \cdot 10^{-5}$	$9.7 \cdot 10^8$
^{15}N	$(I=1/2)$	0.365	$1.2 \cdot 10^{-5}$	$7.0 \cdot 10^9$
^{2}H	$(I=1)$	0.0156	$5.8 \cdot 10^{-6}$	$3.0 \cdot 10^{10}$

NMR signal of the rare spins is very weak, a gain in sensitivity can be obtained by utilizing the reservoir of abundant spins. This can be done in two ways: (i) by the indirect method [1, 2], where the rare spins are detected via the abundant spins and (ii) by the direct method [3], where the rare spins are polarized by the abundant spins. Although the indirect method is of higher sensitivity in principle in the case where the rare spin spectrum contains little spectral information, the direct method has some practical advantages and has so far furnished almost all the high resolution double resonance NMR spectra in solids of rare spins.

In Sect. 4.1 we give a basic account on double resonance experiments, whereas, in Sect. 4.2 we describe the cross-polarization technique of rare spins, which has furnished most of the high resolution spectra in solids to date. In Sect. 4.3 we shall deal with the cross-polarization dynamics and in Sect. 4.4 with the spin-decoupling dynamics, which are both closely interrelated.

4.1 Basic Principles of Double Resonance Experiments

Suppose we have a system of abundant I and rare S spins i.e. $N_I \gg N_S$ where N is the number of spins. Each spin system is coupled to the lattice and approaches the lattice temperature with the spin lattice relaxation time T_{1I} and T_{1S} respectively, as shown schematically in Fig. 4.1. The I and S spins may be coupled by some interaction represented by the cross relaxation time T_{IS}. For the basic understanding let us apply the spin temperature [4, 5] concept, which is discussed in great detail in the standard book by Goldman [6].

In the high temperature approximation we write for the spin density matrix [4, 6].

$$\rho = Z^{-1}(1 - \beta \mathcal{H}) \tag{4.1}$$

where

$$Z = \mathrm{Tr}\{1\} = (2I+1)^{N_I}(2S+1)^{N_S}$$

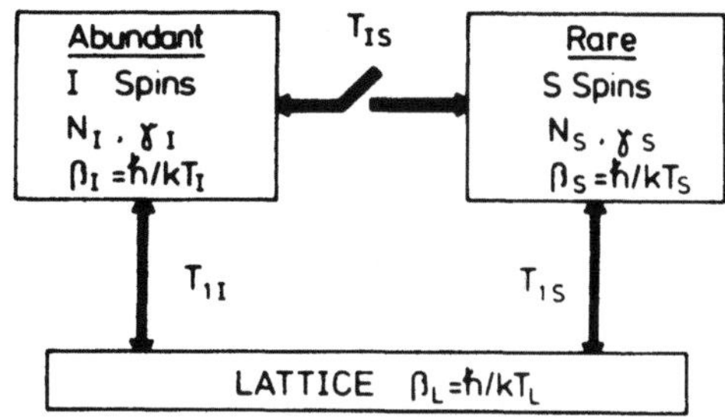

Fig. 4.1. Schematic representation of an abundant I spin reservoir and a rare S spin reservoir, which are coupled to the lattice as expressed by their spin lattice relaxation times T_{1I} and T_{1S}. The coupling between the two reservoirs as represented by the cross relaxation time T_{IS} can be varied at the experimenters discretion by a suitable application of rf fields

and with the inverse temperature

$$\beta = \hbar/kT.$$

Let us define the quantities: [4, 6]

$$\text{magnetization} \quad M_i = \hbar\gamma \, \mathrm{Tr}\{\rho I_i\} \quad i = x, y, z \tag{4.2a}$$

$$\text{energy} \quad E = \hbar \cdot \mathrm{Tr}\{\rho\mathscr{H}\} = -\beta\hbar \, \mathrm{Tr}\{\mathscr{H}^2\}/Z \tag{4.2b}$$

$$\text{entropy} \quad S = -k \, \mathrm{Tr}\{\rho \ln \rho\}. \tag{4.2c}$$

In the case of Zeeman interaction, with

$$\mathscr{H} = -\gamma \boldsymbol{B} \cdot I_i$$

these quantities reduce to

$$M_i = \beta C \cdot B, \tag{4.3a}$$

$$E = -\beta \cdot C \cdot B^2, \tag{4.3b}$$

$$S = kN_I \ln(2I+1) - k\beta^2 \frac{C}{\hbar} B^2, \tag{4.3c}$$

where $C = \frac{1}{3} \cdot N I(I+1)\gamma^2\hbar$.

After waiting several time T_{1I}, T_{1S} when placing a sample into the static magnetic field B_0, the I and S spins reach the magnetization

$$M_{0I} = \beta_L C_I B_0 \quad \text{and} \quad M_{0S} = \beta_L C_S B_0$$

where β_L is the inverse lattice temperature.

The first step in double resonance consists now in

(i) Cooling of the abundant I spin system

This can be achieved e.g. by locking the spins in a field $B_I \ll B_0$. In this case

$$M_{0I} = \beta_L \cdot C_I \cdot B_0 = \beta_I \cdot C_I \cdot B_I \tag{4.4}$$

with

$$\beta_I/\beta_L = B_0/B_I \gg 1$$

or

$$T_I \ll T_L.$$

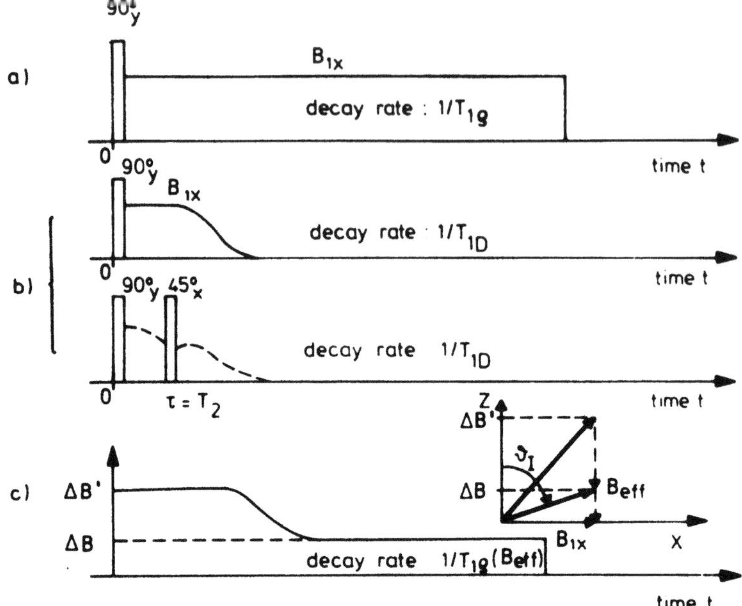

Fig. 4.2a–c. Schematic representation of different means for reducing the spin temperature in the rotating frame. **a** Spin locking, **b** adiabatic demagnetization in the rotating frame (ADRF) and Jeener-Broekaert [9] two pulse experiment, **c** demagnetization in an effective field by sweeping on resonance with a small rf field

There are three basic schemes for achieving a cooling of a spin system, which are represented in the schematical diagram Fig. 4.2.

These are

a) Spin locking [1–7] in the rotating frame with a field B_{1x} $=2B_{1I}\cos\omega_{0I}\cdot t$, which results with $(B_{1I}\gg B_{LI})$ in

$$\beta_I=\frac{B_0}{B_{1I}}\beta_L; \quad M_I=\beta_I\cdot C_I\cdot B_{1I}; \quad E=-\beta_I C_I B_{1I}^2. \tag{4.5}$$

The inverse spin temperature β_I will approach the inverse lattice temperature β_L with the time constant $T_{1\rho}$, the spin lattice relaxation time in the rotating frame.

b) Adiabatic demagnetization in the rotating frame (ADRF) [8] by turning off the B_{1x} field adiabatically, leaving the spins in the "dipolar field" B'_{LI}, where

$$\text{Tr}\{\mathscr{H}_D'^2\}=\gamma^2 B_{LI}'^2 \text{Tr}\{I_z^2\}$$

$$\beta_I=\frac{B_0}{B'_{LI}}\beta_L; \quad E_I=-\beta_I C_I B_{LI}'^2. \tag{4.6}$$

The inverse spin temperature β_I will approach the inverse lattice temperature β_L with the time constant T_{1D}, the spin lattice relaxation time of the dipolar state.

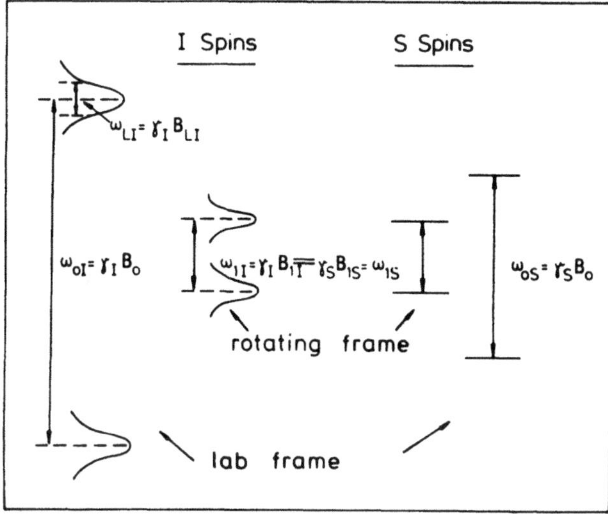

Fig. 4.3. Pictorial representation of level matching in the rotating frame (Hartmann-Hahn condition) for spin 1/2 systems

A different way of achieving a dipolar state was proposed by Jeener and Broekaert, [9], applying a $90^{\circ}_y - \tau - 45^{\circ}_x$ pulse sequence. Since this process is of course not adiabatic, a somewhat smaller inverse temperature β_I is achieved, therefore, [6]

$$\beta_I \cong 0.525 \frac{B_0}{B'_{LI}} \beta_L. \tag{4.7}$$

Nevertheless, this is a very convenient method for cooling the abundant spin system.

c) Adiabatic passage and stop at ΔB. In this case the spins are locked in an effective field $B^2_{\text{eff}} = \Delta B^2 + B^2_{1x}$ which makes an angle $\vartheta_I = \arctan(B_{1x}/\Delta B)$ with the z direction.

$$\beta_I = \frac{B_0}{B_{\text{eff}}} \beta_L; \quad M_I = \beta_I \cdot C_I \cdot B_{\text{eff}}; \quad E_I = -\beta_I \cdot C_I \cdot B^2_{\text{eff}}. \tag{4.8}$$

The next step is

(ii) Bring I and S spins-into contact

Since the I spins are cold and the S spins are hot, there will be a calorimetric effect [2] and energy exchange may proceed with the time constant T_{IS}. Only if $T_{IS} \lesssim T_{1DI}, T_{1\rho I}$ this energy exchange is considerable and can be utilized for a double resonance effect. Rapid energy transfer is possible only under total energy conservation. No such transfer is possible in the laboratory frame. However, in the rotating frame a matching of the energy levels is possible as shown schematically in Fig. 4.3, allowing rapid transfer under energy conservation (in the rotating frame) If the Hartmann-Hahn [1] condition

$$\gamma_S B_{1S} = \gamma_I B_{1I} \quad \text{for } I = S = 1/2 \tag{4.9}$$

or

$$\omega_{1S} = \omega_{1I}$$

is fulfilled, where B_{1S} and B_{1I} are the rf fields in the rotating frame of the I and S spins respectively. Of course, these fields may be effective fields, or correspondingly "effective frequencies" ω_{eS} and ω_{eI}.

As will be shown in more detail in Sect. 4.3, the transfer rate can be expressed as [10, 11]

$$\text{ADRF case:} \quad \left(\frac{1}{T_{IS}}\right)_{\text{ADRF}} = \sin^2 \vartheta_S \, M_2^{IS} \, J_{\text{ADRF}}(\omega_{eS}) \tag{4.10}$$

and in the

$$\text{spin locking case [11]} \quad \left(\frac{1}{T_{IS}}\right)_{SL} = \tfrac{1}{2}\sin^2 \vartheta_S \sin^2 \vartheta_I \, M_2^{IS} \, J_{SL}(\Delta\omega_e)$$

$$\omega_{eS}, \omega_{eI} \gg \omega_{LI} \tag{4.11}$$

where

$$\tan \vartheta_S = B_{1S}/\Delta B_S \quad \text{and} \quad \tan \vartheta_I = B_{1I}/\Delta B_S$$

$$\Delta\omega_e = \omega_{eS} - \omega_{eI}$$

and where M_2^{IS} is the second moment of the I-S coupling Hamiltonian. The spectral distribution function for the cross relaxation process $J(\omega)$ decreases monotonically to zero for increasing ω, i.e. with increasing mismatch of the Hartmann-Hahn condition. This results in a drastic decrease of the transfer-rate $1/T_{IS}$, as is expected. McArthur, Hahn and Walstedt [10] have used an intuitive approach, based on the experimental data to express the functional form of $J_{\text{ADRF}}(\omega_{eS})$ as

$$J_{\text{ADRF}}(\omega_{eS}) = \tfrac{1}{2}\tau_c \exp(-\omega_{eS}\tau_c) \tag{4.12}$$

where τ_c is the correlation time. A more general approach using the memory function technique was applied by Demco, Tengenfeldt and Waugh [11] and will be discussed in more detail in Sect. 4.3.

The spin temperature exchange occuring in a single I-S contact is shown schematically in Fig. 4.4. We suppose again a high inverse spin temperature β_I at the beginning of the contact ($t = 0$) and a zero inverse temperature β_S at $t = 0$. If we neglect spin lattice relaxation, both spin temperatures will finally reach the same value β_f, assuming exponential relaxation as

$$\beta_I(t) = (\beta_I - \beta_f)\, e^{-t/T_{IS}} + \beta_f, \tag{4.13}$$

$$\beta_S(t) = \beta_f(1 - e^{-t/T_{IS}}). \tag{4.14}$$

Assuming energy conservation

$$\beta_I \cdot C_I \cdot B_I^2 + \beta_S C_S \cdot B_S^2 = \beta_f [C_I B_I^2 + C_S B_S^2] \tag{4.15}$$

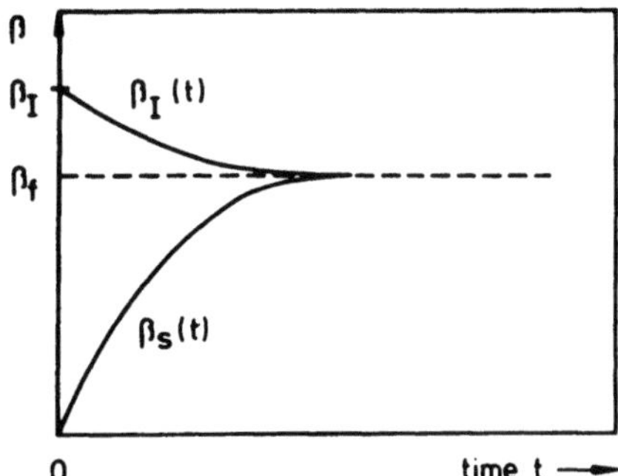

Fig. 4.4. Time evolution of I and S spin inverse temperatures β_I and β_S, when the spin systems are brought into contact at $t=0$. Initial condition: $\beta_I \neq 0$; $\beta_S = 0$. A final inverse spin temperature β_f is reached after several T_{IS}

and with $\beta_S = 0$, Eq. (4.15) leads to

$$\frac{\beta_f}{\beta_I} = \frac{1}{1+\varepsilon'} \tag{4.16}$$

where

$$\varepsilon' = \frac{C_S B_S^2}{C_I B_I^2} \tag{4.17}$$

is the ratio of the heat capacities of the S and I spins.

If the Hartmann-Hahn condition $\gamma_S B_S = \gamma_I B_I$ is fulfilled, we obtain

$$\varepsilon = \frac{N_S S(S+1)}{N_I I(I+1)} \ll 1. \tag{4.18}$$

With $M_I^{(f)} = \beta_f \cdot C_I \cdot B_I$ and $M_S^{(f)} = \beta_f \cdot C_S \cdot B_S$ the final I and S spin magnetization reaches

$$M_I^{(f)} = \frac{1}{1+\varepsilon} M_I^{(i)} \tag{4.19}$$

and

$$M_S^{(f)} = \frac{\gamma_I}{\gamma_S} \cdot \frac{1}{1+\varepsilon} \cdot M_{0S} \tag{4.20}$$

where $M_I^{(i)}$ is the initial magnetization of the I spins and M_{0S} is the Zeeman magnetization of the S spins. Since ε is a very small number of the order of N_S/N_I we may write $1/(1+\varepsilon) \cong 1-\varepsilon$, i.e. according to Eqs. (4.19) and (4.20) the I spin magnetization does not decrease very much in a single contact, however, the S spin magnetization may have been increased if $\gamma_I/\gamma_S > 1$.

In order to achieve a noticable destruction of the I spin magnetization, multiple contacts [2, 3] have to be performed as demonstrated in Fig. 4.5. The I spins are spin-locked in the field B_{1I}, whereas, the S spins become polarized in the field B_{1S} which may be an effective field in the rotating frame. The pulsed B_{1S} field is of duration t_w with a spacing τ_i. Coupling between the I and S spins is achieved, when the B_{1S} field is turned on and I and S are decoupled consecutively, then B_{1S} is turned off. A free precession of the S spins can be observed during this time.

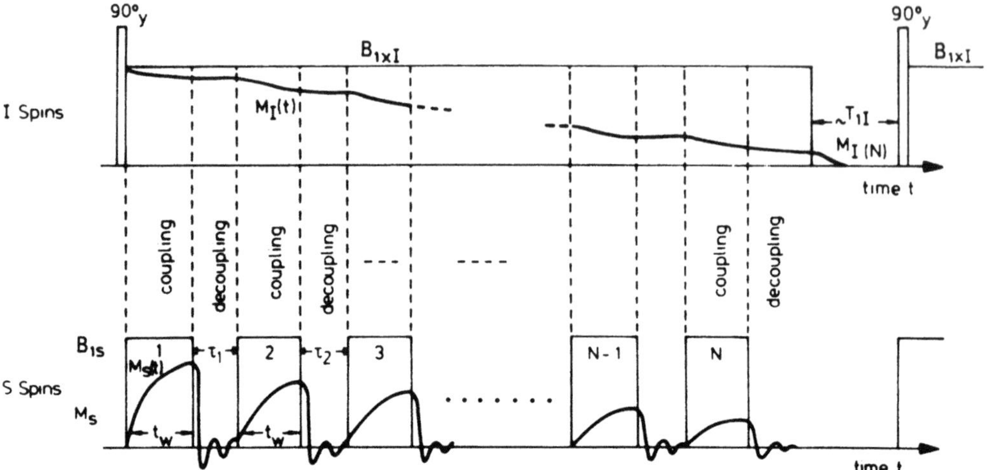

Fig. 4.5. Pulse timing of a typical double resonance experiment in the rotating frame

After the k-th contact we may write for the I spin magnetization [1–3, 12]

$$M_I(k) = \left(\frac{1}{1+\varepsilon}\right)^k M_{0I} \cong (1-\varepsilon)^k M_{0I} \tag{4.21}$$

or

$$M_I(k) \cong \exp(-k\varepsilon) M_{0I} \tag{4.22}$$

and for the S spin magnetization

$$M_S(k) = \frac{\gamma_I}{\gamma_S}(1-\varepsilon)^k M_{0S} \cong \frac{\gamma_I}{\gamma_S}\exp(-k\varepsilon) M_{0S} \tag{4.23}$$

under the condition that the S spin inverse temperature $\beta_S = 0$ at the beginning of each contact. If we sum up all S signals we may write after N contacts

$$M_{SN} = \sum_{k=1}^{N} M_S(k) = \frac{\gamma_I}{\gamma_S} M_{0S} \sum_{k=1}^{N}(1-\varepsilon)^k \tag{4.24}$$

or to a good approxiamtion [3, 12]

$$M_{SN} \cong \frac{\gamma_I}{\gamma_S} M_{0S} \sum_{k=1}^{N} e^{-k\varepsilon}. \tag{4.25}$$

Thus we distinguish between the indirect method, where the I spin magnetization is observed after N contacts and the direct method, where the S spin signals are accumulated.

We are now going to relax the condition $\beta_S = 0$ at the beginning of each contact and discuss the general case. In the direct method, we are interested in $M_I(N)/M_{0I}$ after N contacts i.e.

$$\frac{M_I(N)}{M_{0I}} = \frac{\beta_{NI}}{\beta_{0I}} \tag{4.26}$$

where

$$\beta_{0I} = \beta_L \cdot \frac{B_0}{B_{1x}}.$$

Several authors have discussed the behavior of I and S spin magnetization in a multiple contact double resonance experiment [2, 3, 12–14]. We shall follow here the discussion given by H. Ernst [13] which we find particularly useful for a unified description of all rotating frame double resonance experiments. We shall use the following notation

$\beta_{kI}^{(i)}$: initial inverse temperature of the I spins at the beginning of the k-th B_{1S} pulse,

$\beta_{kI}^{(f)}$: final inverse temperature of the I spins after the k-th B_{1S} pulse.

The expression "temperature" is not to be taken seriously in this context. We simply use it as a thermodynamic parameter which describes the evolution of the energy as the only constant of the motion [6] (see also Sect. 4.3).

We may write:

$$\beta_{(k+1)I}^{(i)} = \beta_{kI}^{(f)} \quad \text{no contact between } I \text{ and } S \text{ spins during } \tau_k \tag{4.27}$$

$$\beta_{(k+1)S}^{(i)} = \beta_{kS}^{(f)} \, G(\tau_k) \tag{4.28}$$

where $G(\tau)$ is the f.i.d. of the S spins. We shall write in short from notation $G(\tau_k) = g_k$ with $|g_k| \leqq 1$.

Energy conservation demands

$$\beta_I^{(i)} + \varepsilon \beta_S^{(i)} = \beta_I^{(f)} + \varepsilon \beta_S^{(f)}. \tag{4.29}$$

If we further assume complete spin temperature exchange within each contact $(t_w \gg T_{IS})$ i.e.

$$\beta_{kI}^{(f)} = \beta_{kS}^{(f)} \tag{4.30}$$

we can combine Eqs. (4.27–4.30) to obtain

$$\beta_{(k+1)I}^{(f)} = \beta_{kI}^{(f)} \frac{1 + \varepsilon g_k}{1 + \varepsilon} \tag{4.31}$$

or

$$\beta_{NI}^{(f)} = \beta_{1I}^{(f)} \cdot \prod_{k=1}^{N-1} \left(\frac{1 + \varepsilon g_k}{1 + \varepsilon} \right). \tag{4.32}$$

Let us keep in mind, that we are looking for $\beta_{NI}^{(f)}/\beta_{0I}$ according to Eq. (4.26) in order to obtain $M_I(N)/M_{0I}$. Using total energy conservation [Eq. (4.29)] and complete spin temperature exchange [Eq. (4.30)], we arrive at

$$\frac{\beta_{1I}^{(f)}}{\beta_{1I}^{(i)}} = \frac{1 + \lambda \varepsilon}{1 + \varepsilon} \tag{4.33}$$

where

$$\lambda = \beta_{1S}^{(i)}/\beta_{1I}^{(i)} = \frac{B_{1xI}}{B_{\text{eff} S}} \cos \vartheta_S. \tag{4.34}$$

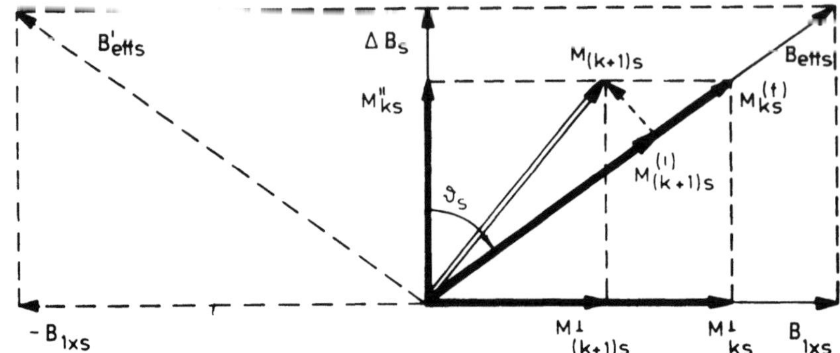

Fig. 4.6. Vector representation of the S spin magnetization at different times during a double resonance experiment according to Fig. 4.5. (H. Ernst [13])

Combining Eqs. (4.26, 4.32, 4.33) leads to

$$\frac{M_I(N)}{M_{0I}}=\frac{1+\lambda\varepsilon}{(1+\varepsilon)^N}\prod_{k=1}^{N-1}(1+\varepsilon g_k). \tag{4.35}$$

With $|g_k|\leqq 1$; $\varepsilon\ll 1$; $\lambda\leqq 1$ and $(1+\varepsilon)^{-N}\cong\exp(-N\varepsilon)$ we may write

$$\frac{M_I(N)}{M_{0I}}=e^{-N\varepsilon}\prod_{k=1}^{N-1}(1+\varepsilon g_k). \tag{4.36}$$

A further approximation is obtained by

$$\ln(M_I(N)/M_{0I})=-N\varepsilon+\sum_{k=1}^{N-1}\ln(1+\varepsilon g_k)\cong -N\varepsilon+\sum_{k=1}^{N-1}\varepsilon g_k$$

or

$$M_I(N)=M_{0I}\,e^{-\delta} \tag{4.37}$$

where the damping constant δ for N different I-S contacts equals

$$\delta=\varepsilon\left[1+\sum_{k=1}^{N-1}(1-g_k)\right]. \tag{4.38}$$

The quantity, still to be determined, is g_k.

In order to get a better understanding of the motion of the S spin magnetization vector, we follow here the vector representation according to H. Ernst [13], as drawn schematically in Fig. 4.6. The notation is self-explanatory and we realize that the magnetization $M_{kS}^{(f)}$ at the end of the k-th contact pulse is polarized, along the effective field $B_{\mathrm{eff}\,S}$ and has the component $M_{kS}^{\perp}$ in the direction of B_{1xS} and the component $M_{kS}^{\parallel}$ parallel to ΔB_S. If the B_{1xS} field is turned off for the time τ_k, only the component $M_S^{\perp}$ is changed to the new value $M_{(k+1)S}^{\perp}$ at the beginning of the $(k+1)$-th contact pulse.

The total S spin magnetization $M_{(k+1)S}$ is no longer parallel to the effective field $B_{\mathrm{eff}\,S}$ and the initial magnetization $M_{(k+1)S}^{(i)}$ is given by the projection of $M_{(k+1)S}$ onto the effective field $B_{\mathrm{eff}\,S}$. We readily realize, that

$$M_{(k+1)S}^{\perp}=M_{kS}^{\perp}g_{xk} \tag{4.39}$$

where

$$g_{xk} = G(\tau_k) \cos \Delta \omega_S \tau_k. \tag{4.40}$$

The quantity wanted is

$$g_k = M^{(i)}_{(k+1)S}/M^{(f)}_{kS}$$

which can be written according to Fig. 4.6 as

$$g_k = \cos^2 \vartheta_S \pm \sin^2 \vartheta_S g_{xk} \tag{4.41}$$

where the minus sign stands for phase alternation of B_{1xS}. We may now write the damping xonstant according to Eqs. (4.37) and (4.38) as [13]

$$\delta = \varepsilon \left[1 + \sin^2 \vartheta_S \sum_{k=1}^{N-1} (1 \mp g_{xk}) \right] \quad \begin{array}{l} \text{``}-\text{'' no phase alternation} \\ \text{``}+\text{'' phase alternation.} \end{array} \tag{4.42}$$

Equation (4.42) is the basic equation which describes virtually all indirect double resonance methods as will be discussed in the following:

a) $\tau_k = 0$ i.e. $g_{xk} = 1$

$$\delta = \begin{cases} \varepsilon \text{ no phase alternation, practically no destruction} \\ \quad \text{of } I \text{ magnetization, since } \varepsilon \ll 1. \\ 2\varepsilon N \sin^2 \vartheta_S + \underbrace{\varepsilon \cos^2 \vartheta_S.}_{\cong 0} \quad \text{phase alternation.} \end{cases} \tag{4.43}$$

With phase alternation of B_{1xS} we arrive at the Hartmann-Hahn experiment [1] where $\delta = 2\varepsilon N$ if irradiation is applied on resonance. The observed line width is determined by the influence of $\sin^2 \vartheta_S$ and the increase of T_{IS} by mismatching the Hartmann-Hahn condition. The line width is typically a few kHz, whereas, the natural line width of the S spins may be a few Hertz. This method has been extensively applied to the investigation of point defects in cubic crystals [14].

b) $\tau_k \gg T^*_{2S}$ i.e. $g_{xk} = 0$

$$\delta = \varepsilon N \sin^2 \vartheta_S + \underbrace{\varepsilon \cos^2 \vartheta_S.}_{\cong 0} \tag{4.44}$$

On resonance ($\vartheta_S = 90°$) this corresponds to the Lurie-Slichter [2] experiment with $\delta = \varepsilon N$. The line width is the same as in a).

c) $\tau_k = \tau = \text{const}$ i.e. $g_{xk} = G(\tau)$ if $\Delta \omega_S = 0$.

It follows

$$\delta = \underbrace{\varepsilon N(1 - G(\tau)) - \varepsilon G(\tau)}_{\cong 0} \tag{4.45}$$

which corresponds to the experiment of McArthur, Hahn and Walstedt [10], who mapped out the S spin FID $G(\tau)$ by varying τ with $0 \leq \tau \leq T^*_{2S}$.

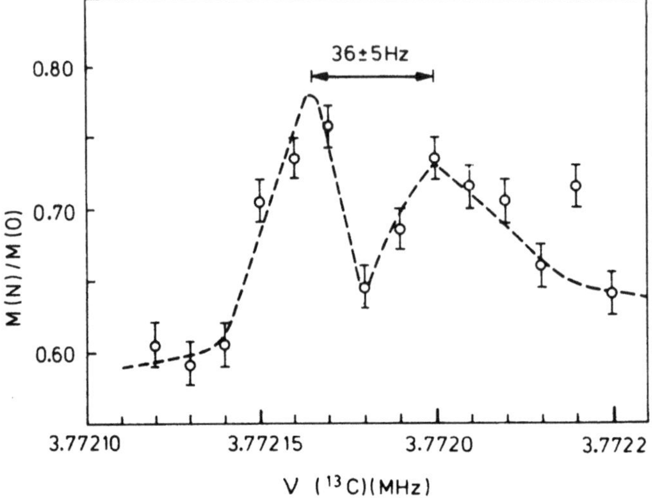

Fig. 4.7. ^{13}C spectrum of solid adamantane obtained by Mansfield and co-workers [12] by a high resolution proton destruction double resonance technique as described in the text

McArthur [10] et al., however, applied ADRF instead of spin locking to the I spins, that is why no decoupling of the I spins was obtained. The S spin FID is determined by the I-S dipolar coupling mainly in this case.

d) same as c) but with a small deviation from resonance

$$\Delta\omega_S \neq 0.$$

It follows

$$\delta = \varepsilon N (1 - G(\tau)\cos\Delta\omega_S\tau) \tag{4.46}$$

where δ is now a function of $\Delta\omega_S$. This represents the experiment performed by Mansfield [12] and co-workers, which allows high resolution spectroscopy to be performed by the indirect method.

In the special case

$$n \cdot 2\pi + \pi/2 < \Delta\omega_S\tau \leqq n2\pi + \tfrac{3}{2}\pi$$

it follows $|\delta| > \varepsilon N$ and the double resonance process becomes very effective. The spectrum $I(\Delta\omega_S)$ of the S spins can be obtained by summing the I signal for different values of τ as

$$\sum_{\tau=\tau_0}^{\tau=n\tau_0} (\delta + \varepsilon N) \cong \frac{N_\varepsilon}{\tau_0} \int_0^\infty d\tau\, G(\tau)\cos\Delta\omega_S\tau = \frac{N_\varepsilon}{\tau_0} I(\Delta\omega_S).$$

A high resolution spectrum of ^{13}C in adamantane was obtained by Mansfield [12] and co-workers with these techniques, as shown in Fig. 4.7. Figure 1.5 in the introduction shows the same spectrum obtained by the direct method of Pines [3] et al., to be discussed in Sect. 4.2.

For completeness we want to discuss briefly some *steady state methods*. Suppose the I spins are in a spin locked state, i.e., they are kept at a low spin

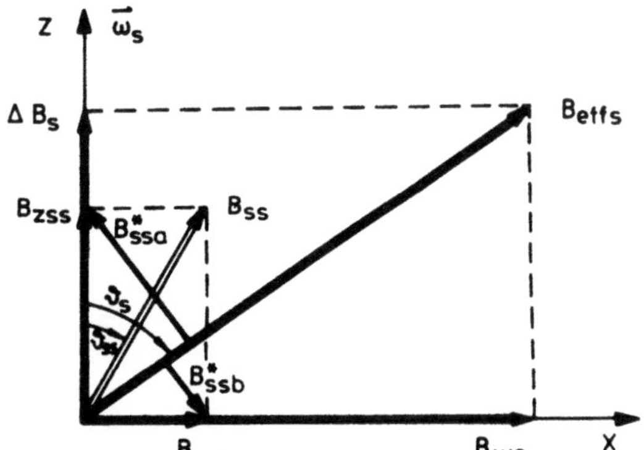

Fig. 4.8. Vector representation of rf field applied in *cw* double resonance experiments. (H. Ernst [13])

temperature. The S spins approach the magnetization

$$M_{0S} = \beta_I \cdot C_S \cdot B_{effS}$$

parallel to B_{effS} with the rate $1/T_{IS}$. In addition another S spin saturation field $B_{1SS}^{(t)} = 2 \cdot B_{SS} \cos \omega_{SS} \cdot t$ is applied to the S spins as shown vectorially in Fig. 4.8.

The angle between the z axis and B_{1SS} may be labelled ϑ_{SS}, whereas the angle between the z axis and B_{effS} may be called ϑ_S as usually. The power absorbed from the saturating field may be expressed as (H. Ernst [13])

$$P_S = -\frac{\omega_{es} M_{0S}}{\gamma_S T_{IS} \left[1 + \left(\dfrac{\Delta \omega_{SS}}{\gamma_S B_{SS}^*}\right)^2\right]} \tag{4.47}$$

where $\Delta \omega_{SS}$ is the deviation from resonance and B_{SS} an effective field according to Fig. 4.8 to be explained later.

The power P_S absorbed by the S spins is transferred to the I spins and changes their magnetization as

$$P_S = \frac{dM_I}{dt} \cdot B_{1xI}. \tag{4.48}$$

With

$$M_I = \beta_I \cdot C_I \cdot B_{1xI}$$

we obtain [13]

$$\delta(\Delta \omega_{SS}) = \ln M_I/M_{0I} = \varepsilon' \frac{T_{1I}}{T_{IS}} \frac{1}{\left[1 + \left(\dfrac{\Delta \omega_{SS}}{\gamma_S B_{SS}^*}\right)^2\right]} \tag{4.49}$$

where we remind the reader of

$$\varepsilon' = \frac{C_S B_{effS}^2}{C_I B_{1xI}^2}.$$

$\delta(\Delta \omega_{SS})$ is a Lorentzian with the half width at half height

$$\Delta \omega = \gamma_S B_{SS}^* < 1/T_{IS}.$$

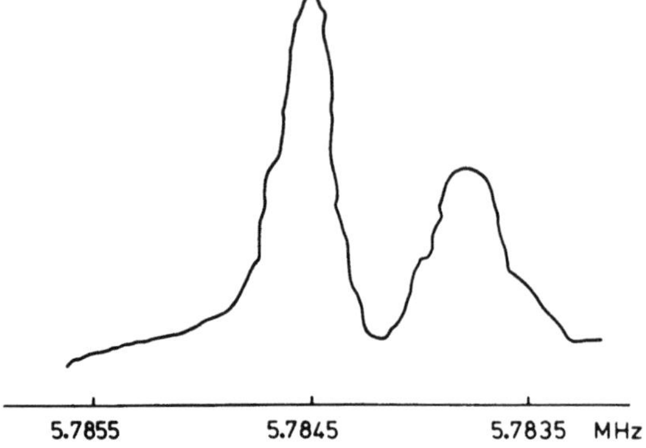

Fig. 4.9. ^{13}C double resonance spectrum of polycrystalline C_6H_6 at $-40°$ C, obtained by C.S. Yannoni and H.E. Bleich [17a] with a *cw* double resonance technique of the Bleich-Redfield [16] type

$B_{1SS}(t)$ may be expressed in the ω_S rotating frame by the vector

$$B_{1SS}(t) = B_{SS} \{\sin \vartheta_{SS} \cos(\omega_{SS} - \omega_S) t, \sin \vartheta_{SS} \sin(\omega_{SS} - \omega_S) t,$$
$$2 \cos \vartheta_{SS} \cos \omega_{SS} \cdot t\}. \tag{4.50}$$

Two possible resonances can occur:

(i) $\omega_{SS} = \omega_{eS}$ i.e. $B^*_{SSa} = B_{zSS} \sin \vartheta_S = B_{SS} \cos \vartheta_{SS} \sin \vartheta_S$.

The saturating field B_{SS} induces transitions in the effective field B_{effS} i.e., it is applied in the audio frequency range. Such an experiment was performed by Walstedt et al. [15] and McArthur et al. [10] by applying an audio-field in the z-direction $\vartheta_{SS} = 0$; $B^*_{SSa} = B_{SS} \sin \vartheta_S$ at the frequency $\omega_{SS} = \omega_{1S} = \gamma_S B_{1S}$.

For $\omega_{SS} > \omega_{1S}$, but $\omega_{SS} = \omega_{eS}$ two resonance minima are observed with varying $\Delta \omega_S$. The condition for the two minima is readily obtained as

$$\Delta \omega_S = \pm (\omega_{SS}^2 - \omega_{1S}^2)^{1/2}. \tag{4.51}$$

Because the Hartmann-Hahn condition is fulfilled, T_{IS} is short (~ 1 m s) and the resulting line width is about 1 kHz.

(ii) $\omega_{SS} = \omega_S + \omega_{eS}$ i.e. $B^*_{SSb} = B_{xSS} \cos \vartheta_S = B_{SS} \sin \vartheta_{SS} \cos \vartheta_S$.

Such an experiment was performed by Bleich and Redfield [16, 17] where B_{SS} lies in the x-direction ($\vartheta_{SS} = 90°$) and $B^*_{SSb} = B_{SS} \cos \vartheta_S$. Bleich and Redfield used $\vartheta_S \cong 5°$ to $10°$ by applying the B_{1xS}-field far off resonance. Already a relative small B_{1xS} produces a large B_{effS} in order to fulfil the Hartmann-Hahn condition. T_{IS} on the other hand becomes very long, because of the factor $\sin^2 \vartheta_S$, which in this case is $\sin^2 \vartheta_S \cong 10^{-2}$.

This leads to a high resolution spectrum, as demonstrated in Fig. 4.9. However, the observed signal is proportional to the quantity $I(\omega)/T_{IS}$ i.e., a strong deformation of the spectrum results, since T_{IS} varies with ϑ_{IS} which is the angle of the *I-S* internuclear vector with respect to the z axis. If ϑ_{IS} equals the magic angle, the next nearest neighbour coupling of ^{13}C to ^{1}H in benzene vanishes and T_{IS} becomes very large, leading to a hole in the spectrum at the corresponding frequency (see Fig. 4.9). These disturbing effects are absent in the direct method of Pines et al. [3] unless they are produced on purpose.

Bleich and Redfield [18] have proposed still another type of Hartmann-Hahn double NMR in solids which is particularly suited for high resolution spectroscopy of spins with low gyromagnetic ratio. They observed the ^{43}Ca resonance of naturally abundant ^{43}Ca (0.145%) in CaF_2 at 16.445 MHz.

Another double resonance technique introduced by Stoll, Vega and Vaughan [19] is based on the spinor character of a two level system. A superposition of states is created by a preparation pulse at an allowed I-spin transition. A connected S-spin transition is then irradiated preferentially by a 2π pulse which changes the sign of the wave-function of the joined state. Since the phase of the joined state is monitored in the coherent decay of the I-spin transition its sign changes also once the 2π pulse is applied on-resonance with the S-spin transition. Such a behaviour is for example shown in Fig. 6.6.

4.2 Cross-Polarization of Dilute Spins

As we have seen in the preceding section, the I spin magnetization decreases, whereas, the S spin magnetization increases during I-S contact, given the I spins are prepared in a low spin temperature state. Pines, Gibby and Waugh [3] have designed a technique to utilize this fact, in order to obtain high resolution spectra of rare spins in solids. They have named their method: "Proton Enhanced Nuclear Induction Spectroscopy". The timing of this technique is the same as schematically shown in Fig. 4.5, where the pulse spacing τ is large to allow the S spin magnetization to decrease fully to zero. The S spin magnetization between the pulses is observed and successive S spin FID's are accumulated in an online computer and Fourier transformed to obtain the S spin spectrum. Since the I-field B_{1xI} is kept on during the S spin FID the abundant I spins are decoupled, resulting in a high resolution spectrum of the S spins. Figure 4.10 shows some representative ^{13}C spectra obtained by this method.

The S spin magnetization after the k-th contact can be expressed according to Eq. (4.23) as [3]

$$M_S(k) = \frac{\gamma_I}{\gamma_S} e^{-k\varepsilon} M_{0S}$$

where

$$\varepsilon = \frac{N_S S(S+1)}{N_I I(I+1)}$$

since the Hartmann-Hahn condition is fulfilled in this experiment. For the same reason the contact time t_w is on the order of milliseconds, whereas, τ may be on the order of a few hundred milliseconds, depending on the wanted spectral resolution. The decrease of I spin magnetization due to $T_{1\rho}$ is not taken into account in this discussion, it may, however, spoil the whole beauty of this experiment, when very high resolution is demanded and also slow motions are present. The total coadded magnetization after N contacts may be

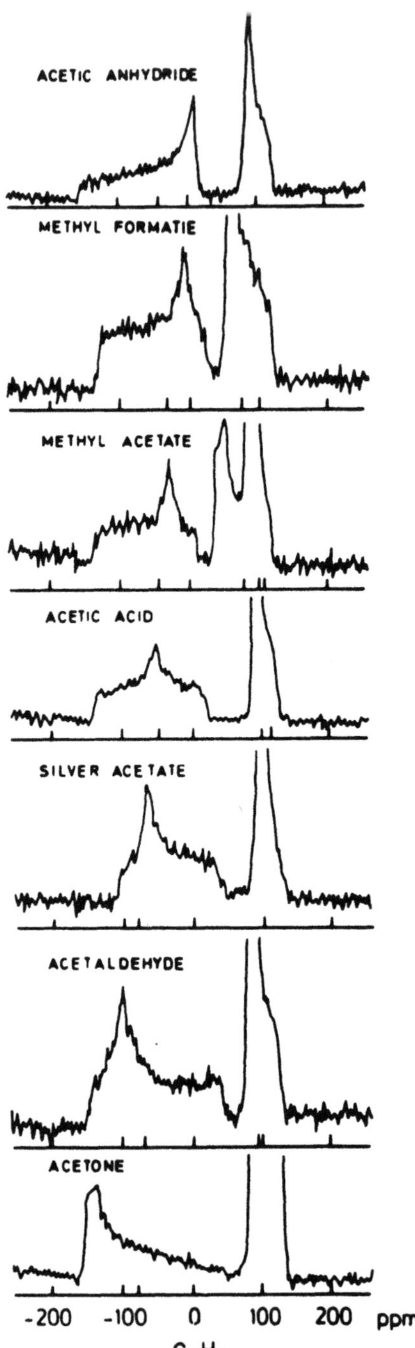

Fig. 4.10. Proton enhanced ^{13}C spectra of polycrystalline compounds containing carbonyl groups (large shielding anisotropy, left side) and methyl groups (small anisotropy, right side) obtained by A. Pines, M.G. Gibby and J.S. Waugh [3b] at liquid nitrogen temperature

expressed according to Eq. (4.25) as

$$M_{SN} = \frac{\gamma_I}{\gamma_S} M_{0S} \sum_{k=1}^{N} e^{-k\varepsilon}.$$

In order to estimate the sensitivity of this method by comparing the enhanced S spin signal with the ordinary Bloch decay, we assume a given quality factor Q of the probe coil, a constant filling factor and a constant detector band-width, producing a constant detector noise, all represented by the constant K. The signal to noise ratio equals

$$S/N = K \cdot \omega_0^{1/2} \cdot M$$

under these conditions as outlined in Sect. 4.

The signal to noise of the FID of the S spins may now be expressed by

$$(S/N)_{\text{FID}} = K \omega_{0S}^{1/2} \cdot M_{0S} \tag{4.52}$$

whereas, the signal to noise ratio of N accumulated multiple contact spectra in a cross-polarization (CP) experiment is given by

$$(S/N)_{\text{CP}} = \frac{1}{\sqrt{N}} K \cdot \omega_{0S}^{1/2} \cdot M_{SN} \tag{4.53}$$

which results in a gain factor G_{CP} of the multiple contact cross-polarization experiment over an ordinary FID as [3, 12]

$$G_{\text{CP}} = \frac{\gamma_I}{\gamma_S} \cdot \frac{1}{\sqrt{N}} \sum_{k=1}^{N} e^{-k\varepsilon}. \tag{4.54}$$

This is maximized for $N\varepsilon \cong 1.3$ from which follows

$$G_{\text{CP}} = 0.64 \frac{\gamma_I}{\gamma_S} \left[\frac{N_I I(I+1)}{N_S S(S+1)} \right]^{1/2} \tag{4.55}$$

or

$$(S/N)_{\text{CP}} = 0.64 \frac{\gamma_I}{\gamma_S} \left[\frac{N_I I(I+1)}{N_S S(S+1)} \right]^{1/2} K \cdot \omega_{0S}^{1/2} M_{0S}. \tag{4.56}$$

For ^{13}C at natural isotopic abundance in organic solids, where $N_I/N_S \cong 150$, we find $G_{\text{CP}} \cong 30$. Notice, that it takes $G_{\text{CP}}^2 \cong 10^3$ accumulations of the FID to reach the same signal to-noise ratio as in the cross-polarization experiment. Another point to be mentioned is the fact that it takes T_{1S} to repeat the FID measurements, whereas, it takes T_{1I} for replicating the cross-polarization experiment. Since usually $T_{1S} > T_{1I}$, a factor T_{1S}/T_{1I} in front of Eq. (4.55) increases further the "gain" of the cross-polarization experiment.

If for technical reasons, or because of $T_{1\rho}$ effects, only a few contacts can be performed, the gain G_{CP} is of course drastically reduced. In these cases it is favourable to reduce T_{1I}. This has been discussed in more detail by Pines et al. [3].

In order to compare the sensitivity of the direct multiple-contact cross-polarization experiment with the indirect method, we have to calculate $(S/N)_{\text{ind}}/$

$(S/N)_{CP}$, where $(S/N)_{CP}$ is given by Eq. (4.56). Whereas, in the cross-polarization experiment the total spectrum is obtained in one shot, each spectral element had to be measured step by step in the indirect method. If we assume that the I magnetization always decreases to $1/e$ of its initial value, $(S/N)_{ind}$ for each spectral element is given by

$$(S/N)_{ind} = \left(1 - \frac{1}{e}\right) \cdot K \cdot \omega_{0I}^{1/2} \, M_{0I}. \tag{4.57}$$

Suppose the S spin spectrum contains n spectral elements, we then define the signal-to-noise ratio of the whole spectrum as

$$(S/N)_{ind} = 0.632 \frac{1}{\sqrt{n}} K \cdot \omega_{0I}^{1/2} \, M_{0I}. \tag{4.58}$$

Remenber, that in the cross-polarization experiment also the whole spectrum is obtained, however in one shot.

If we assume the same quality factor Q of the coil, the same detector bandwidth and noise characteristic in all cases, the factor K is the same in Eqs. (4.56) and (4.58) and we can write [3, 12]

$$\frac{(S/N)_{ind}}{(S/N)_{CP}} \cong \frac{1}{\sqrt{n}} \left(\frac{\gamma_I}{\gamma_S}\right)^{3/2} \left[\frac{N_I \, I(I+1)}{N_S \, S(S+1)}\right]^{1/2} \tag{4.59}$$

for the signal-to-noise ratio of a spectrum, recorded with the indirect method as compared with the cross-polarization method.

This is about 2.85 in the case of ^{13}C and ^{1}H organic compounds ($N_I/N_S \cong 150$) if $n = 1{,}024$.

The advantage in sensitivity of the indirect method over the cross-polarization experiment is even more pronounced if fewer spectral elements are needed to represent the S spin spectrum. The break-even point is reached in the case of ^{13}C and ^{1}H when about 8,300 spectral elements are needed which has so far rarely been the case.

Although the indirect method is of greater sensitivity, almost all of the high resolution spectra of rare spins in solids have been observed by the cross-polarization technique to date. This is because of distortions which may result in the indirect method, and which have been discussed in the last section. On the other hand, very often the indirect method does not produce the total spectrum as fast as the cross-polarization technique.

We are now going to discuss briefly the different steps in the direct detection method of rare spins, following closely the paper of Pines et al. [3]. Four major steps have to be performed:

(1) *Preparation* of the I spins implies the polarization of the I spins in a static magnetic field.
(2) The "*hold*" period keeps the I spin order in some suitable reference frame.
(3) "*Mix*" constitutes the transfer of spin order form the I to the S spin system.
(4) *Observation* of the S spin signal is the final step, where the I spins may be decoupled if high resolution S spectra are desired.

Some suitable procedures to perform the four essential steps are summarized in the following:

(1) *Preparation:*
   (a) polarize I by the spin-lattice relaxation in B_0
   (b) dynamically polarize I by optical or microwave polarization [20, 21]
   (c) polarize I using quadrupolar nucleus with short spin-lattice relaxation time [22–26].
(2) *Hold*
   (a) hold M_I along B_0 in laboratory frame,
   (b) spin lock M_I along B_{1I} in rotating frame,
   (c) hold I order in dipolar state in laboratory or rotating frame [6, 8].
(3) *Mix*
   (a) Hartmann-Hahn [1], matched or unmatched,
   (b) solid effect in laboratory or rotating frames [6],
   (c) adiabatic crossover [3b, 6].
(4) *Observation*
   (a) undecoupled I
   (b) continous I spin decoupling [27, 28]
   (c) pulsed spin decoupling [29].

We want to discuss now some examples of high resolution cross-polarization experiments using an I dipolar state in the rotating frame. Figure 4.11 explains different feasable steps [3b]. The essential part in these experiments is the ADRF of the I spins, followed by turning on the B_{1S}-field and mixing of the I and S spin reservoir. Before the mixing sets in the inverse "dipolar-spin-temperature" β_{DI} of the I spins and the inverse spin-temperature β_S of the S spins are

$$\beta_{DI} = \beta_L \cdot \frac{B_0}{B'_{LI}}; \quad \beta_S = 0$$

where the "local field" of the I spins is defined by

$$\mathrm{Tr}\{\mathcal{H}_{II}'^2\} = \gamma^2 B'_{LI} \, \mathrm{Tr}\{I_z^2\}.$$

The prime labels the secular part of the dipolar Hamiltonian. We assume the mixing to proceed at constant energy i.e., the same final spin-temperature β_f for I and S spins is reached

$$\beta_{DI} \cdot C_I B'^2_{LI} = \beta_f [C_I B'^2_{LI} + C_S B^2_{1S}]$$

resulting in

$$\frac{\beta_f}{\beta_{DI}} = \frac{1}{1 + \varepsilon'} \tag{4.60}$$

where

$$\varepsilon' = \frac{\mathrm{Tr}\{\mathcal{H}_{1S}^2\}}{\mathrm{Tr}\{\mathcal{H}_{II}'^2\}}$$

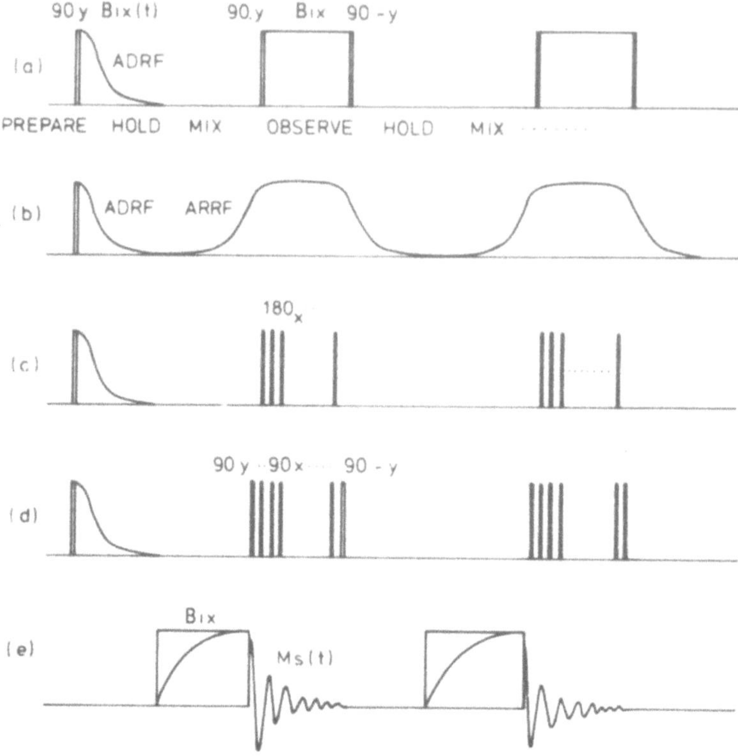

Fig. 4.11a–e. Different double resonance schemes proposed by A. Pines, M.G. Gibby and J.S. Waugh [3b] using an I dipolar state created by ADRF. The following I spin decoupling schemes may be employed: **a** Magic sandwich decoupling. The 90° pulses cause a "spin locking" of the I dipolar state. **b** Adiabatic remagnetization in the rotating frame (ADRF) spin locks the I magnetization along B_{1x} and decouples the S spins. **c** The I spins are decoupled by 180° pulses and maintained in the dipolar state, since $\mathcal{H}'_{II}$ is invariant under 180° rotation (only in the case of δ pulses, see Fig. 3.23). **d** Pulsed version of **a** applying 90° pulses for decoupling. **e** S spin irradiation and magnetization during polarization and observation

or

$$\varepsilon' = \frac{N_S\,S(S+1)}{N_I\,I(I+1)} \cdot \alpha^2 = \varepsilon\alpha^2 \tag{4.61}$$

with

$$\alpha = \frac{\gamma_S B_{1S}}{\gamma_I B'_{LI}}. \tag{4.62}$$

If the Hartmann-Hahn condition is matched i.e., $\alpha = 1$ it follows $\varepsilon' = \varepsilon$. The final magnetization of the S spins is readily obtained using the above equation to be [28]

$$\frac{M_{S\infty}}{M_{0S}} = \frac{\beta_f B_{1S}}{\beta_L B_0} = \frac{\gamma_I}{\gamma_S}\frac{\alpha}{1+\varepsilon\alpha^2}. \tag{4.63}$$

Let us discuss two different cases:

(i) $\alpha = 1$, matched Hartmann-Hahn condition

$$\frac{M_{S\infty}}{M_{0S}} = \frac{\gamma_I}{\gamma_S} \frac{1}{1+\varepsilon}$$

which is the same as in Eq. (4.20) in the case of spin locking of the I spins. The gain in magnetization equals about the ratio of the gyromagnetic moments, however, multiple contacts may be achieved. The transfer time corresponds to milliseconds according to Eq. (4.10).

(ii) $\alpha \gg 1$, unmatched Hartmann-Hahn condition.

The S spin magnetization reaches a maximum according to Eq. (4.63) for $\alpha \cdot \sqrt{\varepsilon}$ $= 1$, with

$$\frac{M_S}{M_{0S_{\max}}} = \frac{1}{2} \frac{\gamma_I}{\gamma_S} \left[\frac{N_I I(I+1)}{N_S S(S+1)} \right]^{1/2} . \tag{4.64}$$

This maximum is reached when the heat capacities of the I and S spins are equal. By comparing Eq. (4.64) with (4.55) we realize that the same total S spin magnetization as in a multiple-contact cross-polarization experiment can be obtained in one shot [21b]. Such a one-shot polarization experiment with unmatched Hartmann-Hahn condition was first performed by A. Pines [31] on adamantane. Figure 4.12 shows schematically how this one-shot cross-polarization experiment can be performed [3b]. The interested reader may readily see by himself that the factor 1/2 vanishes in Eq. (4.64) if the B_{1S} field is turned on adiabatically. In this case entropy conservation according Eq. (4.3c) has to be invoked instead of energy conservation, as used above. Figure 4.13 demonstrates the dependence of M_S on α in the case of the methyl ^{13}C in CF_3COOAg [30]. The experimental data are obtained in a one shot cross-polarization experiment which includes T_{1D} effects to be discussed in the following. T_{1D} effects of the I spin dipole reservoir can severely limit the beauty of the one-shot cross-polarization experiment. Since $\alpha \gg 1$ the transfer time T_{IS} increases exponentially with α, according to Eqs. (4.10, 4.12) and may exceed T_{1D} way before the optimum value of α is reached. This is why in practice maximum S spin magnetization is barely observed.

Since T_{1D} is usually much shorter than T_1 in e.g. polymers it is advantageous to use the spin-locking version (a) with short contact times (1 ms). Tegenfeldt and Haeberlen [32] have recently improved the sensitivity of this experiment by flipping the I-spin magnetization back to the z direction at the end of the decoupling pulse. This allows a rapid repetition of the experiment and I-spin order can be better exploited this way.

In liquids where the isotropic part of $\mathcal{H}_{IS}$, namely the scalar coupling (J) is present, similar double resonance methods have been designed by Ernst and co-workers [33], Morris and Freeman [34] and by Garroway and co-workers [36]. Chingas et al. [36b, 37] have improved the method considerably by an adiabatic scheme which avoids to match the Hartmann-Hahn condition. J order and cross-polarization in scalar coupled spin systems has been discussed in a pictorial way, recently, by Packer and Wright [38].

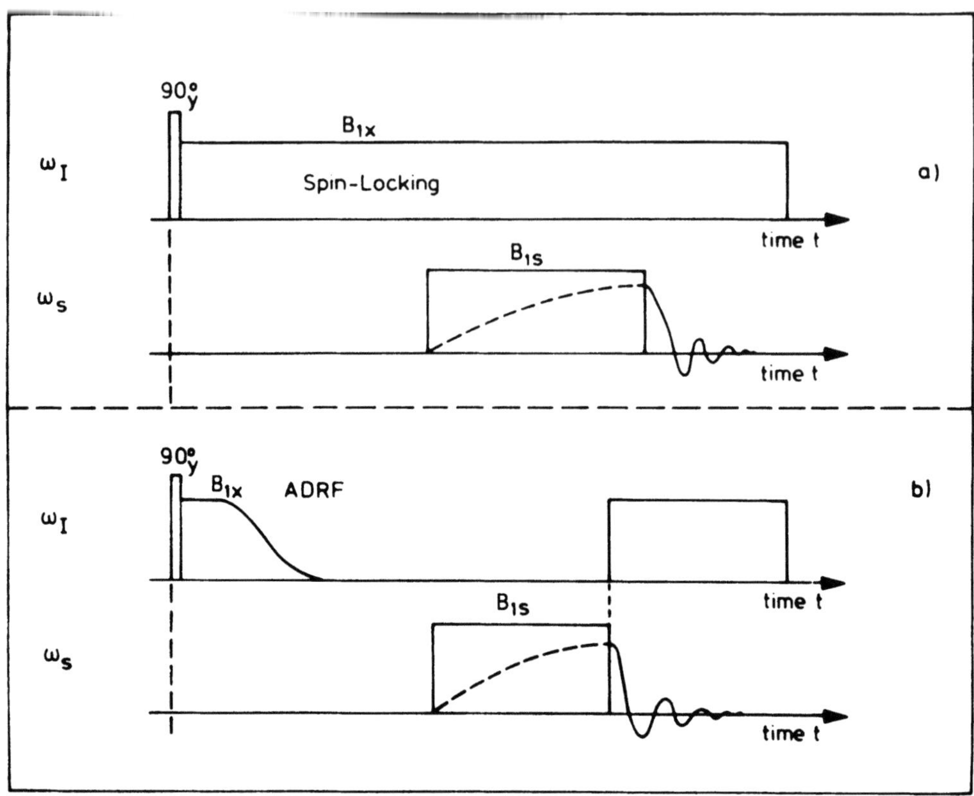

Fig. 4.12. Schematic representation of the one shot cross-polarization experiment according to A. Pines [31]

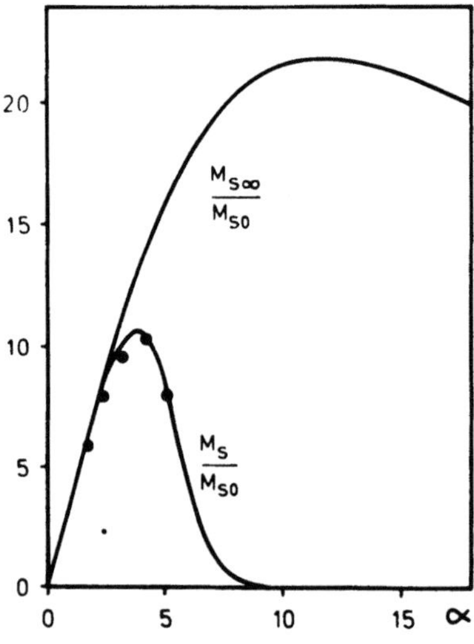

Fig. 4.13. S spin magnetization versus the mismatch parameter α in a one shot cross-polarization experiment employing ADRF. M_{S0} is the Boltzman magnetization and $M_{S\infty}$ corresponds to the equilibrium magnetization for infinite relaxation times [Eq. (4.63)]. In the case of finite relaxation time T_{1D} of the I spins the S spin magnetizations M_S does not reach its full maximum and its dependence on α is modified by T_{1D} and the cross-polarization time $T_{IS}(\alpha)$ according to Eq. (4.68c). The data points correspond to the maximum magnetization of methyl ^{13}C in a single crystal of CF_3COOAg oriented with methyl group C_3-axis at the magic angle with respect to the magnetic field, leaving only intermolecular dipolar interactions effective. (Courtesy of G. Sinning)

The dynamics of the cross-polarization process will be discussed in the next section. Chemical shift tensors which have been determined by this method are summarized in Chap. 5. Applications of the one-shot cross-polarization to relaxation studies will be presented in Chap. 6.

4.3 Cross-Polarization Dynamics

Let us start with a phenomenological approach to describe the variation of the inverse spin temperatures β_I and β_S of the I and S spins respectively in the "mixing" part of the two basic cross-polarization experiments as depicted in Fig. 4.12.

For the initial S spin-temperature, we assume $\beta_S = 0$ and recall for the I spin-temperature according to Sect. 4.1.

(a) $\beta_I = \beta_L \dfrac{B_0}{B_{1I}}$ spin locking $B_{1I} \gg B'_{LI}$

and

(b) $\beta_I = \beta_L \dfrac{B_0}{B_{LI}}$ ADRF.

If we invoke energy conservation in the rotating frame we can write

$$\frac{d}{dt}\beta_I + \varepsilon' \frac{d}{dt}\beta_S = 0 \tag{4.65}$$

where $\varepsilon' = \varepsilon\alpha^2$ is given by Eq. (4.61) of the preceding section, if B'_{LI} is replaced by B_{1I} in the case of spin locking. The S spin temperature β_S is relaxed with the time constant T_{IS} towards the instantanous I spin-temperature which results in the following coupled differential equations [10]

$$\frac{d}{dt}\beta_S = -\frac{1}{T_{IS}}(\beta_S - \beta_I) - \frac{1}{T_{1x}^{(S)}}\beta_S \tag{4.66a}$$

$$\frac{d}{dt}\beta_I = -\frac{\varepsilon'}{T_{IS}}(\beta_I - \beta_S) - \frac{1}{T_{1x}}\beta_I \tag{4.66b}$$

where the last term in Eq. (4.66b) has been added to account for relaxation of the I spins such as $T_{1\rho}$ (spin locking; $x = \rho$) or T_{1D} (ADRF; $x = D$) respectively.

The nature of T_{IS} and its dependence on H_{1S} will be discussed in more detail later in this section. Notice, however, that expressions for $1/T_{IS}$ have been quoted already in Eqs. (4.10) and (4.11). These coupled differential equations (4.66) are straightforwardly solved under the initial conditions:

$$\beta_S(0) = 0 \quad \text{and} \quad \beta_I(0) = \beta_{I0}.$$

$$\beta_S(t) = \beta_{I0}\frac{1}{a_+ - a_-}(e^{-a_- t/T_{IS}} - e^{-a_+ t/T_{IS}}) \tag{4.67a}$$

$$\beta_I(t) = \beta_{I0}\frac{1}{a_+ - a_-}[(1 - a_-)e^{-a_- t/T_{IS}} - (1 - a_+)e^{-a_+ t/T_{IS}}] \tag{4.67b}$$

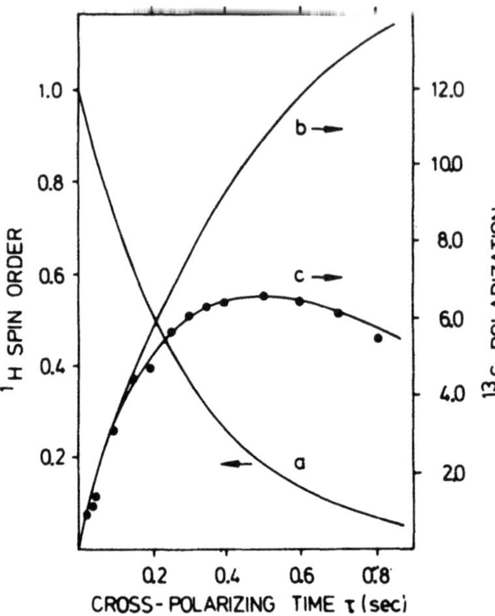

Fig. 4.14. Time evolution of ^{13}C polarization in adamantane at room temperature in a one shot cross-polarization experiment (Pines et al. [39]). The loss of ^{1}H spin order due to T_{1D} has been considered

where $a_\pm = a_0[1 \pm (1 - b/a_0^2)^{1/2}]$

with $a_0 = \frac{1}{2}(1 + \varepsilon\alpha^2 + T_{IS}/T_{1x}^{(I)} + T_{IS}/T_{1x}^{(S)});$ $b = \frac{T_{IS}}{T_{1x}^{(I)}}(1 + T_{IS}/T_{1x}^{(S)}) + \varepsilon\alpha^2 T_{IS}/T_{1x}^{(S)}$

and with $\alpha = \omega_{eS}/\omega_{LI};$ $\varepsilon = \dfrac{N_S S(S+1)}{N_I I(I+1)}.$

The variation of the magnetization $M_S(t)$ of the S spins with the coupling time is

$$\frac{M_S(t)}{M_{S0}} = \frac{\beta_S(t)B_{1S}}{\beta_L B_0} = \alpha \frac{\gamma_I}{\gamma_S} \frac{\beta_S(t)}{\beta_{I0}} \tag{4.68a}$$

where $\beta_S(t)/\beta_{I0}$ must be inserted according to Eq. (4.67a).

The S spin magnetization reaches a maximum at time t_m (see Figs. 4.14 and 4.15):

$$t_m = \frac{T_{IS}}{a_+ - a_-} \ln\left(\frac{a_+}{a_-}\right). \tag{4.68b}$$

Notice, that for large values of T_{1x}/T_{IS} one has to wait a long time compared with T_{IS} for the S spin signal to reach its maximum. However, remember that already 95 % of the maximum signal is reached after $3\,T_{IS}$. Other characteristic cases may be discussed by the interested reader. We turn now to the maximum S spin magnetization M_S obtainable at time t_m for a given value of the "mismatch parameter" α. Applying Eqs. (4.68a, b) we obtain

$$t = t_m: \quad \frac{M_S}{M_{S0}} = \alpha \frac{\gamma_I}{\gamma_S} \frac{1}{a_+ - a_-}\left[\left(\frac{a_-}{a_+}\right)^{\frac{a_-}{a_+ - a_-}} - \left(\frac{a_-}{a_+}\right)^{\frac{a_+}{a_+ - a_-}}\right]. \tag{4.68c}$$

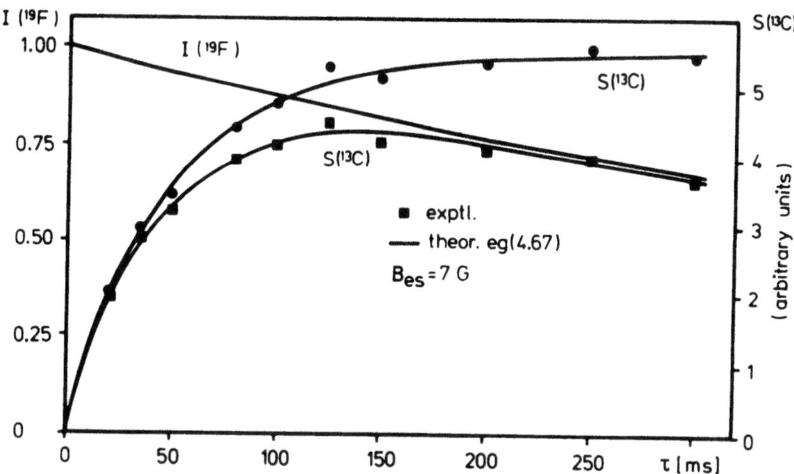

Fig. 4.15. Time evolution of methyl group ^{13}C polarization and ^{19}F order in a CF_3COOAg single crystal oriented as described in Fig. 4.13. The cross relaxation time T_{IS} (see Fig. 4.16a) and the maximum S spin magnetization M_S (see Fig. 4.13) are obtained from this graph, by fitting the data to Eq. (4.67a) (see Ref. [28])

In the case of negligible I spin relaxation ($T_{IS}/T_{1x} \ll 1$, $t_m = \infty$) we obtain the same result as for $M_{S\infty}/M_{S0}$ [see Eq. (4.63)]. This has been schematically drawn in Fig. 4.13. However, in practice, for large rf fields at the S spin resonance ($\alpha \gg 1$), the cross relaxation time T_{IS} becomes comparable to the I spin relaxation time in the rotating frame $T_{1x}^{(I)}$ (whether T_{1D} or $T_{1\rho}$) and Eqs. (4.68a, c) have to be applied. M_S/M_{S0} has been plotted in Fig. 4.13 versus the mismatch parameter α for the case of the methyl ^{13}C in CF_3COOAg.

The data included in Fig. 4.13 are obtained from a series of experiments analogous to the one represented in Fig. 4.15. Similarly one can discuss the I spin order represented by $\beta_I(t)/\beta_{I0}$, which may be deduced straightforwardly for different experimental situations by applying Eq. (4.67b).

Let us now discuss some special cases of Eq. (4.67)

(i) $\varepsilon = 0$; $T_{IS}/T_{1x} = 0$ i.e. vanishing heat capacity of the S spins; negligible T_{1x} relaxation of the I spins and S spins

$$\beta_S(t) = (1 - e^{-t/T_{IS}})\beta_{I0} \tag{4.69a}$$

$$\beta_I(t) = \beta_{I0}.$$

(ii) $\varepsilon = 0$; $T_{IS}/T_{1x}^{(I)} \neq 0$ i.e. same as (i), but $T_{1x}^{(I)}$ relaxation of the I spins.

$$\beta_S(t) = \frac{1}{1-\lambda}(1 - e^{-(1-\lambda)t/T_{IS}})e^{-t/T_{1x}^{(I)}}\beta_{I0} \tag{4.69b}$$

$$\beta_I(t) = e^{-t/T_{1x}^{(I)}} \cdot \beta_{I0}$$

where $\lambda = T_{IS}/T_{1x}^{(I)}$.

This case is usually met under extreme dilution of S spins, and with matched Hartmann-Hahn condition ($\alpha = 1$), but with short $T_{1x}^{(I)}$ relaxation time.

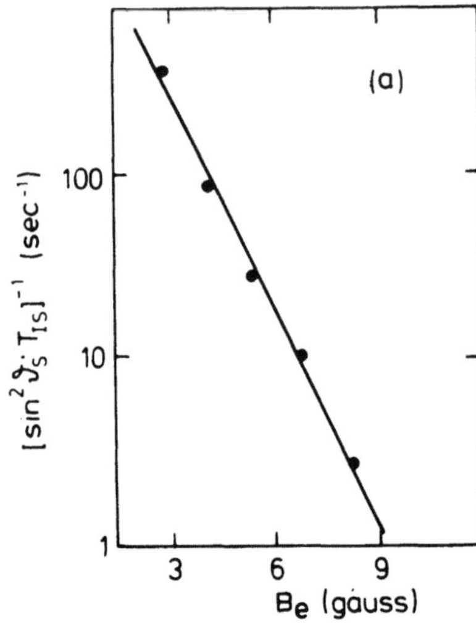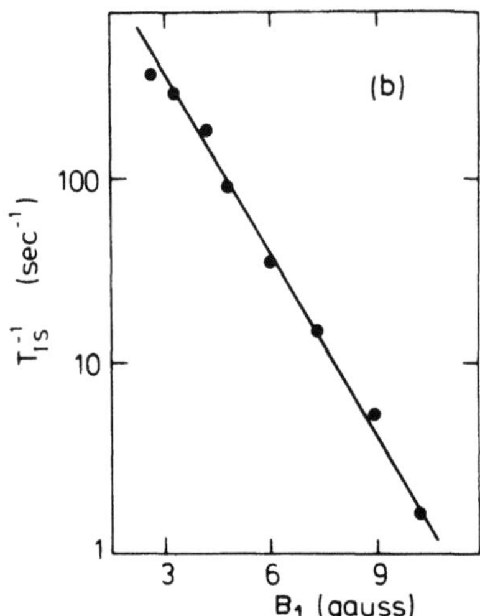

Fig. 4.16a, b. Cross relaxation rate $W_{IS} = 1/T_{IS}$ obtained in an ADRF one shot cross-polarization experiment versus the effective field B_e, B_1 at S spin frequency for (a) the methyl ^{13}C in a CF$_3$COOAg single crystal [20] and (b) ^{13}C in adamantane (A. Pines et al. [39]) as obtained from different sets of Figs. 4.14 and 4.15 with varying rf fields close to the ^{13}C resonance frequency. The straight lines correspond to an exponential cross relaxation spectrum Eq. (4.12) with correlation times $\tau_c = 150\,\mu s$ in the case of CF$_3$COOAg and $\tau_c = 140\,\mu s$ in the case of adamantane. (Fig. 4.16b courtesy of A. Pines)

(iii) $\varepsilon' \neq 0$; $\lambda = 0$ i.e., non-negligible heat capacity of the S spins, but negligible T_{1x} relaxation.

$$\beta_S(t) = \frac{1}{1+\varepsilon'}(1 - e^{-(1+\varepsilon')t/T_{IS}})\beta_{I0} \tag{4.70a}$$

$$\beta_I(t) = \frac{1}{1+\varepsilon'}(1 + \varepsilon' e^{-(1+\varepsilon')t/T_{IS}})\beta_{I0}. \tag{4.70b}$$

This case is usually met, no matter if the Hartmann-Hahn condition is matched or unmatched, as long as $T_{IS} \ll T_{1x}$.

In the one shot cross-polarization experiment, however, ε' usually becomes large and since the Hartmann-Hahn condition is not matched ($\alpha \gg 1$) T_{IS} is comparable with T_{1x} or $\lambda \cong 1$ i.e., Eq. (4.67) has to be used. This case is demonstrated in Fig. 4.14 for ^{13}C in adamantane and in Fig. 4.15 for the methyl ^{13}C in a single crystal of CF$_3$COOAg [30]. By fitting the experimental data to Eq. (4.67) the equilibrium magnetization M_S and the cross-relaxation time T_{IS} can be determined for different coupling fields B_{1S}, or values of ω_{eS} respectively. The cross-relaxation rate $W_{IS} = 1/T_{IS}$ depends strongly on B_{1S} more or less exponentially in the ADRF case, as suggested by McArthur et al. [10] [see Eq. (4.12)]. In Figs. 4.16 and 4.17 the modified cross-polarization rate $1/(T_{IS}\sin^2\vartheta_S)$ is plotted versus B_{1I}, B_{eS}, ν_{eS} according to Eqs. (4.10, 4.12) together with the experimental data, for those completely different systems as ^{13}C in adamantane, methyl ^{13}C in CF$_3$COOAg and ^{43}Ca in CaF$_2$. In all cases

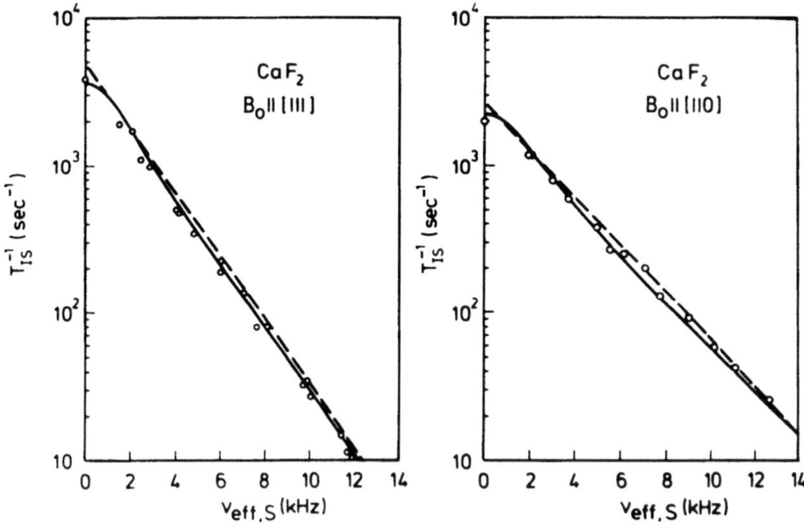

Fig. 4.17a, b. Cross relaxation rate $1/T_{IS}$ versus effective field at S spin resonance (^{43}Ca) under ADRF conditions in ^{43}Ca^{19}F$_2$ for two different orientations (**a**) $B_0\|(110)$, (**b**) $B_0\|(111)$. The data points are obtained by McArthur et al. [10] and the dashed line is their fit to an exponential spectrum Eq. (4.12). The solid lines were calculated by D. Demco, J. Tegenfeldt and J.S. Waugh [11] by using a memory function approach

good agreement between the experimental data and an exponential cross-polarization spectrum is obtained. Demco et al. [11], however, have shown, that this behavior is purely accidental and they have provided a more subtle discussion of cross-polarization spectra. We shall follow briefly the discussion of cross-polarization spectra as initiated by McArthur et al. [10] and improved by Demco, Tegenfeld and Waugh [11].

Let us apply an rf field, with

$$\mathcal{H}_{rf}(t) = -2\omega_{1I}I_x \cos \omega_I t - 2\omega_{1S}S_x \cos \omega_S t \tag{4.71}$$

to the coupled I-S spin system. The total Hamiltonian, neglecting spin lattice relaxation is conveniently expressed in the "tilted rotating frame" (TR), by using the following transformations [11]:

$$\mathbf{R} = \mathbf{R}_I \mathbf{R}_S; \quad \mathbf{R}_I = e^{i\omega_I I_z t}; \quad \mathbf{R}_S = e^{i\omega_S S_z t} \tag{4.72a}$$

$$\mathbf{T} = \mathbf{T}_I \mathbf{T}_S; \quad \mathbf{T}_I = e^{i\vartheta_I I_y}; \quad \mathbf{T}_S = e^{i\vartheta_S S_y}; \tag{4.72b}$$

with $\quad \tan \vartheta_I = \dfrac{\omega_{1I}}{\Delta\omega_I}; \ \tan \vartheta_S = \dfrac{\omega_{1S}}{\Delta\omega_S}$

where $\Delta\omega = \omega_0 - \omega; \ \omega_e^2 = \Delta\omega^2 + \omega_1^2$.

The total Hamiltonian in the "tilted rotating frame" may now be expressed as

$$\mathcal{H} = \mathcal{H}_I + \mathcal{H}_S + \mathcal{H}_{IS} \tag{4.73}$$

where

(i) spin locking case [11]

$$\mathcal{H}_I = -\omega_{eI} I_z + P_2(\cos \vartheta_I) \mathcal{H}_{II}^{(0)} + \mathcal{H}_{II}^{(ns)} \tag{4.74}$$

$$\mathcal{H}_S = -\omega_{eS} S_z \tag{4.75}$$

$$\mathcal{H}_{IS} = \cos \vartheta_I \cos \vartheta_S \mathcal{H}_{IzSz} + \sin \vartheta_I \sin \vartheta_S \mathcal{H}_{IxSx} - \sin \vartheta_I \cos \vartheta_S \mathcal{H}_{IxSz}$$
$$\qquad\qquad - \cos \vartheta_I \sin \vartheta_S \mathcal{H}_{IzSx}. \tag{4.76}$$

(ii) ADRF case [11]

$$\mathcal{H}_I = \mathcal{H}_{II}^{(0)}; \quad \mathcal{H}_S = -\omega_{eS} S_z \tag{4.77}$$

$$\mathcal{H}_{IS} = \cos \vartheta_S \mathcal{H}_{IzSz} - \sin \vartheta_S \mathcal{H}_{IzSx} \tag{4.78a}$$

with

$$\mathcal{H}_{I\mu S\upsilon} = \sum_{i,j} B_{ij} I_{\mu i} S_{\upsilon j} \tag{4.78b}$$

where

$$B_{ij} = \gamma_I \gamma_S \hbar (1 - 3 \cos^2 \vartheta_{ij})/r_{ij}^3. \tag{4.78c}$$

Because of $N_I \gg N_S$ and $\gamma_I \gg \gamma_S$ in the cases considered in this context, we have $\|\mathcal{H}_I\|; \|\mathcal{H}_S\| \gg \|\mathcal{H}_{IS}\|$ and $\mathcal{H}_{IS}$ may be considered as a small perturbation. $\mathcal{H}_I$ and $\mathcal{H}_S$ are orthogonal operators in the sense:

$$(\mathcal{H}_I | \mathcal{H}_S) = 0$$

where the definition

$$(\mathbf{A}|\mathbf{B}) \underset{\text{def}}{=} \mathrm{Tr}\{\mathbf{A}^+ \mathbf{B}\}$$

as in Sect. 3 and Appendix C has been used.

Expressing the total Hamiltonian as

$$\mathcal{H} = \mathcal{H}_1 + \mathcal{H}_2 + \mathcal{H}_p \tag{4.79}$$

where $\mathcal{H}_1 = \mathcal{H}_I$; $\mathcal{H}_2 = \mathcal{H}_S$; $\mathcal{H}_p = \mathcal{H}_{IS}$ will be assumed to be Hermitian in our case. Expanding the density matrix into orthogonal operators leads to

$$\rho = Z^{-1}[1 - \beta_1 \mathcal{H}_1 - \beta_2 \mathcal{H}_2 - \mathbf{R}] \tag{4.80}$$

with

$$(\mathcal{H}_1 | \mathcal{H}_2) = (\mathcal{H}_1 | \mathbf{R}) = (\mathcal{H}_2 | \mathbf{R}) = 0. \tag{4.81}$$

An equation of motion may be derived for the expectation value of the operators $\mathcal{H}_i$; $i = 1, 2$.

$$\langle \mathcal{H}_i \rangle = (\mathcal{H}_i | \rho), \quad i = 1, 2. \tag{4.82}$$

Using the first Born approximation in the time evolution operator

$$\mathbf{S}(t) = \exp[-i(1-P)(\hat{\mathcal{H}}_1 + \hat{\mathcal{H}}_2 + \hat{\mathcal{H}}_p) t] \tag{4.83a}$$

we arrive at

$$\mathbf{S}(t) \cong \mathbf{S}_0(t) = \exp[-i(1-P)(\hat{\mathcal{H}}_1 + \hat{\mathcal{H}}_2) t] \tag{4.83b}$$

where the projector

$$P = \frac{|\mathcal{H}_1)(\mathcal{H}_1|}{(\mathcal{H}_1|\mathcal{H}_1)} + \frac{|\mathcal{H}_2)(\mathcal{H}_2|}{(\mathcal{H}_2|\mathcal{H}_2)} \tag{4.84}$$

has been used and where $\hat{\mathcal{H}}$ denotes a Liouvillian. We arrive at the following integro-differential equation (see Appendix C)

$$\frac{d}{dt}\langle\mathcal{H}_i(t)\rangle = -\left\{\int_0^t dt'\, K_{ii}(t-t')\,\langle\mathcal{H}_i(t')\rangle + \int_0^t dt'\, K_{ij}(t-t')\,\langle\mathcal{H}_j(t')\rangle\right\}$$

$$i,j = 1,2$$
$$i \neq j \tag{4.85}$$

where

$$K_{ii}(t-t') = \frac{(\mathcal{H}_i|\,\hat{\mathcal{H}}_p\,S_0(t,t')\,\hat{\mathcal{H}}_p\,|\mathcal{H}_i)}{(\mathcal{H}_i|\mathcal{H}_i)} \tag{4.86}$$

$$K_{ij}(t-t') = \frac{(\mathcal{H}_i|\,\hat{\mathcal{H}}_p\,S_0(t,t')\,\hat{\mathcal{H}}_p\,|\mathcal{H}_j)}{(\mathcal{H}_j|\mathcal{H}_j)}. \tag{4.87}$$

In the fast correlation limit ($\tau_c \ll t, t-t' = \tau$) this may be expressed as (see Appendix C)

$$\frac{d}{dt}\langle\mathcal{H}_i(t)\rangle = -\left\{\int_0^\infty d\tau\, K_{ii}(\tau)\,\langle\mathcal{H}_i(t)\rangle + \int_0^\infty d\tau\, K_{ij}(\tau)\,\langle\mathcal{H}_j(t)\rangle\right\}. \tag{4.88}$$

Introducing the new thermodynamic coordinates [11]

$$\beta_i = \frac{(\mathcal{H}_i|\rho(t))}{(\mathcal{H}_i|\mathcal{H}_i)} = \frac{\langle\mathcal{H}_i(t)\rangle}{(\mathcal{H}_i|\mathcal{H}_i)} \tag{4.89}$$

and the correlation function $C(\mathcal{H}_i, \mathcal{H}_j, \tau)$, this results readily in the same pair of differential equations as Eq. (4.66), which was obtained by a phenomenological approach:

$$\frac{d}{dt}\beta_i(t) = -\int_0^\infty d\tau\, C(\mathcal{H}_i;\,\mathcal{H}_i;\,\tau)\,\beta_i(t) - \int_0^\infty d\tau\, C(\mathcal{H}_i;\,\mathcal{H}_j;\,\tau)\,\beta_j(t)$$

$$i,j = 1,2$$
$$i \neq j \tag{4.90}$$

where

$$C(\mathcal{H}_i;\,\mathcal{H}_i;\,\tau) = K_{ii}(\tau)$$

and

$$C(\mathcal{H}_i;\,\mathcal{H}_j;\,\tau) = \frac{(\mathcal{H}_j|\mathcal{H}_j)}{(\mathcal{H}_i|\mathcal{H}_i)}\, K_{ij}(\tau).$$

Using energy conservation

$$\frac{d}{dt}[\langle\mathcal{H}_1\rangle + \langle\mathcal{H}_2\rangle] = 0.$$

Equation (4.90) can be casted into the form of Eq. (4.66) for vanishing spin lattice relaxation by using

$$\varepsilon' = \frac{(\mathscr{H}_2|\mathscr{H}_2)}{(\mathscr{H}_1|\mathscr{H}_1)} \tag{4.91}$$

and

$$\frac{1}{T_{IS}} = \int\limits_0^\infty d\tau\, a(\tau) \tag{4.92}$$

where

$$a(\tau) = K_{22}(\tau) = \frac{(\mathscr{H}_2|\,\hat{\mathscr{H}}_p S_0(\tau)\,\hat{\mathscr{H}}_p\,|\mathscr{H}_2)}{(\mathscr{H}_2|\mathscr{H}_2)}. \tag{4.93}$$

The cross relaxation rate $1/T_{IS}$ is the issue of this discussion and we are setting out to calculate it by obtaining an expression for the correlation function $a(\tau)$. A tedious, but straightforward calculation, using Eqs. (4.74–4.78, 4.83) leads to

(i)   spin locking case [11]

$$a(\tau) = \sin^2\vartheta_I \sin^2\vartheta_S\, a_x(\tau)\, \tfrac{1}{2}[\cos(\omega_{eS}-\omega_{eI})\,\tau + \cos(\omega_{eS}+\omega_{eI})\,\tau]$$
$$+ \cos^2\vartheta_I \sin^2\vartheta_S\, a_z(\tau)\cos\omega_{eS}\tau. \tag{4.94}$$

(ii)   ADRF case [11]

$$a(\tau) = \sin^2\vartheta_S\, a_z(\tau)\cos\omega_{eS}\tau \tag{4.95}$$

where in the two different cases (i) and (ii) we define

$$a_z(\tau) = (\sum_i B_i I_{zi}|\, e^{-i P_2(\cos\vartheta_I)\,\hat{\mathscr{H}}_{II}^{(0)}\tau}\,|\sum_i B_i I_{zi}) \tag{4.96a}$$

and

$$a_x(\tau) = (\sum_i B_i I_{xi}|\, e^{-i P_2(\cos\vartheta_I)\,\hat{\mathscr{H}}_{II}^{(0)}\tau}\,|\sum_i B_i I_{xi}). \tag{4.96b}$$

In the ADRF case (ii) we set especially $\cos\vartheta_I = 1$.

It may be convenient to introduce the second moment M_2^{IS} of the heteronuclear dipole coupling as

$$M_2^{IS} = \frac{(S_x|\,\hat{\mathscr{H}}_{IS}^2\,|S_x)}{(S_x|S_x)} = \mathrm{Tr}\{(\sum_i B_i I_{zi})^2\}. \tag{4.97}$$

Let us introduce the correlation functions $C_x(\tau)$ and $C_z(\tau)$ [11]:

$$C_x(\tau) = a_x(\tau)/M_2^{IS}$$
$$C_z(\tau) = a_z(\tau)/M_2^{IS}.$$

i.e.

$$C_\mu(\tau) = \frac{(\sum_i B_i\cdot I_{\mu i}|\, e^{-i P_2(\cos\vartheta_I)\,\hat{\mathscr{H}}_{II}^{(0)}\tau}\,|\sum_i B_i I_{\mu i})}{(\sum_i B_i I_{\mu i}|\sum_i B_i I_{\mu i})} \qquad \mu = x, z. \tag{4.98}$$

Combining Eqs. (4.92–4.98) we are now well equipped to express the cross-polarization rate $1/T_{IS}$ by the cross-polarization spectral densities

$$J_x(\omega) = \int\limits_0^\infty d\tau \cos \omega\tau\, C_x(\tau), \tag{4.99a}$$

$$J_z(\omega) = \int\limits_0^\infty d\tau \cos \omega\tau\, C_z(\tau) \tag{4.99b}$$

as

(i) spin locking case

$$\frac{1}{T_{IS}} = \sin^2 \vartheta_S\, M_2^{IS}[\cos^2 \vartheta_I\, J_z(\omega_{eS})$$
$$+ \sin^2 \vartheta_I\, \tfrac{1}{2}[J_x(\omega_{eS}-\omega_{eI})+J_x(\omega_{eS}+\omega_{eI})]]. \tag{4.100}$$

Since $J_x(\omega)$; $J_z(\omega)$ approach zero for $\omega\tau_c \gg 1$, we find it to be a good approximation in the case $\omega_{eS}, \omega_{eI} \cong 1/T_{2I}$ and $\tau_c \cong T_{2I}$ to write, following DTW [11]

$$\frac{1}{T_{IS}} = \tfrac{1}{2}\sin^2 \vartheta_S \sin^2 \vartheta_I\, M_2^{IS}\, J_x(\Delta\omega_e). \tag{4.101}$$

In the

(ii) ADRF case [10, 11] we obtain

$$\frac{1}{T_{IS}} = \sin^2 \vartheta_S\, M_2^{IS}\, J_z(\omega_{eS}). \tag{4.102}$$

For a full description of the cross-polarization dynamics it is mandatory to calculate the cross-polarization spectral densities $J_x(\omega)$ and $J_z(\omega)$.

Only in exceptional cases it is possible to calculate $C_x(\tau)$ and $C_z(\tau)$ exactly. In order to describe experimental results by the current theory we will have to employ a certain degree of approximation.

This problem was attacked in different ways by "guessing" the functional form of e.g. $C_z(\tau)$ by McArthur, Hahn and Walstedt [10] to be Lorentzian. This special functional form was suggested by the experimental data (CaF_2) and was confirmed theoretically later by Demco, Tegenfeldt and Waugh [11] using the Mori-Zwanzig memory function approach [40, 41].

Approximations of $C_x(\tau)$ and $C_z(\tau)$

For the convenience of the reader we recall Eq. (4.98) here in a slightly different notation

$$C_\mu(\tau) = \frac{\mathrm{Tr}\{(\sum\limits_i B_i I_{\mu i})\, e^{-i P_2(\cos \vartheta_I)\, \mathscr{H}_{II}^{(0)}\tau}(\sum\limits_i B_i I_{\mu i})\}}{\mathrm{Tr}\{(\sum\limits_i B_i I_{\mu i})^2\}} \qquad \mu = x, z \tag{4.103}$$

where in the ADRF case $\cos \vartheta_I = 1$.

A moment expansion of $C_x(\tau)$ and $C_z(\tau)$ can, of course, be done exactly, leading, however, to more and more complicated expressions which become practically unmanagable when moments higher than the sixth moment are involved. Already the fourth moment of $C_{x,z}(\tau)$ contains lattice sums which correspond to the sixth moment of the free induction decay. Low order moments, however, like the second and fourth moment may be conveniently used to adjust approximations to the correct expressions.

A sort of zeroth order approximation would be to guess the functional form of $C_x(\tau)$ and $C_z(\tau)$. Although this may not be very appealing theoretically, it may, however, be very useful, if this method succeeds in representing experimental data. The guessing, of course, has to be strongly supported by the experimental data.

Taking a closer look at the expressions for $C_x(\tau)$ and $C_z(\tau)$, we recognize, that in the case of $B_i = \text{const}$ the correlation function $C_x(\tau)$ would represent the free induction decay of the I spins, where the time scale is scaled by $P_2(\cos \vartheta_I)$. Since the B_i vary slightly in a cubic solid an additional "inhomogeneous broadening" (*cum grano salis*) is introduced which inhibits slightly I spin flip-flop processes. Since free induction decays of abundant spins in cubic solids are often close to a Gaussian, it is the best guess to represent $C_x(\tau)$ by a Gaussian function. This is confirmed by the experimental data and also by higher order approximations to be discussed shortly. $C_z(\tau)$, however, depends merely on the local character of the I-S coupling; note, that $C_z(\tau) = 1$ if $B_i = \text{const}$. Since flip-flop processes among the I spins play a dominant role in $C_z(\tau)$ a sort of Anderson Weiss model would be appropriate, leading to a more or less Lorentzian character of $C_z(\tau)$, which was confirmed experimentally and is also displayed in higher order approximation [11]. Summarizing these arguments we are tempted to write:

$$C_x(\tau) = \exp(-\tau^2/\tau_c^2) \quad \text{with} \quad J_x(\omega) = \frac{\tau_c}{2}\sqrt{\pi}\,\exp(-\omega^2\tau_c^2/4) \tag{4.104a}$$

and

$$C_z(\tau) = [1 + (\tau/\tau_c)^2]^{-1} \quad \text{with} \quad J_z(\omega) = \frac{\tau_c}{2}\exp(-\omega\tau_c) \tag{4.104b}$$

This leads to the following approximate expression for $1/T_{IS}$

(i) spin-locking

$$\frac{1}{T_{IS}} = \frac{\sqrt{\pi}}{4}\sin^2\vartheta_S \sin^2\vartheta_I M_2^{IS}\tau_c \exp(-\Delta\omega_e^2\tau_c^2/4) \tag{4.105a}$$

where

$$\Delta\omega_e = \omega_{eS} - \omega_{eI}.$$

(ii) ADRF

$$\frac{1}{T_{IS}} = \tfrac{1}{2}\sin^2\vartheta_S M_2^{IS}\tau_c \exp(-\omega_{eS}\tau_c). \tag{4.105b}$$

In both cases the correlation time τ_c may be expressed by the second moment N_2 of $C_x(\tau)$ and $C_z(\tau)$, respectively, as

$$\frac{1}{\tau_c^2} = \tfrac{1}{2}N_2. \tag{4.106}$$

Using Eq. (4.103) a straightforward calculation yields for

$$C_x(\tau):\ N_2 = \tfrac{1}{3} I(I+1)\, P_2(\cos\vartheta_I)^2 \cdot \frac{5\sum\limits_{i<j} A_{ij}^2 (B_i^2 + B_j^2) + 8 \sum\limits_{i<j} A_{ij}^2 B_i B_j}{\sum\limits_i B_i^2} \tag{4.107}$$

and for

$$C_z(\tau):\ N_2 = \tfrac{2}{3} I(I+1)\, P_2(\cos\vartheta_I)^2 \frac{\sum\limits_{i<j} A_{ij}^2 (B_i - B_j)^2}{\sum\limits_i B_i^2} \tag{4.108}$$

where

$$B_i = \gamma_I \gamma_S \hbar (1 - 3\cos^2\vartheta_i)/r_i^3$$

and

$$\mathcal{H}_{II} = \sum_{i<j} A_{ij}(3 I_{zi} I_{zj} - \mathbf{I}_i \cdot \mathbf{I}_j)$$

with

$$A_{ij} = \tfrac{1}{2}\gamma_I^2 \hbar (1 - 3\cos^2\vartheta_{ij})/r_{ij}^3.$$

In the ADRF case we set $P_2(\cos\vartheta_I) = 1$.

Equations (4.107) and (4.108) may be expressed in a more compact form, which is more suitable for a numerical computation as

$$C_x(\tau):\ N_2 = \tfrac{1}{24} I(I+1)\, P_2(\cos\vartheta_I)^2 \frac{\gamma_I^4 \hbar^2}{a^6}\, \frac{5S_4 + 18 S_3}{S_1} \tag{4.109a}$$

and

$$C_z(\tau):\ N_2 = \tfrac{1}{12} I(I+1)\, P_2(\cos\vartheta_I)^2 \frac{\gamma_I^4 \hbar^2}{a^6} \cdot \frac{S_4}{S_1} \tag{4.109b}$$

using the lattice sums, defined as follows [11]:

$$
\begin{aligned}
S_1 &= \sum_i b_i^2 & S_7 &= \sum_{i\neq j\neq k} b_{ij}^2 b_{ik}^2 b_i^2 \\
S_2 &= \sum_i b_i^4; \quad S_2^* = \sum_i b_{ij}^4 & S_8 &= \sum_{i\neq j\neq k} b_{ij}^2 b_{ik}^2 b_i b_j \\
S_3 &= \sum_{i\neq j} b_{ij}^2 b_i b_j & S_9 &= \sum_{i\neq j\neq k} b_{ij}^2 b_{ik}^2 b_j b_k \\
S_4 &= \sum_{i\neq j} b_{ij}^2 (b_i - b_j)^2 & S_{10} &= \sum_{i\neq j\neq k} b_{ij}^2 b_{ik} b_{jk} b_k^2 \\
S_5 &= \sum_{i\neq j} b_{ij}^4 b_i b_j & S_{11} &= \sum_{i\neq j\neq k} b_{ij}^2 b_{ik} b_{jk} b_i b_j \\
S_6 &= \sum_{i\neq j} b_{ij}^4 (b_i - b_j)^2 & S_{12} &= \sum_{i\neq j\neq k} b_{ij}^2 b_{ik} b_{jk} b_j b_k
\end{aligned}
\tag{4.110}
$$

where

$$b_{ij} = (1 - 3\cos^2\vartheta_{ij})/r_{ij}^3 \tag{4.111a}$$

$$b_i = (1 - 3\cos^2\vartheta_i)/r_i^3 \tag{4.111b}$$

with r_i being the distance between I and S spins and r_{ij} the distance between the spins I_i and I_j and where both distances are defined in units of the lattice constant a. The angle of the corresponding vector with the static magnetic field B_0 is denoted by $\cos\vartheta_i$ and $\cos\vartheta_{ij}$ respectively.

Different correlation times τ_c calculated according to Eqs. (4.106, 4.109) in the case of CaF_2 are compared with experimental values for different orientations of the magnetic field. The agreement is quite convincing (see Table 4.2).

Nevertheless, Demco et al. [11] went a step further and applied the memory function approach to the calculation of the correlation functions $C_x(\tau)$ and $C_z(\tau)$. Recalling the discussion in Sect. 2.5 and Appendix C, we may write similar to Eq. (2.108)

$$\frac{d}{dt}\,C_\mu(t) = -\int\limits_0^t dt'\, K(t-t')\, C_\mu(t') \tag{4.112}$$

or expressing the corresponding spectral distribution function $J(\omega)$ following Eq. (2.112) directly as

$$J(\omega) = \frac{K'(\omega)}{[\omega - K''(\omega)]^2 + [K'(\omega)]^2} \tag{4.113}$$

where

$$K'(\omega) = \int\limits_0^\infty K(\tau)\cos\omega\tau\, d\tau$$

and

$$K''(\omega) = \int\limits_0^\infty K(\tau)\sin\omega\tau\, d\tau.$$

Approximation of $K(\tau)$ by a Gaussian function which can be rationalized by the calculation of the ratios of higher moments [11] leads to $K'(\omega)$ and $K''(\omega)$ equal to Eq. (2.117). Thus the second and the fourth moment N_2 and N_4 is all that is needed to calculate $J(\omega)$ in this degree of approximation. Expressions for the fourth moments of the correlation function can be obtained from Eq. (4.103) as

$$C_x(\tau): N_4 = [P_2(\cos\vartheta_l)]^4\,\frac{1}{3}\,\frac{I(I+1)}{16S_1}\,\{\tfrac{1}{5}[101\,I(I+1)-\tfrac{49}{2}]\,S_1 S_2^*$$
$$+\tfrac{1}{10}[176\,I(I+1)-32]\,S_5 + \tfrac{1}{3}I(I+1)\,[77S_7 + 88S_8 + 24S_9$$
$$+8S_{10} + 38S_{11} + 8S_{12}]\} \tag{4.114}$$

$$C_z(\tau): N_4 = [P_2(\cos\vartheta_l]^4\,\frac{1}{3}\,\frac{I(I+1)}{8S_1}\,\{\tfrac{1}{5}[16I(I+1)-7]\cdot(S_1\cdot S_2^* - S_5)$$
$$+\tfrac{1}{3}I(I+1)\,[13S_7 - 16S_8 + 3S_9 - 8S_{10} + 16S_{11} - 8S_{12}]\}. \tag{4.115}$$

Figure 4.17 compares those calculations with the experimental data of McArthur et al. [10]. Note, that no adjustable parameter of any kind was used in this theory. The agreement between theory and experiment is quite pleasing. Figure 4.18 shows theoretical cross-polarization rates calculated by Demco et al. [11].

Table 4.2. Correlation times τ_c for the dipolar fluctuation autocorrelation functions $(1+\tau^2/\tau_c^2)^{-1}$ and $\exp(-\tau^2/\tau_c^2)$ for a $^{43}CaF_2$ crystal according to Demco, Tegenfeldt and Waugh (DTW) [11]. The correlation time $\tau_c=(2/N_2)^{1/2}$ is calculated, using Eq. (4.109) and compared with the experimental data of McArthur et al. [10]

Direction of B_0	τ_c (µs)		
	exp (ADRF)[10]	DTW[11] (ADRF)	DTW[11] $\left(SL, \vartheta_I=\dfrac{\pi}{2}\right)$
(100)	–	45	37
(110)	57 ± 0.5	61	64
(111)	78 ± 1; $80+1$	81	102

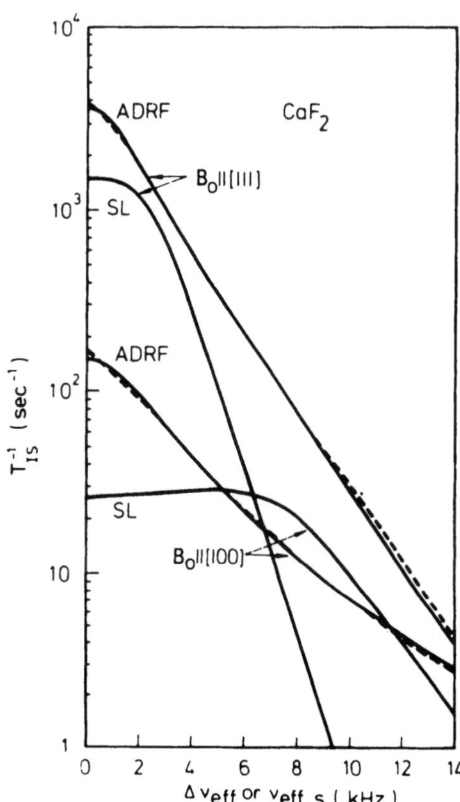

Fig. 4.18. Theoretical cross-polarization rates for different orientations of the CaF_2 crystal as calculated by D.E. Demco et al. [11] for the spin-locking (SL) and ADRF case employing a memory function approach (solid lines) and an information theory approach (dashed line)

Short time behavior:

Two dramatic transient effects have appeared in this regime which have to be accounted for theoretically. The first one to be discussed was observed by Strombotne and Hahn [6, 40] in the laboratory frame and by McArthur, Walstedt and Hahn [10] in the rotating frame (see Fig. 4.19). These effects have been treated by the spin temperature concept in the past, describing them

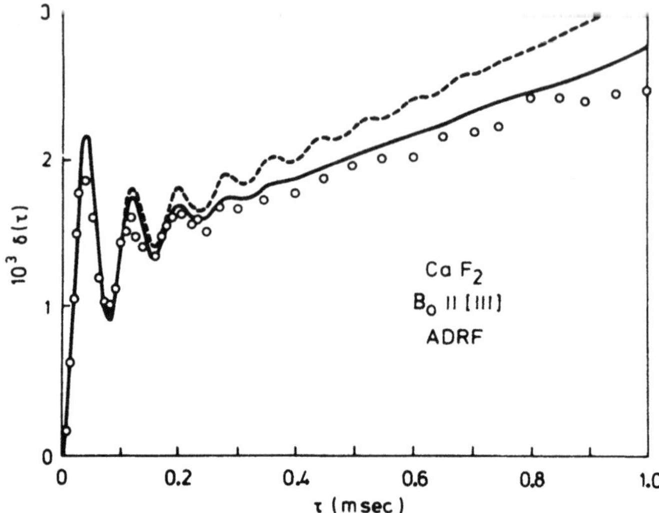

Fig. 4.19. Transient oscillations in the rotating frame in ^{43}Ca^{19}F$_2$ under ADRF conditions as observed by McArthur et al. [10] (open circles) with $v_{eff\,S} = 12.6$ kHz. The fractional decrease $\delta(\tau)$ of I spin (^{19}F) order, when a mixing pulse is applied to the S spins (^{43}Ca) was calculated by McArthur et al. [10] by using a Lorentzian autocorrelation function with $\tau_c = 77\,\mu$s (dashed curve) and by Demco et al. [11] by using their autocorrelation function obtained by a memory function approach (solid curve)

as an adjustment of the spin temperatures between the Zeeman reservoir and the non-secular dipole reservoir. The approach taken here (DTW [11]) is more rigorous, since no explicit spin temperature assumption is involved.

The second transient effect in the short time behavior was observed by R. Ernst [43] and co-workers (see Fig. 4.20) and is due to the discrete I-S dipole spectrum, damped by spin diffusion among the abundant I spins.

Recently J.S. Waugh [44] and co-workers have exploited this effect and designed a sequence which applies rf irradiation at the "magic angle" of the I spins in order to reduce the damping of these oscillations. This sequence is sketched in Fig. 4.21 together with the high-resolution I-S dipole spectrum obtained [44] (Fig. 4.22).

This technique was exploited recently to determine heteronuclear dipole-dipole interactions, leading to quite a precise measurement of bond distances of e.g. C–H bonds [45]. Two-dimensional spectroscopy based on this technique is discussed in more detail in the following chapter.

Let us first try to understand the oscillatory behaviour shown in Fig. 4.19. We start with Eq. (4.85) where we replaced in the right hand side.

$$\langle \mathcal{H}_{i,j}(t') \rangle \quad \text{by} \quad \langle \mathcal{H}_{i,j}(0) \rangle.$$

Integration leads to

$$\langle \mathcal{H}_i(t) \rangle - \langle \mathcal{H}_i(0) \rangle = -\int_0^t dt_2 \int_0^{t_2} dt_1 \, K_{ii}(t_2 - t_1) \langle \mathcal{H}_i(0) \rangle$$

$$-\int_0^t dt_2 \int_0^{t_2} dt_1 \, K_{ij}(t_2 - t_1)\,(i)\,\langle \mathcal{H}_j(0) \rangle. \tag{4.116}$$

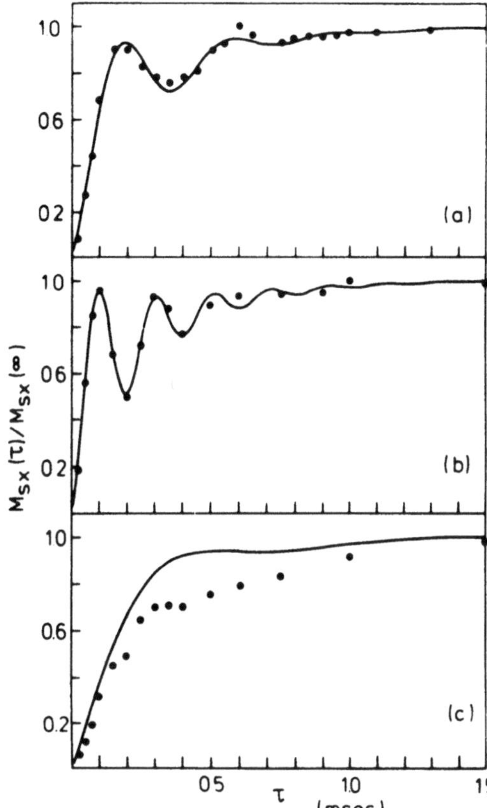

Fig. 4.20. Normalized S spin magnetization $M_{Sx}(\tau)$ of ^{13}C in a ferrocene single crystal for three different orientations versus coupling time τ in a one shot cross-polarization experiment of the spin locking type, with $v_{1I} = v_{1S} = 16\,kHz$. (R.R. Ernst and co-workers [43].) The transient oscillations of the ^{13}C magnetization correspond to the I-S dipolar interaction of next nearest protons. The damping of the oscillations is caused by spin diffusion. For further details see Ref. [43]

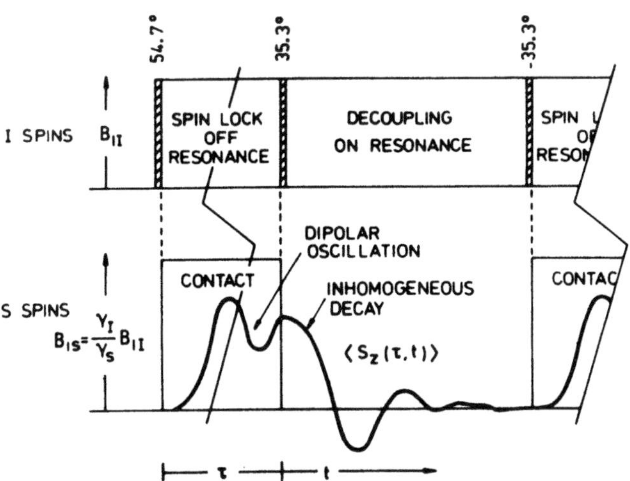

Fig. 4.21. A scheme proposed by J.S. Waugh and co-workers [44] for observing resolved dipolar coupling spectra of dilute nuclear spins in solids

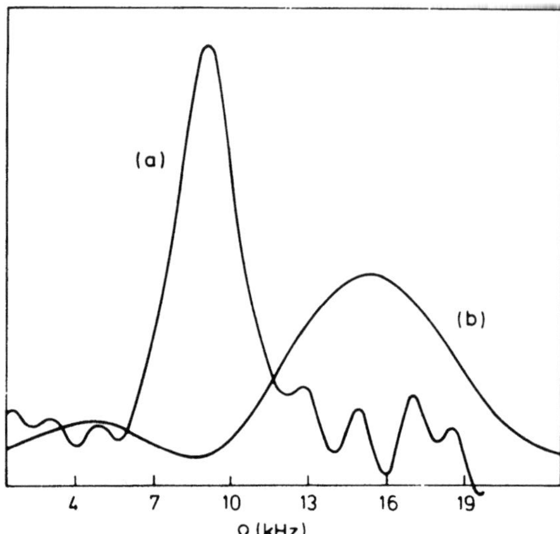

Fig. 4.22a, b. ^{13}C-^{1}H dipolar spectra obtained by J.S. Waugh and co-workers [44] at a certain orientation of an ammonium tartrate single crystal, employing the scheme of Fig. 4.21. The observed line (a) at a dipolar frequency $\Omega = 9.7\,kHz$ arises from the C–H coupling of the CHOH group with off-resonance spin-locking of the protons, whereas, in case (b) resonant spin-locking of the protons is applied

With the initial condition

$$\langle \mathcal{H}_2(0)\rangle = 0; \quad \langle \mathcal{H}_1(0)\rangle \neq 0$$

we obtain

$$\frac{\langle \mathcal{H}_2(t)\rangle}{\langle \mathcal{H}_1(0)\rangle} = -\int_0^t dt_2 \int_0^{t_2} dt_1\, K_{21}(t_2 - t_1).$$

Using energy conservation, the equation above is conveniently expressed as

$$\frac{\langle \mathcal{H}_2(t)\rangle}{\langle \mathcal{H}_1(0)\rangle} = \varepsilon' \int_0^t dt_2 \int_0^{t_2} dt_1\, a(\tau) \tag{4.117}$$

where $a(\tau)$ has been defined above for the two different cases (i) spin locking and (ii) ADRF [Eqs. (4.94–4.98)]. It might be more convenient to express the energy $\langle \mathcal{H}_2(t)\rangle$ in terms of the equilibrium energy $\langle \mathcal{H}_{2,eq}\rangle = \langle \mathcal{H}_2(t=\infty)\rangle$. It is easily recognized that

$$\langle \mathcal{H}_1(0)\rangle = \frac{1+\varepsilon'}{\varepsilon'} \langle \mathcal{H}_{2,eq}\rangle$$

and with

$$\frac{\langle \mathcal{H}_2(t)\rangle}{\langle \mathcal{H}_{2,eq}\rangle} = \frac{M_S}{M_{S,eq}}$$

we rewrite Eq. (4.104) as [11]

$$\frac{M_S}{M_{S,eq}} = (1+\varepsilon') \int_0^t dt_2 \int_0^{t_2} dt_1\, a(\tau). \tag{4.118}$$

As shown above [Eq. (4.94–4.98)] $a(\tau)$ is an oscillatory function, even when the correlation functions $a_x(\tau)$ and $a_z(\tau)$ or $C_x(\tau)$, $C_z(\tau)$ respectively, are monotonic functions due to a more or less isotropic I spin surrounding of the rare S spins, which is e.g. the case in a cubic crystal. The oscillation frequencies are ω_{eS} $\pm\omega_{eI}$ and ω_{eS} as was demonstrated experimentally [10] (see Fig. 4.18). If the I-S interactions is more discrete i.e., a few next neighbour I spins have a distinct dipolar coupling to a single S spin of strength much larger than any other I-S interaction, oscillating correlation functions $C_x(\tau)$ and $C_z(\tau)$ are obtained, producing oscillations in the cross-polarization S spin magnetization. This last effect was observed by R. Ernst and co-workers [43] as was mentioned above and has been demonstrated in Fig. 4.19.

Let us try to understand this experiment on a simple basis. The interesting aspect is the time-evolution of the S spin magnetization $M_S(t)$ during contact time with the I spins. The total Hamiltonian of the problem is given by

$$\mathcal{H} = \mathcal{H}_I + \mathcal{H}_S + \mathcal{H}_{IS} + \mathcal{H}_{II}$$

where

$$\mathcal{H}_I = \omega_{1I} I_x; \quad \mathcal{H}_S = \omega_{1S} S_x.$$

$\mathcal{H}_{IS}$ and $\mathcal{H}_{II}$ are the usual heteronuclear and homonuclear spin interactions. The ratio of the timedependent S spin magnetization $M_S(t)$ and the Boltzmann magnetization M_{SB} can be expressed as

$$M_S(t)/M_{SB} = (\rho(t)|S_x)/(\rho_B|S_z) \tag{4.119}$$

where in the high temperature approximation

$$\rho_B = (1 + \beta_L \omega_{0S} S_z)/\mathrm{Tr}\{1\}$$
$$\rho(t=0) = (1 + \beta_I \omega_{1I} I_x)/\mathrm{Tr}\{1\}$$

with

$$\beta_I = \beta_L \omega_{0I}/\omega_{1I}.$$

Using

$$\rho(t) = e^{-i\mathcal{H}t}\,\rho(t=0)\,e^{i\mathcal{H}t}$$

and $\omega_{0I}/\omega_{0S} = \gamma_I/\gamma_S$ leads to

$$M_S(t)/M_{SB} = \frac{\gamma_I}{\gamma_S}(I_x|\,e^{i\tilde{\mathcal{H}}t}\,|S_x)/(S_z|S_z). \tag{4.120}$$

We now extract $\mathcal{H}_I$ and $\mathcal{H}_S$ by writing

$$e^{-i\mathcal{H}t} = e^{-i\mathcal{H}_It}\,e^{-i\mathcal{H}_St}\,T e^{-i\int_0^t dt'\,[\tilde{\mathcal{H}}_{IS}(t') + \tilde{\mathcal{H}}_{II}(t'')]}$$

where

$$\tilde{\mathcal{H}}(t) = e^{i(\mathcal{H}_I + \mathcal{H}_S)t}\,\mathcal{H}\,e^{-i(\mathcal{H}_I + \mathcal{H}_S)t}.$$

Let us apply average Hamiltonian theory as discussed in Appendix F in zeroth order. In this case we obtain

$$\bar{\mathcal{H}}_{II} = -\tfrac{1}{2}\mathcal{H}_{I_xI_x} \equiv -\tfrac{1}{2}\sum_{i<j} A_{ij}(3I_{xi}I_{xj} - \mathbf{I}_i \cdot \mathbf{I}_j) \tag{4.121}$$

and

$$\bar{\mathcal{H}}_{IS} = 0 \quad \text{if } \omega_{1I} \neq \omega_{1S}.$$

However, when the Hartmann-Hahn condition $(\omega_{1I}=\omega_{1S})$ is invoked we arrive at

$$\bar{\mathcal{H}}_{IS}=\tfrac{1}{2}\sum_{i,j}B_{ij}(I_{zi}S_{zj}+I_{yi}S_{yj}). \tag{4.122}$$

Equation (4.120) can therefore be reduced to

$$M_S(t)/M_{SB}=\frac{\gamma_I}{\gamma_S}(I_x|\,e^{-i(\mathcal{H}_{II}+\mathcal{H}_{IS})t}\,|S_x)/(S_z|S_z). \tag{4.123}$$

In order to treat a simple example, we suppose that only a single I spin interacts with a single S spin, i.e.

$$\bar{\mathcal{H}}_{IS}=\tfrac{1}{2}B(I_zS_z+I_yS_y)$$

and $\mathcal{H}_{II}=0$. In this case Eq. (4.123) results immediately in

$$M_{S_x}(t)/M_{SB}=\frac{1}{2}\frac{\gamma_I}{\gamma_S}(1-\cos\tfrac{1}{2}Bt). \tag{4.124}$$

We see, that an oscillatory behaviour with frequency $B/2$ is observed, which will be damped in practice by other heteronuclear and of course homonuclear dipolar interactions $\mathcal{H}_{II}$. In the situation of off-resonance spin-locking the oscillation frequency $B/2$ is scaled by $\sin\vartheta_I$ depending on the angle ϑ_I of the effective field ω_{eI} in the rotating frame with respect to the z-axis. In the experiment of Ernst and co-workers [43] (Fig. 4.20) $\vartheta_I=\pi/2$, whereas Waugh and co-workers [44] (Fig. 4.21) used $\vartheta_I=54.7°$ ("magic angle") in order to reduce the damping of the oscillations due to $\mathcal{H}_{II}$ (spin diffusion). In this case $\sin\vartheta_I=(2/3)^{1/2}$ resulting in an oscillation frequency $B/\sqrt{6}$.

4.4 Spin-Decoupling Dynamics

Spin decoupling of abundant spins is a prerequisite in order to obtain high resolution NMR spectra in solids, containing different spin species [3]. Several NMR spectra of the cross-polarization type and also multiple-pulse experiments have been reported in the preceding chapters which employed spin decoupling [3, 27–30].

A strong rf field B_{1I} is usually applied in a pulsed or continuous fashion at the resonance frequency of the non-observed abundant spins I which contribute to the coupling represented by $\mathcal{H}_{IS}$. If B_{1I} is strong such that I_z is flipped rapidly compared with the spin-spin interactions, the time average of $\mathcal{H}_{IS}$ vanishes and consequently the excess broadening due to the I spins is zero. The question arises: How strong does B_{1I} have to be in order to average out the heteronuclear coupling? On the first sight it would be sufficient, to make $\omega_{1I}=\gamma_I\cdot B_{1I}$ large compared with $\sqrt{M_2^{IS}}$ where M_2^{IS} is the second moment of the I-S coupling. This would be correct only, if the flip-flop term in the I-I coupling Hamiltonian would be zero. However, the flip-flop "motion" of the I-I spins modulates the I-S interaction which results in a narrowing of the resonance line of the S spins if γ_I/γ_S is large, as was first shown by Abragam and Winter [4a, 46]. In order to narrow the S spin resonance any further, ω_{1I} has to be comparable to the flip-flop rate of the I spins.

This leads to the interesting conclusion that a line broadening instead of a line narrowing will result when the "decoupling" field B_{1I} is applied at the "magic angle", i.e. off resonance by $\Delta\omega = \omega_{1I}/\sqrt{2}$. These aspects are borne out more quantitatively in the following:

We want to calculate the "excess broadening" of the S spin signal due to the I-S interaction.

Let us neglect homonuclear dipolar coupling of the S spins, i.e. we assume the following Hamiltonian in the tilted rotating frame

$$\mathscr{H} = \mathscr{H}_{1I} + \mathscr{H}_{II} + \mathscr{H}_{IS} \tag{4.125}$$

where

$$\mathscr{H}_{1I} = -\omega_{eI} I_z; \quad \mathscr{H}_{II} = P_2(\cos\vartheta_I)\,\mathscr{H}_{II}^{(0)} + \mathscr{H}_{II}^{(ns)}$$
$$\mathscr{H}_{IS} = \cos\vartheta_I\,\mathscr{H}_{I_z S_z} - \sin\vartheta_I\,\mathscr{H}_{I_x S_z}.$$

with

$$\omega_{eI}^2 = \omega_1^2 + \Delta\omega^2 \quad \text{and} \quad \tan\vartheta_I = \omega_1/\Delta\omega.$$

In order to calculate the linewidth of the S spins under the decoupling field we apply the general lineshape theory outlined in Appendix C. We use the simple expression for the linewidth $\delta = \pi\Delta v$ (Δv: FWHH) according to Eq. (C.43b)

$$\delta = \int_0^\infty d\tau\, K(\tau) = K'(\omega = 0) \tag{4.126}$$

where the memory function $K(\tau)$ can be expressed as

$$K(\tau) = \frac{(S_x|\,\hat{\mathscr{H}}_{IS}\,\mathbf{S}(\tau)\,\hat{\mathscr{H}}_{IS}\,|S_x)}{(S_x|S_x)} \tag{4.127}$$

where

$$\mathbf{S}(\tau) = \exp[-i(1-P)\,\hat{\mathscr{H}}\,\tau]$$

with

$$P = \frac{|S_x)(S_x|}{(S_x|S_x)}.$$

Notice, that

$$K(0) = \frac{(S_x|\,\hat{\mathscr{H}}_{IS}^2\,|S_x)}{(S_x|S_x)} = M_2^{IS}. \tag{4.128}$$

Recall from Appendix E that the line shape function

$$G(\omega) = \int_0^\infty dt\,\cos\omega t\,\langle S_x(t)\rangle$$

can be expressed as the Fourier transform of the S spin free induction decay and may be obtained by Laplace inversion from Eq. (4.120) as

$$G(\omega) = \frac{K'(\omega)}{[\omega - K''(\omega)]^2 + [K'(\omega)]^2}$$

where $K'(\omega)$ and $K''(\omega)$ are the cosine and the sine transform of $K(\omega)$ respectively, according to Appendix C. If $K(\omega)$ is approximated by a Gaussian as before the line shape function $G(\omega)$ is straightforwardly calculated following along the lines of Appendix C.

Let us now discuss different limiting cases:

(i) $\omega_{1I}=0$ i.e. $\varDelta\omega=0$ no decoupling

$$\mathbf{S}(\tau)=\exp[-i(1-P)\,\mathscr{H}_{II}+\mathscr{H}_{IS})\,\tau]$$

from which follows for the second moment N_2 of $K(\tau)$

$$N_2=[M_4^{ISIS}+M_4^{ISII}-(M_2^{IS})^2]/M_2^{IS} \tag{4.129}$$

where

$$M_4^{ISIS}=\frac{(S_x|\,\hat{\mathscr{H}}_{IS}^4\,|S_x)}{(S_x|S_x)}$$

and

$$M_4^{ISII}=\frac{(S_x|\,\hat{\mathscr{H}}_{IS}\,\hat{\mathscr{H}}_{II}^2\,\hat{\mathscr{H}}_{IS}\,|S_x)}{(S_x|S_x)}$$

which may be expressed in terms of the lattice sums, as defined in the preceding section by

$$M_4^{ISIS}=\frac{1}{3}I(I+1)\left(\frac{\gamma_I\gamma_S\hbar}{a^3}\right)^4\left[\frac{1}{5}(3I^2+3I-1)S_2+I(I+1)(S_1^2-S_2)\right], \tag{4.130a}$$

$$M_4^{ISII}=\left[\frac{1}{3}I(I+1)\right]^2\cdot\frac{1}{4}\left(\frac{\gamma_I}{\gamma_S}\right)^2\left(\frac{\gamma_I\gamma_S\hbar}{a^3}\right)^4 S_4, \tag{4.130b}$$

$$M_2^{IS}=\frac{1}{3}(I(I+1))\left(\frac{\gamma_I\gamma_S\hbar}{a^3}\right)\cdot S_1. \tag{4.130c}$$

In the following we shall use the moment ratios

$$\mu_1=\frac{M_4^{ISIS}}{(M_2^{IS})^2}=3-\frac{1}{5}\left[\frac{1}{3}I(I+1)\right]^{-1}(2I(I+1)+1)\frac{S_2}{S_1^2} \tag{4.131a}$$

and

$$\mu_2=\frac{M_4^{ISII}}{(M_2^{IS})^2}=\frac{1}{4}\left(\frac{\gamma_I}{\gamma_S}\right)^2\frac{S_4}{S_1^2}. \tag{4.131b}$$

The ratio μ_1 is usually very close to three, whereas μ_2 may become very large, depending on the ratio of γ_I/γ_S. The second moment of $K(\tau)$ may now be expressed as

$$N_2=M_2^{IS}(\mu_2+\mu_1-1). \tag{4.132}$$

The linewidth δ may be calculated as outlined in Appendix C if the functional form of $K(\tau)$ is known or can be represented by a sufficiently reasonable approximate expression. For simplicity and a qualitative discussion let us assume $K(\tau)$ to be Gaussian in the following.

The line width of the undecoupled spectrum can now be calculated, using Eqs. (4.128, 4.132) and

$$\delta_0 = \sqrt{\frac{\pi}{2}} \left[\frac{K^2(0)}{N_2} \right]^{1/2}$$

as derived in Appendix C. As mentioned before this line width is already reduced by the flip-flop "motion" of the I spins as expressed by the large value of N_2 [4a, 47]. A similar approach has been taken by R.E. Walstedt [47] invoking the short correlation limit, resulting in a Lorentzian line shape

(ii) $\omega_{1I} \gg \omega_{LI}$ strong decoupling $\Delta\omega \neq 0$

$$S(\tau) = S_1(\tau) S_0(\tau)$$

where

$$S_1(\tau) = e^{-i(1-P)\hat{\mathscr{H}}_{1I}\tau}$$

and

$$S_0(\tau) = T \exp\left[-i \int_0^\tau dt'(1-P)(\hat{\mathscr{H}}_{II}(t') + \hat{\mathscr{H}}_{IS}(t')) \right]$$

with

$$(1-P)(\hat{\mathscr{H}}_{II}(t') + \hat{\mathscr{H}}_{IS}(t')) = S_1(t')^{-1}(1-P)(\hat{\mathscr{H}}_{II} + \hat{\mathscr{H}}_{IS}) S_1(t').$$

Due to the strong decoupling field, we neglect non-secular terms in $\mathscr{H}(t')$ and write

$$S_0(\tau) = \exp[-i(1-P)\hat{\mathscr{H}}^{(0)}\tau]$$

where

$$\hat{\mathscr{H}}^{(0)} = P_2(\cos\vartheta_1)\hat{\mathscr{H}}_{II} + P_1(\cos\vartheta_I)\hat{\mathscr{H}}_{IS}.$$

Here ϑ_I is the angle of the effective field in the rotating frame with respect to B_0. A tedious, but straightforward calculation evaluating $K(\tau)$ according to Eq. (4.127) leads to

$$K(\tau) = K_z(\tau) + K_x(\tau)\cos\omega_{eI}\tau \tag{4.133}$$

where

$$K_\mu(\tau) = \frac{(S_x| \hat{\mathscr{H}}_{I\mu Sz} S_0(\tau) \hat{\mathscr{H}}_{IS} |S_x)}{(S_x|S_x)} \times \begin{cases} \cos\vartheta_I; & \mu = z \\ -\sin\vartheta_I; & \mu = x \end{cases} \tag{4.134}$$

with

$$K_z(0) = \cos^2\vartheta_I \, M_2^{IS} \tag{4.135a}$$

$$N_{2z} = [P_1^2(\cos\vartheta_i)(\mu_1 - 1) + P_2^2(\cos\vartheta_I)\mu_2] \, M_2^{IS} \tag{4.135b}$$

where M_2^{IS}, μ_1, and μ_2 are expressed in terms of lattice sums according to Eqs. (4.130, 4.131). In the case of large ω_{eI}, the case considered here, $K_x(\tau)$ does not contribute appreciably to the spectral distributions $K'(\omega)$ and $K''(\omega)$ and will therefore be neglected.

Using Eqs. (4.133) and (4.134) the line shape and the corresponding "excess" linewidth can be calculated, using the memory function approach as discussed in Appendix C [48].

With

$$\delta = \sqrt{\frac{\pi}{2}} \left[\frac{K^2(0)}{N_2} \right]^{1/2} \tag{4.136}$$

we obtain [48]

$$\delta = \sqrt{\frac{\pi}{2}} \left[\frac{\cos^4 \vartheta_I \, M_2^{IS}}{P_2^2(\cos \vartheta_I) \, \mu_2 + \cos^2 \vartheta_I (\mu_1 - 1)} \right]^{1/2} . \tag{4.137}$$

A similar expression for the excess line width results if a cutoff Lorentzian line shape of the S spin signal is assumed [4a, 6] and the corresponding second and fourth moments are inserted. The only difference is the factor μ_1 instead of $\mu_1 - 1$ in Eq. (4.137) and a different prefactor, namely $\pi/(2\sqrt{3})$.

Equation (4.137) describes qualitatively the behavior of the S spin linewidth δ under off-resonance decoupling of the I spins. Notice, that the linewidth δ does not follow the simple $\cos \vartheta_I$ variation of the second moment, but reflects rather sensitively the scaling of the correlation time of the I spin flip-flops, represented by $P_2^2(\cos \vartheta_I) \cdot \mu_2$. Figure 4.23 demonstrates this behavior very clearly in the case of ^{13}C in adamantane under proton decoupling and in the case of ^{109}Ag in AgF under ^{19}F decoupling. No decoupling is obtained far off-resonance i.e., for $\vartheta_I = 0$. The S spin resonance line is already narrowed by the unscaled fast fluctuation of the I spins due to their flip-flop motion as mentioned before [4a, 46].

Approaching the "magic angle" with $P_2(\cos \vartheta_I) = 0$, the correlation time of the I spins is slowed down and is finally quenched at the magic angle, resulting in a "broadened" S spin line which is solely due to the I-S interaction. The phenomenon may be considered as a freezing of the motion in spin space, resulting in a similar linewidth broadening as is observed with the freezing of lattice motion. However, the I-S interaction is also scaled by $\cos \vartheta_I$ which is an inevitable consequence of the I spin irradiation. At the "magic angle" the S spin line shape can be calculated exactly as the Fourier transform of

$$G_{IS}(t) = \langle S_x(t) \rangle = \prod_j \cos \left(\frac{1}{2\sqrt{3}} B_j t \right) \quad \text{for } I = \frac{1}{2} \tag{4.138}$$

where B_j is defined by Eq. (4.78c). Equation (4.135a) is exact in the sense, that no memory function approach or moment expansion is involved and the corresponding lineshape can be calculated readily as a sum of δ functions over all combinations of $P_j = \pm 1$

$$F(\omega) = \frac{1}{2^n} \sum_{\text{comb. } P_j} \delta \left(\omega - \sum_{j=1}^{n} \frac{1}{2\sqrt{3}} B_j P_j \right); \quad P_j \pm 1. \tag{4.139}$$

Notice, that this is the same spectrum as observed in the experiment of Waugh and co-workers [44] (preceding section) for vanishing chemical shift.

When ϑ_I approaches $90°$, the correlation time of the I spin flip-flops decreases towards half its natural value. However, since the I-S interactions is drastically reduced by $\cos \vartheta_I$ in this region, the fluctuation rate of the I spins becomes very large with respect to the I-S interaction and the line narrowing becomes very effective. This general behavior is borne out in Fig. 4.23 and qualitatively described by Eq. (4.137). For a more thorough discussion one would have to resort to a more realistic memory function, rather than simply assuming a Gaussian as was done above (see Ref. [48]).

Recall that $K_z(\tau)$ [Eq. (4.134)] would be sufficiently close approximated by a Lorentzian, if $S_0(\tau)$ would contain only the Hamiltonian $\mathcal{H}_{II}$, as was shown in the preceding section. On the other hand a Gaussian approximation would be appropriate if only the term $(1-P)\,\mathcal{H}_{IS}$ would be present in $S_0(\tau)$. We may combine those approximations by writing:

$$K_z(\tau) = K_{II}(\tau) \cdot K_{IS}(\tau) \tag{4.140}$$

where the second moment N_{2z} of $K_z(\tau)$ is given by Eq. (4.135b). Following the arguments above, we write

$$K_{II}(\tau) = \frac{1}{1 + (\tau/\tau_{II})^2} \tag{4.141a}$$

with

$$\frac{1}{\tau_{II}^2} = \frac{1}{2} P_2(\cos \vartheta_I)^2 \cdot \mu_2 \cdot M_2^{IS}$$

and

$$K_{IS}(\tau) = \cos^2 \vartheta_I \, M_2^{IS} \exp[-\tau^2/\tau_{IS}^2] \tag{4.141b}$$

where

$$\frac{1}{\tau_{IS}^2} = \frac{1}{2} \cos^2 \vartheta_I (\mu_1 - 1) \, M_2^{IS}.$$

If an even higher degree of approximation is desired one has to express $K_z(\tau)$ itself by an integro-differential equation as was done by Demco, Tegenfeldt and Waugh [11] in the case of cross-polarization dynamics. Notice, however, that this involves higher order lattice sums and demands considerable computational effort. On the other hand, $K_{IS}(\tau)$ can be calculated numerically from

$$K_{IS}(\tau) = -\int_0^\tau dt \, K_{IS}(t) \, G'_{IS}(\tau - t) - G''_{IS}(\tau) \tag{4.141c}$$

where $G'_{IS}(t)$ and $G''_{IS}(t)$ are the first and second time derivative of $G_{IS}(t)$.

Using this memory function in the mixed memory function approach above leads to the exact lineshape at the magic angle and a decent approximation for the other values of ϑ_I. We finally remark, that the exact calculation of the linewidth $\Delta v = \delta/\pi$ at the magic angle $\vartheta_I = 54.7°$ according to Eq. (4.137) yields

^{13}C in adamantane $\Delta v = 1{,}223$ Hz

^{109}Ag in AgF $\qquad \Delta v = \phantom{1{,}}453$ Hz.

The reason that the experimental spectra do not reach these values, is the inhomogeneity of the rf field at the I spin resonance. This may be circumvented by using a four-pulse sequence instead. For further details consult reference [48].

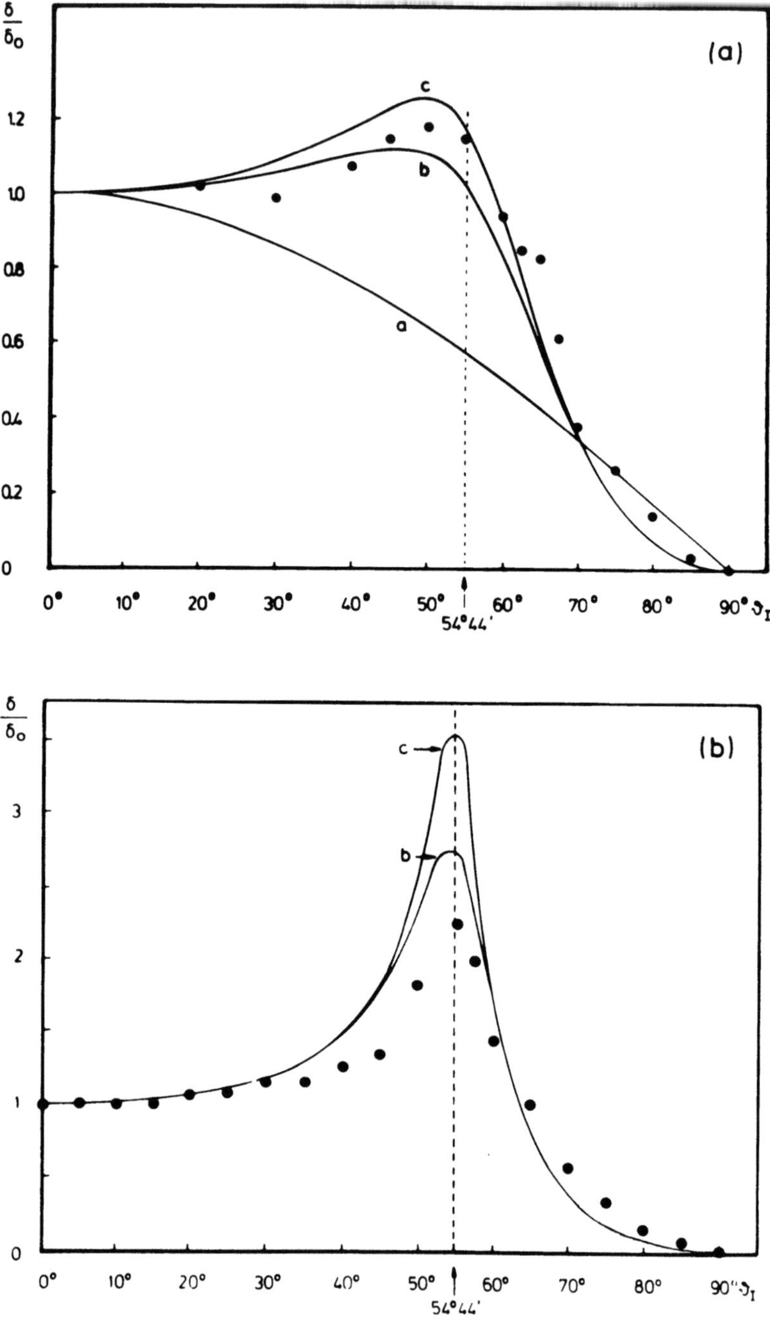

Fig. 4.23. a Normalized excess line width of the ^{13}C spectrum (S spins) in adamantane versus the angle ϑ_I of a strong effective field in the rotating frame of the I spins (protons) [39]. a Variation of the second moment M_2^{IS} with ϑ_I. b Calculated line width using the second and fourth moment assuming a truncated Lorentzian line. c Linewidth calculated according to Eq. (4.137) using a memory function approach. **b** ^{109}Ag excess line width versus ϑ_I in AgF with ^{19}F being the I spin [48]

Let us consider the situation of strong near resonance decoupling, i.e.

$$\omega_1 \gg (M_2^{IS})^{1/2}, \quad (M_2^{II})^{1/2} \quad \text{and} \quad \Delta\omega.$$

We are interested in the line broadening caused by the deviation from resonance $\Delta\omega$. Under this condition we can apply Eq. (4.137) with

$$P_2(\cos\vartheta_I) \cong -\tfrac{1}{2} \quad \text{and} \quad \cos\vartheta_I \cong \Delta\omega/\omega_1.$$

The resulting linewith $\Delta v = \delta/\pi$ is

$$\Delta v = \frac{1}{\sqrt{2\pi}} \frac{\Delta\omega^2}{\omega_1^2} (M_2^{IS})^{1/2} \left[\mu_2/4 + (\mu_1 - 1)\,\Delta\omega^2/\omega_1^2\right]^{-1/2} \tag{4.142a}$$

which may be further approximated to give

$$\Delta v \cong \sqrt{\frac{2}{\pi}} \frac{\Delta\omega^2}{\omega_1^2} (M_2^{IS})^{1/2}/\sqrt{\mu_2}. \tag{4.142b}$$

Van der Haart et al. [49] have recently applied these arguments to discuss in a detailed investigation the different broadening mechanisms responsible for ^{13}C resonance lines in solids under high resolution conditions. The frequency offset $\Delta\omega$ can in this case e.g. be caused by a distribution of chemical shift values in the S spin spectrum.

(iii) *On resonance Decoupling* $(\Delta\omega = 0)$

In high resolution NMR experiments in solids as discussed in the preceding sections the decoupling rf fields is usually applied on resonance in order to make the spin decoupling most effective. This is highly recommended, as has been demonstrated in Fig. 4.23, since for small deviations from the resonance frequency $(\vartheta_I \neq 90°)$, the line broadening becomes appreciable.

However, as will be shown subsequently, the line narrowing due to on-resonance decoupling becomes very effective already for small rf fields which are comparable to the "local field" of the spins to be decoupled (here I spins). If we neglect the S spin dipolar coupling, we may write for the total Hamiltonian in the rotating frame

$$\mathcal{H} = \mathcal{H}_{II} + \mathcal{H}_{IS} + \mathcal{H}_{1I}$$

where

$$\mathcal{H}_{1I} = -\omega_{1I} I_x.$$

Using the memory function approach as outlined in Eqs. (4.126, 4.127), we obtain

$$K(\tau) = K_0(\tau) \cos\omega_{1I}\tau \tag{4.142a}$$

with

$$K_0(\tau) = \frac{(S_x|\,\mathcal{H}_{IS}\, S_0(\tau)\,\mathcal{H}_{IS}\,|S_x)}{(S_x|S_x)} \tag{4.142b}$$

where we approximate the propagator $S_0(\tau)$ for small ω_{1I} as

$$S_0(\tau) = \exp[-i(1-P)(\hat{\mathcal{H}}_{II} + \hat{\mathcal{H}}_{IS})\tau] \tag{4.142c}$$

using the same projector P as above. A moment expansion of $K_0(\tau)$ results in:

$$K_0(0) = M_2^{IS}$$

and

$$N_2 = M_2^{IS}(\mu_2 + \mu_1 - 1).$$

The approximation implied in Eq. (4.142c) is reasonable only for $\omega_{1I} \to 0$ i.e. $\omega_{1I} \ll \|\mathcal{H}_{II}\|$, $\|\mathcal{H}_{IS}\|$. The relation is therefore not expected to hold for large values of $\omega_{1I} \gg \|\mathcal{H}_{II}\|$. In this case the operator $(1-P)(\hat{\mathcal{H}}_{II} + \hat{\mathcal{H}}_{IS})$ in Eq. (4.142c) has to be replaced by $-\frac{1}{2}\hat{\mathcal{H}}_{I_xI_x}$. The corresponding second moment N_2 of the memory function $K_0(\tau)$ is than given by Eqs. (4.107) and (4.109a) with $P_2(\cos\vartheta_I) = -\frac{1}{2}$.

For the following discussion it is irrelevant which form of N_2 is used.

In order to obtain numerical values for the linewidth of the S spin resonance under on-resonance decoupling of the I spins, we assume a functional form of the memory function $K_0(\tau)$ in Eq. (4.142). The simplest functional form of $K_0(\tau)$ would be a Gaussian, which corresponds to the first term of a cumulant expansion of $K_0(\tau)$. The corresponding line width of the S spin signal for different values of ω_{1I} can be calculated as discussed before by using the lattice parameters of the system under investigation. There is no adjustable parameter in the calculation and the only approximation made concerns the assumption of the functional form of the memory function $K_0(\tau)$.

Notice, however, that the Gaussian approximation is "legitimate" only in the case $\omega_{1I} \gg \|\mathcal{H}_{II}\|$ where $K_0(\tau)$ corresponds to the cross-correlation function of the spin-locking case. For $\omega_{1I} \ll \|\mathcal{H}_{II}\|$, however, the "memory function" $K_0(\tau)$ may be better approximated by $K_z(\tau)$ of Eq. (4.141a). In order to obtain a closed form expression for qualitative arguments we shall assume a Gaussian "memory function" $K_0(\tau)$ for the sake of simplicity.

The line shape and line width of the resonance line is readily obtained, using the corresponding equations of Appendix C.

The line width δ may be obtained by iteration from

$$\delta = K''(\delta) + [2K'(\delta=0)K'(\delta) - [K'(\delta)]^2]^{1/2}. \tag{4.143}$$

Where because of the cosine in Eq. (4.141) $K'(\delta)$ and $K''(\delta)$ have to be replaced by

$$\begin{aligned} K'(\delta) &= \tfrac{1}{2}[K'(\delta+\omega_{1I}) + K'(\delta-\omega_{1I})] \\ K''(\delta) &= \tfrac{1}{2}[K''(\delta+\omega_{1I}) + K''(\delta-\omega_{1I})]. \end{aligned} \tag{4.144}$$

Using Gaussian approximation of $K_0(\tau)$ [Eq. (4.142)], the line width δ has been calculated in the case of ^{13}C in adamantane under proton on resonance decoupling [48]. This is compared with the experimental data in Fig. 4.24.

The agreement between theory and experiment is quite pleasing. Notice, that the decoupling is already very effective for fields comparable with the local field $H'_L(I) = [M_2^{II}/3]^{1/2}$ of the I spins. This was realized soon in practice [3] when already small decoupling fields sufficed to produce a high resolution spectrum.

For an estimate of the linewidth $\Delta v = \delta/\pi$ we may simply apply Eq. (4.126) where $K(\tau)$ is given by Eq. (4.142) and by using a Gaussian approximation for

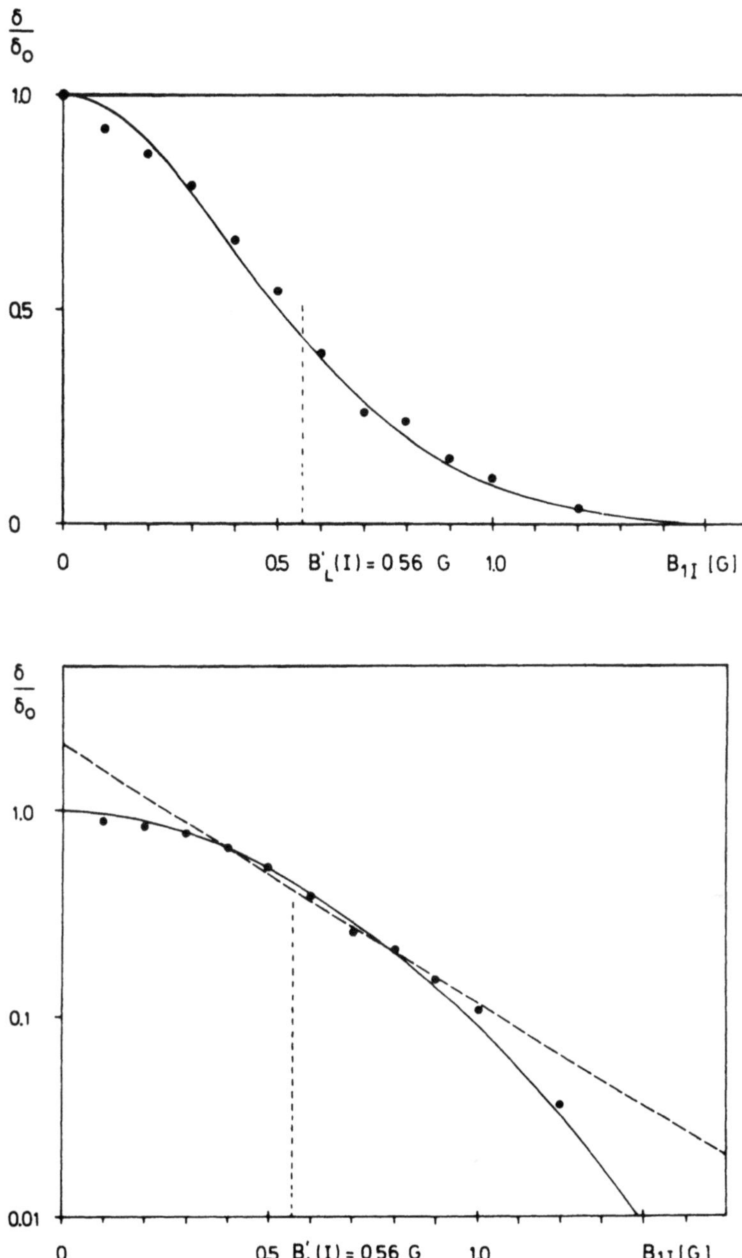

Fig. 4.24. Top: Normalized excess linewidth of the ^{13}C spectrum in adamantane applying an rf field B_{1I} at the proton resonance frequency (on-resonance decoupling [48c]). Bottom: Logarithmic plot of top figure. The local field of the I spins (protons) is indicated. The solid line is calculated using a memory function approach as described in the text

$K_0(\tau)$. This leads to

$$\Delta v = \frac{1}{\sqrt{2\pi}}\,\frac{M_2^{IS}}{\sqrt{N_2}}\,\exp[-\omega_{1\,I}^2/2N_2].\tag{4.145}$$

Depending on the strength of $\omega_{1\,I}$, N_2 is either given by Eqs. (4.107) and (4.109a) ($\omega_{1\,I}\gg\|\mathscr{H}_{II}\|$) or by Eq. (4.132) ($\omega_{1\,I}\ll\|\mathscr{H}_{II}\|$). Van der Haart et al. [49] have discussed this situation also using a slightly different approach.

In order to get a clear apprehension of the effectiveness of on-resonance decoupling we have plotted δ/δ_0 in Fig. 4.25 versus the dimensionless parameter

$$\Omega_1 = \frac{\omega_1}{\sqrt{M_2^{IS}}}\tag{4.145}$$

for different values of the moment ratio μ_2 and where μ_1 is kept constant, namely, $\mu_1=3$. These plots are generally applicable, since any specific spin system may be represented by μ_2. However, Fig. 4.25 serves only a demonstrative purpose in the light of the approximations involved in Eq. (4.142).

In a similar fashion we may treat the on-resonance pulse-decoupling. Let us apply the same approach as above, by representing the rf field Hamiltonian $\mathscr{H}_1(t)$ in Eq. (4.142) as

$$\mathscr{H}_1(t) = \pi I_x\,\delta(t-(2n+1)\,t_c/4)\qquad n=0,1,2,\ldots\tag{4.146}$$

describing the application of π pulses at the I spin resonance twice every cycle (cycle time t_c). The memory function $K(\tau)$ can now be expressed by

$$K(\tau) = K_0(\tau)\,F(t)\tag{4.147}$$

where $K_0(\tau)$ is defined in Eq. (4.142) and where $F(t)$ is represented by

$$\begin{aligned}
F(t) &= \frac{4}{\pi}\left[\cos\omega_c t - \frac{1}{3}\cos 3\omega_c t + \frac{1}{5}\cos 5\omega_c t - \ldots\right]\\
&= \frac{4}{\pi}\sum_{k=0}^{\infty}\frac{(-1)^k}{2k+1}\cos(2k+1)\,\omega_c t;\qquad \omega_c = 2\pi/t_c.
\end{aligned}\tag{4.148}$$

Correspondingly, the function $K'(\delta)$ and $K''(\delta)$ in Eq. (4.143) have to be replaced by

$$K(\delta) = \frac{2}{\pi}\sum_{k=0}^{\infty}\frac{(-1)^k}{2k+1}\,[K(\omega+(2k+1)\,\omega_c)+K(\omega-(2k+1)\,\omega_c)]\tag{4.149}$$

where K is either K' or K'', respectively. The iteration procedure of Eq. (4.143) using the K-functions as given by Eq. (4.149) leads to the lower graph in Fig. 4.25, where δ/δ_0 is plotted versus

$$\Omega_c = \frac{\omega_c}{\sqrt{M_2^{IS}}}\tag{4.150}$$

for different values of the moment ratio μ_2 and where $\mu_1=3$. The continuous decoupling procedure is in general more effective than the pulsed decoupling for the same cycle time ($\Omega_1=\Omega_c$), as is demonstrated in Fig. 4.25.

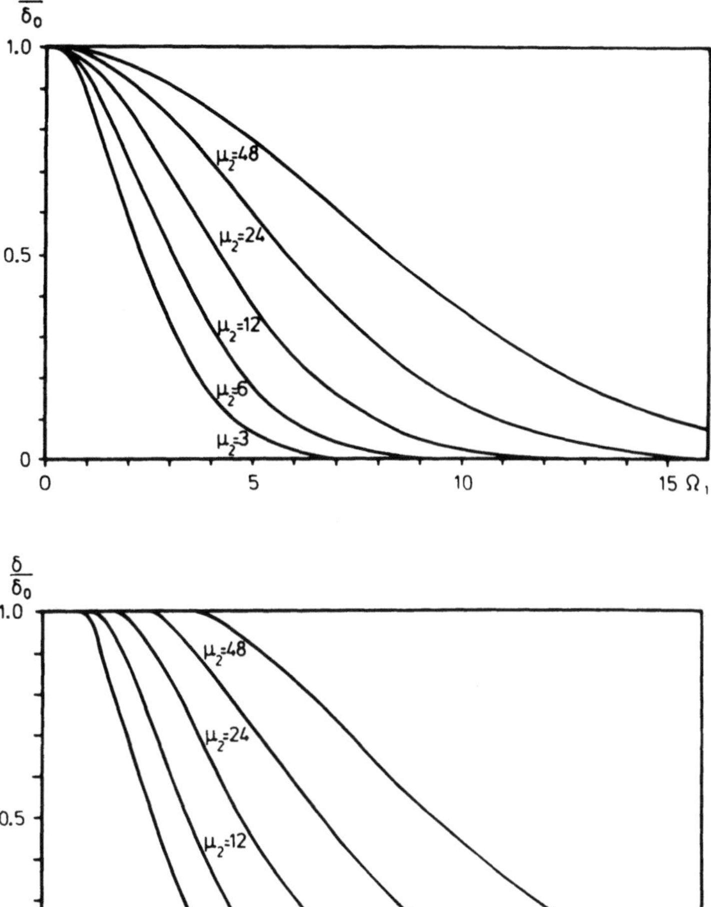

Fig. 4.25. Normalized excess linewidth of a dilute S spin system embedded in an abundant I spin system under I spin on-resonance decoupling conditions. Top: Continuous decoupling of I spins with a rate $\Omega_1 = \omega_1/(M_2^{IS})^{1/2}$. Bottom: Pulsed decoupling with a rate $\Omega_c = \omega_c/(M_2^{IS})^{1/2}$, where ω_c is the cycle frequency of the decoupling sequence (two π pulses per cycle in this case). The moment ratio μ_2 is explained in the text

Deuteron Decoupling

A further approach towards proton high resolution NMR in solids has been proposed by Pines et al. [50]. If most of the protons are exchanged by deuterons ($I=1$) the main contribution to the linewidth of the protons is caused by the dipolar coupling of the protons to the nearby deuterons. Due to the different gyromagnetic ratios of deuterons and protons this coupling is reduced by a factor of about six, compared with the linewidth of the fully

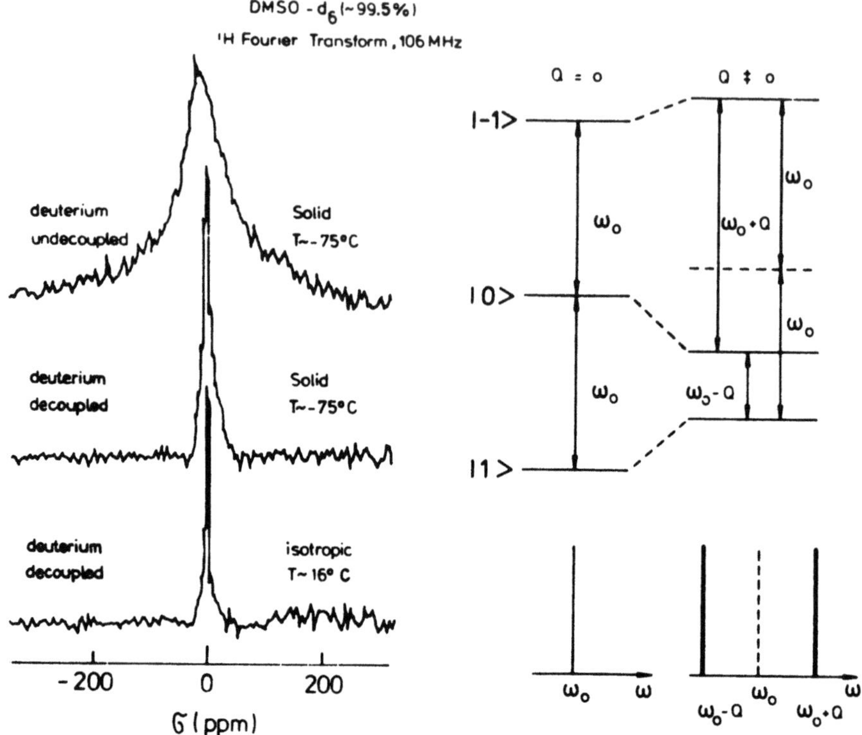

Fig. 4.26. Proton spectra of 99.5% deuterated dimethyl sulfoxide (DMSO) in the solid and liquid state respectively, with and without deuteron decoupling [50]. The independence of the double quantum transition on the quadrupole interaction of the deuterons is indicated in the energy level diagram of the deuterons. (Courtesy of A. Pines)

protonated compound. If strong rf irradiation is now applied to the deuteron resonance frequency, the deuterons are decoupled from the protons and a high resolution spectrum of the protons results. This is demonstrated in Fig. 4.26. If the quadrupole interaction of the deuterons vanishes, a medium size rf field with

$$\omega_1^2 \cong D^2 \tag{4.151}$$

is needed for the onset of line narrowing, if D is the size of the proton-deuteron coupling.

However, at first sight the deuteron decoupling seems to be impossible, since the deuteron spectrum is "inhomogeneously" broadened by a strong quadrupole interaction Q on the order of about 100 kHz. A first sighted argument would than imply that at least the spectral width Q of the deuteron spectrum has to be covered by the decoupling field, i.e.

$$\omega_1^2 \cong Q^2 \tag{4.152}$$

in order to observe the onset of narrowing. This corresponds to a B_1 field of about 155 G and to a much higher value for obtaining a reasonable resolution.

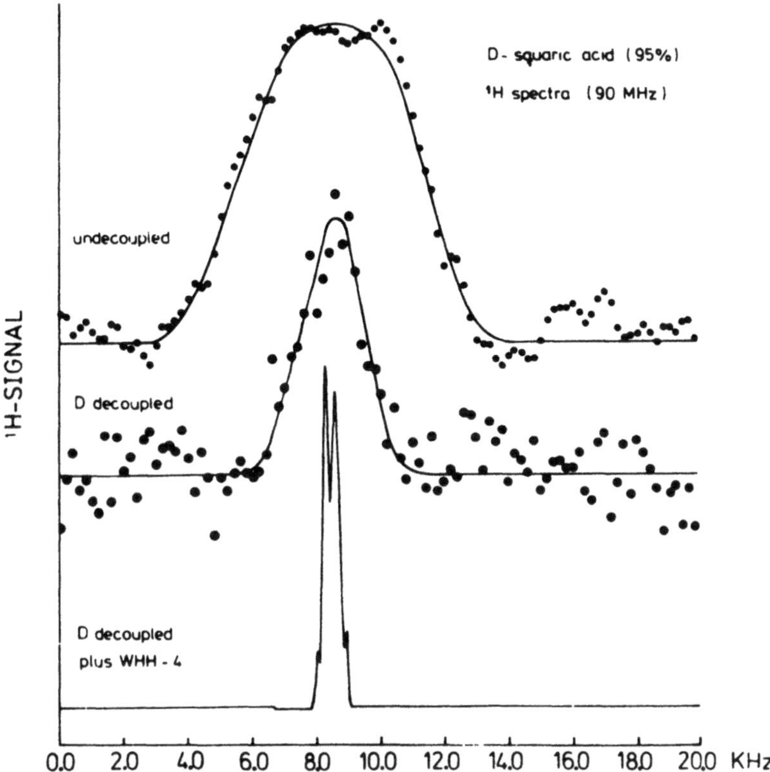

Fig. 4.27. Proton spectra of a 95 % deuterated single crystal of squaric acid at 90 MHz at arbitrary orientation. A 20 G decoupling field is sufficient to decouple the deuterons from the protons, however, the dipolar broadening of the residual protons is still too large to resolve any fine structure (middle). If a WHH-4 sequence ($t_c = 72\,\mu s$) is applied additionally, a high resolution proton spectrum results. (Courtesy of D. Suwelack)

This is certainly a difficult task beyond technical practicability. Snyder and Meiboom [51], however, realized in the deuteron decoupling of liquid crystal spectra that a "double quantum transition" becomes very effective if the decoupling field is applied exactly at the resonance of the deuterons. Although all the three energy levels of the deuteron states in a magnetic field are shifted by the quadrupole interaction, the difference between the $|+1\rangle$ and $|-1\rangle$ states equals $2\omega_0$, independent of the quadrupole interaction in first order.

Since these levels are not coupled by rf fields, no transitions ($\Delta m = 2$) can be induced by a single photon. If, however, two photons are applied at exact resonance, rapid transitions among the $|\pm 1\rangle$ levels may be introduced. The first important point is, that the frequency is independent of the quadrupole interaction in first order and the second point, that the necessary rf field for the onset of decoupling needs to be much less than according to Eq. (4.152), namely [50]

$$\omega_1^2 \cong D \cdot Q.$$

This is why it is possible to decouple the deuterons with medium size rf fields, as is demonstrated in Fig. 4.26. The residual broadening of the proton line which is due to the proton-proton interaction, in a say 5% protonated compound, may be effectively reduced by applying a WHH-4 sequence in addition to the deuteron decoupling field (see Fig. 4.27). Since the dipolar coupling among the protons is already reduced considerably, no severe conditions apply to the multiple pulse sequence.

Especially the cycle time may become rather large, reducing the influence of accumulative pulse errors. A more detailed discussion may be found in Ref. [50]. Further aspects of double quantum decoupling are discussed in Chap. 6.

4.5 Application of Cross-Polarization Experiments

It is beyond the scope of this monograph to cover all possible applications of cross-polarization (CP) NMR. A large variety of molecular systems have been investigated by this technique and we concentrate here only on some specific examples.

a) Chemical shift tensor and the structure of molecules

The chemical shift tensor is the primary result of CP-NMR. Some representative examples are collected in Chap. 7.5. Besides this the structure of molecules may be determined from the heteronuclear dipole-dipole interaction. This goal is best be achieved by two-dimensional spectroscopy to be discussed in Chap. 5.2. However, also from standard CP-NMR these informations can be obtained simultaneously by curve fitting with chemical shift data and dipole-dipole couplings [52–56]. Zilm et al. [56, 57] have recently used the matrix isolation technique to determine these parameters for simple molecules at 15 K.

b) Molecular motion

As discussed extensively already in Chap. 2.8 the chemical shift tensor is a versatile intrinsic parameter for studying molecular motion. Not only the static orientation of a molecule in a solid can be determined from chemical shift tensor data evaluation, but also the molecular dynamics in the solid state once it falls into the kHz-range can be detected. These aspects have been discussed in more detail in Chap. 2.8. Of course, also other anisotropic spin interactions, like dipole-dipole interaction and quadrupole interaction can be exploited in the same way.

c) Bio-molecules

It were Griffin and co-workers [59] who realized, that the chemical shift tensor obtained from cross-polarization experiments can be a versatile tool for monitoring molecular dynamics in bio-molecules. There also the quadrupole interactions of deuterons labelled at particular sites have been utilized [60, 61]. CP-NMR has been used very successfully in the determination of structure and motion of bio-molecules [62, 63] in the past and will continue to do so in the future.

d) Polymers

Polymers show a wealth of structure and molecular dynamics and are particularly suited candidates for CP-NMR. It were Schäfer and Stejskal [64] who advanced this area of research by combining magic angle spinning (MAS) with CP-NMR. The appearance if liquid-like spectra (isotropic shift) in solid polymers was highly appreciated at the time by NMR spectrocopists. This theme has been followed by others [65–67] and is currently a standard tool in polymer research. Note, that the cross-polarization dynamics in MAS CP-NMR is not as simple as in ordinary CP-NMR. Stejskal, Schaefer and Waugh [68] have addressed this aspect.

Nevertheless, even more information about the structure and dynamics can be obtained when the full chemical shift tensor is exploited. This has been realized by Opella and Waugh [69] in their investigation on oriented polyethylene. If the spinning sideband technique as outlined in Chap. 2.6 is combined with CP-NMR the full chemical tensor shift information for different nuclear sites in the polymer can be determined as is demonstrated in Fig. 4.28 in the case of cis and trans-polyacetylene.

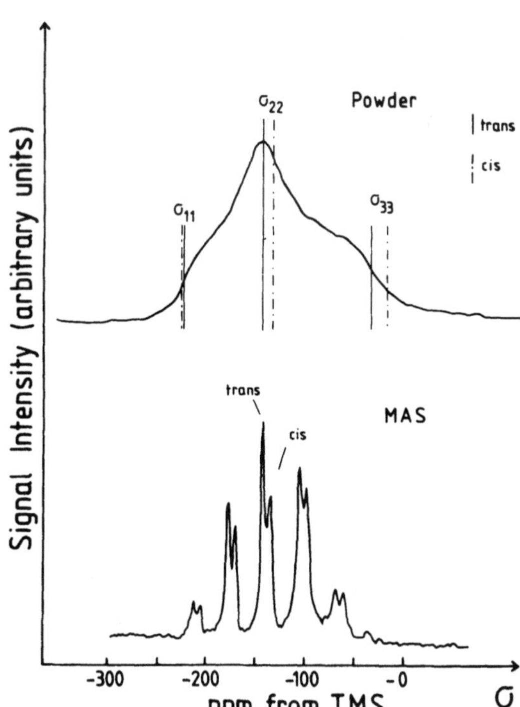

Fig. 4.28. Spinning sideband (bottom) and powder (top) pattern of ^{13}C in cis- and trans-poly-acetylene [70]. The chemical shift tensor has been evaluated using the Herzfeld-Berger method discussed in Sect. 2.6 and is included in Table 7.5. It is comparable with the data obtained elsewhere [76]

e) Coal, clay and food

About any organic material, which always contains protons and some ^{13}C or ^{31}P can be used as a sample in a CP-NMR experiment. Maciel and co-workers [71] have shown that coal, clay, wood etc. can be analyzed by ^{13}C MAS CP-NMR. The differentiation between aromatic and aliphatic content of the sample becomes possible this way. Since there is no other spectroscopic way to determine these parameters MAS CP-NMR has become the working horse in this area [72].

f) Phase transitions

The orientation or electronic structure usually differs in different structural phases of molecular solids. The chemical shift tensor may therefore be the appropriate monitor for the order parameter in different phases of the solid. This has been demonstrates in the case of squaric acid ($H_2C_4O_4$) [73] where an ordered phase exists at room temperature and a phase transition occurs at 374 K. It could be shown that the linesplitting $\Delta\sigma$ due to the ^{13}C chemical shift tensor difference of different nuclear sites is directly related to the order parameter. Fig. 4.29 displays the variation of the linesplitting (order parameter) with temperature [73]. Moreover it was concluded from the chemical shift tensor evaluation, that the electronic structure is switching between the four possible states above the transition temperature.

The evidence comes e.g. from the change in the isotropic shift which would be zero if only molecular reorientation is involved. This, however, is the general case in phase transitions e.g. in KDP ($K_2H_2PO_4$) where Blinc and co-workers [74] detected a change in the chemical shift tensor but not in the isotropic shift at the phase transition. CP-NMR will prove to be very valuable for phase transition investigations in the future.

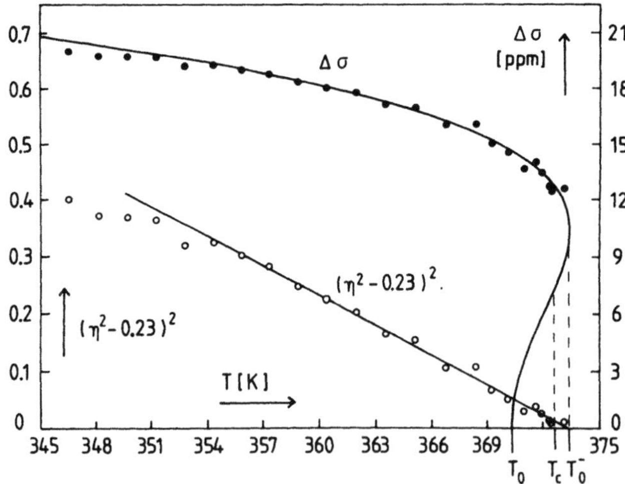

Fig. 4.29. Linesplitting $\Delta\sigma$ of the ^{13}C spectra in squaric acid ($H_2C_4O_4$) versus temperature [73d]. The theoretical lines (full and dashed) are obtained from a Landau theory for first-order phase transition

g) Others

It is beyond the scope of this monograph to discuss all possible applications of CP-NMR. There is such a variety of molecular aggregates in nature which represent good samples for CP-NMR, that the supply of samples will be unlimited. Only a few other classes of substances might be worth mentioning like (i) Resins [75, 76], (ii) surfaces [77] and other chemicals [78].

5 Two-Dimensional NMR Spectroscopy

Spectroscopy usually employs one frequency axis along which the signal intensity is plotted. In high resolution NMR spectroscopy in liquids a wealth of overlapping lines of different nuclear species is commonly observed. The same statement applies to solid state NMR especially if high resolution techniques are applied as described in this monograph. The reader should be aware, that so far only simple highly resolved solid state spectra have been discussed in order to present simple examples. In practice, however, lines of different nuclear species with possibly different near neighbour couplings do often overlap especially in powder samples. A technique is therefore desirable, which employs a second frequency axis representing a different interaction of the spins than the first one. The spectra would be disentangled and correlations could be visualized. Such a technique was proposed for liquids by Jeener [1] at a summer school in Jugoslavia and has been further developed by Ernst and co-workers [2–4] and by Freeman and co-workers [5, 6]. Waugh and co-workers [7–10], Stoll, Vega and Vaughan [11] and Alla and Lippmaa [12] designed 2D-experiments specifically tailored for solid state applications. Since then numerous variations of these basic experiments have been proposed. Some of these we will deal with in Sect. 5.4. Space does not permit a complete review of 2D-spectroscopy. We therefore restrict ourselves to the basic principles (Sect. 5.1) and some applications in solid state NMR. The reader who is interested in liquid state 2D-NMR is referred to the papers by Ernst et al. [2–4] and Freeman et al. [5, 6]. A review on 2D-NMR spectroscopy has been published recently by Freeman and Morris [13].

5.1 Basic Principles of 2D-Spectroscopy

The basic aspects of 2D-spectroscopy are most easily understood if time is separated into four different domains or periods as shown in Fig. 5.1. Preparation, evolution and detection periods are mandatory. Only certain experiments, however, require the mixing period.

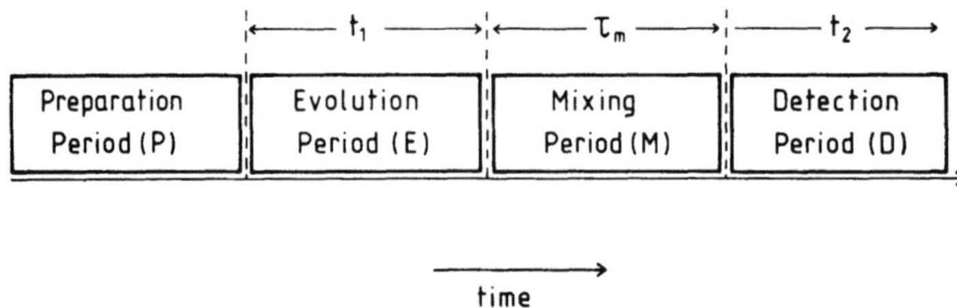

Fig. 5.1. Timing blocks of two-dimensional spectroscopy (Ernst et al. [17])

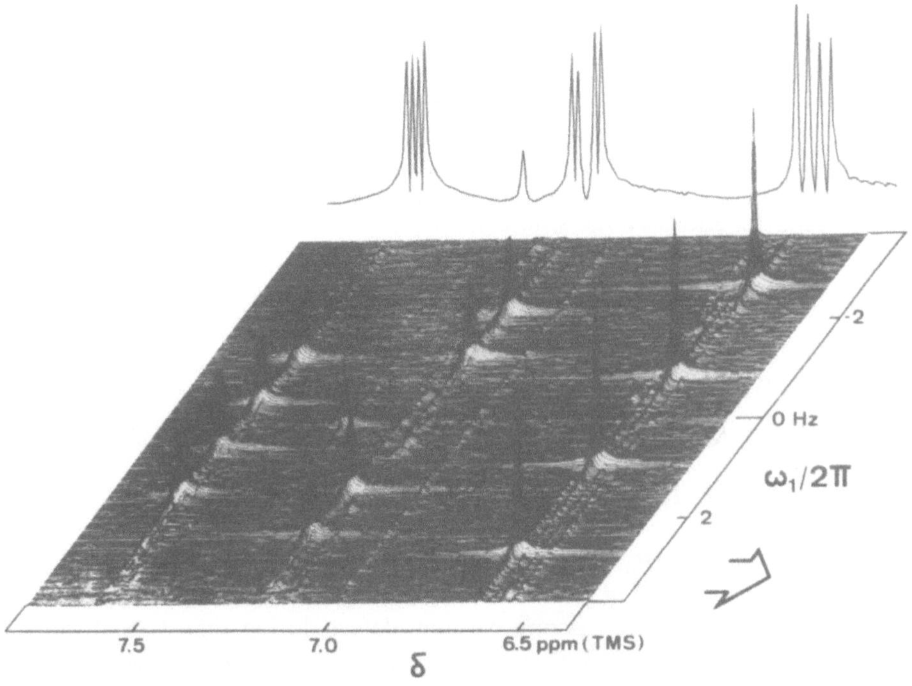

Fig. 5.2. Two-dimensional resolved spectrum of 2-furancarboxylic acid methyl ester in hexafluoro-
benzene and deuterated chloroform according to Aue et al. [14]. The ω_1-axis corresponds to the J-
coupling, whereas the ω_2-axis represents J-couplings and isotropic shifts. The ordinary 1 D-NMR
spectrum shown in the upper part corresponds to the projection onto the ω_2-axis

During the preparation period the system is converted from a Boltzman
equilibrium state into a non-equilibrium state. Subsequent evolution of the
system propagates with time t_1 under the action of some internal Hamiltonian
[1]. The spectral response $\omega_{jk}^{(1)}$ during time t_1 labels the transitions $j \leftrightarrow k$
involved. After evolution for times t_1 detection sets in beginning with a new
time axis t_2. The eigenfrequencies during t_2 are labelled $\omega_{jk}^{(2)}$.

Suppose now that we observe the signal response $M(t_2)$ during detection
period for different values of t_1. A 2D-free induction decay $M(t_1, t_2)$ is obtained, which after 2D-Fouriertransform results in a 2D-spectrum $S(\omega_1, \omega_2)$.
Such a spectrum is shown in Fig. 5.2 [14]. Note that the frequency pair $\omega_{jk}^{(1)}$,
$\omega_{jk}^{(2)}$ determines the coordinates in the 2D-spectrum. If, however, mixing of
different transitions (e.g. coherence transfer) is employed during the mixing
period for some time τ_m peaks from pairs like $\omega_{jk}^{(1)}$, $\omega_{lm}^{(2)}$ appear in the 2D-
spectrum and cross-correlations can be determined.

(i) homonuclear spinsystems
Suppose our sample contains a number of spins I_j with eigenfrequencies ω_j
due to different chemical shifts for example. The spins are coupled by a scalar
interaction constant J_{jk}. The Hamiltonian of the system can then be expressed

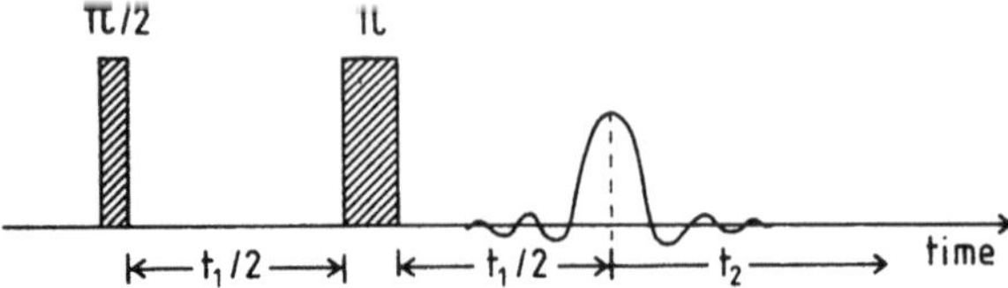

Fig. 5.3. Simple pulse sequence (Hahn echo) for two-dimensional spectroscopy of homonuclear coupled spins. During the evolution period (t_1) shift interactions are eliminated "on the average" by the π-pulse. The detection period starts at the echo peak

as

$$\mathcal{H} = \mathcal{H}_{ZI} + \mathcal{H}_{II} \tag{5.1a}$$

$$\mathcal{H}_{ZI} = \sum_j \omega_j I_{Zj} \tag{5.1b}$$

$$\mathcal{H}_{II} = \sum_{j<k} J_{jk} \mathbf{I}_j \cdot \mathbf{I}_k \tag{5.1c}$$

where $\omega_j = -\gamma_I(1-\sigma)B_0$. The 2D-experiment we want to apply to this situation is sketched in Fig. 5.3. This is nothing but the Hahn-echo sequence [15]. In view of 2D-spectroscopy we now detect during time t_2 half the echo response for different t_1. Both pulses are considered to be non-selective, i.e. they cover the full spectral width. The π-pulse refocusses all linear spin interactions during t_1, i.e. chemical shift interactions are eliminated during t_1. The scalar interaction, however, is invariant under π rotation, i.e. $\mathcal{H}(1) = \mathcal{H}_{II}$.

Sampling only the echo amplitude at $t_2 = 0$ for different t_1 would lead to $M(t_1)$ which after Fourier-transform results in a J-resolved spectrum $S(\omega_1)$ which only reflects the J couplings. During t_2 on the other hand the full Hamiltonian is active, i.e. $\mathcal{H}(2) = \mathcal{H}_{ZI} + \mathcal{H}_{II}$. In a 2D-experiment according to Fig. 5.3 we therefore obtain for the transverse magnetization M_j of a resonance line ω_j within a multiplet due to the coupling J_{jk} to k magnetically equivalent nuclei [2–6]

$$M_j(t_1, t_2) = M_j(0,0) \cos\{J_{jk} m_k t_1 + (\omega_j + J_{jk} m_k) t_2\} \tag{5.2}$$

where m_k is the quantum number of the nuclear spin k. We have neglected so far relaxation effects and B_0 inhomogeneity. In order to simulate real spectra these aspects can be included phenomenologically as [4]

$$M_j^*(t_1, t_2) = M_j(t_1, t_2) \exp[-(t_1 + t_2)/T_{2j}]$$
$$\cdot \exp[-(t_2/T_2^*)^2] \exp[-D\gamma_I^2 G^2(t_1 + t_2)^3/12] \tag{5.3}$$

where the first relaxation term is caused by the transverse relaxation time T_{2j}, whereas the second and third relaxation terms take the B_0 inhomogeneity (T_2^*) and diffusion (D) in the magnetic field gradient G into account, respectively.

After 2D-Fourier-transform of Eq. (5.3) a 2D-spectrum results where the projection onto the ω_1-axis yields a J-resolved spectrum, whereas projection onto the ω_2-axis results in the ordinary 1D-NMR-spectrum. An example of this type of experiment is displayed in Fig. 5.2 [14].

(ii) heteronuclear spinsystems

Suppose our sample contains I-spins (e.g. ^{1}H) and S-spins (e.g. ^{13}C) which are coupled by scalar interactions J_{IS}. The Hamiltonian of the spin system is readily expressed as

$$\mathcal{H} = \mathcal{H}_{ZI} + \mathcal{H}_{II} + \mathcal{H}_{ZS} + \mathcal{H}_{IS} \tag{5.4a}$$

$$\mathcal{H}_{ZS} = \omega_S S_Z \tag{5.4b}$$

$$\mathcal{H}_{IS} = \sum_j J_{ISj} I_{Zj} S_Z \tag{5.4c}$$

and where $\mathcal{H}_{ZI}$ and $\mathcal{H}_{II}$ are given by Eq. (5.1). We want to observe the S spin resonance only and aim at a complete separation of J_{IS}-coupling and chemical shift ω_S. Indeed this is possible in a heteronuclear 2D-experiment as follows. During the evolution period (t_1) we want to eliminate S and I chemical shifts which is possible by combined π-pulses at ω_I and ω_S at $t_1/2$ in the same spirit as before (i), i.e. $\mathcal{H}(1) = \mathcal{H}_{IS} + \mathcal{H}_{II}$. During detection (t_2) we decouple the I spins by application of a strong B_1-field at the resonance frequency ω_I of the I spins, i.e. $\mathcal{H}(2) = \mathcal{H}_{ZS}$. The pulse sequence of two possible experiments of this type are shown in Fig. 5.4. In both cases (a) and (b) saturation of the $I(^1$H)-spins is performed in the preparation period to create $S(^{13}$C)-spin polarization by means of the Overhauser effect. Experiments according to Fig. 5.4a, however, are known to show fake lines, when I-spin couplings are strong. The modification according to Fig. 5.4b remmedies this defect. Note, however, that now $\mathcal{H}(1)$ has to be replaced by $\mathcal{H}(1) = \frac{1}{2}(\mathcal{H}_{IS} + \mathcal{H}_{ZI} + \mathcal{H}_{II})$ whereas $\mathcal{H}(2) = \mathcal{H}_{ZS}$ as before.

We readily obtain the magnetization $M_j(t_1, t_2)$ in the heteronuclear 2D-experiment according to Fig. 5.4a as [16]

$$M_j(t_1, t_2) = M_j(0, 0) \cos(J_{ISk} m_k t_1 + \omega_{Sj} t_2). \tag{5.5}$$

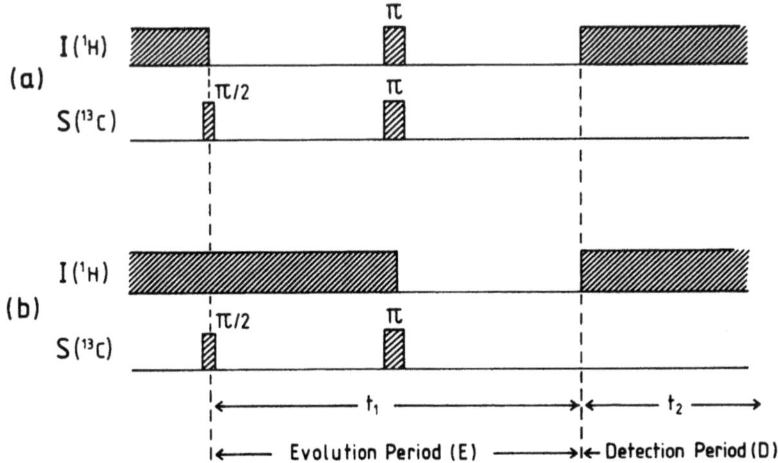

Fig. 5.4a, b. Pulse timing for heteronuclear J-resolved two-dimensional spectroscopy [13]. **a** Heteronuclear Hahn echo may produce fake lines if coupling among the I spins is strong. **b** Improved version has not the drawback of **a**

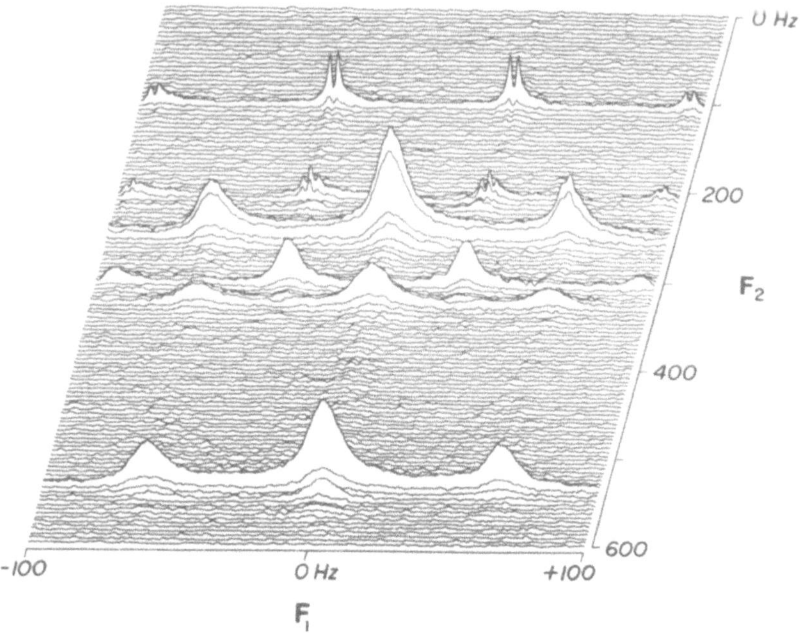

Fig. 5.5. Two-dimensional J versus shift spectrum of ^{13}C in 2(1-methylcyclohexyl)-4,6-dimethylphenol, according to Freeman and Morris [13]. Note the fine structure on two of the methyl group quartets, which identifies them as the aromatic 4-methyl (triplet structure) and 6-methyl (doublet structure)

Relaxation terms may be included as in Eq. (5.3). The corresponding 2D-spectrum displays the J_{IS}-coupling along the ω_1-axis and the chemical shift ω_{S_j} along the ω_2-axis [13]. The modification according to Fig. 5.4b results in half the J_{IS}-coupling along the ω_1-axis. A representative example is shown in Fig. 5.5.

Ernst et al. have constructed building blocks for a 2D-spectroscopy kit and established guide lines for the design of 2D-experiments [17]. We shall follow here closely their philosophy. Two different spin systems I (abundant) and S (rare) are considered. The corresponding interaction Hamiltonian are expressed as

liquid phase: $\quad \mathcal{H} = \mathcal{H}_I + \mathcal{H}_{II} + \mathcal{H}_{IS} + \mathcal{H}_S$ (5.6a)

solid phase: $\quad \mathcal{H} = \mathcal{H}_I^{\text{iso}} + \mathcal{H}_I^{\text{aniso}} + \mathcal{H}_{II} + \mathcal{H}_{IS} + \mathcal{H}_S^{\text{iso}} + \mathcal{H}_S^{\text{aniso}}$ (5.6b)

where only Zeeman and scalar or dipolar interactions are taken into account.

Figure 5.6 shows a set of building blocks for 2D-experiments which have been collected by Ernst et al. [17].

The three experiments we have discussed above in Figs. 5.3 and 5.4 can be expressed in terms of these building blocks as 5.3: Pa, Eb, Da in the homonuclear case (i) and 5.4a: Pd, Eh, Dd or 5.4b: Pd, Ei, Dd in the heteronuclear case (ii). A wealth of 2D-experiments can and has been designed by combination and permutation of these building blocks.

1. *Preparation period:*

Pa. Non-selective excitation of single quantum transitions: Single 90° pulse (Selective excitation of single or multiple quantum transition is also feasible)

Pb. Non-selective excitation of multiple quantum transitions: 90° pulse pair

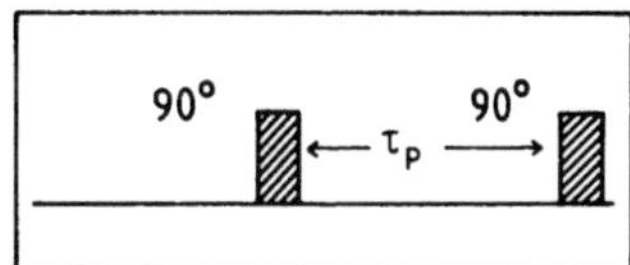

Pc. Selective excitation of double and zero quantum transitions: selective 90°/180° pulses

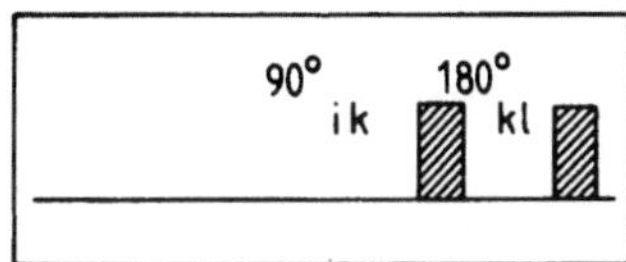

Pd. Heteronuclear Overhauser polarization enhancement and 90° pulse:

Pe. Hartmann-Hahn cross-polarization in the rotating frame: $\gamma_I B_I = \gamma_S B_S$

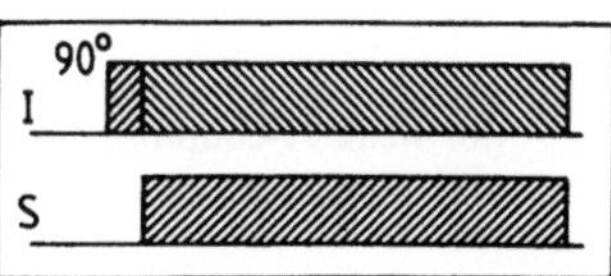

Pf. Spin inversion for relaxation experiments:

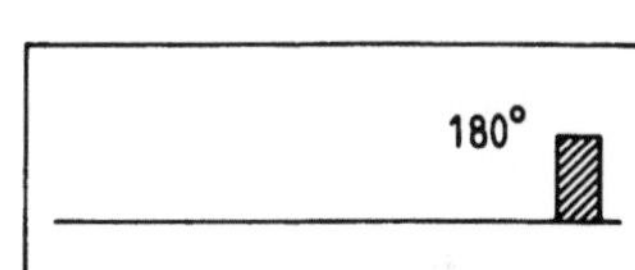

Fig. 5.6a

2. *Evolution period:* The following average Hamiltonians $\bar{\mathscr{H}}^{(1)}$ can be created to effect the frequency spread in the ω_1 direction:

Homonuclear systems in liquids:

Ea. Full unperturbed Hamiltonian:
$$\bar{\mathscr{H}}^{(1)} = \mathscr{H}$$

Eb. Elimination of Zeeman interaction (weak coupling):
$$\bar{\mathscr{H}}^{(1)} = \mathscr{H}_{II}$$

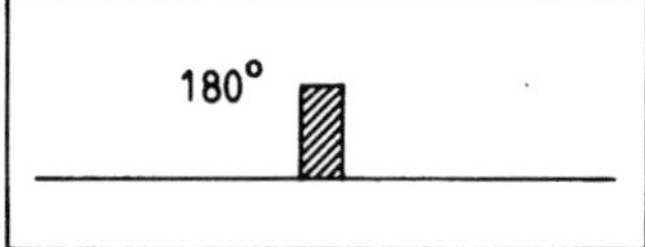

Ec. Elimination of Zeeman interaction (weak coupling):
$$\bar{\mathscr{H}}^{(1)} = \mathscr{H}_{II}$$

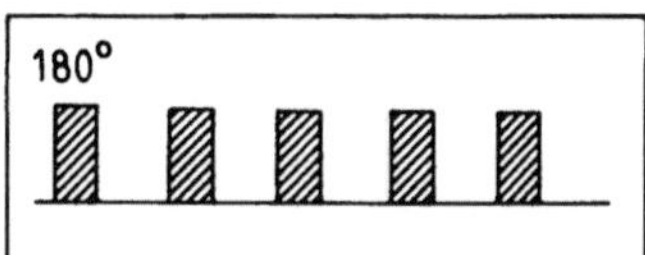

Fig. 5.6b

Heteronuclear systems in liquids:

Ed. Heteronuclear decoupling
(coherent or noise-modulated):
$$\bar{\mathcal{H}}^{(1)} = \mathcal{H}_S$$

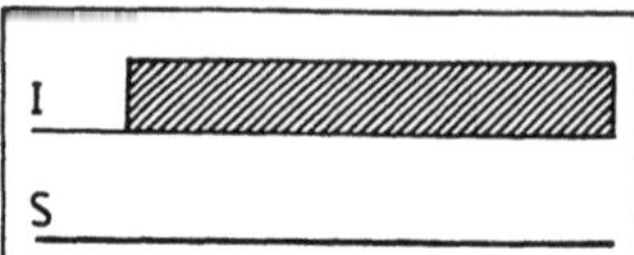

Ee. Heteronuclear decoupling
(coherent or noise-modulated):
$$\bar{\mathcal{H}}^{(1)} = \mathcal{H}_I + \mathcal{H}_{II}$$

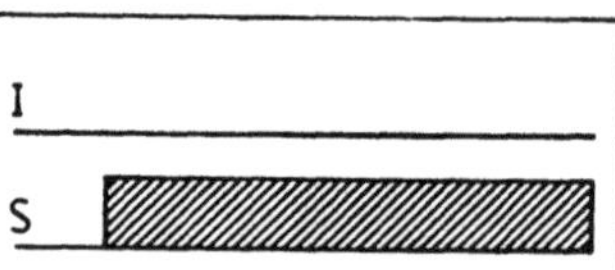

Ef. Heteronuclear refocusing (for weak II-coupling only):
$$\bar{\mathcal{H}}^{(1)} = \mathcal{H}_S + \mathcal{H}_{II}$$

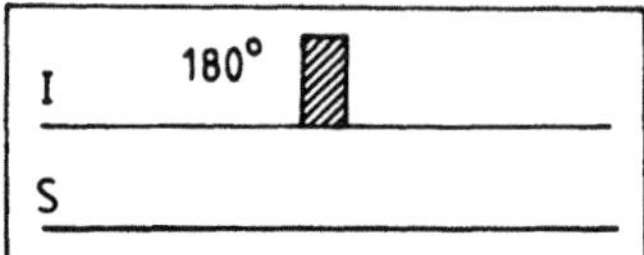

Eg. Heteronuclear refocusing (for weak II-coupling only):
$$\bar{\mathcal{H}}^{(1)} = \mathcal{H}_I + \mathcal{H}_{II}$$

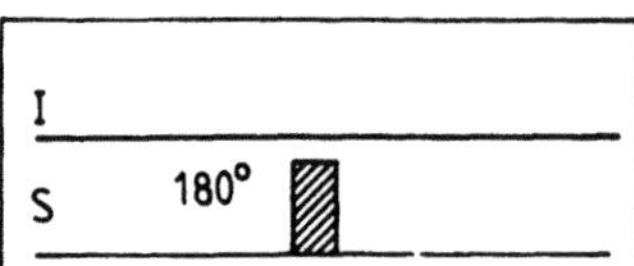

Eh. Elimination of Zeeman interactions
(for weak II-coupling only):
$$\bar{\mathcal{H}}^{(1)} = \mathcal{H}_{IS} + \mathcal{H}_{II}$$

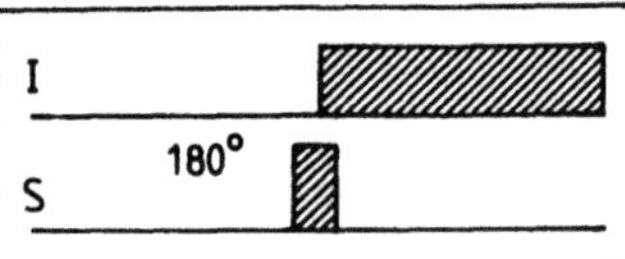

Ei. Elimination of S-Zeeman interaction:
$$\bar{\mathcal{H}}^{(1)} = \tfrac{1}{2}(\mathcal{H}_{IS} + \mathcal{H}_{II} + \mathcal{H}_I)$$

Fig. 5.6b

By off-center pulses at $t \neq t_1/2$ or by using decoupling during a part of the evolution period, only, scaling of spin interactions can be effected as well.

3. *Mixing period:*

Ma. No coherence transfer:

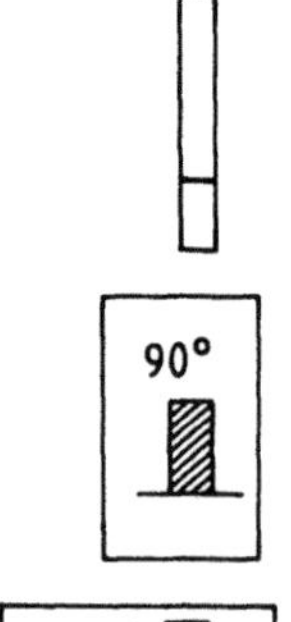

Mb. Complete homonuclear mixing (including multiple quantum transitions). (Partial mixing by pulses with rotation angles $<90°$ may also lead to informative 2D spectra)

Mc. Differential heteronuclear coherence transfer:

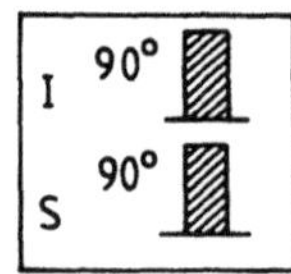

Fig. 5.6c

Md. Net heteronuclear coherence transfer by delayed pulses:

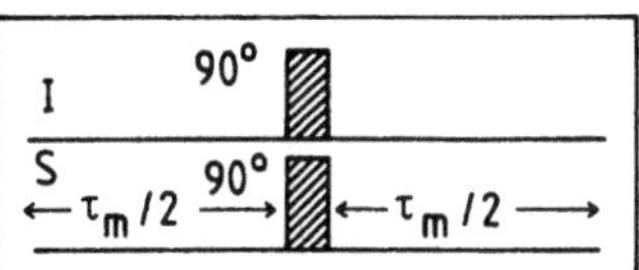

Me. Net heteronuclear coherence or polarization transfer:
$$\gamma_I B_I = \gamma_S B_S$$

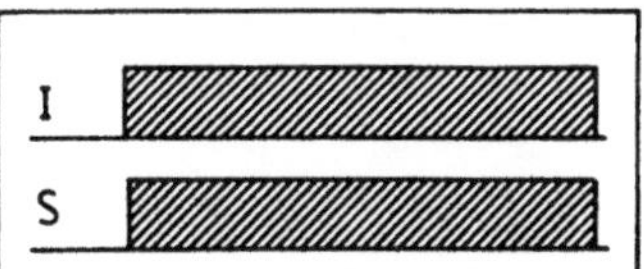

Mf. Study of cross-relaxation and of chemical exchange processes with optional field gradient pulse g_y:

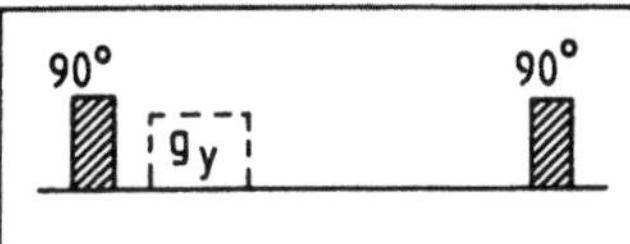

Fig. 5.6c

4. *Detection period:* During detection again a modified Hamiltonian $\bar{\mathcal{H}}^{(2)}$ may be used to obtain a suitable frequency spread in the ω_2 direction. The Hamiltonians, described for evolution, may be used whenever they involve exclusively continuous or periodic perturbations, i.e.

$Da = Ea, Dc = Ec, Dd = Ed, De = Ee, Dk = Ek, Dl = El, Dm = Em, Dn = En, Do = Eo,$
$Dp = Ep$ and $Dq = Eq.$

Homonuclear systems in solids or in liquid crystals:

Ek. Elimination of dipolar interactions, multipulse sequences:
$$\bar{\mathcal{H}}^{(1)} = \mathcal{H}_I^{iso} + \frac{1}{\sqrt{3}} \mathcal{H}_I^{aniso}$$

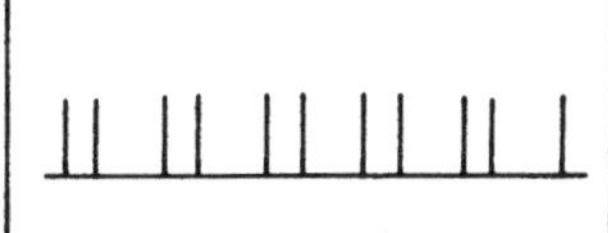

El. Elimination of dipolar interactions, off-resonance rf field at magic angle:
$$\bar{\mathcal{H}}^{(1)} = \mathcal{H}_I^{iso} + \frac{1}{\sqrt{3}} \mathcal{H}_I^{aniso}$$

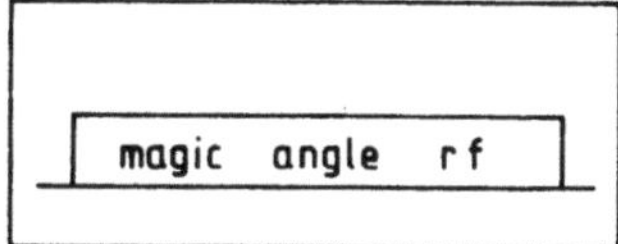

Em. Elimination of anisotropic interaction, magic angle rotation:
$$\bar{\mathcal{H}}^{(1)} = \mathcal{H}_I^{iso}$$

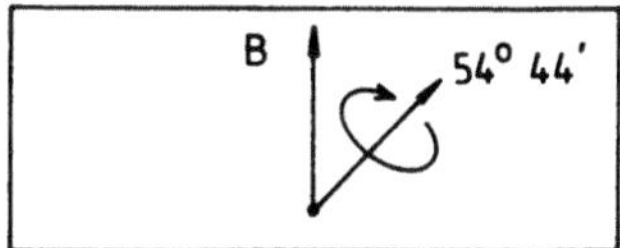

Heteronuclear systems in solids or in liquid crystals:

En. Heteronuclear decoupling: $\bar{\mathcal{H}}^{(1)} = \mathcal{H}_S^{iso} + \mathcal{H}_S^{aniso}$

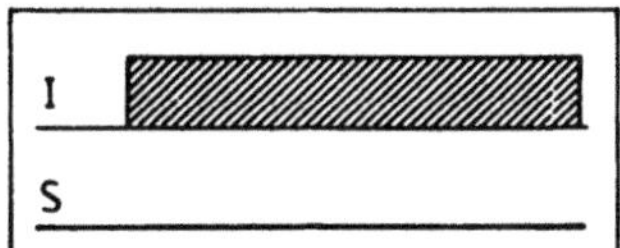

Eo. Elimination of homonuclear dipolar interactions by multipulse sequences:
$$\bar{\mathcal{H}}^{(1)} = \mathcal{H}_S^{iso} + \mathcal{H}_S^{aniso} + \frac{1}{\sqrt{3}} \mathcal{H}_{IS} + \mathcal{H}_I^{iso} + \frac{1}{\sqrt{3}} \mathcal{H}_I^{aniso}$$

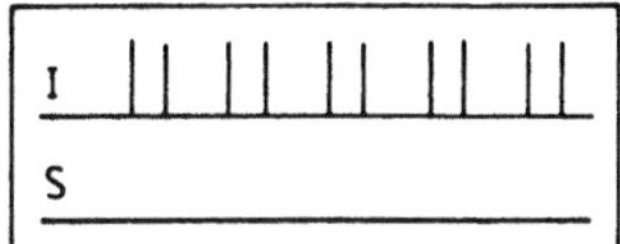

Fig. 5.6d

Ep. Elimination of homonuclear dipolar interactions by off-resonance rf at magic angle:

$$\bar{\mathcal{H}}^{(1)} = \mathcal{H}_S^{iso} + \mathcal{H}_S^{aniso} + \frac{1}{\sqrt{3}}\,\mathcal{H}_{IS} + \mathcal{H}_I^{iso} + \frac{1}{\sqrt{3}}\,\mathcal{H}_I^{aniso}$$

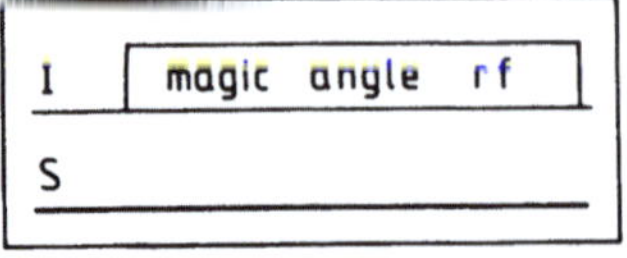

Eq. Elimination of anisotropic interactions by magic angle rotation and decoupling:

$$\bar{\mathcal{H}}^{(1)} = \mathcal{H}_S^{iso}$$

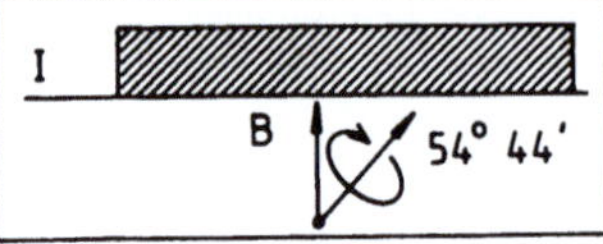

Fig. 5.6d

In addition, pulse sequences synchronized to magic angle spinning lead to further possibilities to tailor the average Hamiltonian differently in evolution and detection periods.

Fig. 5.6a–d. Building blocks of 2D spectroscopy kit. (Ernst et al. [17])

Before discussing some aspects of 2D-spectroscopy to solids let us briefly summarize some peculiarities which arise during 2D-Fourier-transformation. The 2D-Fourier-transform of the 2D-FID $M(t_1, t_2)$ reads [4, 13]

$$S(\omega_1, \omega_2) = \int_0^\infty dt_1 \exp(-i\omega_1 t_1) \int_0^\infty dt_2 \exp(-i\omega_2 t_2)\, M(t_1, t_2) \tag{5.7}$$

where

$$M(t_1, t_2) = 0 \quad \text{for } t_1, t_2 < 0$$

i.e. the FID only exists for $t_1, t_2 \geqq 0$.

Equation (5.7) results in 4 terms

$$
\begin{aligned}
S(\omega_1 > 0, \omega_2) &= S^{CC}(\omega_1, \omega_2) - S^{SS}(\omega_1, \omega_2) &&\text{real part}\\
&\quad - i\{S^{CS}(\omega_1, \omega_2) + S^{SC}(\omega_1, \omega_2)\} &&\text{imaginary part}
\end{aligned}
\tag{5.8a}
$$

and

$$
\begin{aligned}
S(\omega_1 < 0, \omega_2) &= S^{CC}(\omega_1, \omega_2) + S^{SS}(\omega_1, \omega_2) &&\text{real part}\\
&\quad - i\{S^{CS}(\omega_1, \omega_2) - S^{SC}(\omega_1, \omega_2)\} &&\text{imaginary part}
\end{aligned}
\tag{5.8b}
$$

where

$$S^{CS}(\omega_1, \omega_2) = \int_0^\infty dt_1 \cos(\omega_1 t_1) \int_0^\infty dt_2 [-\sin\omega_2 t_2]\, M(t_1, t_2) \tag{5.9}$$

etc.

In order to discuss the lineshape problems which occur in 2D-spectroscopy we consider a model FID which mainly consists of two different eigenfrequencies Ω_1, Ω_2 with corresponding line broadenings $1/T_2$ and $1/T_2^*$.

$$M(t_1, t_2) = M(0,0) \cos(\Omega_1 t_1 + \Omega_2 t_2) \exp(-t_1/T_2) \exp(-t_2/T_2^*) \tag{5.10}$$

resulting in

$$
\begin{aligned}
\operatorname{Re} S(\omega_1, \omega_2) &= S^{CC}(\omega_1, \omega_2) - S^{SS}(\omega_1, \omega_2)\\
&= \tfrac{1}{2} M(0,0)\,\{a(\omega_1)\, a(\omega_2) - d(\omega_1)\, d(\omega_2)\}
\end{aligned}
\tag{5.11}
$$

where

$$a(\omega) = \frac{1/T_2}{(\omega - \Omega_1)^2 + (1/T_2)^2} \tag{5.12}$$

with $a(\omega)$ being a Lorentzian lineshapes of the absorption type, and

$$d(\omega) = \frac{\omega - \Omega_1}{(\omega - \Omega_1)^2 + (1/T_2)^2} \tag{5.13}$$

being the corresponding dispersion like lineshape, which is mixed into the real part of $S(\omega_1, \omega_2)$.

The total lineshape is therefore a mixture of absorption and dispersion like lineshapes which make the spectrum look very unpleasant. Moreover this leads to linebroadening and linedistortion which make the disentanglement of close lines difficult [4, 13].

Several schemes have been proposed to cope with this problem. The simplest one suggests to calculate the absolute value spectrum, i.e. [4, 13]

$$|S(\omega_1 > 0, \omega_2)| = \{[S^{CC} - S^{SS}]^2 + [S^{CS} + S^{SC}]^2\}^{1/2} \tag{5.14a}$$

$$|S(\omega_1 < 0, \omega_2)| = \{[S^{CC} + S^{SS}]^2 + [S^{CS} - S^{SC}]^2\}^{1/2}. \tag{5.14b}$$

The line distortion is eliminated by this procedure, however, linebroadening is still severe.

An elegant method has been proposed by Ernst and co-workers [14] in the heteronuclear case. It is based on the general principle of phase alternating before the detection period to eliminate the sin-terms. If only one frequency term is inverted

$$\bar{M}(t_1, t_2) = M(0, 0) \cos(-\Omega_1 t_1 + \Omega_2 t_2). \tag{5.15}$$

When several FID's are accumulated using phase alternation an average FID of the form

$$\begin{aligned}
\langle M(t_1, t_2) \rangle &= M(t_1, t_2) + \bar{M}(t_1, t_2) \\
&= M(0, 0) \cos \Omega_1 t_1 \cos \Omega_2 t_2
\end{aligned} \tag{5.16}$$

results. The 2D-Fourier-transform of $\langle M(t_1, t_2) \rangle$ has the same lineshape features as the usual 1D-spectrum.

5.2 2D-Spectroscopy of ^{13}C–^{1}H Interactions in Solids

Separated local field spectroscopy (SLF) of rare ^{13}C nuclei in solids was introduced by Waugh [8] and Waugh and co-workers [7–10], and Alla and Lippmaa [12] to uncover heteronuclear dipole-dipole interactions in solids. These experiments were soon converted to 2D-spectroscopy to display heteronuclear dipole-dipole interaction (^{13}C–^{1}H) versus chemical shift (of ^{13}C) in a 2D-spectrum [9, 10]. The corresponding pulse scheme is shown in Fig. 5.7. During the preparation period T_{IS} a Hartmann-Hahn type cross-polarization experiment as described in Chap. 5 is performed to polarize the ^{13}C spins. After termination of the rf fields evolution of the $S(^{13}$C)-spin magnetization sets in under the action of the full Hamiltonian $\mathscr{H}(1) = \mathscr{H}_{II} + \mathscr{H}_{IS} + \mathscr{H}_S$. In the case of not too large magnetic fields B_0 the evolution period (t_1) is governed

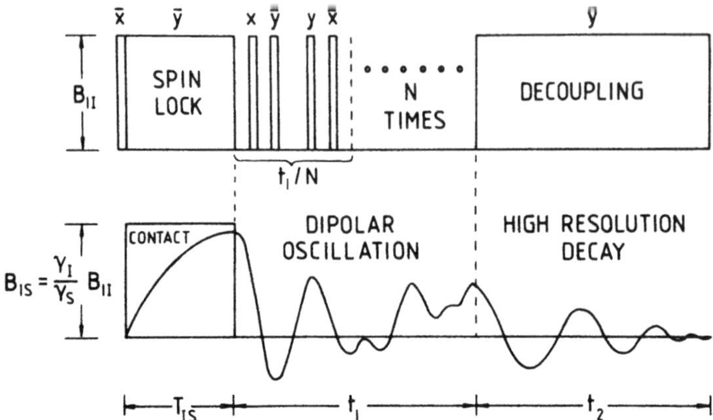

Fig. 5.7. Pulse scheme for 2D spectroscopy of ^{13}C–^1H interactions according to Rybaczewski et al. [10]. During evolution period (t_1) mainly heteronuclear spin interactions $\mathscr{H}_{IS}$ are effective, whereas shift interactions govern the time evolution during the detection period (t_2) as explained in the text

mainly by heteronuclear dipole interactions $\mathscr{H}_{IS}$, which may be "dressed" by strong homonuclear interactions $\mathscr{H}_{II}$ among the $I(^1H)$-spins (see Sect. 4.4). In order to eliminate $\mathscr{H}_{II}$ during t_1 "magic angle decoupling" or a 4-pulse experiment may be applied to the $I(^1H)$-spin resonance as proposed by Waugh and co-workers [7]. One should be aware, however, that according to the arguments used in Sect. 4.4 the spectral lines will become broader in this case instead of narrower. Detection period (t_2) is initiated by turning on the decoupling field at the I-spin resonance, to make $\mathscr{H}(2)=\mathscr{H}_S$. The time evolution of the density matrix after the cross-polarization (preparation period) can be described in Liouville-space notation as introduced in Appendix C as

$$|\rho(t_1,t_2))=e^{-it_2\,\hat{\mathscr{H}}(2)}\,e^{-it_1\,\hat{\mathscr{H}}(1)}\,|\rho(0)) \tag{5.17}$$

which results in the 2D-FID (with $\rho(0)=S_x$)

$$g_\alpha(t_1,t_2)=(S_\alpha|\,e^{-it_2\,\hat{\mathscr{H}}(2)}\,e^{-it_1\,\hat{\mathscr{H}}(1)}\,|S_x)/(S_\alpha|S_\alpha)\qquad \alpha=x,y. \tag{5.18}$$

After a 2D-Fourier transform this results in a 2D-spectrum $S(\omega_1,\omega_2)$. In order to be specific we consider a non-trivial example, namely a CH_2-group. This situation among others has been discussed by Rybaczewski et al. [10] and is nicely demonstrated in Fig. 5.8.

Starting with the Hamiltonian

$$\mathscr{H}(1)=\omega_S S_Z+2(B_1 I_{Z1}+B_2 I_{Z2})S_Z+A_{12}(3I_{Z1}I_{Z2}-\mathbf{I}_1\cdot\mathbf{I}_2) \tag{5.19}$$

where ($j=1,2$)

$$B_j=(\gamma_I\gamma_S\hbar/r_{IS}^3)\,P_2(\cos\vartheta_j) \tag{5.20a}$$

$$A_{12}=(\gamma_I^2\hbar/r_{II}^3)\,P_2(\cos\varphi) \tag{5.20b}$$

with ϑ_j being the angle between the radius vector from the $S(^{13}C)$-spin to spin I_j and the magnetic field B_0, whereas φ is the angle between r_{II} and B_0.

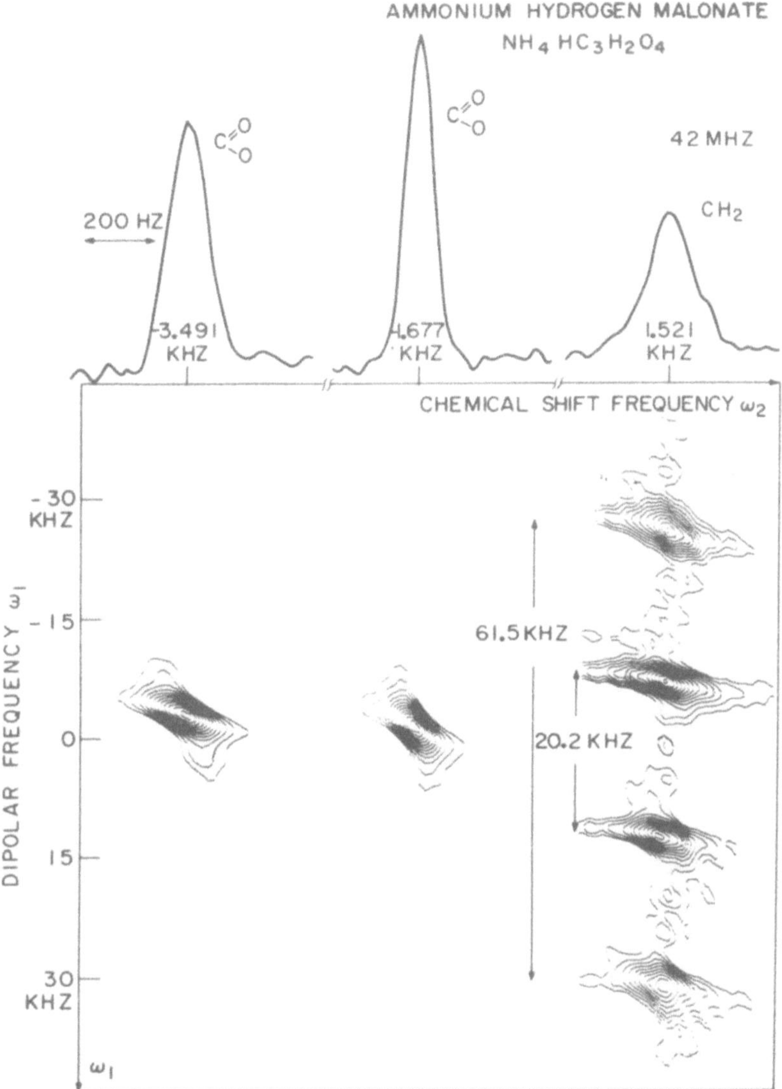

Fig. 5.8. Two-dimensional ^{13}C spectrum of NH$_4$OOC CH$_2$COOH (ammonium hydrogen malonate) according to Rybaczewski et al. [10]. The spectrum was obtained by using the pulse scheme shown in Fig. 5.7 which allows to plot the heteronuclear dipolar interaction (ω_1) versus chemical shift frequency (ω_2). The projection onto the ω_2-axis is shown at the top and represents the ordinary ^{13}C high-resolution NMR spectrum in the solid state. (Courtesy of J.S. Waugh)

During detection period (t_2) the $I(^1$H)-spins are completely decoupled, i.e.

$$\mathcal{H}(2) = \omega_S S_z. \tag{5.21}$$

If I-I interaction is neglected in Eq. (5.19) we obtain

$$g_x(t_1, t_2) = \frac{(S_x|\, e^{-i\omega_S(t_1+t_2)\,\hat{S}_z}\, e^{-it_1\, 2(B_1\hat{I}_{z1} + B_2\hat{I}_{z2})\,\hat{S}_z}\, |S_x)}{(S_x|S_x)} \tag{5.22}$$

which reduces immediately to

$$g_x(t_1,t_2)=\cos\omega_S(t_1+t_2)\frac{(S_x|\cos[2(B_1I_{z1}+B_2I_{z2})t_1]|S_x)}{(S_x|S_x)}. \tag{5.23}$$

In Eq. (5.23) spin I can be any integer or half-integer spin. We now restrict ourselves to $I=1/2$ and take the trace to obtain

$$g_x(t_1,t_2)=\cos\omega_S(t_1+t_2)\{2\cos(B_1+B_2)t_1+2\cos(B_1-B_2)t_1\} \tag{5.24}$$

and after 2D-Fourier-transformation a four-line spectrum results along the ω_1-axis with the frequencies

$$\Omega_{1,4}=\omega_S\mp(B_1+B_2)$$
$$\Omega_{2,3}=\omega_S\mp(B_1-B_2)$$

and equal intensity. Figure 5.8 displays this feature for the CH_2-group, whereas the other carbons with no near neighbour protons show up as single lines at ω_S along the ω_1-axis. The ω_2-axis separates the different chemical shifts ω_S [10].

However, the neglect of A_{12} in Eq. (5.19) is not justified. If A_{12} is included, Rybaczewski et al. [10] obtain also a four line spectrum along ω_1, however with different intensities and splittings

$$\Omega_{1,4}=\omega_S\mp(B_1+B_2)\qquad\qquad\begin{array}{c}\text{Intensity}\\1\end{array}$$
$$\Omega_{2,3}=\omega_S\mp[(B_1-B_2)^2+A_{12}^2]^{\frac{1}{2}}\quad \lambda^4(1-\varepsilon^2) \tag{5.25}$$

where

$$\lambda=\frac{\Omega_a-\Omega_b}{[(\Omega_a-\Omega_b)^2+A_{12}^2]^{\frac{1}{2}}} \tag{5.26}$$

and

$$\varepsilon=\frac{-A_{12}}{\Omega_a-\Omega_b}$$

with

$$\Omega_a=[(B_1-B_2)^2+A_{12}^2]^{\frac{1}{2}}\quad\text{and}\quad\Omega_b=B_1-B_2.$$

In this section we have discussed the 2D-spectroscopy of $^{13}C-^{1}H$ interactions in single crystals only. In practice, however, very often only powder samples are available. Opella and Waugh [9] have demonstrated recently that $^{13}C-^{1}H$ 2D-spectroscopy can be performed in polymers, where especially interesting lineshapes are observed when the polymer is partially oriented as was discussed already in Sect. 2.5. Stoll, Vega and Vaughan [11] have observed time resolved ^{13}C chemical shift spectra in benzene and have proposed schemes for observing 2D-spectra applying multiple-pulse cycles. Linder, Höhener and Ernst [18] have treated the general problem of how to determine the orientation of tensorial interactions from two-dimensional NMR powder spectra. We shall briefly discuss their results.

Remember that the 1D-powder spectrum can be expressed in the form [18]

$$S(\omega) = \frac{1}{4\pi} \sum_j \int_0^{2\pi} \int_0^\pi a_j(\vartheta, \varphi)\, g_j[\omega - \omega_j(\vartheta, \varphi)] \sin\vartheta\, d\vartheta\, d\varphi \qquad (5.27)$$

where ϑ, φ are the polar angles which describe the orientation of the randomly distributed crystallites with respect to the static magnetic field B_0. The eigenfrequencies $\omega_j(\vartheta, \varphi)$ are the transition frequencies of the orientation dependent Hamiltonian $\mathscr{H}(\vartheta, \varphi)$. The line-shape function $g_j(\omega)$ may be orientation dependent (e.g. by $T_2(\vartheta, \varphi)$ or homonuclear dipole-dipole interaction) but will often be assumed to be universal, i.e. independent of j, ϑ and φ. The line intensities $a_j(\vartheta, \varphi)$ are given by the transition matrix elements. Summation is performed over all allowed transitions. In 2D-spectroscopy two Hamiltonian $\mathscr{H}(1)$ with transition frequencies ω_j and $\mathscr{H}(2)$ with ω_k are created during the evolution period (1) and the detection period (2). The 2D-spectrum consists of contributions from all possible cross-peaks (jk): [18]

$$S(\omega_1, \omega_2) = \frac{1}{4\pi} \sum_{j,k} \int_0^{2\pi} \int_0^\pi a_{jk}(\vartheta, \varphi)\, g_j[\omega_1 - \omega_j^{(1)}(\vartheta, \varphi)]$$

$$\cdot g_k[\omega_2 - \omega_k^{(2)}(\vartheta, \varphi)] \sin\vartheta\, d\vartheta\, d\varphi. \qquad (5.28)$$

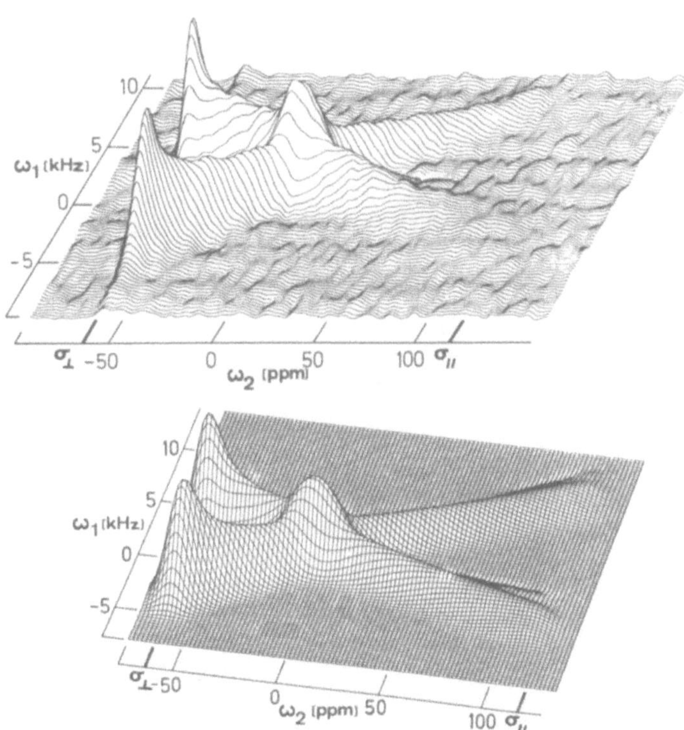

Fig. 5.9a, b. ^{13}C Chemical shift resolved dipolar powder spectrum of benzene (C_6H_6) according to Linder et al. [18]. **a** Top: The experimental ^{13}C spectrum was obtained at 25 MHz and at 148 K. **b** Bottom: Computer calculation of the 2D spectrum by assuming that the symmetric chemical shift and heteronuclear dipole tensor are coaxial. (Courtesy of R.R. Ernst)

In simple cases (usually encountered in ^{13}C-^1H NMR) all $a_{jk}(\vartheta,\varphi)$ are equal and independent of ϑ and φ.

To be specific let us treat the case of ^{13}C-^1H spectroscopy where only the chemical shift $\mathscr{H}_{zS}$ of the $S(^{13}C)$-spins and the heteronuclear dipole-dipole interaction $\mathscr{H}_{IS}$ between protons (I-spins) and the ^{13}C nuclei is taken into account. Interaction among the protons is assumed to be eliminated by a "magic angle irradiation" or a corresponding multiple pulse sequence. The experimental scheme of such an experiment is shown in Fig. 5.7. The effective Hamiltonian during

$$\text{evolution:} \quad \mathscr{H}(1)=\frac{1}{\sqrt{3}}\,\mathscr{H}_{IS}+\mathscr{H}_{zS}$$

and

$$\text{detection:} \quad \mathscr{H}(2)=\mathscr{H}_{zS}$$

periods now determine the 2D-spectrum. The ω_2-axis represents solely the chemical shift interaction, whereas the ω_1-axis contains chemical shift as well as scaled ^{13}C-^1H dipole interaction. In not too large magnetic fields, however, the dipole-dipole interaction dominates the shift interaction.

As an illustrative example Fig. 5.9 shows the 2D-spectrum of ^{13}C in benzene (C_6H_6) as obtained by Linder et al. [18]. The simulated spectra compare very well with the experimental ones. The situation here is fairly

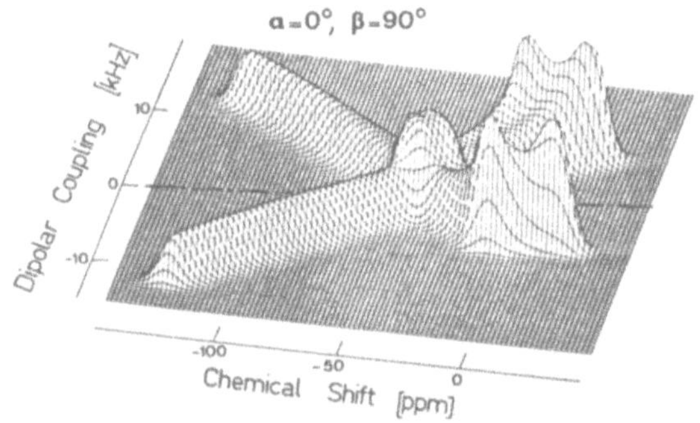

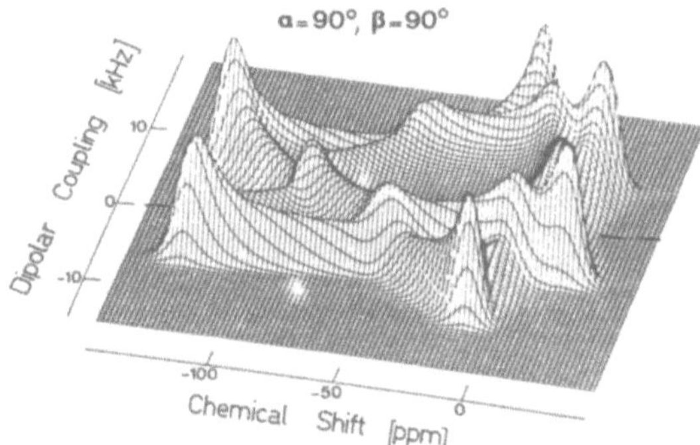

Fig. 5.10. Computer calculated ^{13}C chemical shift resolved dipolar powder spectra of methyl formate due to Linder et al. [18]. Two different sets **(a)**, **(b)** of angles (α, β) have been chosen, where α, β are the Euler angles of C–H vector in the principal axis frame of the chemical shift tensor. Top: C–H parallel to the 2-axis. Bottom: C–H is parallel to the 1-axis. (Courtesy of R.R. Ernst)

simplified due to the rotational averaging of the chemical shift and the dipolar tensor. Both have their main principal axis parallel to the rotation axis (C_6-axis). The two chemical shift pattern cross at the magic angle position, where the internuclear dipolar interaction disappears.

If the orientation between the two interaction tensors is arbitrary complicated pattern can arise. Some of these cases have been discussed by Linder et al. [18] and an example of their calculations is shown in Fig. 5.10. The reader is referred to reference [18] for further details.

5.3 Applications of 2D-Spectroscopy

Two-dimensional NMR spectroscopy was originally developed for and applied to liquids [1–6]. The interested reader is referred to a review article by Freeman and Morris [13]. Soon solid state applications were proposed [10–12] and have improved structural investigations. Recently coupling networks in molecular systems of biological relevance were unravelled by 2D-J-resolved spectroscopy [19, 20]. An example which shows the complexity of 2D-spectra in biological compounds is presented in Fig. 5.11.

A homonuclear 2D-experiment for resolving dipole-dipole interaction with respect to chemical shift interaction has been applied to ^{1}H in polycrystalline CCl_3COOH by Stoll et al. [11b]. After preparing the initial state by a 90°-pulse the system evolves during time t_1 due to the homonuclear dipole-dipole interaction, $\mathscr{H}_{II}$ and chemical shift interaction $\mathscr{H}_S$, i.e. $\mathscr{H}(1) = \mathscr{H}_{II} + \mathscr{H}_S$. Detection during t_2 is governed solely by $(\sqrt{2}/3)\,\mathscr{H}_S$ due to the application of an 8-pulse (MREV-8, see Sect. 3.4) which eliminates $\mathscr{H}_{II}$ on the average. Spectra $S(\omega_2, t_1)$ are shown for different t_1 values of ^{1}H in polycrystalline CCl_3COOH in Fig. 5.12. Oscillations of different parts of the chemical shift powder spectrum are due to different dipolar couplings.

An interesting indirect detection scheme of heteronuclear 2D-spectroscopy has been discussed by Maudsley and Ernst [21]. The method relies on a coherent transfer of transverse magnetization of nuclei of low gyromagnetic ratio to nuclei with high gyromagnetic ratio. Bodenhausen et al. [22] have separated dipolar and quadrupolar splittings in single crystal ^{14}N NMR by a heteronuclear 2D-experiment called DISQUO.

2D-rotational spin echo NMR in solids was introduced recently by Manowitz and Griffin [23]. There dipolar interaction is separated with respect to the sideband pattern of a slow spinning powder sample (DIPSHIFT). The separation of different powder spectra with respect to their trace was recently achieved by Griffin and co-workers [24] by means of 2D-spectroscopy under spin rotational echo conditions. A typical spectrum obtained by Griffin et al. [24] was already shown in Sect. 2.6. The basic idea of their experiments is to start the detection period (t_2) at the n-th echo peak, where $t_1 = n\,2\pi/\omega_r$ with $n = 0, 1, 2, \ldots$. Since the echo peak envelope is only due to the isotropic shift the ω_1-axis corresponds to the isotropic shift only, whereas the ω_2-axis represents the full anisotropic sideband pattern.

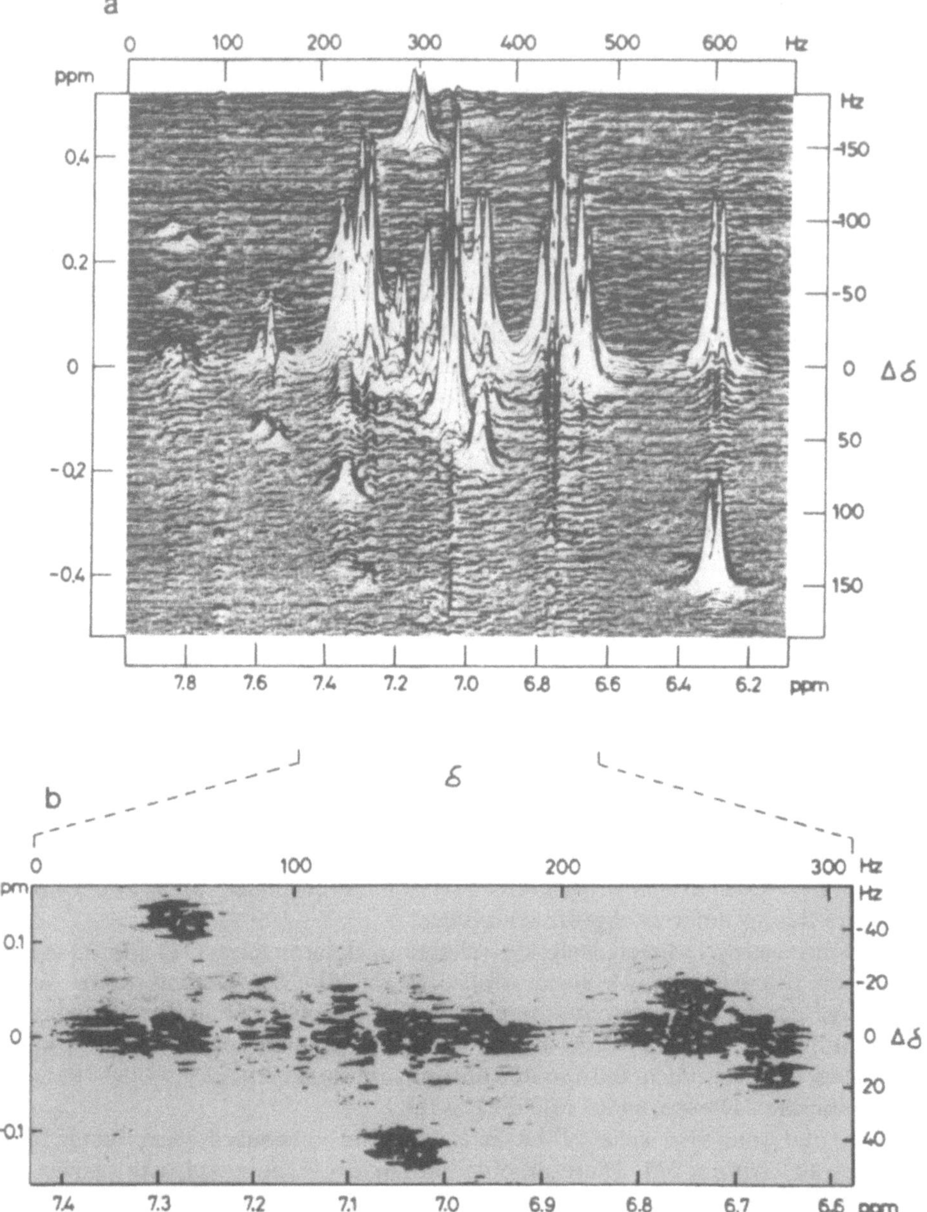

Fig. 5.11. Three-dimensional presentation of a 360 MHz spin-echo correlated ^{1}H NMR spectrum of a protein according to Nagayama et al. [20]. The chemical shift δ on the horizontal axis corresponds to that in conventional one-dimensional NMR. The vertical axis ($\Delta\delta$) represents the frequency difference between correlated nuclei

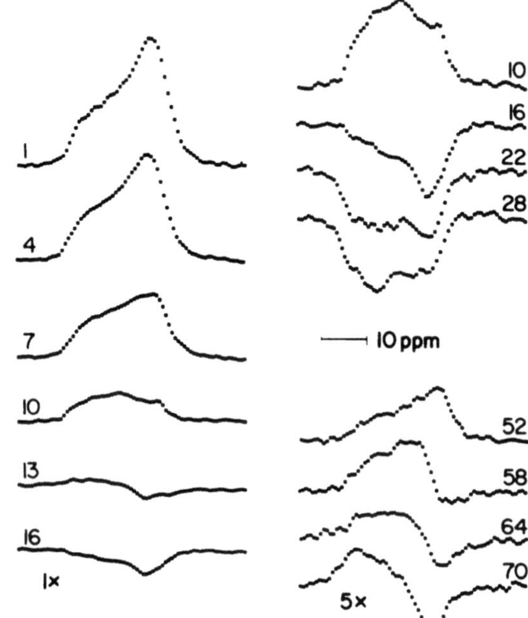

Fig. 5.12. Homonuclear dipolar-modulated proton chemical shift spectra in polycrystalline CCl₃COOH according to Stoll et al. [11b]. Dipolar evolution has taken place for times τ, in units of 4.17 μs, indicated by the numbers. The right hand spectra are enlarged by a factor 5

Disentenglement of chemical exchange networks by using 2D-spectroscopy was recently proposed by Jeener et al. [25]. Here the spins evolve during time t_1 after a 90° pulse under the action of their chemical shift Hamiltonian. At time t_1 after the initial 90° pulse a second 90° pulse is applied to create z-magnetization whoose magnitude depends on the phase of the spins at time t_1, i.e. on their chemical shift. After the second 90° pulse exchange may take place for a time τ_m allowing for mixing of the different z-magnetizations. Following this mixing period (τ_m) the detection period (t_2) starts by transferring the z-magnetization at $t_1 + \tau_m$ into observable x, y-magnetization with the help of a third 90° pulse. The corresponding 2D-spectrum of heptamethylbenzenonium ion is shown as a contour plot in Fig. 5.13. The four different lines due to the chemical shifts of the protons A, B, C and D with the intensity ratio 2:2:2:1 are found on the diagonal of the $f_1 - f_2$ contour plot (Fig. 5.13). Without chemical exchange or for $\tau_m = 0$ only these diagonal lines would be present. Under the action of chemical exchange during τ_m, off-diagonal lines appear which show the connectivity of the exchange process. It is evident from Fig. 5.13 that lines A–C, B–C, B–D are connected by exchange and not e.g. A–B or C–D [25].

Although this experiment was applied to liquids only it has, however, potential applications to solids. With slight modifications it can be used to study chemical exchange in solids, spin-diffusion, relaxation pathways etc. to name a few.

Finally I should remark, that the 2D-concept was exploited also in multiple-quantum spectroscopy by Pines and co-workers [26] and Wokaun and Ernst [27]. These aspects, however, will be discussed in the next chapter.

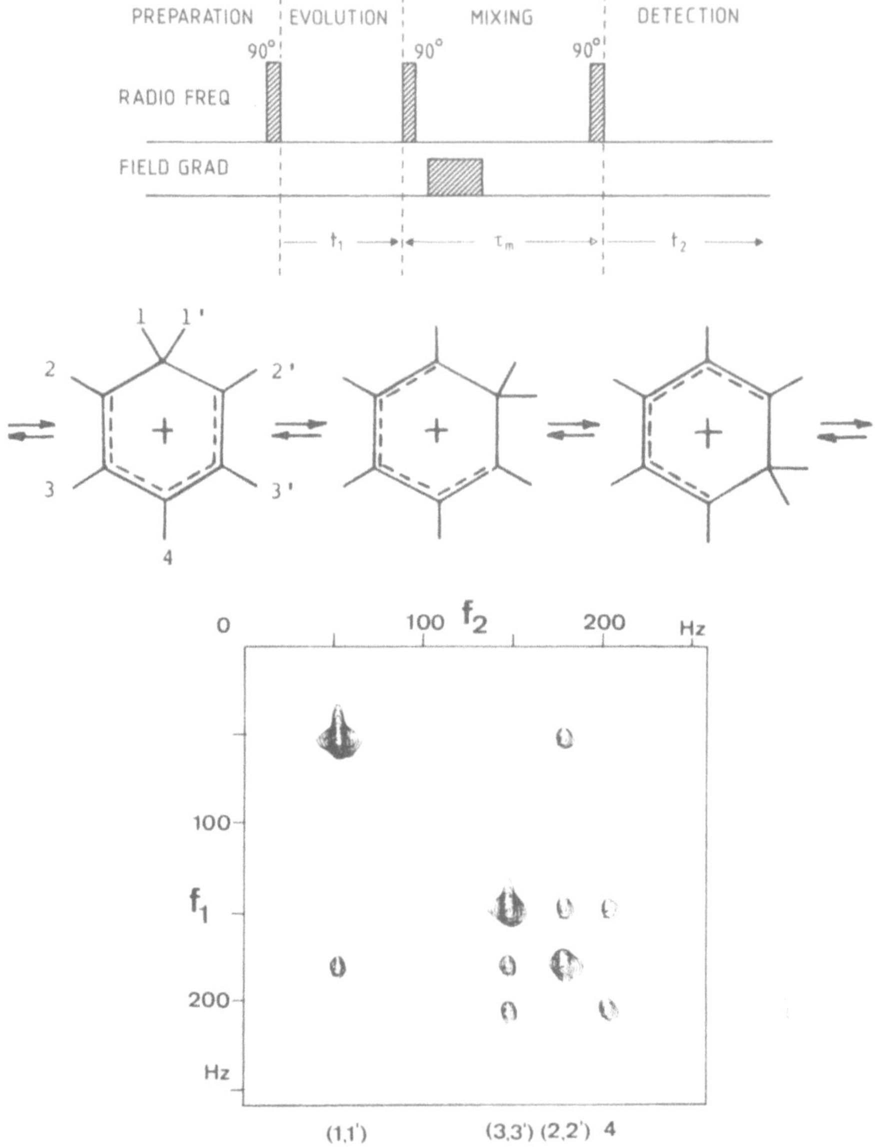

Fig. 5.13. Pulse sequence (top) and 2D exchange spectrum contour plot (bottom) of heptamethyl-benzenonium ion in 9.4 M H_2SO_4 obtained by Jenner et al. [25]. The spectral lines (diagonal) which are correlated by exchange possess off-diagonal lines

6 Multiple-Quantum NMR Spectroscopy

In the preceding chapters we have dealt with single quantum transitions in the sense that a single "quantum of energy" $\hbar\omega_{jk}$ is necessary and sufficient to create a transition (superposition of states) between the two eigenstates E_j and E_k of the systems Hamiltonian. In NMR we usually employ large magnetic fields B_0 which ensures that the magnetic quantum number m is a good quantum number. Magnetic dipole transitions $\Delta m = 1$ occur under the action of a radio-frequency field B_1. In NMR therefore an n-quantum transition usually involves $\Delta m = n$.

In optical spectroscopy, ESR and NMR in low fields, however, m may no longer be a good quantum number, i.e. a single quantum transition with e.g. $\Delta m = 2$ can be observed, due to an admixture of the wavefunctions.

A single quantum $\Delta m = 2$ transition may be even induced in a spin 1 system in a large magnetic field B_0 with the help of a single phonon induced quadrupolar transition. Still we shall use the notion of a single quantum transition in these cases. The literature is full of confusion in these respects. We only speak of an n-quantum transition if really n-quanta are absorbed by the system in contrast to the situation where a single quantum is absorbed in a $\Delta m = n$ transition.

Double- and multiple quantum transitions have been discussed already in the early days of quantum mechanics and spectroscopy [1]. In optical spectroscopy multiple quantum transitions are well established and comprise techniques like Raman-spectroscopy, two-photon absorption, multi-photon absorption etc. [2–5]. Also in magnetic resonance multiple-quantum transitions were observed at a very early stage [6, 7]. The technique applied was the continous wave spectroscopy (cw), where higher quantum transitions are observed with increasing radio-frequency field strength. The corresponding perturbation theory was derived by Yatsiv [8] and more recently in second quantization language by Bucci et al. [9]. In the following sections, however, we are interested in transient and coherent multiple-quantum phenomena and the theory of Yatsiv [8] and Bucci et al. [9] is not readily applicable to these cases. A special theory which is tailored to these needs based on fictitious spin 1/2 operators was applied by Vega and Pines [10] and by Wokaun and Ernst [11].

In Sect. 6.1 we discuss a technique of proton high resolution spectroscopy in highly deuterated solids by double quantum decoupling of the deuterons. This leads us to the question of double quantum coherence in the three-level system ($I = 1$) in Sect. 6.2. Based on these principles we then advance to multiple quantum coherence in Sect. 6.3 continuing to the recent selective multiple quantum experiments by Pines and co-workers [12] in Sect. 6.4. Double-quantum cross-polarization is covered in Sect. 6.5.

6.1 Double-Quantum Decoupling

Heteronuclear dipolar coupling $\mathcal{H}_{IS}$ between two different spin species I (abundant) and S (rare) is the major broadening mechanism of the S spin spectral lines in solids. Spin-decoupling is therefore a pre-requisite to observe highly resolved S spin resonance as was discussed in Sect. 4.4. Here we want to apply this concept to $^1\text{H}(S) - {}^2\text{D}(I)$, i.e. we are observing the proton resonance in highly deuterated solids. The dipole-dipole interaction $\mathcal{H}_{II}$ among the protons is assumed to be sufficiently reduced by the dilution, whereas the deuteron-proton coupling $\mathcal{H}_{IS}$ is responsible for the linebroadening of the proton resonance.

The deuteron with spin $I = 1$ has a quadrupole moment Q which couples to the electric field gradient resulting in the three-level system shown in Fig. 6.1. Two transitions ($\Delta m = \pm 1$) are allowed at the frequencies $\omega_0 \pm \omega_Q$, whereas the $\Delta m = \pm 2$ transition is dipole forbidden. The spectrum therefore consists of two lines symmetric to the Larmorfrequency ω_0 with splitting $2\omega_Q$. In a polycrystalline sample mirror-symmetric powder pattern equivalent to those for the chemical shift tensor (see Sect. 2.4) with a spectral width of typically 100 kHz result. Thus, it might appear that the deuterium decoupling rf field intensity must be sufficient to cover the entire quadrupolar spectrum, i.e.

$$\omega_1 \gtrsim \omega_Q \tag{6.1}$$

where ω_1 is the intensity of the rf field. It was realized by Meiboom and co-workers [13] in liquid crystal work that deuterium spin decoupling takes place via double quantum transitions between the $m = \pm 1$ levels. In this case it was shown that the decoupling condition [14]

$$\omega_1 \gtrsim (\omega_D \omega_Q)^{1/2} \tag{6.2}$$

applies, where ω_D is the strength of the heteronuclear dipolar interaction. This condition is less severe than Eq. (6.1) since ω_D is typically 2π times a few kHz. As a result deuteron decoupling by a double quantum process is easily performed with standard rf power.

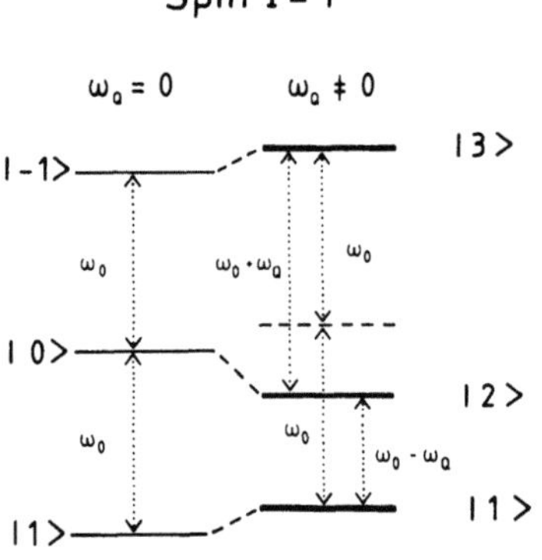

Fig. 6.1. Three-level diagram for a spin $I = 1$ in a magnetic field $B_0 = \omega_0/\gamma$ and with quadrupole interaction ω_Q. Allowed $\Delta m = 1$ transitions (1-2; 2-3) and the forbidden transition $\Delta m = 2$ (1-3) are indicated

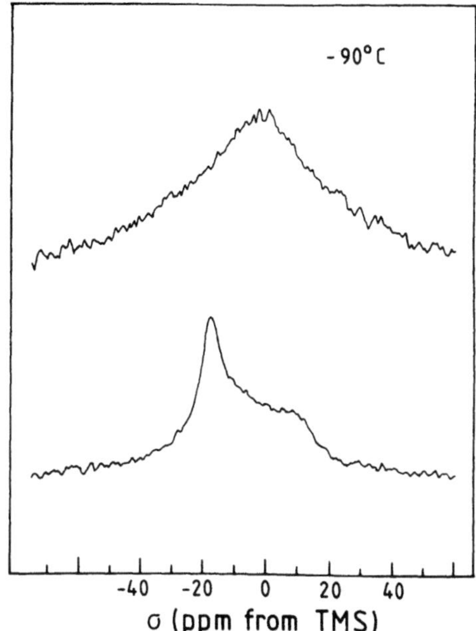

Fig. 6.2. Powder pattern due to the ^{1}H chemical shift tensor in solid D_2O (ice) without (top) and with (bottom) deuterium double-quantum spin decoupling according to Pines et al. [14]

A typical example was shown already in Figs. 4.26 and 4.27. Application of this technique to solids by Pines et al. [14, 15] resulted among others in the observation of the proton chemical shift tensor in ice (see Fig. 6.2) and in squaric acid [16].

Before discussing some details of the double-quantum decoupling mechanism let us estimate the reduction of the homonuclear linewidth δ due to dilution. Since the linewidth δ due to homonuclear dipolar interaction cannot be calculated rigorously we have to resort to approximate expressions using e.g. the second (M_2) and fourth (M_4) moment [17]. The halfwidth at halfheight δ in rad s^{-1} can be calculated for a Gaussian and truncated Lorentzian lineshape as [17]

Gaussian
$$\delta = 1.18(M_2)^{1/2} \tag{6.3a}$$

truncated Lorentzian
$$\delta = \frac{\pi}{2\sqrt{3}\mu}(M_2)^{1/2} \tag{6.3b}$$

where
$$\mu = M_4/M_2^2.$$

A universal linewidth formular for $\mu \geqq 3$ can be derived, as shown in Sect. 2.5 and in Appendix C, which covers the two limiting cases Gaussian ($\mu = 3$) and truncated Lorentzian ($\mu \gg 3$), namely

$$\delta = \left(\frac{\pi}{2}\right)^{1/2}\left(\frac{M_2}{\mu - 1.87}\right)^{1/2}. \tag{6.3c}$$

In the following we shall use this formula to calculate the linewidth δ from the knowledge of the second and the fourth moment. We shall use the simple statistical argument that each nuclear site is occupied with a probability

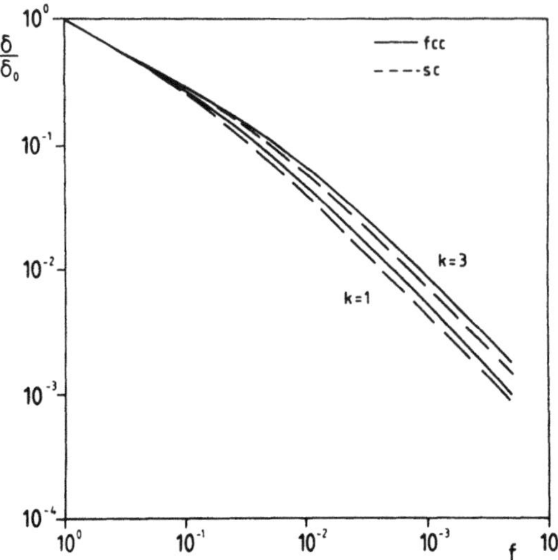

Fig. 6.3. Normalized linewidth δ/δ_0 versus dilution factor f for a simple cubic (sc) and a face centered cubic (fcc) lattice. The parameter k (1, 3) represents the number of spins per lattice site. (Courtesy of D. Suwelack)

$f\,(0 \leqq f \leqq 1)$ where f is called the dilution factor [17]. In the calculation of lattice sums it is convenient to count lattice sites rather than nuclear sites, i.e. the bunching factor k is introduced which counts the number of nuclei at a particular lattice site. In adamantane ($C_{10}H_{16}$), for example, $k=10$ for ^{13}C dipole-dipole interaction. Introducing now the modified dilution factor $F=k \cdot f$ we obtain for the second and fourth moment [17]

$$M_2 = \tfrac{1}{3} S(S+1)\, \gamma^4 \hbar^2 F \sum_k b_{jk}^2 \tag{6.4}$$

and

$$M_4 = [\tfrac{1}{3} S(S+1)\, \gamma^4 \hbar^2]^2\, F^2 \{ \tfrac{7}{3} \sum_{k \neq l} b_{jk}^2 b_{jl}^2$$

$$+ \tfrac{2}{3} \sum_{k \neq l} b_{kl}^2 b_{jk} b_{jl} + \tfrac{1}{5}(7 - \tfrac{3}{2}(S^2 + S)^{-1}) \sum_k b_{jk}^4 / F \} \tag{6.5}$$

where

$$b_{jk} = \tfrac{3}{2}(1 - 3\cos^2 \vartheta_{jk})\, r_{jk}^{-3}.$$

The dependence of the linewidth on dilution according to Eqs. (6.3c, 6.4, 6.5) is shown in Fig. 6.3. With increasing dilution $(F \to 0)$ the second moment decreases proportional to F whereas $\mu = M_4 / M_2^2$ increases due to the last term in the fourth moment. As a consequence the normalized linewidth $\delta/\delta_0 = \delta(f)/\delta(f=1)$ decreases proportional to the square root of f for medium dilution $(0.1 \leqq f \leqq 1)$ and only for high dilution $(f < 0.1)$ the lineshape becomes more Lorentzian $(\mu > 3)$ and narrows proportional to the dilution [18].

Let us discuss now qualitatively the double quantum decoupling mechanism [13-16]. Irradiation at the Larmorfrequency ω_{0I} of the deuterons [see Eq. (6.1)] cannot induce transitions from $|1\rangle$ to $|-1\rangle$ in first order. In second order, however, this transition rate is given by

$$W_2 = (2\omega_1^2/\omega_Q) |\langle 1| I_x |0\rangle \langle 0| I_x |-1\rangle|. \tag{6.6}$$

Evaluating the matrix elements in Eq. (6.6) results in

$$W_2 = \omega_1(\omega_1/\omega_Q) \tag{6.7}$$

which clearly demonstrates the scaling of the rf field ω_1 by the factor ω_1/ω_Q.

We now turn to some quantitative aspects of double quantum decoupling [15, 16]. The appropriate Hamiltonian

$$\mathcal{H} = \omega_1 I_x + \omega_Q[I_z^2 - I(I+1)/3] + 2BI_z S_z \tag{6.8}$$

in the doubly rotating frame is comprized of the decoupling field strength ω_1, the quadrupole interaction ω_Q of the I spins and the heteronuclear coupling strength B. Without the decoupling field ($\omega_1 = 0$) the S spin spectrum consists of three lines located at ω_{0S} ($m_{zI} = 0$) and $\omega_{0S} \pm B(m_{zI} = \pm 1)$. Note, that the $m_{zI} = 0$ level does not lead to line splitting or broadening of the S spins. In order to decouple the I spins it is therefore only necessary to exchange the $m_{zI} = \pm 1$ levels rapidly enough. This, however, corresponds to a double quantum transition ($\Delta m_{zI} = 2$) at the Larmor-frequency ω_{0I} of the I spins.

Using the fictitious spin 1/2 operators [10, 11] applied to quadrupolar interactions and multiple-quantum transitions by Vega [10b] and Wokaun-Ernst [11] recently, the Hamiltonian Eq. (6.6) can be rewritten as [16]

$$\mathcal{H} = \sqrt{2}\,\omega_1(I_x^{1-2} + I_x^{2-3}) + \tfrac{2}{3}\omega_Q(I_z^{1-2} - I_z^{2-3}) + 4BI_z^{1-3}S_z. \tag{6.9}$$

The fictitious spin 1/2 operators will be discussed in more detail in the next section. Here it might suffice to note, that I_α^{r-s} denotes a spin 1/2 operator with polarization ($\alpha = x, y, z$) of the transition $r - s(r, s = 1, 2, 3)$. After some algebra [16] and under the assumption $\omega_1 \ll \omega_Q$ Eq. (6.7) can be expressed as

$$\mathcal{H} = (\omega_1^2/\omega_Q)\,I_x^{1-3} + 4BI_z^{1-3}S_z + \tfrac{2}{3}\omega_Q(I_z^{1-2} - I_z^{2-3}). \tag{6.10}$$

Now the double-quantum operators I_α^{1-3} commute with $(I_z^{1-2} - I_z^{2-3})$ and the last part can be ignored, when calculating the S spin spectrum, i.e. the problem reduces to a "double-quantum rotation" of the I spins with the effective frequency

$$\omega_e = [(2B)^2 + (\omega_1^2/\omega_Q)^2]^{1/2}. \tag{6.11}$$

The rf decoupling field appears to be renormalized by the scaling factor ω_1/ω_Q. In fact the double-quantum rotation has been observed as a satellite line in the S spin spectrum [16].

Off-resonance effects can be treated in a similar fashion as in single quantum decoupling [15] (see Sect. 4.4). Again the relative linewidth is proportional to the projection of the effective field onto the z-axis, i.e. [15]

$$\delta/\delta_0 = \cos\vartheta_D = \frac{2\Delta\omega}{[(2\Delta\omega)^2 + (\omega_1^2/\omega_Q)^2]^{1/2}} \tag{6.12}$$

where $\Delta\omega = \omega_{0I} - \omega$ is the frequency deviation from the double quantum frequency ω_{0I}. For small offset $\Delta\omega \ll \omega_1^2/\omega_Q$

$$\delta/\delta_0 \sim \Delta\omega\,\omega_Q/\omega_1^2 \tag{6.13}$$

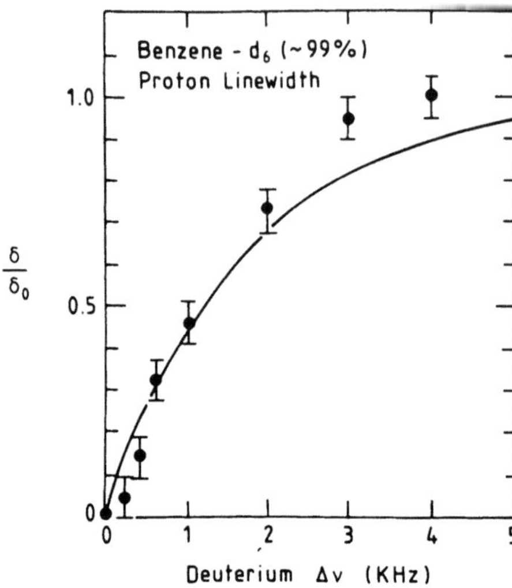

Fig. 6.4. Normalized linewidth δ/δ_0 of the residual protons in perdeuterated benzene (C_6D_6) versus off-set frequency of the double-quantum decoupling field. (Pines et al. [15])

i.e. the residual linewidth is very sensitive to frequency offset as was noted already by Meiboom et al. [13]. A demonstration of this behaviour is shown in Fig. 6.4. Further details are discussed in references [15].

6.2 The Three-Level System; Double Quantum Coherence

The three-level system is the celebrated level scheme to discuss double quantum transitions and coherence in optical [19-23] as well as NMR spectroscopy [24-29]. In fact any multi-level system can be truncated to a three level system to a certain degree of approximation by connecting just three levels only by selective application of radiation fields. Based on these ideas a number of transient phenomena have been observed in optical spectroscopy [23] and NMR [23-29]. Space does not permit to go into detail here. Especially the publications on NMR by Hashi and co-workers [24, 25] are recommended to the interested reader.

Brewer and Hahn [19] have solved the equation of motion of the density matrix in the three-level case and coherent transients may be calculated using their results. We, however, will take the opportunity to use this simple level scheme to introduce fictitious spin 1/2 operators [10, 11] in the notation of Vega [10b] and Wokaun and Ernst [11].

We shall restrict ourselves to the spin $I = 1$ case whose level scheme is shown in Fig. 6.1. The Hamiltonian of this three-level system is readily expressed as

$$\mathscr{H}_0 = -\omega_0 I_z + \omega_Q [I_z^2 - \tfrac{1}{3} I(I+1)] \tag{6.14}$$

with $I = 1$ and where ω_Q is the quadrupolar interaction. The different single quantum ($\Delta m = \pm 1$; $\omega_0 \pm \omega_Q$) and double quantum ($\Delta m = \pm 2$; $2\omega_0$) transitions are indicated in Fig. 6.1.

The fictitious spin 1/2 operators in the Zeeman basis $|+1\rangle$, $|0\rangle$ and $|-1\rangle$ are expressed in matrix form as [10b, 11].

$$I_x^{1-2}=\frac{1}{2}\begin{pmatrix}0 & 1 & 0\\ 1 & 0 & 0\\ 0 & 0 & 0\end{pmatrix};\quad I_y^{1-2}=-\frac{i}{2}\begin{pmatrix}0 & 1 & 0\\ -1 & 0 & 0\\ 0 & 0 & 0\end{pmatrix};\quad I_z^{1-2}=\frac{1}{2}\begin{pmatrix}1 & 0 & 0\\ 0 & -1 & 0\\ 0 & 0 & 0\end{pmatrix}$$

$$\text{(6.15a)}$$

$$I_x^{2-3}=\frac{1}{2}\begin{pmatrix}0 & 0 & 0\\ 0 & 0 & 1\\ 0 & 1 & 0\end{pmatrix};\quad I_y^{2-3}=-\frac{i}{2}\begin{pmatrix}0 & 0 & 0\\ 0 & 0 & 1\\ 0 & -1 & 0\end{pmatrix};\quad I_z^{2-3}=\frac{1}{2}\begin{pmatrix}0 & 0 & 0\\ 0 & 1 & 0\\ 0 & 0 & -1\end{pmatrix}$$

$$\text{(6.15b)}$$

$$I_x^{1-3}=\frac{1}{2}\begin{pmatrix}0 & 0 & 1\\ 0 & 0 & 0\\ 1 & 0 & 0\end{pmatrix};\quad I_y^{1-3}=-\frac{i}{2}\begin{pmatrix}0 & 0 & 1\\ 0 & 0 & 0\\ -1 & 0 & 0\end{pmatrix};\quad I_z^{1-3}=\frac{1}{2}\begin{pmatrix}1 & 0 & 0\\ 0 & 0 & 0\\ 0 & 0 & -1\end{pmatrix}$$

$$\text{(6.15c)}$$

where

$$I_z^{1-3}=I_z^{1-2}+I_z^{2-3}.$$

Note, that I_α^{1-2} and $I_\alpha^{2-3}(\alpha=x,y,z)$ are "single quantum operators", whereas the I_α^{1-3} are "double quantum operators". For the general three-level subsystem $r,s=1,2,3$ of a multilevel system, these fictitious spin 1/2 operators can be written as [10b, 11]

$$I_x^{r-s}=\tfrac{1}{2}\{|r\rangle\langle s|+|s\rangle\langle r|\},$$

$$I_y^{r-s}=-\frac{i}{2}\{|r\rangle\langle s|-|s\rangle\langle r|\},\qquad(6.16)$$

$$I_z^{r-s}=\tfrac{1}{2}\{|r\rangle\langle r|-|s\rangle\langle s|\}$$

with

$$I_x^{r-s}=I_x^{s-r};\quad I_y^{r-s}=-I_y^{s-r};\quad I_z^{r-s}=I_z^{s-r}.\qquad(6.17)$$

In the case of $I=1$ we obtain for the total spin operators [10b, 11]

$$I_{x,y}=\sqrt{2}(I_{x,y}^{1-2}+I_{x,y}^{2-3});\quad I_z=2(I_z^{1-2}+I_z^{2-3})=2I_z^{1-3}.\qquad(6.18)$$

Other bilinear combinations of total spin operators can be conveniently expressed as linear combinations of fictitious spin 1/2 operators as

$$\begin{aligned}
I_x^2-\tfrac{1}{3}I(I+1)&=I_x^{1-3}-\tfrac{1}{3}(I_z^{1-2}-I_z^{2-3})\\
I_y^2-\tfrac{1}{3}I(I+1)&=-I_x^{1-3}-\tfrac{1}{3}(I_z^{1-2}-I_z^{2-3})\\
I_z^2-\tfrac{1}{3}I(I+1)&=\tfrac{2}{3}(I_z^{1-2}-I_z^{2-3})\\
I_xI_z+I_zI_x&=\sqrt{2}(I_x^{1-2}-I_x^{2-3})\\
I_yI_z+I_zI_y&=\sqrt{2}(I_y^{1-2}-I_y^{2-3})\\
I_xI_y+I_yI_x&=2I_y^{1-3}.
\end{aligned}\qquad(6.19)$$

We are now able to rewrite the Hamiltonian Eq. (6.14) in terms of fictitious spin 1/2 operators by using Eqs. (6.18, 6.19) as

$$\mathscr{H}_0=-2\omega_0 I_z^{1-3}+\tfrac{2}{3}\omega_Q(I_z^{1-2}-I_z^{2-3}).\qquad(6.20)$$

Coherent transients must be derived from the equation of motion

$$\frac{d}{dt}I_\alpha = i[\mathcal{H}, I_\alpha] \qquad \alpha = x, y, z \tag{6.21}$$

or the corresponding equation for the density matrix.

Since the Hamiltonian $\mathcal{H}$ can always be expressed as a linear combination of fictitious spin 1/2 operators, e.g.

$$\mathcal{H} = \omega_a I_a^{r-s} + \omega_b I_b^{s-t} + \omega_c I_c^{t-r} \tag{6.22}$$

the equation of motion (6.21) leads to Bloch-type equations [28]

$$\frac{d}{dt}\mathbf{I} = k\boldsymbol{\Omega} \times \mathbf{I} \tag{6.23}$$

where $\mathbf{I} = (I_a^{r-s}, I_b^{s-t}, I_c^{t-r})$ and $\boldsymbol{\Omega} = (\omega_a, \omega_b, \omega_c)$ and the factor $k = 1/2$, 1 or 2 depends on the chosen frame defined by $\mathbf{I}$ [28].

This implies a rotation of the vector $\mathbf{I}$ about the vector $\boldsymbol{\Omega}$ with the effective rotational frequency

$$k|\boldsymbol{\Omega}| \equiv \omega_e = k[\omega_a^2 + \omega_b^2 + \omega_c^2]^{1/2}. \tag{6.24}$$

In order to derive these Bloch-type equations we need to know the commutators which are summarized below [10, 11]

$$[I_x^{r-s}, I_y^{r-s}] = i I_z^{r-s} \quad \text{and cyclic permutation} \tag{6.25}$$

and

$$[I_x^{r-t}, I_x^{s-t}] = [I_y^{r-t}, I_y^{s-t}] = \frac{i}{2} I_y^{r-s} \tag{6.26a}$$

$$[I_z^{r-t}, I_z^{s-t}] = 0 \tag{6.26b}$$

$$[I_x^{r-t}, I_y^{s-t}] = \frac{i}{2} I_x^{r-s} \tag{6.26c}$$

$$[I_x^{r-t}, I_z^{s-t}] = -\frac{i}{2} I_y^{r-t} \tag{6.26d}$$

$$[I_y^{r-t}, I_z^{s-t}] = \frac{i}{2} I_x^{r-t}. \tag{6.26e}$$

A change of the transition index (e.g. $s-t$ by $t-s$) changes the sign of the commutator according to Eq. (6.17), e.g.

$$[I_x^{r-t}, I_y^{t-s}] = -\frac{i}{2} I_x^{r-s}. $$

Operators which do not have a common level commute

$$[I_a^{r-s}, I_b^{u-t}] = 0 \qquad a, b = x, y, z. \tag{6.27}$$

If operators belong to connected transitions the triangular relation

$$I_z^{r-s} + I_z^{s-t} + I_z^{t-r} = 0 \tag{6.28}$$

holds for the z-operators.

Suppose we irradiate the three-level system by an rf field of strength B_1 $=\omega_1/\gamma$ in a frame rotating with the frequency ω around the z-direction. With the secular part of the rf Hamiltonian

$$\mathcal{H}_1 = -\omega_1 I_{x,y} = -\sqrt{2}\,\omega_1(I_{x,y}^{1-2} + I_{x,y}^{2-3}) \tag{6.29}$$

and the three-level Hamiltonian according to Eq. (6.20) we obtain for the total Hamiltonian

$$\mathcal{H} = -2(\omega_0 - \omega)\,I_z^{1-3} + \tfrac{2}{3}\omega_Q(I_z^{1-2} - I_z^{2-3}) - \sqrt{2}\,\omega_1(I_{x,y}^{1-2} + I_{x,y}^{2-3}). \tag{6.30}$$

To be specific let us now treat some representative examples [10, 11].

(i) irradiation near a satellite transition $\omega_0 \pm \omega_Q$

With $\omega = \omega_0 + \omega_Q - \delta\omega$ (2–3 transition, see Fig. 6.1) where $\delta\omega \ll \omega_Q$ the Zeeman term of Eq. (6.30) may be rewritten by using Eq. (6.28) as

$$\mathcal{H}_0 = -\delta\omega\, I_z^{2-3} + \mathcal{H}_2$$

where

$$\mathcal{H}_2 = (\tfrac{4}{3}\omega_Q - \delta\omega)(I_z^{1-2} + I_z^{1-3})$$

with the eigenvalues in the rotating frame [10, 11]

$$(\mathcal{H}_0)_{11} = \tfrac{4}{3}\omega_Q - \delta\omega$$
$$(\mathcal{H}_0)_{22} = -\tfrac{2}{3}\omega_Q$$
$$(\mathcal{H}_0)_{33} = -\tfrac{2}{3}\omega_Q + \delta\omega.$$

If $\omega_1 \ll \omega_Q$ the term $2\omega_1 I_{x,y}^{1-2}$ in Eq. (6.30) can be neglected since it connects levels (1–2) which are separated by $2\omega_Q$. Under this assumption we arrive at

$$\mathcal{H} = \mathcal{H}_1 + \mathcal{H}_2 \tag{6.31a}$$
$$\mathcal{H}_1 = -\delta\omega\, I_z^{2-3} - \sqrt{2}\,\omega_1 I_{x,y}^{2-3} \tag{6.31b}$$

where $[\mathcal{H}_1, \mathcal{H}_2] = 0$ and any action on the transition 2–3 is mediated by $\mathcal{H}_1$, i.e. we observe the usual spin precession around an effective field $\omega_e = (\delta\omega^2 + 2\omega_1^2)^{1/2}$ which is tilted with respect to the z-axis by an angle $\vartheta = \arctan$ $(\sqrt{2}\omega_1/\delta\omega)$. The two-level subsystem (2–3) behaves like an ordinary spin 1/2, however, with the modified Rabi oscillation frequency $\sqrt{2}\omega_1$. If we turn on the rf field for a time t at the 2–3 transition, such that $\sqrt{2}\omega_1 t = \pi/2$ we have applied a 90°-pulse to the 2–3 transition, i.e. we have converted $\rho^{2-3}(0)$ $= I_z^{2-3}$ into $I_{x,y}^{2-3}$. The on-resonance precession follows immediately from the commutators Eq. (6.26) as

$$e^{i\sqrt{2}\,\omega_1 t\, I_x^{2-3}}\, I_x^{2-3}\, e^{-i\sqrt{2}\,\omega_1 t\, I_x^{2-3}}$$
$$= I_x^{2-3}\cos(\sqrt{2}\,\omega_1 t) + I_y^{2-3}\sin(\sqrt{2}\,\omega_1 t). \tag{6.32}$$

Corresponding arguments hold for irradiation near the 1–2 transition. In other words if we connect a two-level subsystem in a multi-level system by selective irradiation with a weak rf field the two-level subsystem behaves to a good approximation like a spin 1/2. All transient phenomena like Hahn spin-

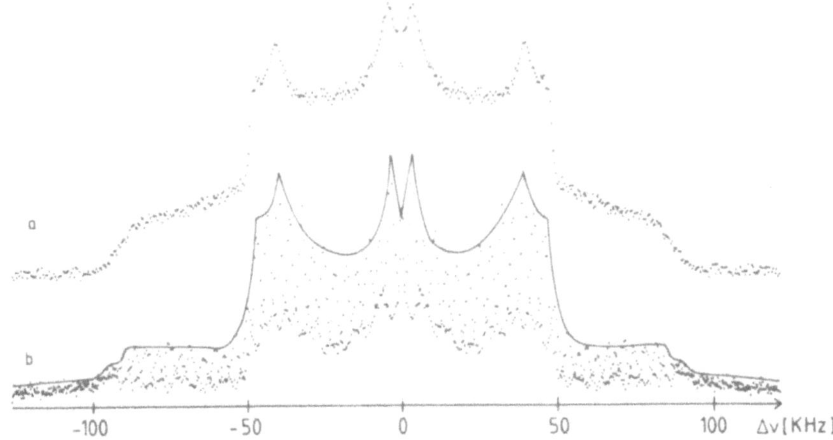

Fig. 6.5a, b. Spin-echo spectrum (**a**) and MFE spectrum (**b**) of ^{2}H in deuterated α-oxalic acid powder. Several frequency shifted MFE spectra are displayed in (**b**) with a line drawn through the peak points to guide the eye (Wemmer et al. [30])

echo etc. can be observed. Note, that irradiation of a spin 1 system with two weak rf fields $\sqrt{2}\omega_1$ at the two transitions $\omega_0 \pm \omega_Q$ (1−2; 2−3) is equivalent to applying a field ω_1 to the total spin without quadrupolar interaction, because $\sqrt{2}\omega_1(I_x^{1-2}+I_x^{2-3})=\omega_1 I_x$. i.e. spin precession of the form

$$e^{i\omega_1 t I_x} I_z e^{-i\omega_1 t I_x}=I_z \cos\omega_1 t+I_y \sin\omega_1 t$$

is observed. This has been utilized in multiple frequency excitation (MFE [30]), where a wide quadrupolar broadened spectrum is excited with a double sideband multiline rf signal. An example of such an MFE-spectrum [30] is shown in Fig. 6.5.

Suppose now, that we apply a 90°-pulse at the $1-2$ transition in order to create the initial state I_x^{1-2} [31]. We ask now, what happens to this initial state if we perform a rotation by β_{23} at the transition $2-3$. Since both transitions have level 2 in common, a phase change of level 2 due to the rotation β_{23} should be felt by I_x^{1-2}. Rotation of the $2-3$ transition proceeds as

$$e^{-i\beta_{23}I_z^{2-3}} I_z^{2-3} e^{i\beta_{23}I_z^{2-3}}=I_z^{2-3} \cos\beta_{23}+I_x^{2-3} \sin\beta_{23}$$

i.e. for $\beta_{23}=\pi$ we obtain $-I_z^{2-3}$ and $\beta_{23}=0$ and 2π are equivalent as usually. However, the $1-2$ transition varies according to the commutator relations Eq. (6.26) as

$$e^{-i\beta_{23}I_z^{2-3}} I_x^{1-2} e^{i\beta_{23}I_z^{2-3}}=I_x^{1-2} \cos(\beta_{23}/2)+I_x^{1-3} \sin(\beta_{23}/2) \tag{6.33}$$

i.e. for $\beta_{23}=2\pi$ the sign of I_x^{1-2} is changed and only for $\beta_{23}=4\pi$ the original state is reached [32]. This "spinor behaviour" of a two level quantum system is usually stated in a general way as

$$\exp(-i2\pi \mathbf{I}\cdot\mathbf{n})|\psi\rangle=-|\psi\rangle \tag{6.34}$$

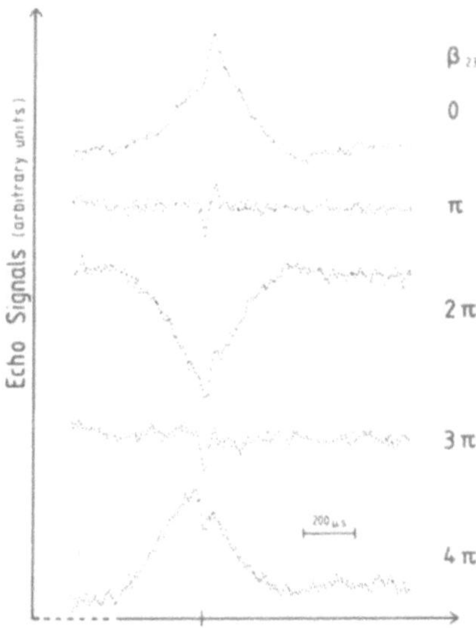

Fig. 6.6. Demonstration of spinor behaviour of a two-level subsystem (2–3 in Fig. 6.1) according to Stoll et al. [31a]. Coherence of the 1–2 transition is monitored by a spin-echo signal shown in the figure for different rotation angles β_{23} due to an rf field at the 2–3 transition. Note the inversion under $\beta_{23} = 2\pi$ rotation

where **n** is a unit vector in direction of the axis of rotation, **I** is the spin operator, and $|\psi\rangle$ the "spinor wavefunction". This behaviour is demonstrated in Fig. 6.6. Stoll et al. [32] have also demonstrated this in a four-level system of two coupled spin 1/2 nuclei. Also the double-quantum transition $(1-3)$ in a spin 1 system behaves in the same way [31b].

(ii) irradiation near the double-quantum transition ω_0 [10, 11, 14–16, 24, 25]

This problem was already briefly discussed in Sect. 6.1 in connection with double-quantum decoupling. Suppose we irradiate slightly off the double-quantum transition at $\omega = \omega_0 - \delta\omega$, with $\delta\omega \ll \omega_Q$ Eq. (6.30) results in

$$\mathscr{H} = -2\delta\omega I_z^{1-3} + \tfrac{2}{3}\omega_Q(I_z^{1-2} - I_z^{2-3}) - \sqrt{2}\,\omega_1(I_x^{1-2} + I_x^{2-3}). \tag{6.35}$$

By using a sequence of unitary transformations $\mathscr{H}$ can be brought into the form [10, 11]

$$\begin{aligned}
\mathscr{H}^* = &-2\delta\omega\cos(\vartheta/2)\,I_z^{1-3} + \tfrac{1}{2}(\omega_e - \omega_Q)\,I_x^{1-3} \\
&+ [\tfrac{2}{3}\omega_Q + \tfrac{1}{2}(\omega_e - \omega_Q)](I_z^{1-2} - I_z^{2-3}) \\
&- \sqrt{2}\,\delta\omega\sin(\vartheta/2)\,(I_x^{1-2} - I_x^{2-3})
\end{aligned} \tag{6.36}$$

where

$$\sin\vartheta = 2\omega_1/\omega_e; \quad \cos\vartheta = \omega_Q/\omega_e \tag{6.37a}$$

and

$$\omega_e = [\omega_Q^2 + 4\omega_1^2]^{1/2}. \tag{6.37b}$$

Under the assumption of weak irradiation ($\omega_1 \ll \omega_Q$; $\vartheta \cong 0$) we obtain

$$\mathscr{H}^* = -2\delta\omega I_z^{1-3} + (\omega_1^2/\omega_Q)\,I_x^{1-3} + [\tfrac{2}{3}\omega_Q + (\omega_1^2/\omega_Q)](I_z^{1-2} - I_z^{2-3}). \tag{6.38}$$

Note, that $I_z^{1-2} - I_z^{2-3}$ commutes with $I_\alpha^{1-3}(\alpha = x, y, z)$ and spin precession of the double-quantum subsystem $(1-3)$ evolves under the action of

$$\mathcal{H}^{1-3} = -2\delta\omega I_z^{1-3} + (\omega_1^2/\omega_Q)\, I_x^{1-3} \tag{6.39}$$

with

$$e^{-i\beta_{13} I_x^{1-3}}\, I_y^{1-3}\, e^{i\beta_{13} I_x^{1-3}} = I_y^{1-3}\cos\beta_{13} + I_z^{1-3}\sin\beta_{13}$$

and cyclic permutations. Two aspects are to be noted: (1) The rf field is scaled by ω_1/ω_Q and (2) the resonance shift $\delta\omega$ is transformed into $2\delta\omega$, i.e. chemical shifts in the double-quantum frame are doubled [10]. Waugh [33] has derived this relation by using coherent averaging theory (see Sect. 3.3). The argument proceeds by noting that the quadrupolar term in Eq. (6.35) is the largest contribution and

$$e^{it(\mathcal{H}_Q+\mathcal{H}_1)} = e^{it\mathcal{H}_Q}\, T e^{i\int_0^t dt'\,\tilde{\mathcal{H}}(t')}$$

with

$$\tilde{\mathcal{H}}(t) = e^{-it\mathcal{H}_Q}\,\mathcal{H}_1\, e^{it\mathcal{H}_Q}$$

may be a useful separation. With

$$\mathcal{H}_Q = \tfrac{2}{3}\omega_Q(I_z^{1-2} - I_z^{2-3})$$

and

$$\mathcal{H}_1 = -2\delta\omega I_z^{1-3} - \sqrt{2}\omega_1(I_x^{1-2} + I_x^{2-3})$$

one calculates by using the commutator relations Eq. (6.26)

$$\tilde{\mathcal{H}}(t) = -2\delta\omega I_z^{1-3} - \sqrt{2}\omega_1\{(I_x^{1-2} + I_x^{2-3})\cos\omega_Q t + (I_y^{1-2} - I_y^{2-3})\sin\omega_Q t\}$$

with

$$\bar{\mathcal{H}}^{(0)} = -2\delta\omega I_z^{1-3}$$

$$\bar{\mathcal{H}}_S^{(1)} = \frac{\omega_1^2}{\omega_Q}\{(I_z^{1-2} - I_z^{2-3}) + I_x^{1-3}\}. \tag{6.40}$$

Since $(I_z^{1-2} - I_z^{2-3})$ commutes with I_α^{1-3} we obtain

$$e^{it(\mathcal{H}_Q+\mathcal{H}_1)} = e^{it(\mathcal{H}_Q + \bar{\mathcal{H}}^{(0)} + \bar{\mathcal{H}}_S^{(1)})}$$

with

$$\mathcal{H}_Q + \bar{\mathcal{H}}^{(0)} + \bar{\mathcal{H}}_S^{(1)} = \mathcal{H}^*$$

as given by Eq. (6.38).

Note, however, that the average Hamiltonian of Haeberlen and Waugh [34] is not identical with Eq. (6.40) if $\delta\omega \neq 0$. Instead we have used the secular average Hamiltonian as discussed in Appendix G where the higher order terms are calculated from the commutators of $\tilde{\mathcal{H}}(t) - \bar{\mathcal{H}}^{(0)}$ rather than $\tilde{\mathcal{H}}(t)$.

Let us now turn to a third class of double quantum experiments in solids which use non-selective pulses. Pines and co-workers have initiated this field in solids and oriented molecules and performed a number of interesting experiments. On the other hand there is abundant work of this sort in liquids. Space, however, does not allow to discuss this here.

(iii) Non-selective excitation of double quantum transitions [10, 27, 35–37]

We want to discuss now a basic and interesting experiment performed by Pines and co-workers [10, 27, 36] to observe the chemical shift tensor in

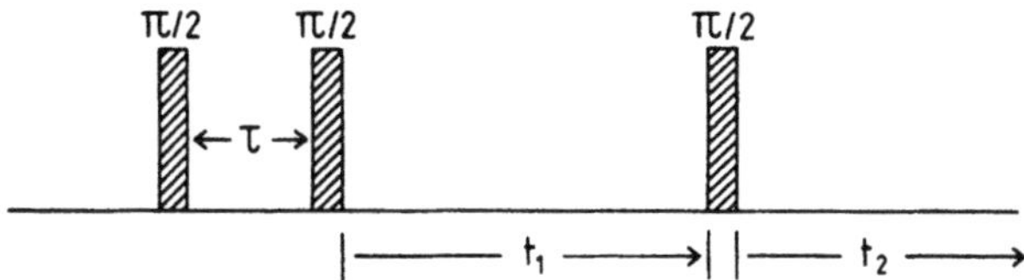

Fig. 6.7. Three-pulse sequence for excitation and detection of multiple-quantum coherence

quadrupolar distorted deuterons in solids. The chemical shift tensor is usually masked by the large quadrupole interaction in these systems and double-quantum coherence experiments allow very elegantly to separate both contributions.

The pulse scheme resembles that used in 2D-spectroscopy (see Chap. 5). First the double-quantum state has to be prepared during the preparation period τ. Pines et al. [10, 27], Wokaun-Ernst [11, 37] and others [29, 38] have proposed different schemes to achieve this goal. The pulse sequence which is most commonly used consists of two 90°-pulses separated by τ as shown in Fig. 6.7. After the second pulse the double-quantum evolution period of duration t_1 begins and double-quantum oscillations (unobservable) with frequency shift $2\delta\omega$ occur. In order to transfer the unobservable double-quantum oscillation $2\delta\omega$ during t_1 into observable magnetization a third 90°-pulse is applied and magnetization is measured a time t_2 after the third pulse. Consider the Hamiltonian [10]

$$\mathcal{H} = -\delta\omega I_z + \omega_Q[I_z^2 - \tfrac{1}{3}I(I+1)]$$
$$= -2\delta\omega I_z^{1-3} + \tfrac{2}{3}\omega_Q(I_z^{1-2} - I_z^{2-3}) \tag{6.41}$$

where $\delta\omega \ll \omega_Q$. The evolution of the density matrix is readily obtained by using the different transformation properties. During time τ we neglect $\delta\omega$ and obtain after the first 90°-pulse

$$\rho(\tau) = e^{-i\tau\tilde{\mathcal{H}}_Q}|I_x)$$
$$= I_x \cos\omega_Q\tau + \sqrt{2}(I_y^{1-2} - I_y^{2-3})\sin\omega_Q\tau.$$

After the second pulse

$$\rho(t_1 = 0, \tau) = e^{-i\frac{\pi}{2}\hat{I}_y}|\rho(\tau))$$
$$= -I_z \cos\omega_Q\tau + 2I_y^{1-3}\sin\omega_Q\tau.$$

Note that $[\mathcal{H}_Q, \rho(t_1 = 0, \tau)] = 0$, i.e. evolution of $\rho(t_1, \tau)$ proceeds only due to $2\delta\omega I_z^{1-3}$

$$\rho(t_1, \tau) = e^{it_1 2\delta\omega \hat{I}_z^{1-3}}|\rho(t_1 = 0, \tau))$$
$$\rho(t_1, \tau) = -I_z \cos\omega_Q\tau + 2\sin\omega_Q\tau[I_y^{1-3}\cos 2\delta\omega t_1 - I_x^{1-3}\sin 2\delta\omega t_1]. \tag{6.42}$$

The third pulse transforms this into

$$\rho(t_2 = 0, t_1, \tau) = e^{-i\frac{\pi}{2}\hat{I}_y}|\rho(t_1, \tau))$$
$$= -\{I_x \cos\omega_Q\tau + \sqrt{2}(I_y^{1-2} - I_y^{2-3})\sin\omega_Q\tau \cos 2\delta\omega t_1$$
$$+ (I_z^{1-2} - I_z^{2-3} + I_x^{1-3})\sin\omega_Q\tau \sin 2\delta\omega t_1\}. \tag{6.43}$$

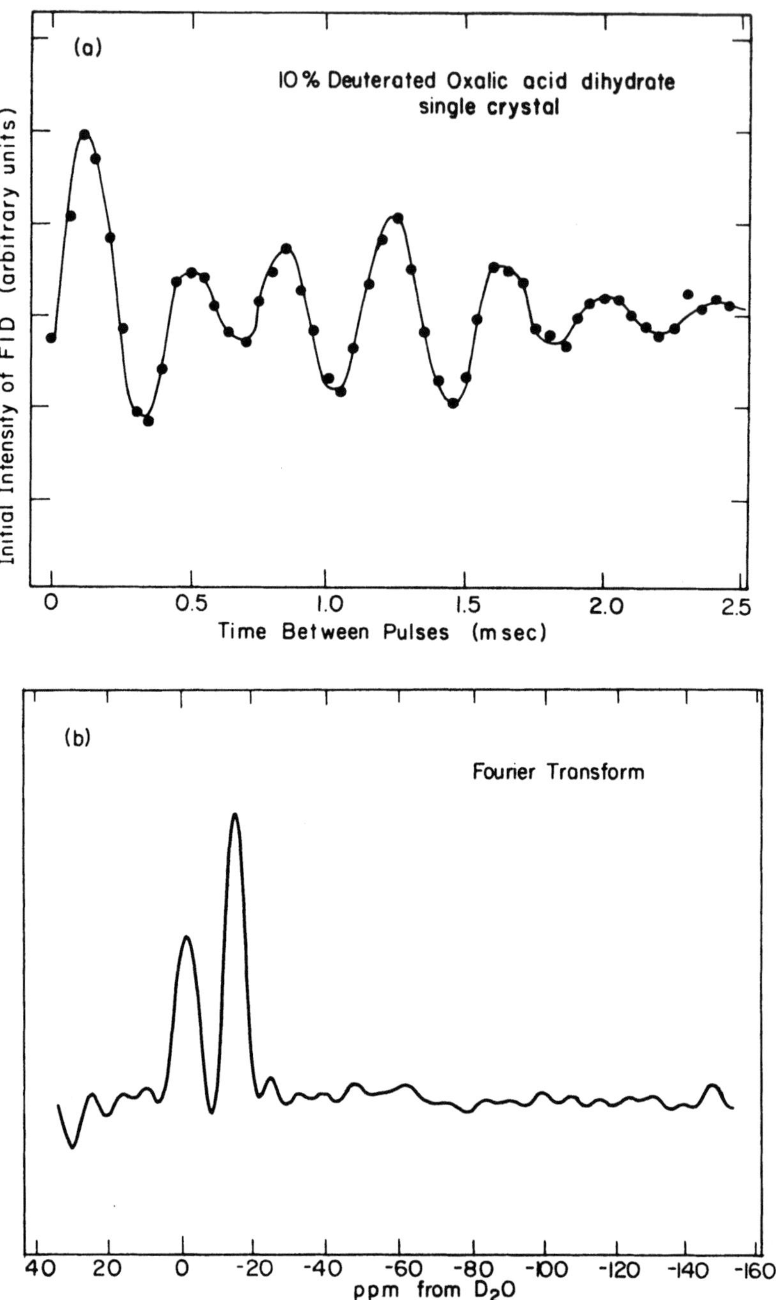

Fig. 6.8a, b. Double-quantum FID (**a**) and spectrum (**b**) of ^{2}H in oxalid acid dihydrate according to Vega et al. [10, 27]. (Courtesy of A. Pines)

Only the I_x term produces observable magnetization. It does, however, not yet depend on $\delta\omega$. Waiting a time t_2 (on the order of τ) in fact results in

$$\rho(t_2, t_1, \tau) \cong e^{-it_2 \hat{\mathscr{H}}_Q} \,|\rho(t_2 = 0, t_1, \tau))$$

or

$$\begin{aligned}
\rho(t_2, t_1, \tau) = I_x(&\cos 2\delta\omega t_1 \sin \omega_Q \tau \sin \omega_Q t_2 - \cos \omega_Q \tau \cos \omega_Q t_2) \\
&- \sqrt{2}(I_y^{1-2} - I_y^{2-3}) \cos \omega_Q \tau \sin \omega_Q t_2 \\
&- (I_z^{1-2} - I_z^{2-3} + I_x^{1-3}) \sin \omega_Q \tau \sin 2\delta\omega t_1.
\end{aligned} \tag{6.44}$$

The spectrometer now detects the normalized magnetization

$$\begin{aligned}
\langle I_x(t_2, t_1, \tau)\rangle &= \mathrm{Tr}\{I_x\,\rho(t_2, t_1, \tau)\}/\mathrm{Tr}\{I_x^2\} \\
&= \cos 2\delta\omega t_1 \sin \omega_Q \tau \sin \omega_Q t_2 - \cos \omega_Q \tau \cos \omega_Q t_2.
\end{aligned} \tag{6.45}$$

By keeping τ and $t_2 \cong \tau$ fixed and varying t_1 step by step the double quantum oscillations $2\delta\omega$ can be mapped out as demonstrated by Pines and co-workers [10, 27]. Their experimental results are shown in Fig. 6.8.

Note, however, that the second order quadrupolar interaction not taken into account in our treatment also shifts the double quantum transition by an amount of order ω_Q^2/ω_0. This effect has to be subtracted from the data before determining the true chemical shift interaction.

In a spin 3/2 system three single quantum transitions are observed at ω_0 and $\omega_0 \pm 2\omega_Q$. The two double quantum transitions $(1-3)$ and $(2-4)$ at frequency $\omega_0 \pm \omega_Q$ (level spacing $2(\omega_0 \pm \omega_Q)$) are driven by an effective field [10b, 39]

$$\omega_1^{1-3} = \omega_1^{2-4} = 7\omega_1^2/(4\omega_Q). \tag{6.46}$$

Besides the single quantum transition at ω_0 also a triple quantum transition $(1-4;\ 3\omega_0)$ exists with an effective transition rate [39]

$$\omega_1^{1-4} = 3\omega_1^3/(8\omega_Q^2). \tag{6.47}$$

It is straightforward to calculate the corresponding matrix elements in systems with spin $I > 3/2$ but we will not persue this here.

6.3 Multiple-Quantum Coherence

In the preceding section we saw that a sequence of two 90°-pulses separated by τ is capable of preparing a double-quantum coherent state. It has been shown [37, 40, 41], that in fact all orders of multiple-quantum coherent states can be produced by this sequence (if τ is large enough). Before discussing this in more detail let us make some general remarks concerning multi-level systems and multiple-quantum transitions. To be specific we shall consider a group of N interacting spins $I = 1/2$ in a large magnetic field B_0. The total z-component of the angular momentum $M = \sum_j m_j$ can have any of $2N$ values, with $-N/2 \leqq M \leqq +N/2$. For a fixed value of M there are $N!/[(N/2 - M)!/$

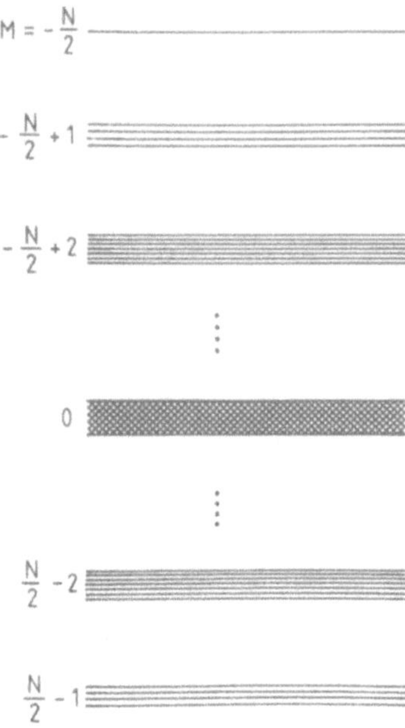

Fig. 6.9. Energy level diagram for a general system with N spin 1/2. Each value of M has degeneracy $\binom{N}{M+N/2}$, corresponding to a binomial distribution (see Warren et al. [40, 41])

$(N/2+M)!]$ energy levels as shown schematically in Fig. 6.9 which are all degenerate if there is no interaction among the spins. The spin interactions which are usually much smaller than the Zeeman interaction break this degeneracy as shown in Fig. 6.9 but leave M to be a good quantum number, i.e. levels of different M do not cross. In a system of N spin 1/2 there are 2^N distinct energy levels with a maximum number of $2^N 2^{N-1}$ transitions if transitions between any two levels are allowed.

If the systems Hamiltonian has no symmetry element whatsoever, all of the levels can be non-degenerate and the number of distinct n-quantum transitions is $(2N)!/[(N+n)!(N-n)!]$. The situation is more complicated if the spins are magnetically equivalent as was discussed by Wokaun and Ernst [37] and by Warren et al., Weitekamp and Pines [40, 41]. The most simplest transition in an N spin-1/2 system is the $\Delta M = N$ (N-quantum) transition which results in a single line at frequency $\omega = \omega_0 + N\,\Delta\omega - \sum_j \sigma_j \omega_0$ [40, 41]. Therefore the N-quantum transition is shifted off resonance by $N\,\Delta\omega$ and by the sum of the N chemical shifts $\sigma_j \omega_0$. The dipolar interaction $\mathcal{H}_D$ does not contribute here, since it is a bilinear interaction and its eigenvalue is invariant to flipping all the spins. Therefore N-quantum spectroscopy fulfills the goal of high resolution NMR in solids. On the other hand dipolar interaction can be a valuable tool for structural analysis as was already demonstrated in Chap. 5. Only because of the tremendous amount of single-quantum transitions in dipolar coupled systems this tool is not very sharp if many spins are involved. Pines and co-

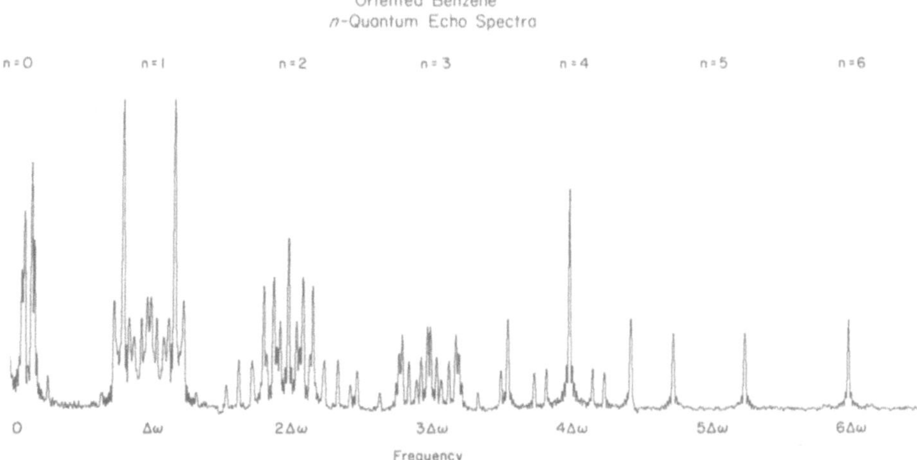

Fig. 6.10. Multi-quantum ^{1}H spectrum of oriented benzene (C_6H_6) in a liquid crystalline phase according to Pines and co-workers [12]. All n-quantum spectra up to $n=6$ are displayed

workers [12, 41–44] therefore drew attention to the fact that especially the $N-1$ and $N-2$ quantum transitions are very useful for structural and motional analysis. We shall follow their arguments throughout this section.

The $(N-1)$- and $(N-2)$-quantum spectra are far simpler than the single-quantum spectrum, but still contain enough useful information as is evident from Fig. 6.10 which shows the n-quantum spectra ($n=0,1,2,3,4,5,6$) of benzene dissolved in a nematic liquid crystal [12]. There are N states with $M = \pm(N/2-1)$ and $N(N-1)/2$ states with $M = \pm(N/2-2)$. The number of allowed transitions depends on the symmetry of the Hamiltonian [40]. Here we consider only the influence of dipolar interaction $\mathcal{H}_D$. In this case the $(N-1)$-quantum spectrum has N pairs of lines $(-N/2+1 \leftrightarrow N/2; N/2-1 \leftrightarrow -N/2$ see Fig. 6.9). In the case of benzene $(N=6)$ the $(N-1)$-quantum spectrum $(n=5)$ should have 6 pairs of lines. However, all six protons in benzene are equivalent resulting in just a single pair of lines as shown in Fig. 6.10. Pines and co-workers [43] have pointed out that the number of line pairs in the $(N-1)$-quantum spectrum is equal to the number of different sites a single spin can occupy in the system. In this sense the $(N-1)$-quantum spectrum corresponds to isotope labelling with a single spin.

The $(N-2)$-quantum spectrum is also symmetric and consists of $N(N-1)$ pairs of lines $(N/2 \leftrightarrow -N/2+2; N/2-1 \leftrightarrow -N/2+1; N/2-2 \leftrightarrow -N/2$; see Fig. 6.9) plus a highly degenerate peak at $(N-2)\Delta\omega$ due to flipping all the spins in the transition $N/2-1 \leftrightarrow -N/2+1$. This feature is also evident in the $(N-2)$-quantum spectrum of benzene $(n=4)$ where $N(N-1)=30$ pairs of lines are expected. Figure 6.10, however, shows clearly that the number of lines is drastically reduced due to symmetry. Invoking the spin labelling picture of Pines et al. [43, 44] again, we note that each possible site a pair of spins can occupy in the system contributes a triplet or quartet to the $(N-2)$-quantum spectrum. In the case of benzene we can clearly place a pair of protons in three distinct ways on the molecule, namely in ortho-, meta- and para-position leading to three pairs of lines as shown in Fig. 6.10.

How are the n-quantum spectra generated? As mentioned in the introduction to this section a pair of 90°-pulses separated by τ is capable of generating all orders of multiple-quantum coherence, i.e. we can use the same non-selective pulse sequence (Fig. 6.7) as discussed in the context of double-quantum coherence. The density matrix after the two preparation pulses may be expressed as

$$\rho(t_1=0,\tau)=e^{-i\frac{\pi}{2}\hat{I}_y}\,e^{-i\tau\hat{\mathcal{H}}_z}\,e^{-i\frac{\pi}{2}\hat{I}_y}|I_z)\tag{6.48}$$

where we denote the interaction Hamiltonian by $\mathcal{H}_z$. Equation (6.48) may be rewritten as

$$\rho(t_1=0,\tau)=-e^{-i\tau\hat{\mathcal{H}}_x}|I_x)\tag{6.49}$$

where

$$\mathcal{H}_x=e^{-i\frac{\pi}{2}I_y}\,\mathcal{H}_z\,e^{i\frac{\pi}{2}I_y}$$

e.g. in the case of dipolar interaction

$$\mathcal{H}_z\equiv\mathcal{H}_{Dz}=\sum_{i<j}A_{ij}(3I_{zi}I_{zj}-\mathbf{I}_i\cdot\mathbf{I}_j)$$

$$\mathcal{H}_x\equiv\mathcal{H}_{Dx}=\sum_{i<j}A_{ij}(3I_{xi}I_{xj}-\mathbf{I}_i\cdot\mathbf{I}_j).$$

The minus sign in Eq. (6.49) may be ignored or eliminated by using phase alternated pulses. The density matrix is a $2^N\times2^N$ matrix as is I_z. The time evolution of the density matrix during τ from the initial diagonal state I_z to non-diagonal coherent states is mediated by the bilinear Hamiltonian $\mathcal{H}_x$. In order to calculate the density matrix element ρ^{j-k} of the $j-k$ two-level sub-system we have to project $\rho(t_1=0,\tau)$ onto this state. By using the projector

$$P=\frac{|I_\alpha^{j-k})\,(I_\alpha^{j-k}|}{(I_\alpha^{j-k}|I_\alpha^{j-k})}\tag{6.50}$$

we obtain

$$\rho^{j-k}=|I_\alpha^{j-k})\frac{(I_\alpha^{j-k}|\,e^{-i\tau\hat{\mathcal{H}}_x}\,|I_z)}{(I_\alpha^{j-k}|I_\alpha^{j-k})}\tag{6.51}$$

or

$$\rho^{j-k}=M_\alpha^{j-k}(\tau)\,|I_\alpha^{j-k})\tag{6.52a}$$

where

$$M_\alpha^{j-k}(\tau)=\frac{(I_\alpha^{j-k}|\,e^{-i\tau\hat{\mathcal{H}}_x}\,|I_z)}{(I_\alpha^{j-k}|I_\alpha^{j-k})}\equiv\frac{\mathrm{Tr}\{(I_\alpha^{j-k})^+\,e^{-i\tau\mathcal{H}_x}I_z e^{i\tau\mathcal{H}_x}\}}{\mathrm{Tr}\{(I_\alpha^{j-k})^+\,I_\alpha^{j-k}\}}\tag{6.52b}$$

with $\alpha=0$ $(I_0^{j-k}=I_z^{j-k})$ for on-diagonal elements and $\alpha=+(-)$ for the off-diagonal elements above (below) the diagonal. The calculation of the multiple-quantum coefficients $M_\alpha^{j-k}(\tau)$ is a formidable task and we are not persuing this here. Inspection of Eq. (6.52b) tells us immediately that for $\tau=0$, $M_\pm^{j-k}(0)=0$; i.e. multiple-quantum coherence evolves with increasing τ. In fact the interaction Hamiltonian $\mathcal{H}_x$ carries the diagonal initial density matrix (I_z) into higher and higher quantum coherent states when time proceeds. The usual expansion techniques and moment analysis may be applied to $M_\alpha^{j-k}(\tau)$ to elucidate the τ-dependence of the different n-quantum states.

We want to continue by assuming that all quantum coherent states have been created and evolve under the action of the Hamiltonian $\mathcal{H}_z$ for some time t_1 after the two 90°-pulses (see Fig. 6.7).

$$\rho(t_1, \tau) = e^{-it_1 \hat{\mathcal{H}}_z} e^{-i\tau \hat{\mathcal{H}}_x} |I_z) \tag{6.53}$$

where we have dropped the minus sign of Eq. (6.49). The multiple-quantum oscillations are not observable and have to be read out with the help of a third pulse plus a delay t_2 to create observable $\langle I_x \rangle$ or $\langle I_y \rangle$ magnetization

$$\langle I_x(t_2, t_1, \tau) \rangle = \frac{(I_x | e^{-it_2 \hat{\mathcal{H}}_z} e^{-i\frac{\pi}{2} \hat{I}_y} e^{-it_1 \hat{\mathcal{H}}_z} e^{-i\tau \hat{\mathcal{H}}_x} |I_z)}{(I_z|I_z)} \tag{6.54}$$

which can be written in a more symmetric form

$$\langle I_x(t_2, t_1, \tau) \rangle = \frac{(I_z | e^{it_2 \hat{\mathcal{H}}_x} e^{-it_1 \hat{\mathcal{H}}_z} e^{-i\tau \hat{\mathcal{H}}_x} |I_z)}{(I_z|I_z)}. \tag{6.55}$$

Inserting now complete sets of fictitious spin-1/2 operators

$$\mathbb{1} = \sum_{\substack{j \neq k \\ \alpha}} \frac{|I_\alpha^{j-k})(I_\alpha^{j-k}|}{(I_\alpha^{j-k}|I_\alpha^{j-k})} \tag{6.56}$$

leads to

$$\langle I_x(t_2, t_1, \tau) \rangle = (I_z|I_z)^{-1} \sum_{\substack{l \neq m; j \neq k \\ \beta; \alpha}} \frac{(I_z | e^{it_2 \hat{\mathcal{H}}_x} |I_\beta^{l-m})}{(I_\beta^{l-m}|I_\beta^{l-m})}$$

$$\cdot (I_\beta^{l-m}| e^{-it_1 \hat{\mathcal{H}}_z} |I_\alpha^{j-k}) \frac{(I_\alpha^{j-k}| e^{-i\tau \hat{\mathcal{H}}_x} |I_z)}{(I_\alpha^{j-k}|I_\alpha^{j-k})}. \tag{6.57}$$

Using the multiple-quantum coefficients $M_\alpha^{j-k}(\tau)$ as defined in Eq. (6.52b) and their symmetry relations, Eq. (6.57) can be reformulated as

$$\langle I_x(t_2, t_1, \tau) \rangle$$

$$= (I_z|I_z)^{-1} \sum_{\substack{l \neq m; j \neq k \\ \beta; \alpha}} M_\beta^{l-m}(t_2)^* \, M_\alpha^{j-k}(\tau) (I_\beta^{l-m}| e^{-it_1 \hat{\mathcal{H}}_z} |I_\alpha^{j-k}). \tag{6.58}$$

The multiple-quantum oscillations are mapped out step by step by varying the evolution time t_1, whereas τ and t_2 (on the order of τ) are kept fixed. The product of the multiple-quantum matrix elements in Eq. (6.58) determines the intensity of the corresponding line. This intensity varies with t_2 and τ and spectra with different sets of (t_2, τ) have to be averaged (ensemble average). By noting, that

$$(I_\beta^{l-m}| e^{-it_1 \hat{\mathcal{H}}_z} |I_\alpha^{j-k}) \equiv C_{\beta\alpha}(lm; jk) \, e^{-it_1 \alpha \omega_{rs}} \tag{6.59}$$

where ω_{rs} is the $r-s$ subsystem transition frequency with (rs) being any index pair out of the set $(jklm)$, Eq. (6.58) may be further compressed to

$$\langle I_x(t_2, t_1, \tau)\rangle = \sum_{\substack{l \neq m; j \neq k \\ \beta; \alpha}} K_{\beta\alpha}(lm; jk)\, e^{-it_1 \alpha \omega_{rs}} \qquad rs \in jklm$$

$$\alpha\beta = 0, \pm \tag{6.60}$$

where

$$K_{\beta\alpha}(lm, jk) = M_\beta^{l-m}(t_2)^*\, M_\alpha^{j-k}(\tau)\, C_{\beta\alpha}(lm; jk)/(I_z|I_z). \tag{6.61}$$

Equation (6.60) expresses the fact, that in principle all multiple-quantum transitions $r-s$ are observable where the intensity is given by the coefficient $K_{\beta,\alpha}(lm; jk)$. The sixfold sum is taken over all indices $l \neq m$ and $j \neq k$ as well as over α, $\beta = 0$, $+$, and $-$. So far we have neglected relaxation effects. They may be phenomenologically included by multiplying the exponential function in Eq. (6.60) by $\exp(-t_1/T_2^{r-s})$ if τ, $t_2 \ll T_2$. The investigation of spin-lattice relaxation effects in multiple-quantum transitions has its own fascination and complexities, and the interested reader is referred to the literature [29, 45].

Multiple-quantum spectra like the one shown in Fig. 6.10 utilize a further subtlety which is worth mentioning. As discussed in the introduction the n-quantum spectrum is shifted by an amount $n\Delta\omega$ off-resonance if the single quantum spectrum has a frequency shift $\Delta\omega$ with respect to the reference signal. Only by choosing $\Delta\omega$ large enough all the multiple-quantum spectra are separated. Note, also that magnetic field inhomogeneities become increasingly more severe with increasing n. In the next section we will turn to the question how to select a specific n-quantum spectrum.

6.4 Selective Multiple-Quantum Coherence

In the preceding section we learned how to excite all multiple-quantum spectra of a given system. Pines and co-workers [12] have recently demonstrated in a series of experiments that it is indeed possible to excite selectively a specific n-quantum transition with non-selective pulses by using the full orchestra of multiple-pulse techniques [40–44]. Before introducing their ideas we turn to the simpler question of how to detect selectively a specific n-quantum spectrum. Schemes for doing this have been discussed in references [37, 42].

Suppose we change the phase of the first two pulses in the basic experiment shown in Fig. 6.7. We also could change of course only the phase of the third pulse. This corresponds to a rotation about the z-direction of the rotating frame by an angle Φ. The density matrix at $t_1 = 0$ $\rho(t_1 = 0, \tau)$ (Eq. (6.49)) is then replaced by

$$\rho_\Phi(t_1 = 0, \tau) = e^{-i\Phi \hat{I}_z}\,|\rho(t_1 = 0, \tau)). \tag{6.62}$$

This modifies Eq. (6.59) as

$$(I_\beta^{l-m}|\, e^{-it_1 \mathscr{H}_z}\, e^{-i\Phi \hat{I}_z}\, |I_\alpha^{j-k}) = C_{\alpha\beta}^{(\Phi)}(lm; jk)\, e^{-it_1 \alpha \omega_{rs}}\, e^{-i\alpha n\Phi} \tag{6.63}$$

with $n = \Delta M$ corresponding to the n-quantum transition and where $\alpha = 0, \pm$; i.e. the n-quantum spectrum is shifted by $n\Phi$ and $\langle I_x(t_2, t_1, \tau, \Phi)\rangle$ appears to be

Table 6.1. Some possibilities for selective detection of multiple quantum spectra [37]

Phase values Φ								Observed orders n of multiple-quantum spectra									
0°	60°	90°	120°	180°	240°	270°	300°										
x								0	1	2	3	4	5	6	7	8	9
x				+				0		2		4		6		8	
x				−					1		3		5		7		9
x		+		+		+		0				4				8	
x		−		+		−				2				6			
x	−		+	−	+		−				3						9

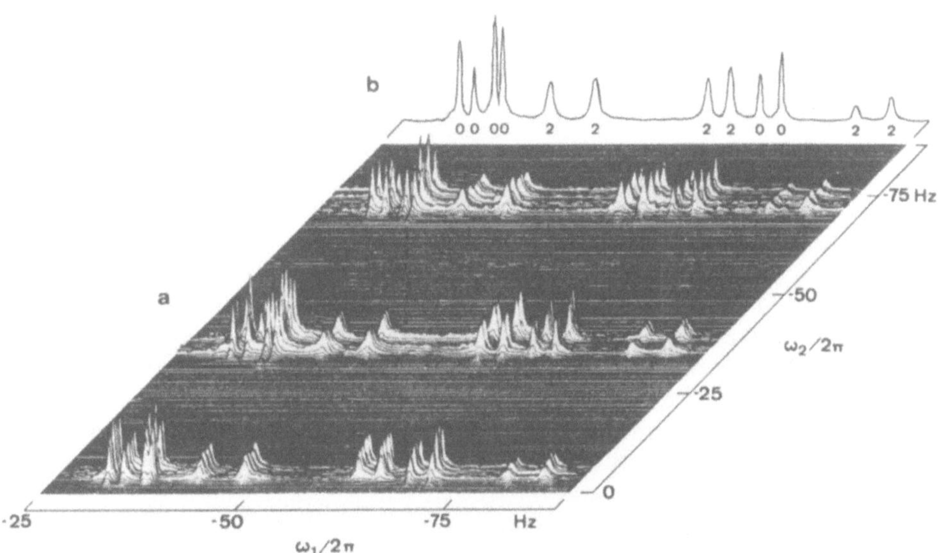

Fig. 6.11a, b. Zero-quantum and double-quantum proton spectrum of 2-furancarboxylic acid methyl ester according to Wokaun and Ernst [37]. **a** Two-dimensional representation of zero- and double-quantum transitions along the ω_1-axis and single-quantum transitions along the ω_2-axis. **b** Projection of the 2D spectrum onto the ω_1-axis

modulated with $\cos n\Phi$ and $\sin n\Phi$ terms. Suppose we take multiple-quantum spectra with $\Phi = 0$ and $\Phi = \pi$ and add those, only even-quantum spectra with $n = 0, 2, 4, \ldots$ survive, whereas odd-quantum spectra vanish. If we subtract, however, the $\Phi = 0$ and $\Phi = \pi$ spectra only the odd-quantum spectra $n = 1, 3, 5, \ldots$ survive.

Table 6.1 summarizes a number of possible cases [37]. A demonstration of this behaviour by Wokaun and Ernst [37] in case of the liquid AMX spectra of 2-furncarboxylic acid methyl ester is shown in Fig. 6.11. In this case $\Phi = 0, \pi$ spectra were coadded and correspondingly only zero- and double-quantum spectra are recorded in a 2D-representation. The $n\Phi$-dependence of the n-quantum spectrum according to Eq. (6.63) has found another useful application by Drobny et al. [42]. Suppose we choose the phase Φ of the first two pulses

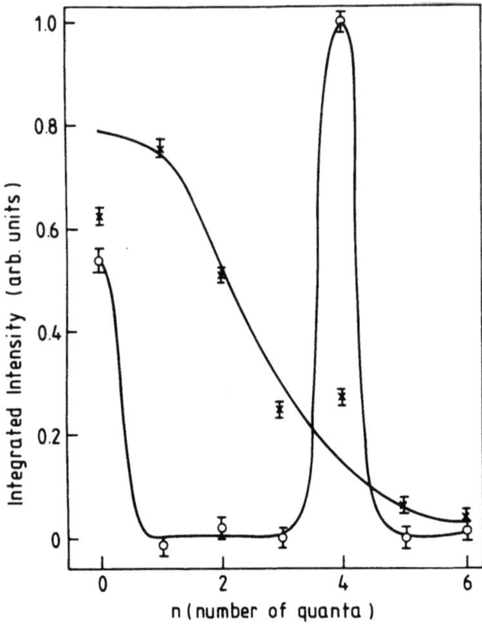

Fig. 6.12. Integrated intensities of the multiple quantum spectra of benzene for different Δm according to Drobny et al. [44]. In a non-selective experiment (X) a monotonic decrease (near Gaussian) of the intensity is observed with increasing Δm, whereas the intensity is enhanced at $\Delta m = 4$ for a $n = 4$ selective experiment

as $\Phi = \Delta \omega t_1$, i.e. for each step in t_1 we advance also Φ proportionally in the data taking process. We have created artificially an off-set $\Delta \omega$ which separates the n-quantum spectra as $n\Delta \omega$, but is independent of the time evolution during t_1. A π-pulse e.g. may be placed between the second and the third pulse to cancel magnetic field inhomogeneity effects. Applying time proportional phase increments (TPPI [42]) still allows then to separate all the n-quantum spectra without using any spectrometer off-set and selective detection is still feasible.

It is evident, that selective detection does not enhance a specific n-quantum spectrum but rather selects it from others. With increasing order n, however, the number of possible transitions decreases drastically resulting in a reduced overall intensity of an n-quantum transition ($n > 2$) with respect to the single-quantum transition. Drobny et al. [40, 44] have estimated the overall intensity of multiple-quantum spectra assuming different kind of models. Figure 6.12 shows their result for benzene, where a Gaussian distribution seems to fit the data pretty well. Note the drastic intensity loss for the 5- and 6-quantum spectra. This supplied motivation to invent sequences which selectively pump a specific n-quantum transition. Pines and co-workers [12] achieved this goal by phase-cycling and nesting certain multiple-pulse cycles. Their success in achieving this goal is displayed in Fig. 6.13 which shows the selectively pumped $n = 4$ quantum spectrum of oriented benzene dissolved in a liquid crystal [12]. In the case of selective excitation as discussed here, the selectively pumped n-quantum transition is enhanced with respect to the others and the intensity loss as displayed in Fig. 6.11 is absent. This is in contrast to selective detection. The basic idea of Pines and co-workers [12] to produce selective multiple-quantum sequences is to replace the two-pulse sequence ($90_y - \tau - 90_{\bar{y}}$) in the preparation period by phase-incremented multiple-pulse cycles which allow

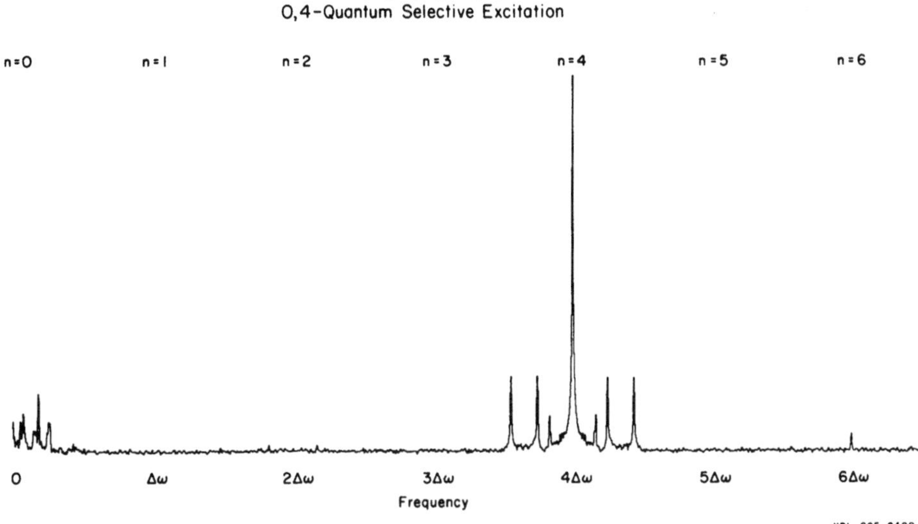

Fig. 6.13. $n=4$ quantum selective spectrum of ^1H in oriented benzene according to Warren et al. [12]. This spectrum should be compared with the non-selective multiple quantum spectrum shown in Fig. 6.10

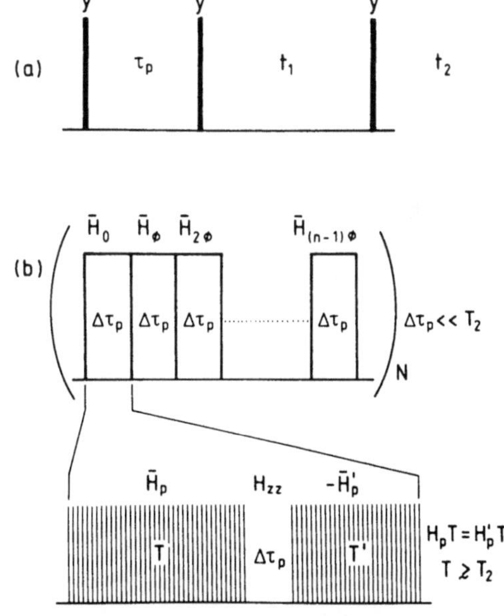

Fig. 6.14a, b. Multiple-quantum pulse sequences due to Warren et al. [12, 41]. Sequence (**a**) is a version of the non-selective multiple-quantum sequence shown in Fig. 6.7. Sequence (**b**) is an n-quantum selective sequence if $\Phi=2\pi/n$ as explained in the text

destructive interference of all unwanted multiple-quantum coherences and constructive interference of a specific n-quantum coherence. Such a scheme is shown in Fig. 6.14. In discussing how these sequences work we follow closely the reasoning of Warren et al. [12, 40, 41].

The sequence to create multiple-quantum spectra so far is shown again in

Fig. 6.14a. Multiple-quantum coherence of all orders is excited by the first two $90°$-pulses separated by τ which is on the order of T_2 where T_2 is the single quantum coherence time, i.e. $T_2^{-1} \cong \|\mathcal{H}_z\|$. Figure 6.14b shows schematically how selective wide band excitation is performed. Two steps of coherent averaging [34] (see also Chap. 3) are performed.

(i) Subcycle averaging.

An effective Hamiltonian $\bar{H}_0$ containing all multiple quantum matrix elements is produced under which the system evolves during time $\Delta\tau_p$ such that $\|\bar{H}_0\| \Delta\tau_p \ll 1$. In the next subcycle all the pulses are then phase shifted by an angle Φ to create a new subcycle of duration $\Delta\tau_p$, with a new effective Hamiltonian

$$\bar{H}_\Phi = e^{-i\Phi I_z} \bar{H}_0\, e^{i\Phi I_z}. \tag{6.64}$$

As shown in Eq. (6.63) every n-quantum term is then multiplied by $e^{-in\Phi}$. The phase increment used is $\Phi = 2\pi/n$ if n subcycles are repeated to produce a full cycle.

(ii) Cycle averaging.

The whole cycle made up from n phase incremented subcycles is repeated N times, such that $\tau_p = N\, n\Delta\tau_p > T_2$. The total average Hamiltonian for the cycle is readily obtained as [12, 41]

$$\bar{H}^{(0)} = \frac{1}{n} \sum_{k=0}^{n-1} \bar{H}_{k\Phi} = \frac{1}{n} \sum_{k=0}^{n-1} e^{-ik\Phi I_z} \bar{H}_0\, e^{ik\Phi I_z}. \tag{6.65}$$

Since $\Phi = 2\pi/n$ interference destroys all $m \neq n$ terms and only the n-quantum term survives so that $\bar{H}^{(0)}$ can be visualized as an n-quantum operator to lowest order in $\Delta\tau_p$. Warren et al. [12, 41] have discussed also higher order correction terms and the interested reader is referred to their original work.

At least one open question still remains, i.e. how can one achieve multiple quantum excitation is a time $\Delta\tau_p$ (which supposedly should be on the order of T_2) and simultaneously fulfill the condition $N\, n\Delta\tau_p \cong T_2$.

How Warren et al. [12] achieved this is shown in the lower part of Fig. 6.14 schematically. The idea is based on nested cycles [46] (see Fig. 3.36 p. 124) as used in magic angle high resolution NMR, where the effective evolution time of the system is much shorter than the cycle time. In the case discussed here the effective cycle time $\Delta\tau_p$ is much shorter than the real cycle time $T + T' + \Delta\tau_p$. The multiple-quantum excitation evolves under the action of H_p during the time $T \simeq T_2$, where a multiple-pulse sequence is applied. After the window $\Delta\tau_p$ another multiple-pulse sequence is applied for time T' which inverts H_p [47] to cancel the effect of the rather long times T and T'. The effective subcycle Hamiltonian may therefore be written as

$$\begin{aligned}
\bar{H}_0 &= e^{iH'_p T'} H_{zz}\, e^{-iH_p T} \\
&= e^{iH_p T} H_{zz}\, e^{-iH_p T} \tag{6.66}
\end{aligned}$$

acting effectively only during $\Delta\tau_p$.

Rhim, Pines and Waugh [47] have proposed several ways to produce time-reversed $\bar{H}_p$. Two simple pulse sequences used by Warren et al. [12] are mentioned:

a) $\bar{H}_p = H_{xx}$ by using $90_y - T - 90_{\bar{y}}$ and $\bar{H}'_p = -\frac{1}{2}H_{xx}$ created by the time-reversing sequence $(90_x - \tau - 90_x - \tau - 90_{\bar{x}} - \tau - 90_{\bar{x}} - \tau - 90_{\bar{x}} - \tau - 90_{\bar{x}} - \tau - 90_x - \tau - 90_x - \tau)_m$ which is repeated m times to make a total interval $T' = 2T$. Note that τ must fulfill the condition $\tau \ll \Delta \tau_p$.

b) $\bar{H}_p = 1/2(H_{yy} - H_{xx}) = 1/2(2H_{yy} + H_{zz})$ produced by the sequence $(90_x - 2\tau - 90_x - \tau - 90_{\bar{x}} - 2\tau - 90_{\bar{x}} - \tau - 90_{\bar{x}} - 2\tau - 90_{\bar{x}} - \tau - 90_x - 2\tau - 90_x - \tau)_m$ repeated m times to make the interval T. Reversal is simply achieved by phase shifting each pulse by $\pi/2$, giving $-\bar{H}'_p = -1/2(H_{yy} - H_{xx})$ and a time $T' = T$. The $4k$-quantum spectrum $(k = 0, 1, 2, \ldots)$ of benzene shown in Fig. 6.13 was obtained by using this sequence.

Warren et al. [12, 41] have discussed other schemes and higher order corrections. Treatment of these details is beyond the goal of this monograph.

A few applications of multiple-quantum coherence experiments such as conformational studies [48a], methyl group relaxation [48b–d] and high resolution NMR in inhomogeneous fields [48e] are worth mentioning.

6.5 Double-Quantum Cross-Polarization

Double-resonance and cross-polarization of rare spins has been discussed extensively in Chap. 4. Here we want to restrict ourselves to the specific problem of double-quantum cross-polarization in solids, namely the direct or indirect detection of quadrupolar distorted $S = 1$ spins e.g. ^{2}D and ^{14}N. There are only four publications which have dealt with these aspects so far, namely Shattuck [48], Vega, Shattuck and Pines [49], Brunner, Reinhold and Ernst [50] and Reinhold, Brunner and Ernst [51]. We shall follow closely their argumentation. The spin dynamics involved is actually very similar to the one discussed in Chap. 4 with the main difference that polarization transfer to the "rare spins" S evolves in the effective field $\omega_{1S}(\omega_{1S}/\omega_Q)$ instead of ω_{1S} as in the single-quantum transfer. The scaling of the S spin transition rate by ω_{1S}/ω_Q is therefore more severe the larger the quadrupolar interaction ω_Q is.

Nevertheless the double-quantum cross-polarization is still feasible [48–51] even in the case of large quadrupolar interaction like ^{14}N in glycine as was demonstrated by Reinhold et al. [51]. Their ^{14}N double-quantum spectrum is shown in Fig. 6.15.

Let us consider a single spin $S = 1$ surrounded by a number of spins $I = 1/2$. Two rf fields with strength ω_{1I} and ω_{1S} are applied at the corresponding frequencies ω_S and ω_I respectively. The full Hamiltonian in the rotation frame can be written as

$$\mathscr{H} = \mathscr{H}_{SZ} + \mathscr{H}_{SQ} + \mathscr{H}_{IS} + \mathscr{H}_{IZ} + \mathscr{H}_{II} + \mathscr{H}_{1S} + \mathscr{H}_{1I} \tag{6.67}$$

where

$$\mathscr{H}_{SZ} = \Delta \omega_S S_Z; \quad \mathscr{H}_{IZ} = \sum_j \Delta \omega_{Ij} I_{zj} \tag{6.68}$$

with

$$\Delta \omega_S = \omega_{0S} - \omega_S; \quad \Delta \omega_{Ij} = \omega_{0j} - \omega_I.$$

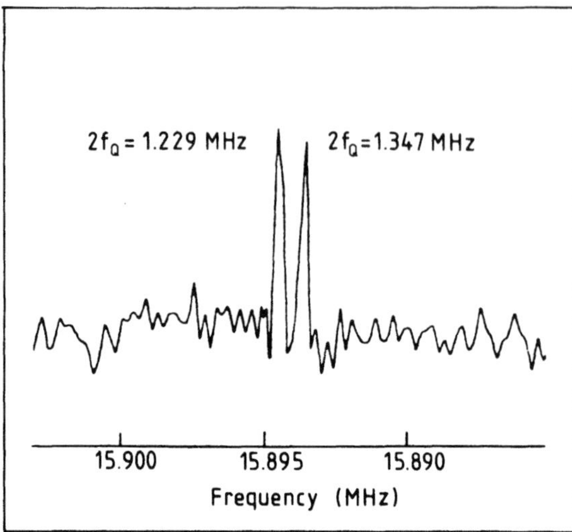

Fig. 6.15. ^{14}N double quantum spectrum of glycine with the large quadrupolar splittings 1.229 MHz and 1.347 MHz respectively. The spectrum was obtained by double-quantum cross polarization via dipolar order by Reinhold et al. [51]

The quadrupolar interaction $\mathscr{H}_{SQ} = \mathbf{S} \cdot \tilde{Q} \cdot \mathbf{S}$ (see Chap. 2) is usually smaller than the Zeeman interaction, resulting in a first order splitting

$$2\omega_Q = \left(\frac{3e^2qQ}{4\hbar}\right)[(3\cos^2\beta - 1) - \eta\sin^2\beta\cos 2\gamma] \tag{6.69}$$

where the Euler angles $(\alpha\beta\gamma)$ specify the orientation of quadrupole interaction with respect to the magnetic field B_0 and where the asymmetry factor $\eta = (V_{22} - V_{11})/V_{33}$ describes the deviation from axial symmetry. However, in double-quantum spectroscopy this first order quadrupole interaction does usually not appear in the spectrum and second order contributions [50, 51]

$$\delta_Q^{(2)} = \frac{1}{8\omega_{0S}}\left(\frac{3e^2qQ}{4\hbar}\right)^2\{\sin^2\beta(3\cos^2\beta + 1) + \tfrac{1}{2}\eta\cos 2\gamma\sin^2\beta(3\cos^2\beta - 1)$$
$$+ \tfrac{1}{9}\eta^2[4 - 3\cos^2 2\gamma(\cos^2\beta - 1)^2]\} \tag{6.70}$$

become increasingly important. The dipolar interactions are

$$\mathscr{H}_{IS} = \sum_j B_j I_{zj}S_z; \quad \mathscr{H}_{II} = \sum_{i<j} A_{ij}(3I_{zi}I_{zj} - \mathbf{I}_i \cdot \mathbf{I}_j)$$

as defined before. The rf field terms are given by

$$\mathscr{H}_{1S} = \omega_{1S}S_x; \quad \mathscr{H}_{1I} = \omega_{1I}\sum_j I_{xj}. \tag{6.71}$$

As shown before in this chapter it is possible to transform the Hamiltonian $\mathscr{H}$ according to Eq. (6.67) into a double-quantum Hamiltonian within second-order as [49–51]

$$\mathscr{H}^{(2Q)} = \mathscr{H}_{SZ}^{(2Q)} + \mathscr{H}_{SQ}^{(2Q)} + \mathscr{H}_{IS}^{(2Q)} + \mathscr{H}_{IZ} + \mathscr{H}_{II} + \mathscr{H}_{1S}^{(2Q)} + \mathscr{H}_{1I} \tag{6.72}$$

where

$$\mathscr{H}_{SZ}^{(2Q)} = 2\Delta\omega_S S_Z^{1-3}; \quad \mathscr{H}_{SQ}^{(2Q)} = 2\delta_Q^{(2)}S_Z^{1-3}$$

$$\mathscr{H}_{IS}^{(2Q)} = \sum_j 2B_j I_{zj}S_z^{1-3}; \quad \mathscr{H}_{1S}^{(2Q)} = \frac{\omega_{1S}^2}{\omega_Q}S_x^{1-3}. \tag{6.73}$$

Double-quantum cross-polarization proceeds now similar to the single-quantum cross-polarization as treated in Chap. 4. The I and S spin systems have to be matched in the rotating frame, i.e. their effective fields

$$\omega_{eS}^{(2Q)} = [(\omega_{1S}^2/\omega_Q)^2 + (2\Delta\omega_S)^2]^{1/2} \tag{6.74a}$$

$$\omega_{eI} = [\omega_{1I}^2 + \Delta\omega_I^2]^{1/2} \tag{6.74b}$$

must be equal to fulfill the Hartmann-Hahn condition ($\omega_{eS}^{(2Q)} = \omega_{eI}$). Note that the effective field $\omega_{eS}^{(2Q)}$ depends on the quadrupolar interaction, i.e. Hartmann-Hahn condition can be matched only for a certain range of ω_Q values. Moreover, the mismatch sensitivity to frequency offset is twice that of single quantum transitions.

The cross-relaxation rate $1/T_{IS}$ for single quantum cross-relaxation was derived in Chap. 4. Here we can apply the same equations by just inserting the appropriate effective fields according to Eq. (6.74) and taking the doubled frequency offset $2\Delta\omega_S$ into account. In addition $1/T_{IS}$ is proportional to the second moment [49–51]

$$M_2^{IS(2Q)} = \mathrm{Tr}\{(\sum_j 2B_j I_{zj})^2\} = 4M_2^{IS} \tag{6.75}$$

which is four times larger than the single quantum second moment. The double-quantum cross-relaxation rate proceeds therefore four times faster than the single-quantum rate [48–51]

$$1/T_{IS}^{(2Q)} = 4/T_{IS}^{(1Q)}. \tag{6.76}$$

The result can be generalized to n-quantum cross-relaxation rate as [50, 51]

$$1/T_{IS}^{(nQ)} = n^2/T_{IS}^{(1Q)}. \tag{6.77}$$

This behaviour is demonstrated in Fig. 6.16 where the single-quantum ($1QT$) and double-quantum ($2QT$) magnetization transfer of ^{14}N–^{1}H in $(NH_4)_2SO_4$ is shown according to Brunner et al. [50].

As was already discussed in Chap. 4 cross-polarization of rare spins from a dipolar state of abundant spins is often advantageous, because much smaller rf fields ω_{1S} suffice. Since S spin rf fields are scaled by ω_{1S}/ω_Q this advantage becomes severe in double-quantum cross-polarization as was noticed by Reinhold et al. [51]. Their pulse scheme is shown in Fig. 6.17 and was actually used to obtain the spectrum shown in Fig. 6.15.

An I spin dipolar state is first created by ADRF (adiabatic demagnetization in the rotating frame) and the S spins are coupled to this dipolar state for a time T_p. During this time double-quantum cross-polarization takes place as described above. Consecutively the double-quantum state evolves for some time t_1 under decoupling of the I spins and where I spin dipolar order was destroyed before by application of a "magic pulse" (54.7°) [52]. After the evolution period (t_1) the S spin double quantum coherence is locked again for some time T_m and mixing with the I spins occurs to transfer leftover S spin polarization into a dipolar ordered I spin state. This transfered I spin dipolar order is monitored at the I spin resonance (indirect detection) by the sequence ARRF (adiabatic remagnetization in the rotation frame) and a 90°-pulse to

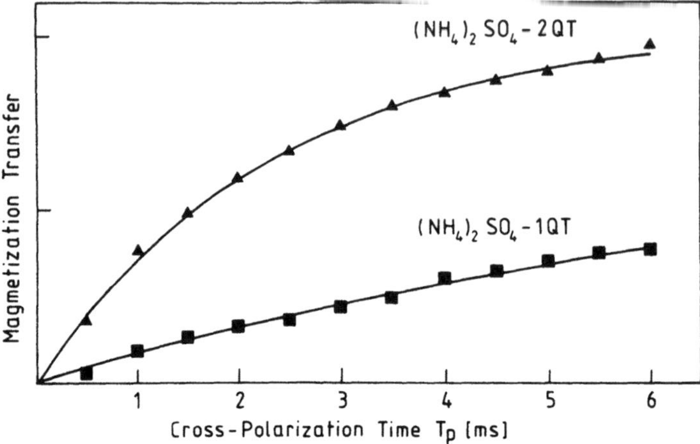

Fig. 6.16. Comparison of single-quantum (1 QT) and double-quantum (2 QT) cross-polarization rates according to Brunner et al. [50]

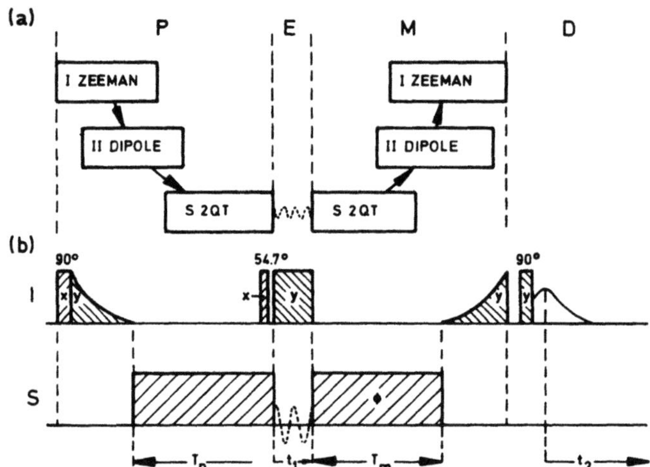

Fig. 6.17. Pulse sequence for double-quantum cross polarization via dipolar order by Reinhold et al. [51] (see text)

produce a solid echo. The signal intensity of the solid echo is then plotted for different values of t_1 to map out the double-quantum free induction decay of the S spins. After Fourier-transform the frequency spectrum as shown in Eq. (6.15) results. This method is quite sensitive due to the indirect detection scheme used (see Chap. 4).

A large variety of experiments of this sort also in connection with two-dimensional spectroscopy (see Chap. 5) can be invented by permutation of the different pulse schemes. The technique discussed here might be useful in ^{14}N spectroscopy in powder material.

7 Magnetic Shielding Tensor

As has been demonstrated in the preceding chapters multiple pulse NMR and cross-polarization of dilute spins are the most prominent techniques for evaluating the chemical shielding tensor σ of nuclei in solids. Other methods for determining parts of the chemical shielding tensor like the "second moment" analysis, the NMR of partially oriented molecules dissolved in clathrates or liquid crystals will not be reviewed in this monograph because the shift tensors determined by these methods are mostly unreliable. A review of these methods including theoretical calculations has been given recently by Appleman and Dailey [1]. The interested reader is referred to their review article also for additional reference of chemical shielding data.

The interest in the magnetic shielding tensor can be traced back to the early days of NMR and generations of chemists have utilized the isotropic part of this quantity which is observed in a liquid, as a fingerprint of the chemical bond. In fact, the measurement of chemical shift has been developed into a tool for the investigation of molecular structure [2, 3]. Although the trace of this tensor is only one out of six numbers [4] (in case the tensor is symmetric), considerable effort has been devoted to relating the molecular electronic structure to this isotropic shift theoretically [5, 6]. Since now the full symmetric second rank shift tensor becomes available with the advent of the techniques described in this book, it is hoped that a better understanding of the underlying electronic structure will result and the "valid" theory will be discernable from the "invalid" theory.

Beside the actual size of the shielding tensor, its symmetry is an interesting quantity, since it is related to the molecular and crystal structure. In contrast to high resolution NMR in liquids, where the equivalence of nuclei is defined with respect to molecular symmetry alone (taking scalar coupling into account), in crystals we call those nuclei which are related by one of the symmetry elements of the space group *crystallografically equivalent*. On the other hand nuclei are called *magnetically equivalent*, if they are related by translation or/and inversion, which are elements of the space group. The tensor elements of magnetically equivalent nuclei are identical, whereas nuclei which are crystallografically equivalent are related by the point symmetry operations (rotation-reflection) of the space group, i.e. they are not necessarily magnetically equivalent [7].

It should be noted, that even from a single crystal study determining the complete shielding tensor including the principal axis system in general, no complete unambiguous assignment of the shielding tensor to a particular nuclear site in the unit cell can be made. This is readily visualized if two crystallografically equivalent nuclei are considered which are magnetically non-equivalent. These two nuclei result in two lines in the NMR spectrum and no symmetry argument can be applied to make an assignment. One has to rely on other arguments, based on theoretical grounds or consistency checks with related compounds. A more detailed account of symmetry can be found in Ref. [7].

7.1 Ramsey's Formula [4]

It is convenient to express the shift tensor as a sum of a "diamagnetic part" $\sigma^{(d)}$ and a "paramagnetic part" $\sigma^{(p)}$ [4]:

$$\sigma = \sigma^{(d)} + \sigma^{(p)}. \tag{7.1}$$

Let us consider a rigid molecule with no net electronic orbital or spin angular momentum in the absence of the magnetic field. As shown by Ramsey [4] for such as system in the state $|0\rangle$ the components $\sigma_{\alpha\beta}^{(d)}$ and $\sigma_{\alpha\beta}^{(p)}$ of the shielding tensor take the form

$$\sigma_{\alpha\beta}^{(d)} = \frac{e^2}{2mc^2} \langle 0| \sum_i \frac{r_i^2 \delta_{\alpha\beta} - r_{i\alpha} r_{i\beta}}{r_i^3} |0\rangle \tag{7.2}$$

and

$$\sigma_{\alpha\beta}^{(p)} = \frac{e^2}{2mc^2} \sum_k (E_0 - E_k)^{-1} \langle 0| \sum_i L_{\alpha i} |k\rangle \langle k| \sum_i L_{\beta i}/r_i^3 |0\rangle$$
$$+ \langle 0| \sum_i L_{\beta i}/r_i^3 |k\rangle \langle k| L_{\beta i} |0\rangle \tag{7.3}$$

where r_i is the vector from the nucleus whose shielding is to be determined to the ith electron; $L_{\alpha i}$ is the α component of the angular momentum operator of electron i about the nucleus under consideration. E_k is the energy of the state $|k\rangle$.

The "diamagnetic" term $\sigma_{\alpha\beta}^{(d)}$ contains only matrix elements involving the ground state wavefunctions, whereas the "paramagnetic" term, resulting from second order perturbation theory displays the "mixing in" of excited states into the ground state due to the vector potential of the magnetic field. The axial character of the magnetic fields leads to a "dequenching" of the orbital angular momentum.

The calculation of the paramagnetic term is rather involved and in principle the knowledge of all the excited states is needed. Note, even though this term is of second order, it is by no means small compared with the diamagnetic term.

7.2 Approximate Calculations of the Shielding Tensor

A great amount of *ab initio* calculations exist, employing all kinds of approximated wave functions [1, 8, 9]. However, only simple molecules like H_2, F_2, HF etc. and molecular fragments have been treated in this fashion using Hartree-Fock functions as a suitable compromise between accuracy and simplicity [6, 9]. However, as Reid [10] states "the complexity of paramagnetic shielding is apparently sufficiently deterring that, for highly accurate calculations, molecules other than hydrogen tend to lose their appeal". Other so-called ab initio calculations using SUFO's (*SU*itably *F*udged *O*rbitals) have been applied to numerous simple molecules.

Although the isotropic shift calculations seem to theoretically reproduce the experimental values, there is usually no resemblance between the calculated anisotropies and the recently accurately measured shielding tensor elements. One approximation, usually employed, is the average energy or closure approximation [6] where $(E_0 - E_k)$ in the denominator of Eq. (7.3) is expressed by an average energy ΔE. This can be valid only in the case if one energy separation is small compared with all others. On the contrary, it has been demonstrated recently by Kempf et al. [11] that transitions between different states are responsible for different tensor elements. A review of these theoretical methods may be found in Refs. [1, 7].

We turn now to the approximate method of Gierke and Flygare [12, 13] (GF), which has met with some success. Especially in the case of proton shielding the GF method has been successfully applied by Haeberlen [14-16], and Spiess and co-workers [11] to molecular fragments. The diamagnetic part of the shielding tensor is approximated in the GF-method by a multipole expansion up to the molecular quadrupole moment, whereas the paramagnetic part is mainly related to the spin rotation interaction tensor C, which is by no means easier to evaluate than the paramagnetic term itself, but may be obtained experimentally by other means. The shielding tensor in this approximation may be written as a sum of six terms with [12]

$$\sigma_{xx}^{(d)} = I + II + III + IV$$

where

$$I = \sigma_{\text{atom}}^{(d)}$$

$$II = \frac{e^2}{2mc^2} \sum_i Z_i r^{-3}(y_i^2 + z_i^2) \tag{7.4}$$

$$III = \frac{e^2}{2mc^2} \sum_i [2r_i^{-3}(y_i\langle y\rangle_i + z_i\langle z\rangle_i) - 3r_i^{-5}(y_i^2 + z_i^2)\,\mathbf{r}_i \cdot \langle \vec{\rho}\rangle_i]$$

$$IV = \frac{e^2}{2mc^2} \sum_i r_i^{-3}\langle \tfrac{1}{3}\rho^2\rangle_i [2 - 3(y_i^2 + z_i^2)/r_i^2]$$

and

$$\sigma_{xx}^{(p)} = V + VI$$

where

$$V = \frac{e^2}{2mc^2} \cdot \frac{C_{xx} I_{xx} c}{e\hbar \mu_0 g_I}$$

$$VI = \frac{e^2}{2mc^2} \sum_i Z_i r_i^{-3}(y_i^2 + z_i^2) = -II. \tag{7.5}$$

Here $r_i = (x_i, y_i, z_i)$ is the vector from the shielded nucleus to a neighbour nucleus i and $\langle \vec{\rho}\rangle_i$ is the expectation value of the vector from nucleus i to the electrons surrounding it

$$\langle \vec{\rho} \rangle_i = \sum_{ki}^{zi} \langle 0| \rho_{ki} |0\rangle. \tag{7.6}$$

The components of $\langle \vec{\rho} \rangle_i$ are $\langle x \rangle_i$, $\langle y \rangle_i$ and $\langle z \rangle_i$.

In the same way $\langle \rho^2 \rangle_i$ is defined.

$C_{\mu\nu}$ is the spin-rotation interaction tensor and $I_{\mu\nu}$ is the moment of inertia tensor. The other constants have their usual meaning. The summation is performed with the shielded nucleus omitted.

Term I represents the free atom contribution, whereas II gives the contribution of the electronic point charges centered at the other nuclei. This term II cancels with term VI. If the point charges are not centered at the i-th nucleus, but are displaced by a distance $\langle \vec{\rho} \rangle_i$ the third term III arises, which is called the dipolar term [13]. This term is in general quite small with respect to the others. On the contrary, term IV, the quadrupole term, may become rather large [13], because the electronic charge density at nucleus i is spatially extended. Note, that the quadrupole part is traceless, i.e. it does not contribute to the isotropic shift tensors for simple molecules and molecular fragments. Combination with known spin-rotation tensors leads to the full shielding tensor in a semi-empiric fashion. Haeberlen, Spiess and co-workers [11, 14–16] have applied this method with some success to the explanation of proton shielding tensors (see also Ref. [7] for a detailed discussion).

The GF approach cannot really be applied to solids, since the σ values diverge logarithmically with increasing volume. Vaughan and co-workers [17] have therefore proposed a different approach which does not separate into "diamagnetic" and "paramagnetic" parts, which are gauge dependent, but rather separates into "geometrical" and "chemical" parts, which are separately gauge independent. Another different semiclassical approach has been proposed by Schmiedel and others [18].

7.3 Proton Shielding Tensors

In this most difficult domain of high resolution NMR in solids a breakthrough has been brought about by the multiple-pulse methods. Other methods such as the liquid crystal method, which claim to have obtained decent proton shift anisotropies give completely unreliable results in this field.

Multiple-pulse experiments were first applied to powder samples containing OH-groups, which show the largest proton shielding anisotropies (~ 20 ppm) observed to date [7, 14–26]. An example of such a spectrum was presented in Fig. 3.11, Sect. 3.2. As can be seen, the anisotropy range extends over the total isotropic shift observed for protons. Haeberlen and co-workers [7, 14–16] have pushed this field and determined a large amount of shielding tensors from powder spectra and performed very meticulous single crystal studies. A sample of such a single crystal spectrum was given in Fig. 3.9, Sect. 5.2. Haeberlen [7] and co-workers also got a clear physical apprehension of protonic shielding tensors by applying the Gierke-Flygare approach.

Since Haeberlen has dealt with these different aspects in his excellent review article extensively, we refer the interested reader to Ref. [7].

We shortly summarize different facts observed in protonic shielding tensors:

a) Protons in Hydrogen Bonds

(i)   The shielding tensor is approximately axially symmetric, with the unique axis close to the vector joining the next nearest neighbours i.e. the "hydrogen bond direction".

(ii) The anisotropy is the largest observed among the proton shieldings and tends to be larger in ionic system than in neutral systems. Its size is about 20 ppm, with the most shielded axis being the unique axis.

(iii) In a planar structure always the least shielded axis is orthogonal to the plane.

Haeberlen [7, 14] and co-workers have shown, that these features can be explained by the quadrupolar term (IV) in Eq. (5.4). Because of its $1/r^3$ dependence only next nearest neighbours have to be considered, leading to a simple physical explanation.

A single crystal spectrum of protons in hydrogen bonds and its orientation dependence has been shown in Figs. 3.9 and 3.10 in Sect. 3.2.

b) Methylene Protons ($-CH_2$)

Methylene protons have been studied by Haeberlen [7] and co-workers [14]. This shielding tensor is known to be very small from powder spectra. On the other hand methylene protons have a very large dipolar coupling due to the small distance (1.7 Å), which makes the detection of shift ansisotropies a formidable task. Haeberlen and co-workers solved this problem by utilizing the fact, that the dipolar interaction of the two protons vanishes when their internuclear vector is placed at the "magic angle" ($\vartheta_m = 54° 44'$) with respect to the magnetic field [7]. The residual dipolar broadening due to other protons can be eliminated easily by multiple pulse techniques. Notice, that a conical 2π rotation of the magnetic field can be performed around the internuclear vector keeping the magic angle condition, but tracing out part of the shielding tensor. In fact, five of the six tensor elements are determined by this 2π rotation (see Sect. 2.3). The sixth number has been obtained by Kohlschuetter [27] by a rotation of the magnetic field perpendicular to the internuclear vector. It is claimed [7], that the shift tensor of these methylene protons is axially symmetric about the $C\text{-}H$ bound direction with $\Delta\sigma \approx 5\text{-}6$ ppm.

$$\text{H}\quad\text{H}$$

c) Olefinic Protons ($-C{=}C-$)

These protons have been studied in three compounds so far [7, 15, 16, 28] (see Table 7.1). The conclusions about the shielding tensor which can be drawn tentatively are:

(i)   complete non-axial symmetry,
(ii) total anisotropy $\sigma_{33} - \sigma_{11} \cong 6$ ppm,
(iii) the least shielded direction is orthogonal to the molecular plane,
(iv) the in-plane principal axes are not always related to the bond directions.

Table 7.1. Proton shielding tensor σ_{11}, σ_{22}, and σ_{33} in ppm with respect to its average $\bar{\sigma}$ for different compounds. The reference for the average $\bar{\sigma}$ is σ_{TMS}. (PW, powder spectra; SC, single crystal; CW, wide line; MP, multiple pulse; DD, deuteron decoupling; RT, room temperature; $\sigma_{TMS}-\sigma_{adamantane}=1.74\,ppm$ has been used)

Compound	Formula		T (°K)	σ_{11}	σ_{22}	σ_{33}	$\bar{\sigma}$	Method	Refs.
Carboxyl									
Oxalic acid	$(COOH^*)_2$		RT	− 6.6	− 4.5	11.1	−12.6	SC, MP	26, 19
Squaric acid	$C_4O_4H_2^*$		RT	−11.3	− 5.0	16.2	−15.2	SC, MP	30
Butynedioic acid	$^*HOOCC_2COOH^*$		RT	− 6.7	− 6.7	13.3	−11.8	PW, MP	20
Trichloro acetic acid	CCl_3COOH^*		RT	− 6.6	− 6.6	13.3	−12	PW, MP	22
Trichloro acetic acid	CCl_3COOH^*		RT	− 8	− 6	13	−13.8	SC, CW	23, 31
Malonic acid	$CH_2(COOH^*)_2$	(a)	RT	− 7.9	− 5.3	13.2	−15.6	SC, MP	14, 39b
		(b)	RT	− 8.1	− 4.1	12.2	−14.9	SC, MP	14, 39b
Maleic acid	$(CHCOOH^*)_2$		RT	− 8.5	− 4.2	12.7	−14.5	SC, MP	15
Maleic acid	$(CHCOOH^*)_2$		RT	−10	− 5.1	15.1	−16.6	SC, MP	15
Succinic acid	$(CH_2CO_2H^*)_2$		RT	− 6.3	− 6.3	12.6	−15.4	PW, MP	19
Fumaric acid	$(CHCO_2H^*)_2$		RT	− 5.6	− 5.6	11.2	−16.9	PW, MP	19
Phtalic acid	$C_6H_4(COOH^*)_2$		RT	− 7.3	− 7.3	14.6	−15	PW, MP	19
Potassium hydrogen maleate	$H^*OOCC_2H_2COOK$		RT	−11.6	− 8.7	20.2	−21	SC, MP	16
Potassium hydrogen malonate	H^*OOCCH_2COOK		RT	−14.5	− 3.8	18.3	−20.5	SC, MP	29
Terephthalic acid	$p\text{-}C_6H_4(COOH^*)_2$		RT	−11.6	− 8.5	20.3	−11.5	SC, MP	43b
Pyromellitic acid	$C_6H_2(COOH^*)_4$	(a)	RT	− 8.9	− 3.8	12.8	−12.6	SC, MP	42
		(b)		− 8.5	− 4.1	12.6	−11.9	SC, MP	42
Methylene									
Malonic acid	$CH_2^*(COOH)_2$	(a)	RT	− 8.9	− 3.8	12.8	−12.6	SC, MP	14, 39b
		(b)	RT	− 8.5	− 4.1	12.6	−11.9	SC, MP	14, 39b
Succinic acid	$(CH_2^*COOH)_2$		RT	− 2.9	− 2.9	5.7	− 5.7	PW, MP	19
Succinic acid anhydride	$C_4H_4O_3$		RT	− 3.4	− 3.4	6.8	− 5.4	PW, MP	19
Terephtalic acid	$p\text{-}C_6H_4^*(COOH)_2$	(a)	RT	− 2.4	− 0.7	3.2	− 8.6	SC, MP	43b
		(b)	RT	− 3.0	− 1.0	4.0	− 9.3	SC, MP	43b
		(c)	RT	− 4.0	0.2	4.2	− 9.4	SC, MP	43b
		(d)	RT	− 3.5	− 0.3	3.7	− 9.4	SC, MP	43b
Durol	$C_6H_2^*(CH_3)_4$	(a)	RT	− 4.4	1.3	3.1	−	SC, MP	43c
		(b)	RT	− 5.6	1.2	4.4	−	SC, MP	43c
Pyromellitic acid	$C_6H_2^*(COOH)_4$		RT	− 3.1	0.8	2.3	− 8.0	SC, MP	42
Pyromellitic acid dianhydride	$C_{10}H_2^*O_6$		RT	− 6.1	− 2.7	8.8	− 4.3	SC, MP	41

Table 7.1 (continued)

Compound	Formula		T (°K)	σ_{11}	σ_{22}	σ_{33}	$\bar{\sigma}$	Method	Refs.
Olefinic									
Potassium hydrogen maleate	$KOOCC_2H_2^*COOK$		RT	$-$ 4.6	$-$ 3.5	8.0	$-$ 2.8	SC, MP	16, 40
Maleic acid	$(CH^*COOH)_2$		RT	$-$ 3	$-$ 0.3	3.3	$-$ 7.6	SC, MP	15
Fumaric acid	$(CH^*CO_2H)_2$		RT	0	0	0	$-$ 8.8	PW, MP	19
Aromatic									
Ferrocene	$(C_5H_5^*)_2Fe$		RT	$-$ 4.4	2.2	2.2	$-$ 6.1	SC, MP	7
Phtalic acid	$C_6H_4^*(COOH)_2$		RT	0	0	0	$-$ 9	PW, MP	19
Phtalic acid anhydride	$C_6H_4^*C_2O_3$		RT	0	0	0	$-$ 9	PW, MP	19
Others									
Ice	H^*DO		180	-11.5	-11.5	23	$-$ 5.1	PW, DD	29
	H_2^*O		173	-11.4	-11.4	22.8	$-$ 7.5	PW, MP	33
			77	$-$ 9.5	$-$ 9.5	19.0	$-$ 5.5	SC, MP	34
Hydrogen sulfide	H_2^*S		173	$-$ 3.7	$-$ 3.7	7.4	$-$ 9.0	PW, MP	33
Gypsum	$CaSO_4 \cdot 2H_2^*O$		RT	$-$ 6.2	0.6	5.6	-11.5	SC, MP	35
Trans-diiodoethylene	$(CH^*I)_2$		RT	$-$ 2.4	0.3	2.1	$-$ 6.2	SC, MP	37
Potassium hydrogen-carbonate	KH^*CO_3		RT	-10.7	$-$ 7.1	17.8	-16.1	SC, MP	28c
Ca-Formiate	$Ca(HCOO)_2$	(a)	RT	$-$ 3.3	0.2	3.1	$-$ 8.9	SC, MP	43a
		(b)	RT	$-$ 3.5	0.9	2.7	$-$ 9.1	SC, MP	43a
Pb-Formiate	$Pb(HCOO)_2$	(a)	RT	$-$ 6.0	2.2	3.9	-11.1	SC, MP	43a
		(b)	RT	$-$ 6.1	1.7	4.5	-10.7	SC, MP	43a
KDP	$KH_2^*PO_4$		RT	-12	-12	24	-16.2	SC, MP	25
Potassium hydrogen sulfate	KH^*SO_4	(a)	RT	$-$ 8.6	$-$ 7.6	16.2	-12.7	SC, MP	7, 28c
		(b)	RT	$-$ 9.8	$-$ 8.2	18	-14.5	SC, MP	7, 28c
	KH^*F_2		RT	-18.6	-11.4	30	-21.1	SC, MP	28b
Calcium hydroxide	$Ca(OH^*)_2$		RT	$-$ 4.7	$-$ 4.7	9.3	$-$ 4.6	SC, MP	21
Magnesium sulfate	$MgSO_4H_2^*O$		RT	$-$ 6.5	$-$ 6.5	13	$-$ 9.3	SC, MP	24

d) Aromatic Protons

have been studied for the first time in ferrocene, $Fe(C_5H_5)_2$ by the Heidelberg group [7, 28]. The C_5H_5 rings are rapidly rotating about their 5-fold axes, leading to a motionally averaged axially symmetric shielding tensor $\Delta\sigma = -6.5 \pm 0.1$ ppm with the unique axis being the rotation axis. Because of structural phase transitions at low temperature which destroy the single crystal, it is not possible to obtain the complete shielding tensor. However, this result is remarkable, because it was demonstrated in this case that reliable shielding tensors can only be obtained when both the *shape* and the *intrinsic* anisotropy of the bulk susceptibility are properly taken into account [28].

An excellent discussion of proton shielding tensors has been given by Haeberlen and the interested reader is referred to Ref. [7] for further reading on this subject. Table 7.1 lists a number of proton shielding tensors mostly determined by multiple-pulse methods.

Very detailed investigations of proton shielding tensors on different molecular systems have been obtained by Haeberlen and co-workers [38–43] in addition to the ones listed in Table 7.1. Ususally different sets of crystallografically non-equivalent protons can be differentiated. Therefore these data have not been included in Table 7.1. Fine details of shielding tensors were determined and could unravel subtleties of the corresponding crystal structure.

The compounds investigated are: 4,4'-difluorobiphenyl [38], α-oxalic acid dihydrate [39], potassium hydrogen maleate [40], pyromellitic acid anhydride [41], pyromellitic acid dihydrate and malonicacid [42], calcium and lead formate [43].

Oxalic acid dihydrate was also investigated by Ernst, Fenzke and Heinz [44]. Voigtberger and Rosenberger [45] determined the chemical shift anisotropies of the H_3N^+-group and the CH_2-group in α glycine. Although one would expect a chemical shift anisotropy in alkaline earth hydrides, Nicol and Vaughan [46a] were not able to resolve any anisotropy at their low fields (1.3 T), but could determine the isotropic shift of CaH_2 (-4.5 ± 3.0 ppm), SrH_2 (-6.7 ± 1.0 ppm), BaH_2 (-8.7 ± 1.0 ppm) relative to TMS.

Of particular interest are protons bound to heavy metals. Nicol and Vaughan [46b] made an investigation of $H_2Ru_4(CO)_{12}$ and $H_4Os_4(Co)_{12}$. The anisotropy for the bridging protons was less than 30 ppm. Dubois, Murphy and Gerstein [47] have made some detailed investigation on ^1H shielding tensors in zirconium halide hydrides using multiple pulse techniques. It is found that the proton shielding tensors exhibit a large anisotropy in these "sandwichlike" compounds compared to those found in hydrocarbons and salts.

These examples show the potential usefulness of ^1H multiple-pulse NMR in different systems. More data of this sort have been obtained but it is beyond the scope of this monograph to list them all.

7.4 ^{19}F Shielding Tensors

The first application of multiple pulse techniques was performed by Waugh, Huber and Haeberlen [48] on ^{19}F in CaF_2. No shift anisotropy was to be

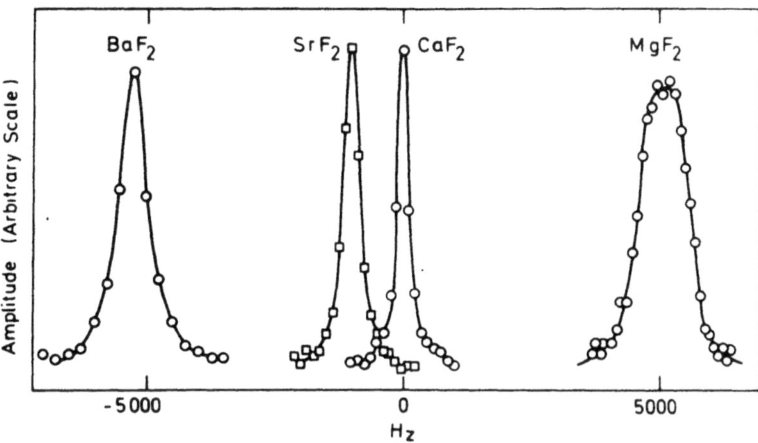

Fig. 7.1. Multiple pulse (WHH-4) spectra of group II difluorides obtained by R.W. Vaughan et al. [49a] at 56.4 MHz

expected, of course, since CaF_2 has cubic symmetry. However, not only the feasibility of this approach to line narrowing was demonstrated by this experiment, also valuable information was obtained about the isotropic shift. Before proceeding to non degenerate shielding tensors, we take a brief look at the isotropic shift in cubic crystals thus far determined.

Vaughan and co-workers [49] have obtained a number of highly resolved spectra in group *II* difluorides such as cubic: CaF_2, SrF_2, BaF_2, CdF_2, HgF_2 and noncubic MgF_2 and ZnF_2. Figure 7.1 gives a sample of these spectra. All these compounds are closely related with nearly completely ionic bonding. Their isotropic shift, however, can be correlated with electronegativities and a covalency parameter, calculated from electron spin resonance superhyperfine interaction parameters as shown in Fig. 7.2. Sears [50] has contributed to this investigation the shift of BeF_2. Since BeF_2 appears in different structures, this shift value is not expected to fit into this scheme. It is, however, a long standing discussion, which electronegativity scale should be used [51], but we are not going to discuss this point here.

Vaughan et al. [49] have also accounted for the shielding tensor in MgF_2 and ZnF_2 theoretically. The corresponding data may be found in Table 7.4. H. Ackermann and co-workers [52] have determined the quadrupole coupling constant of radioactive ^{20}F in MgF_2. In order to investigate the dependence of the chemical shift on the volume, Lau and Vaughan [17] performed a pressure experiment on CaF_2. They observed a shift of -1.7 ppm/kbar, indicating greater overlap of the wavefunctions or "covalency" as the pressure is increased.

However, Burum, Elleman and Rhim [53] have repeated these pressure experiments by using a multiple pulse zero crossing NMR technique. They obtained data with higher precision on CaF_2 than Lau and Vaughan [17] and extended these measurements to BaF_2. Burum et al. [53] obtained -0.29 ±0.02 ppm/kbar for ^{19}F in CaF_2 and -0.62 ± 0.05 ppm/kbar for BaF_2. Apply-

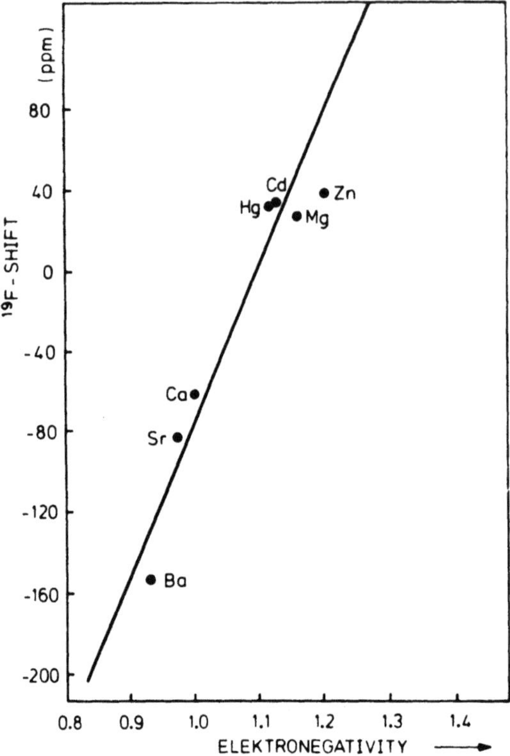

Fig. 7.2. ^{19}F shielding versus cation electronegativity on the Gordy scale [51b] for cubic and tetragonal group II difluorides obtained from multiple pulse spectra as demonstrated in Fig. 7.1 by R.W Vaughan et al. [49a]

ing the multiple zero crossing NMR technique Burum et al. [53] determined a number of isotropic ^{19}F shifts in different solids.

For completeness we represent in Fig. 7.3 the isotropic shift of ^{19}F in alkalifluorides [54, 55], which have been partially determined by multiple pulse techniques, employing spin decoupling. Here the same statement concerning the electronegativities applies as before. We only remark here, that the correlation with the Phillips [51a] ionicity scale is even worse.

Representative ^{19}F powder spectra as obtained by multiple pulse techniques have been shown already in Figs. 2.6 and 3.8. The first reliable ^{19}F shift tensor were obtained from those spectra [56]. Although no information about the principal axes was furnished by these experiments, still some assignments could be made by the virtue of known anisotropic molcular reorientation (see Sect. 2.7). Such a case is C_6F_6, in which an axially symmetric ^{19}F shielding tensor was observed at 200°K, where the molecules are rotating about their 6-fold axis.

The immediate conclusion to be drawn, assigns the unique axis of the shielding tensor to the 6-fold axis (see Sect. 2.7).

If the rotation is slowed down (much less than $\Delta\sigma$ at about 40°K) the complete non averaged shielding tensor is expected to be observed. The unexpected result, however, was that the powder spectrum did not change within the limits of resolution. This leaves as the only conclusion, that the shielding

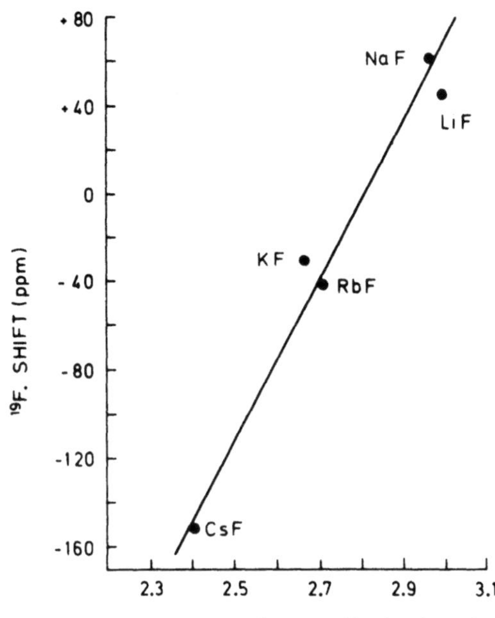

Fig. 7.3. ^{19}F shielding versus overlap repulsion energy between ions/mean excitation energy of F^-, for alkali fluorides according to R.E.J. Sears [54]

tensor is axially symmetric within the limit of resolution and that the unique axis is the 6-fold axis [56b].

A similar conclusion was drawn from the high temperature and low temperature powder spectra of CF$_3$COOAg [56b]. An assignment of the principal axes could be made, (although not unambiguously) which turned out to be correct later as proved by a single crystal study [58]. The same arguments have been applied to the high and low temperature spectra of partially fluorinated benzenes [59], whose low temperature tensor elements are shown in Fig. 7.4. The assignment of the principle axes was supported by liquid crystal studies [60].

However, one should be aware, that this assignment is very preliminary, especially for the in-plane components of the shielding tensor. However, single crystal measurements on potassium tetrafluorophtalate seem to support this assignment [61]. Since the protons in these fluorinated benzenes lead to substantial broadening, pulsed spin decoupling had to be applied in this case (see Sects. 3.2 and 4.4). The beauty of these data is reflected in the individual change of the different tensor elements due to the bonding

(i) The so-called *"ortho-effect"* is very pronounced, it shifts the σ_{33}-component ($\perp$ plane) by $\sim +50$ ppm if there is another fluorine in the ortho position, whereas leaving the other two components (σ_{11}, σ_{22}) more or less unchanged. *Meta-* and *para*-substitution does not lead to a similar, pronounced effect.

(ii) The tensor element in the bond direction (σ_{22}) is not very sensitive at all to any kind of substitution which is also true for the σ_{33} component besides the *"ortho-effect"*.

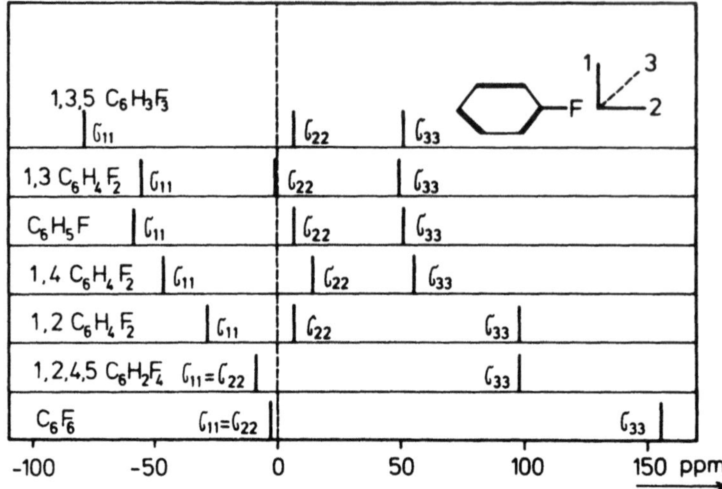

Fig. 7.4. Principal elements of the ^{19}F shielding tensor in partially fluorinated benzenes obtained from proton decoupled multiple pulse powder spectra [59]. The assignment of the principal axes is tentative, but in agreement with liquid crystal studies [60]

(iii) The only component which changes gradually with substitution is the σ_{11} component perpendicular to the bond direction and perpendicular to the plane.

Up to now, there is no reliable theory to account for these observations. In the early days of multiple-pulse line narrowing NMR we have tried to analyse our data along the lines of the theory of Karplus and Das [6, 56b]. This led to completely unacceptable conclusions about such quantities as "bond characters" and "hybridization parameters". The main outcome was to demonstrate, that this theory is completely unapplicable. But also more empiric approaches which have been successful in proton shielding, like the Gierke-Flygare method [12], failed completely.

This can be easily demonstrated by comparing the measured values of the shift tensor in monofluorobenzene ($\sigma_{11} = -58$ ppm, $\sigma_{22} = 8$ ppm, $\sigma_{33} = 53$ ppm) with the value calculated by Gierke and Flygare [12] ($\sigma_{11} = -56$ ppm, $\sigma_{22} = 123$ ppm, $\sigma_{33} = -68$ ppm). Besides the accidental agreement in the σ_{11}-component there is no resemblance between the theoretical and the experimental values. However, this can not be taken as a disproof of the GF approach, since most of the discrepancy, I believe, stems from the inaccurate values of the paramagnetic term, where one has to resort to the values of the spin rotation interaction tensor [57]. Also *ab initio* calculations employing SUFO's have not contributed much to the understanding of the shift tensors. Anyway, these few remarks should suffice to give pleasure to the theorist and challenge further theoretical effort towards a better understanding of this molecular quantity. Especially the paramagnetic part of the shielding tensor, which seems to be an extremely sensitive tool for studying the structure of molecular orbitals is still lacking a simple theoretical interpretation.

Leaving the complete theoretical understanding of the ^{19}F shielding tensor

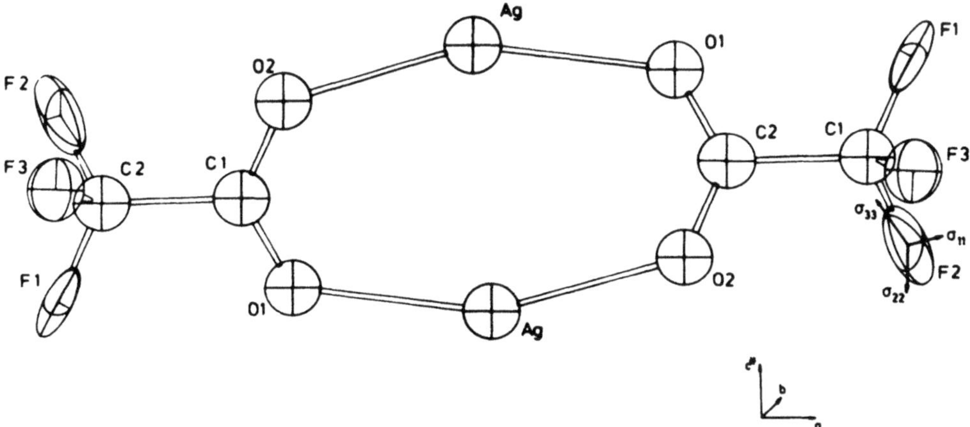

Fig. 7.5. Pictorial representation of the ^{19}F shielding tensors (aestheticized shielding tensor representation according to the prescription in Sect. 2.2) of the CF_3COOAg dimer as viewed along the b-axis of the monoclinic crystallografic unit cell. The chemical shielding tensors were determined from orientation plots of a single crystal study similar to Fig. 3.9 by Griffin et al. [58]. (Courtesy of D. Suwelack)

to the future, we now turn to another interesting example of a single crystal study, namely, the ^{19}F shielding tensor in

Silver-trifluoroacetate (CF_3COOAg) [58]

The first extensive single crystal study using multiple-pulse line narrowing techniques was performed on the ^{19}F spins in the methyl group of this compound [58]. As part of the same investigation the complete crystal structure was determined by X-rays. The monoclinic unit cell contains four (CF_3COOAg)$_2$-dimers ($Z = 8$) as shown in Fig. 7.5. The space group determined is Cc or C_2/c, which cannot be distinguished by the X-ray analysis nor by the NMR data. The most pronounced result is, that the apparent 3-fold axes of all CF_3-groups is parallel and approximately along the a-axis of the unit cell. At room temperature the CF_3-group is rapidly rotating about its 3-fold axis, leading to a single line. The shielding tensor as determined from the room temperature data, however, is not axially symmetric as expected, because it displays the average of three incongruent tensors (see Table 7.2).

This is borne out by the low temperature spectrum (see Fig. 3.8) which displays all the six lines of the six magnetically inequivalent ^{19}F nuclei. Three incongruent shielding tensors are observed for the three crystallografically inequivalent fluorine atoms in the methyl group. This is consistent with the space group C_2/c which is a special case of the space group Cc. A slight deviation of the dimer away from the two-fold axis would lead to the space group Cc with four magnetically inequivalent CF_3-groups, resulting in twelve lines at low temperature. This has not been observed, but cannot be excluded because of the limited resolution. On the contrary, the ^{13}C spectra we have observed in the cross-polarization experiments (see Sect. 4.2) seem to support

Table 7.2. Tensor elements σ_{11}, σ_{22}, σ_{33} in ppm and direction cosines with respect to the a, b, c^* crystal axis system of ^{19}F in CF_3COOAg at room temperature from evaluation of orientation plots like Fig. 2.4 (Griffin et al. [58]). The average of the three low temperature tensors performing random jumps about the C_3-axis is included

	σ_{11}	Direction cosines	σ_{22}	Direction cosines	σ_{33}	Direction cosines	Ref. $\bar{\sigma}(C_6F_6)$
Room		-0.024		0.9985		0.0495	
temperature	-43.8	0.0203	15.7	-0.0490	29.1	0.9986	-95
		0.9995		0.0253		-0.0192	
Low temper-		0.0039		0.9999		0.0169	
ature (40°)	-44.0	0.0027	13.0	0.0169	31.0	0.9999	-92
rapid jump average		1.0000		0.0038		0.0027	

the space group Cc. However, this point is not completely settled at the time of writing.

Notice, that the symmetries of the unit cell render the fluorines, labelled F_1, F_2, and F_3 in the methyl group to be crystallografically inequivalent. However, this does not necessarily allow any conclusion to be drawn about the shielding tensor of these nuclei. In other words, the symmetries of the unit cell permit the fluorines to have incongruent shielding tensors, but symmetry is only a necessary, not a sufficient condition for the three shielding tensors to be different. Table 7.3 summarizes the low temperature single crystal results of the three incongruent shielding tensors of the CF_3-group. We may summarize the results as follows:

(i)   The shielding tensor is "complete non-axially symmetric" with average values ($\sigma_{11} = -71$ ppm, $\sigma_{22} = 0.0$ ppm, $\sigma_{33} = 71$ ppm).
(ii) The most shielded direction lies approximately along the CF-bond, with a deviation of about 10°.
(iii) The least shielded axis is perpendicular to the bond and in the CCF plane.
(iv) The intermediate direction is perpendicular to the bond and the CCF plane.

A pictorial representation of these shielding tensors according to the rules stated in Sect. 2.2 is given in Fig. 7.5.

The difference of the three fluorine tensors is pronounced and by no means negligible. It is imposed by the crystal structure and is very likely due to the quadrupolar term of the diamagnetic shielding.

Scalar coupling (J-coupling) could have been observed in principle, since the three fluorine of the methyl group are magnetically inequivalent. However, the resolution was not sufficient to detect any isotropic part of the indirect spin-spin coupling. Notice, that the symmetric part of the indirect spin-spin coupling tensor is averaged by the multiple pulse experiment applied. This is so, because it has the same spin symmetry as the dipolar coupling Hamiltonian, if antisymmetric parts are neglected. We conclude this section by summarizing in Table 7.4 ^{19}F shielding tensors, which have to our knowledge thus far been determined.

We have not included in this table, however, those values, which did not

Table 7.3. ^{19}F shielding tensor in solid CF_3COOAg at 40°K according to Griffin et al. [58]. The direction cosines refer to the a, b, c^* unit cell axes. The labelling of the fluorine corresponds to Fig. 5.5, where a pictorial representation of the tensors is given. The errors are ± 2 ppm and $\pm 2°$ respectively

Fluorine	σ_{11}	Direction cosine	σ_{22}	Direction cosine	σ_{33}	Direction cosine	$\bar{\sigma}$ (Ref. C_6F_6)	CCF bond angle	CF bond[a] direction cosines
F_1	-64.3	0.8791 -0.0638 0.4724	4.3	0.2545 0.8645 -0.4335	60.0	-0.4031 0.4986 0.7674	-89.4	114.1°	-0.3658 0.3780 0.8512
F_2	-75.0	0.9138 0.3103 -0.2622	2.0	0.0156 -0.6559 -0.7548	73.0	-0.4059 0.6883 -0.6014	-90.0	110.3°	-0.3481 0.5639 -0.7491
F_3	-73.6	0.8864 -0.4614 0.0400	-6.6	-0.0357 0.0236 0.9992	80.0	-0.4617 -0.8870 0.0152	-96.5	110.3°	-0.3875 -0.9144 -0.1182
average	-71.0		-0.1		71.0		-92.0		
Powder pattern 32b	-67.0		-3.0		70.0		-95.5		

[a] The direction cosines of the C–C bond are: 0.9990, 0.0405, and 0.0343.

Table 7.4. Fluor (^{19}F) shielding tensor σ_{11}, σ_{22}, σ_{33} in ppm with respect to its average $\bar{\sigma}$ for different compounds. The reference for the average is $\sigma_{C_6F_6}$. (PW, powder spectra; SC, single crystal; MF, molecular frame; LC, liquid crystal; CT, clathrate; CW, wide line; FID, free induction decay; FM, first moment; SM, second moment; MP, multiple pulse; RT, room temperature; NP, nematic phase; $\sigma_{C_6F_6} - \sigma_{CaF_2} = 57$ ppm has been used)

Compound	Formula	T (°K)	σ_{11}	σ_{22}	σ_{33}	$\bar{\sigma}$	Method	Refs.
Aromatic								
Fluorobenzene	C_6F_6	40	− 51.7	− 51.7	103.4	7	PW, MP	56b
Fluorobenzene	C_6F_6	NP	− 53	− 53	106	0	MF, LC	60b
	$1,3,5\text{-}C_6H_3F_3$	77	− 74	7	68	− 57	PW, MP	59
	$1,3,5\text{-}C_6H_3F_3$	NP	− 94	27	67	− 53	MF, LC	60a
	$1,3\text{-}C_6H_4F_2$	77	− 69	11	57	− 52	PW, MP	59
	$1,3\text{-}C_6H_4F_2$	NP	− 98	29	69	− 53	MF, LC	60a
	C_6H_5F	77	− 58	7	51	− 50	PW, MP	59
	C_6H_5F	NP	− 94	27	67	− 50	MF, LC	60a
	$1,4\text{-}C_6H_4F_2$	77	− 63	7	56	− 42	PW, MP	59
	$1,4\text{-}C_6H_4F_2$	NP	− 60	0	60	− 43	MF, LC	60a
	$1,2\text{-}C_6H_4F_2$	77	− 67	− 17	85	− 24	PW, MP	59
	$1,2\text{-}C_6H_4F_2$	NP	− 34	− 3	107	− 24	MF, LC	60a
	$1,2,4,5\text{-}C_6H_2F_4$	77	− 58	− 24	80	− 24	PW, MP	59
	$C_6F_3Br_3$	NP	− 75	37	37		MF, LC	66
Fluoranil	$C_6F_4O_2$	300	− 84	− 33	117	− 20	PW, MP	56
Perfluoro naphtalene	$C_{10}F_8$	300	− 53	− 53	106	− 19	PW, MP	56b
Perfluoro naphtalene	$C_{10}F_8$	83	− 50	− 50	100	− 16	PW, MP	56b
Perfluoro biphenyl	$(C_6F_5)_2$	83	− 48	− 48	96	− 16.3	PW, MP	56b
Perfluoro benzophenone	$C_6F_5COC_6F_5$	83	− 66	− 31	97	− 29	PW, MP	56b
Potassium-tetrafluorophtalate	$F_2^*C_6F_2(CO_2K)_2$	77	− 55	− 43	98	− 3.5	SC, MP	61
Potassium-tetrafluorophtalate	$F_2C_6F_2^*(CO_2K)_2$	77	− 52	− 33	85	− 22	SC, MP	61
4,4′ Difluorobiphenyl	$C_{12}H_8F_2^*$	RT	− 54.9	3.2	51.8	− 52.5	SC, MP	38

Table 7.4 (continued)

Compound	Formula	T (°K)	σ_{11}	σ_{22}	σ_{33}	$\bar{\sigma}$	Method	Refs.
Methyl								
Silver trifluoro acetate	CF_3COOAg	40	-71	0	71	-92	SC, MP	58
Silver trifluoro acetate	CF_3COOAg	RT	-44	15	29	-95	SC, MP	58
	$(CF_3CO)_2O$	182	-46	23	23	-93	PW, MP	56b
		40	-75	-1	76	-72	PW, MP	56b
	CHF_3	4	-35	-35	70	-78	CT, CW	64b, 66
	CF_2BrCF_2Br	85	-87	-87	173		SM, CW	65, 66
	$CFCl_2CFCl_2$	85	-80	-80	160		SM, CW	65, 66
Others								
Fluorine	F_2	29	-350	-350	700	-593	SM, CW, FM	62
	NF_3	4	-130	-130	260	-282	CT, CW	64b, 66
	KHF_2	RT	-36	-20	56	-23.5	SC, FID	28b
	MgF_2	RT	-15	0	15	28	SC, MP	49
	ZnF_2	RT	-22	0	22	37	SC, MP	49a
	XeF_4		-190	-190	280	-107	SM, CW	70, 71
	XeF_2		-35	-35	70	61	SM, CW	72
	$BaFPO_2$		-61	-61	122		SM, CW	73
Fluoroapatite	$Ca_5F(PO_4)_3$	298	-56	28	28	-83	SC, FID	63
	CH_3F	1.3	-44	22	22	116	CT, CW	64a, 66
	$CF_2S_2CF_2$	77	-109	54	54	-111	PW, MP	68
Teflon	$(F_2C{=}CF_2)_n$	153	-71	20	51	-41	PW, MP	56b
Teflon	$(F_2C{=}CF_2)_n$	81	-80	21	59	-40	PW, MP	56b
Teflon	$(F_2C{=}CF_2)_n$	77	-80	10	7	-35	PW, MP	56b
Teflon	$(F_2C{=}CF_2)_n$		-114	57	57		SM, CW	69

seem to be very reliable because of the method used. Only those data from liquid crystal studies are included which can be compared directly with solid state measurements. Hull and Sykes [67] have utilized some of the listed ^{19}F shielding tensors to investigate the spin lattice relaxation of fluor labelled protein molecules at high fields (5 Tesla).

Even before the advent of multiple pulse experiments ^{19}F shielding tensors were estimated from first moments of resonance lines [64, 65, 70-73]. However, these data have inferior accuracy compared with the multiple pulse data. Nevertheless some of them are included in Table 7.4 for comparison.

7.5 ^{13}C Shielding Tensors

Some ^{13}C shielding tensors have been observed in solids already before the advent of line narrowing techniques, by the virtue of the small gyromagnetic ratio of ^{13}C, its low natural abundance and its relative shielding anisotropy in noncubic surrounding [74, 75]. The brute force technique of determining shielding tensors directly from the NMR spectrum by applying sufficiently high magnetic fields (6-8 Tesla) and exchanging the protons by deuterons has undergone a renaissance recently [11, 58].

Nevertheless, the biggest supply of ^{13}C shielding tensors has been furnished by Pines, Waugh and co-workers [79-85] applying cross-polarization techniques or the so-called PENIS experiment (Proton Enhanced Nuclear Induction Spectroscopy) as proposed by Pines et al. [79] (see Chap. 4).

As representative examples we have shown in Fig. 4.10 of Sect. 4.2 some powder spectra of related compounds as determined by Waugh and co-workers [81]. ^{13}C shielding tensors as determined from these spectra are listed among others in Table 7.5.

Let us take a closer look at a representative single crystal investigation of ^{13}C shielding tensors in durene (1,2,4,5-tetramethylbenzene), which was performed by Waugh and co-workers [80, 84]. The crystal structure of durene is monoclinic with space group C_{2h}^{5} $(P2_1/a)$. The asymmetric structural unit consists of one-half of a molecule. On expects therefore at most five incongruent ^{13}C shielding tensors, namely two from the methyl carbons, two from the ring carbons, which are bonded to the methylgroup and one unsubstituted ring carbon. The other tensors are related to these by the symmetry operations of C_{2h}^{5}.

Since the unit cell contains four asymmetric units, and at most five incongruent ^{13}C shielding tensors are allowed for each asymmetric unit, the maximum number of lines to be observed would be twenty. However, the two halves of each molecule are related by a center of inversion, leaving only ten lines, because of the invariance of the symmetric shielding tensor to inversion. That the ten lines are actually observed at some crystal orientations is demonstrated in Fig. 7.6.

Orientation plots of the shift data were obtained for three orthogonal rotation axes, from which the ^{13}C shielding tensors could be determined.

A complete pictorial representation of these tensors using the aesthetizised

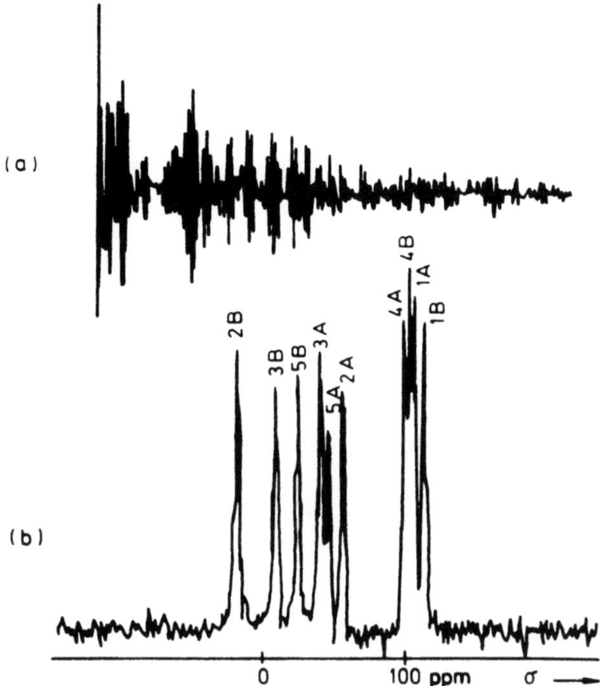

Fig. 7.6. ^{13}C spectrum in durene obtained by the cross-polarization technique by S. Pausak et al. [80]. All possible ten lines, allowed by the crystal structure are borne out at this specific crystal orientation

shielding tensor representation as proposed in Sect. 2.2 is given in Fig. 7.7. Numerical values may be obtained from Table 7.5. Let us summarize some conclusions which can be drawn from these investigations:

(i)   Methyl carbons display almost axially symmetric shielding tensors, with $\Delta\sigma \cong 24$ ppm and the unique axis being the 1-axis almost parallel to the C-C bond. The shift tensor is most likely averaged due to tunelling motion or hindered rotation about the C_3-axis.

(ii) The ring carbons possess three distinct tensor elements with the most shielded axis perpendicular to the plane and the least shielded axis bisecting almost the C-C-C angle of the ring carbons.

(iii) The isotropic average of the shielding tensor is closely related to the isotropic shift in liquids, showing only little dependence on packing in the solid state.

The last symptom is also found in the eigenvalues and the principal axes of the shielding tensors, being closely related to molecular symmetry, rather than crystal symmetry. This is in contrast with ^{19}F shielding tensors which reflect subtle changes in the crystalline structure. This fact may be caused by the marginal location of the fluoratoms at the outskirts of molecular bonds, whereas the carbon atoms are buried in the interior of the molecule.

Numerous attempts have been made to relate the observed shielding tensors to the molecular orbitals involved. An interesting approach has been

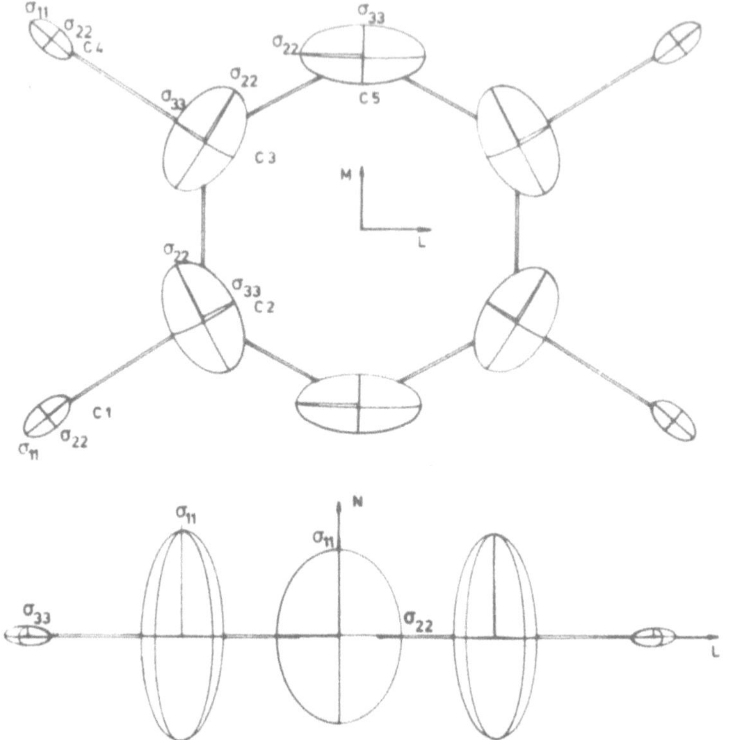

Fig. 7.7. Aestheticized chemical shielding tensor representation of ^{13}C in durene obtained by S. Pausak et al. [80]. The dimensions of the ellipsoids are directly proportional to the shielding with respect to a certain reference, according to the prescription in Sect. 2.2. The methyl ^{13}C ellipsoids are magnified by a factor of two. (Courtesy of D. Suwelack)

taken by Spiess and co-workers [11], relating this quantity to the different orbital transitions as known from optical absorption frequencies. Others have tried to apply *ab initio* calculations, mainly to the paramagnetic part of the shielding tensor with varying success [86]. Rather than going into those details, we would like to summarize some common features [11, 78]:

(i)   The most shielded direction is:
      a) perpendicular to the ring in aromatic carbons,
      b) the C_3-axis for methyl carbons,
      c) perpendicular to the sp^2 plane for carbonyl and carboxyl carbons.
(ii) The least shielded direction is:
      a) in the plane for ring carbons, bisecting the C-C-C angle,
      b) for methyl carbons perpendicular to the C_3-Axis and perpendicular to a plane of symmetry to which the methyl group is connected (if not averaged),
      c) in the sp^2 plane for carboxyl and carbonyl carbons.
(iii) The intermediate shielded direction is:
      a) tangent to the ring for aromatic carbons,

Table 7.5. Examples of carbon (^{13}C) shielding tensor σ_{11}, σ_{22}, σ_{33} in ppm with respect to its average $\bar{\sigma}$ for different compounds. The reference for the average is C_6H_6. (PW, powder spectra; SC, single crystal; FID, free induction decay; CP, cross polarization; RT, room temperature; $\sigma_{C_6H_6} - \sigma_{CS_2} = 65$ ppm has been used)

Compound	Formula	T (°K)	σ_{11}	σ_{22}	σ_{33}	$\bar{\sigma}$	Method	Refs.
Aromatic								
Benzene	$C_6^*H_6$	223	-60	-60	120	-3	PW, CP	81
Hexafluor benzene	$C_6^*F_6$	233	-13	-13	27	-29	PW, CP	81
Hexamethyl benzene	$C_6^*(CH_3)_6$	296	-56	-56	112	-5	PW, CP	81
			-56	-56	112	-3	SC, CP	84
Hexamethyl benzene	$C_6^*(CH_3)_6$	87	-95	-17	113	-9	PW, CP	81
Toluene	$C_6^*H_5CH_3$	87	-104	-17	121	11	PW, CP	81
Durene	$C_2H_2(C^*CH_3)_4$	RT	-94	-22	116	-5	SC, CP	80
	$C_2^*H_2(CCH_3)_4$	RT	-92	8	85	-2	SC, CP	80
Hexaethyl benzene	$C_6^*(CH_2CH_3)_6$	RT	-85	-29	114	-6.5	SC, CP	84
Pentamethyl benzene	$CHC_5^*(CH_3)_5$	RT	-84	-29	112	-4	SC, CP	84
Pentamethyl benzene	$CHC_5^*(CH_3)_5$	RT	-84	1	83	-24	SC, CP	84
Benzene	C_6H_6	14	-97	-21	118	9	PW, CP	90
p-Xylene	$C_4^*H_4(CCH_3)_2$	103	-98.7	-9.3	108	0.6	SC, CP	91, 92
	$C_4H_4(C^*CH_3)_2$	103	-99	-28	127	-5.0	SC, CP	91, 92
Acetophenone	$C_5^*H_5(CCOCH_3)$	103	-107.2	-22.3	129.5	-4.7	SC, CP	92, 93
	$C_5H_5(C^*COCH_3)$	103	-89.5	-24.1	113.6	-5.1	SC, CP	92, 93
Ferrocene	$(C_5H_5)_2Fe$	RT	-25.3	-25.3	50.6	59.3	PW, CP	94
Ruthenocene	$(C_5H_5)_2Ru$	RT	-26.3	-26.3	52.6	55.3	PW, CP	94
Magnesocene	$(C_5H_5)_2Mg$	RT	-44	-44	88	20	PW, CP	94
Methyl								
Methanol	C^*H_3OH	87	-21	-21	42	76	PW, CP	81
Ethanol	$C^*H_3CH_2OH$	103	-11	-2	13	115	PW, CP	81
Methyl formate	$HCOOC^*H_3$	87	-27	-15	41	80	PW, CP	81
Methyl acetate	$HC_3COOC^*H_3$	133	-33	-17	51	70	PW, CP	81
Dimethyl carbonate	$(C^*H_3O)_2CO$	87	-26	-20	45	76	PW, CP	81
Dimethyl oxalate	$(C^*H_3OCO)_2$	87	-27	-15	43	79	PW, CP	81
Dimethyl acetylene	$(C^*H_3C)_2$	87	-5	-5	9	118	PW, CP	81
Methyl Urea	$C^*H_3NHCONH_2$	77	-17	-5	22	-127	PW, MP	10

Table 7.5 (continued)

Compound	Formula	T (°K)	σ_{11}	σ_{22}	σ_{33}	$\bar{\sigma}$	Method	Refs.
Dimethyl oxalate	$(C^*H_3OCO)_2$	RT	-86	15	71	43	SC, CP	83c
Hexamethyl benzene	$C_6(C^*H_3)_6$	RT	-1.1	0.1	0.9	111.8	SC, CP	84
Dimethyl sulfoxide	$(C^*H_3)_2SO$	226	-23	-1	25	86	PW, CP	81
Octomethyl cyclotetrasiloxane	$[(C^*H_3)_2SiO]_4$	87	0	0	0	128	PW, CP	81
Hexamethyl disiloxane	$[(C^*H_3)_3Si]_2O$	87	0	0	0	124	PW, CP	81
Diethyl ether	$[(C^*H_3CH_2)]_2O$	133	-11	-1	13	111	PW, CP	81
Acetaldehyde	C^*H_3CHO	87	-21	-9	31	95	PW, CP	81
Acetone	$(C^*H_3)_2CO$	87	-17	-17	33	98	PW, CP	81
Acetic acid	C^*H_3COOH	87	-12	-12	23	95	PW, CP	81
Acetic anhydride	$(C^*H_3CO)_2O$	91	-13	-10	25	103	PW, CP	81
Methyl acetate	$C^*H_3COOCH_3$	133	-8	-5	12	94	PW, CP	81
Toluene	$C_6H_5C^*H_3$	87	-13	-2	15	109	PW, CP	81
Durene	$C_6H_2(C^*H_3)_4$	RT	-12	-5	16	110	SC, CP	80
Hexamethyl benzene	$C_6(CH_2C^*H_3)_6$	RT	-12	-3	16	115	SC, CP	84
Hexamethyl dewar benzene	$(CC^*H_3)_4(CC^*H_3)_4$	87	0	0	0	118	PW, CP	81
Silver acetate	C^*H_3COOAg	87	-16	-10	26	110	PW, CP	81
	C^*F_3COOAg	296	-6	-6	13	6	PW, CP	81
Dimethyl dimethoxy silane	$(C^*H_3)_2(CH_3O)_2Si$	87	0	0	0	132	PW, CP	81
Thioacetic acid	C^*H_3COSH	87	-21	-14	35	94	PW, CP	81
Trifluor acetic anhydride	$(C^*F_3CO)_2O$	109	-6	-6	11	16	PW, CP	81
Dimethyl dimethoxy silane	$(CH_3)_2(C^*H_3O)_2Si$	87	-27	-22	50	82	PW, CP	81
Tetramethyl silane	$(C^*H_3O)_4Si$	87	-25	-19	45	82	PW, CP	81
Dimethyl disulfide	$(C^*H_3)_2S_2$	87	-24	6	18	105	PW, CP	81

Table 7.5 (continued)

Compound	Formula		T (°K)	σ_{11}	σ_{22}	σ_{33}	$\bar{\sigma}$	Method	Refs.
Methylene									
Ethanol	$CH_3C^*H_2OH$		103	$-$ 19	$-$ 19	38	75	PW, CP	81
Diethylether	$(CH_3C^*H_2)_2O$		133	$-$ 30	$-$ 18	49	61	PW, CP	81
Hexaethyl benzene	$C_6(C^*H_2CH_3)_6$		RT	$-$ 7	$-$ 5	13	108.5	SC, CP	84
Ammonium hydrogen malonate	methylene C		RT	$-$ 18	$-$ 8	26	84	SC, CP	83b
ADIMU	$(CHC^*H_2COOH)_2$		RT	$-$ 12	$-$ 10	22	100.5	SC, CP	95
Glycine	$H_3^+NC^*H_2COOH$		RT	$-$ 19	0	19	82.4	SC, CP	96
Polyethylene	$(CH_2)_n$			$-$ 21.5	4.5	17	34.4 (TMS)	PW, CP	97–100
Diacetylene	$(...C^*H_2...)$	(a)	RT	$-$ 37.7	10.0	27.6	32.0 (Adam)	SC, CP	43c
		(b)	RT	$-$ 33.7	14.2	19.6	40.5 (Adam)	SC, CP	43c
		(c)	163	$-$ 72.7	72.7	145.4	63	PW, CP	104
Carboxyl									
Acetic acid	CH_3C^*OOH		87	$-$ 81	4	78	$-$ 59	PW, CP	81
Ammonium-D tartrate	carboxyl C		RT	$-$ 61	$-$ 12	72	$-$ 50	SC, CP	85, 83a
Ammonium hydrogen malonate	carboxyl C		RT	$-$ 68.5	$-$ 1.5	70	$-$ 44	SC, CP	83b
Benzoic acid	$C_6H_5C^*OOH$		RT	$-$ 57	$-$ 14	71	$-$ 46	SC, FID	11, 78
ADIMU	$(CHCH_2C^*OOH)_2$		RT	$-$ 71	$-$ 6	77	$-$ 45	SC, CP	95
Glycine	$H_3^+NCH_2C^*OOH$		RT	$-$ 70	$-$ 4	74	$-$ 48.6	SC, CP	96
Squaric Acid	$H_2C_4O_4$	(a)	RT	$-$ 72.3	$-$ 17.7	90	$-$ 63.5	SC, CP	101
		(b)	RT	$-$ 55.7	$-$ 45.3	101	$-$ 57.1	SC, CP	101
DAOX	$(NH_4)_2C_2O_4 \cdot H_2O$		RT	$-$ 74.4	12.3	62.1	$-$ 42.2	SC, CP	102
HOX	$H_2C_2O_4 \cdot 2H_2O$		RT	$-$ 85.5	31.1	54.4	$-$ 34.3	SC, CP	102
AHOX	$(NH_4)HC_2O_4 \cdot \frac{1}{2}H_2O$	(a)	RT	$-$ 81.3	24.8	21.9	$-$ 39.2	SC, CP	103
		(b)	RT	$-$ 86.2	31.2	55	$-$ 37.7	SC, CP	103
Olefinic									
Acetylene	$(CH)_2$		NP	$-$ 84.3	$-$ 84.3	168.6	$-$	LC	105
Polyacetylene	$(CH)_n$	cis	RT	-100	$-$ 10	110	$-$ 1.0	PW, CP-MAS	107
		trans	RT	$-$ 97	$-$ 8	105	7.0	PW, CP-MAS	107
ADIMU	$(C^*HCH_2COOH)_2$		RT	$-$ 95	10	85	13.1	SC, CP	95

Table 7.5 (continued)

Compound	Formula	T (°K)	σ_{11}	σ_{22}	σ_{33}	$\bar{\sigma}$	Method	Refs.
Carbonyl								
Acetone	$(CH_3)_2C^*O$	87	-71	-57	129	-79	PW, CP	81
Acetaldehyde	CH_3C^*HO	87	-77	-35	112	-70	PW, CP	81
Thioacetic acid	CH_3C^*OSH	87	-72	-28	101	-73	PW, CP	81
Silver acetate	CH_3C^*OOAg	87	-62	-25	87	-48	PW, CP	81
Silver trifluoro acetate	CF_3C^*OOAg	296	-77	39	39	-39	PW, CP	81
Acetic anhydride	$(CH_3C^*O)_2O$	91	-110	55	55	-42	PW, CP	81
Methyl formate	HC^*OOCH_3	87	-88	29	58	-36	PW, CP	81
Methyl acetate	$CH_3C^*OOCH_3$	133	-85	22	62	-53	PW, CP	81
Dimethyl carbonate	$(CH_3O)_2C^*O$	87	-81	40	40	-21	PW, CP	81
Dimethyl oxalate	$(CH_3OC^*O)_2$	87	-98	49	49	-19	PW, CP	81
Trifluoro acetic anhydride	$(CF_3C^*O)_2O$	109	-108	46	62	-23	PW, CP	81
Benzoic acid anhydride	$C_6H_5C^*O_3C^*H_5C_6$	RT	-83	9	73	-24	PW, FID	11, 78
Silver benzoate	$C_6H_5C^*OOAg$	RT	-75	-4	79	-43	PW, FID	11, 78
Benzophenone	$C_6H_5C^*OC_6H_5$	RT	-72	-29	101	-72	SC, FID	11, 78
Others								
Nickelcarbonyl	$Ni(CO)_4$	4.2	-132	-132	264	-66	PW, FID	88
Ironcarbonyl	$Fe(CO)_5$	4.2	-142	-142	284	-85	PW, FID	88
	$K_2Pt(CN)_4Br_{0.3}3H_2O$	RT	-116	-78	193	19	SC, FID	87
Thiobenzophenone	$C_6H_5C^*SC_6H_5$	RT	-148	-39	187	-107	PW, FID	11, 78
Dimethyl acetylene	$(CH_3C^*)_2$	87	-67	-67	135	38	PW, CP	81
Hexamethyl dewar benzene	$(C^*CH)_3)_4(CCH_3)_2$	87	-101	-6	107	-14	PW, CP	81
Hexamethyl dewar benzene	$(CCH_3)_4(C^*CH_3)_2$	87	-15	0	14	72	PW, CP	81
Ammonium-D tartrate	hydroxyl C	RT	-13	-7	19	55	SC, CP	85, 83a
Acetonitrile	CH_3C^*N	83	-68	-68	136		PW, CP	89
Calcite	$CaCO_3$	RT	25	25	-51	-15	PW, FID	74a, 75
Carbondisulfide	CS_2		-144	-144	288	-65	PW, FID	75

    b) for non averaged methylgroups perpendicular to the C_3-axis in the plane of symmetry,

    c) in the sp^2 plane, but perpendicular to the C-C bond for carboxyl groups.

For further reference of ^{13}C shielding tensors and their principal axes, we list a number of data in Table 7.5.

A more detailed discussion of ^{13}C shielding tensors may be found in the review of Spiess [89]. In complex molecular solids usually several non-equivalent nuclear sites are encountered. It is therefore difficult to include these data in Table 7.5. Among these is the detailed investigation of ^1H and ^{13}C magnetic shielding in solid pyromellitic acid dihydrate and malonic acid by Tegenfeldt et al. [108]. Up to eight different ^{13}C shielding tensors were evaluated in ammonium tartrate by Pines et al. [109]. A detailed discussion of the direction of the principal axes of the ^{13}C shielding tensor in carboxyl and hydroxyl groups is found in Ref. 109. Van der Hart [110] has investigated the ^{13}C chemical shift tensor of methylene groups in alkane n-$C_{20}H_{42}$. This single crystal investigation is of major importance since it furnishes also the directions of the principal axes.

Zilm, Grant and co-workers [111] dissolved small molecules in an argon matrix. By this dilution technique they were able to investigate the ^{13}C chemical shift tensor as well as the ^{13}C-^1H dipolar interaction at 15 K without any intermolecular interaction.

7.6 Other Shielding Tensors

Shielding tensors of other nuclei like ^{15}N, ^{29}Si, ^{31}P, ^{77}Se, ^{113}Cd, ^{125}Te, ^{207}Pb etc. have also been determined [47, 53–57]. As demonstrated in Figs. 3.12 and 3.13 (Sect. 3.2) line narrowing techniques have been applied also to metals such as ^{27}Al and ^9Be. We are not going to discuss the details of the magnetic shielding of these compounds, but would rather like to summarize the data available in the following Tables 7.6, 7.7, and 7.8.

Table 7.6. Nitrogen (^{15}N) shielding tensor σ_{11}, σ_{22}, σ_{33} in ppm with respect to its average $\bar{\sigma}$ for different compounds. (PW, powder spectra; SC, single crystal; LC, liquid crystal; MF, molecular frame; CP, cross polarization; FID, free induction decay; NP, nematic phase; RT, room temperature)

Compound	Formula		T (°K)	σ_{11}	σ_{22}	σ_{33}	$\bar{\sigma}$	Reference	Method	Refs.
Ammonium nitrate	$NH_4N^*O_3$		RT	-83	-57	140	352	NO_3^-	PW, CP	114
Ammonium sulfate	$(NH_4)_2SO_4$		RT	0	0	0	352	NO_3^-	PW, CW	114
Glycine			RT	0	0	0	352	NO_3^-	PW, CW	114
Nitrous oxide	NN^*O		NP	-170	-170	340	135	HNO_3	MF, LC	115
Nitrous oxide	N^*NO		NP	-123	-123	246	219	HNO_3	MF, LC	115
Nitrobenzene	$C_6H_5NO_2$		170	-265	102	164	-8	liquid	PW, FID	113
Pyridine			105	-334	-115	448	-21	liquid	PW, FID	116
Acetonitrile	CH_3CN		77	-326	163	163	311	NO_3^-	PW, CP	112
Imidazole	C_3N_2H	(a)	RT	-127	32	95	165.8	$(NH_4)_2SO_4$	SC, CP	125
in L-His·HCl·H$_2$O		(b)	RT	-117	18	99	152	$(NH_4)_2SO_4$	SC, CP	125

Table 7.7. Phosphorous (^{31}P) shielding tensor σ_{11}, σ_{22}, σ_{33} in ppm with respect to its average $\bar{\sigma}$ for different compounds. (PW, powder spectra; SC, single crystal; MF, molecular frame; FM, first moment; FID, free induction decay; MP, multiple pulse; LC, liquid crystal; NP, nematic phase; RT, room temperature)

Formula		T (°K)	σ_{11}	σ_{22}	σ_{33}	$\bar{\sigma}$	Reference	Method	Refs.
P_4		25	-270	135	135			PW, FID	88
Zn_3P_2		RT	-80	40	40	195	Ortho phosphoric acid	PW, FID	117
Mg_3P_2		RT	-75	-75	150	-95	Ortho phosphoric acid	PW, FID	117
KH_2PO_4		77	-5.1	-5.1	10.3	-11.2	H_3PO_4	SC, FM	121
$P^*S_3P_3$		RT	-150	64	86	-89	Ortho phosphoric acid	SC, FID	117
$P^*S_3P_3$		NP	-162	81	81			MF, LC	122
$PS_3P_3^*$		RT	-188	-131	319	87	Ortho phosphoric acid	SC, FID	117
$PS_3P_3^*$		NP	-177	-177	335			MF, LC	122
$P(CN)_3$			-43	-43	86	136	H_3PO_4	SM, CW	123
P_4O_{10}			-109	-109	218	150	H_3PO_4	SM, CW	123
P_4S_{10}			-63	-63	126	-45	H_3PO_4	SM, CW	123
$BaFPO_3$			-96	48	48			SM, CW	73
KH_2PO_4		RT	-12	-12	24			SC, MP	126
α-$Ca_2P_2O_7$	(a)	RT	-60.4	13.7	46.7	18.3	H_3PO_4	SC, MP	127
	(b)		-69.5	23.5	46	21	H_3PO_4	SC, MP	127

Table 7.8. Silicon (^{29}Si), Selenium (^{77}Se), Cadmium (^{113}Cd), Tellurium (^{125}Te) and Lead (^{207}Pb) shielding tensor σ_{11}, σ_{22}, σ_{33} in ppm with respect to its average $\bar\sigma$ for different compounds. The average $\bar\sigma$ is referenced to TMS (^{29}Se), selenic acid (^{77}Se) and TeCl$_2$ solution (^{125}Te). (PW, powder spectra; SC, single crystal; CW, wide line; CP, cross polarization; RT, room temperature)

Compound	Formula	T (°K)	σ_{11}	σ_{22}	σ_{33}	$\bar\sigma$	Method	Refs.
TMS	$(CH_3)_4Si$	87	0	0	0	0	PW, CP	118
Trimethyl methoxy silane	$(CH_3)_3SiOCH_3$	87	− 13.7	− 13.7	27.3	− 19	PW, CP	118
Dimethyl dimethoxy silane	$(CH_3)_2Si(OCH_3)_2$	87	− 16	− 16	32	4	PW, CP	118
Methyl trimethoxy silane	$CH_3Si(OCH_3)_3$	87	− 19	− 7	26	42	PW, CP	118
Tetramethoxy silane	$Si(OCH_3)_4$	87	0	0	0	80	PW, CP	118
	$[(CH_3)_3Si]_3CH$	87	− 13	0	13	5	PW, CP	118
	$(CH_3)_3SiC_6H_5$	87	− 12	− 6	19	10	PW, CP	118
Hexamethyl disiloxane	$[(CH_3)_3Si]_2O$	87	− 13	− 5	17	− 3	PW, CP	118
	$[(CH_3)_2SiO]_3$	87	− 26	− 18	44	18	PW, CP	118
Actamethyl cyclotetra siloxane	$[(CH_3)_2SiO]_4$	87	− 16	− 16	33	20	PW, CP	118
	^{77}Se	77	− 420	− 420	840	33	SC, CW	119, 120
	^{125}Te	RT	−1190	570	620	620	SC, CW	124
	^{111}Cd	RT	− 160	− 160	320	− 3221 (absolute)	PW, CW	128
	^{113}Cd	RT	− 160	− 160	320	− 3224 (absolute)	PW, CW	128
	$^{113}CdCl_2$	RT	− 38	− 38	76	− 1311 (absolute)	PW, CW	129
	$^{113}CdBr_2 \cdot 4H_2O$	RT	− 85	42.5	42.5	− 1141 (absolute)	PW, CW	129
	^{113}CdS	RT	− 36	18	18	− 1792 (absolute)	PW, CW	129
	$^{113}CdSe$	RT	− 42	21	21	− 1650 (absolute)	PW, CW	129
	$^{113}CdSO_4 \cdot H_2O$	RT	− 92	46	46	− 40 $(Cd(NO_3)_2 \cdot 4H_2O)$	PW, CP	130
	$^{113}Cd(NO_3)_2 \cdot 4H_2O$	RT	− 126	63	63	0 $(Cd(NO_3)_2 \cdot 4H_2O)$	PW, CP	130
	$^{207}Pb(NO_3)_2$	RT	− 17.7	− 17.7	35.4	601 (absolute)	SC, FID	131

8 Spin-Lattice Relaxation

In this chapter we want to account briefly for spin-lattice relaxation, i.e. the loss of order or the increase in entropy of the spin system due to interactions with the lattice. The general relaxation theory is outlined in Appendix I. For additional information the reader is referred to the fundamental books by Abragam [1] and Goldman [2] and to the more recent review by Spiess [3].

Here we emphasize more the context with high resolution NMR techniques in the solid state.

We define the different relaxation times T_1, $T_{1\rho}$ and T_2 as

$$\frac{d}{dt}\langle I_z(t)\rangle = -\frac{1}{T_1}\langle I_z(t)\rangle \tag{8.1a}$$

$$\frac{d}{dt}\langle I_x(t)\rangle = -\frac{1}{T_2}\langle I_x(t)\rangle \tag{8.1b}$$

$$\frac{d}{dt}\langle I_\rho(t)\rangle = -\frac{1}{T_{1\rho}}\langle I_\rho(t)\rangle \tag{8.1c}$$

where the z-axis is parallel to the static magnetic field B_0 and $\langle I_z(t)\rangle$ is the deviation from the equilibrium value. The ρ-axis in a spin-locking or multiple-pulse experiment is considered to be the quantization axis in the rotating frame. The relaxation rate $1/T_{1\rho}$ describes therefore the relaxation along this quantization axis, which may be tilted by an angle β with respect to the z-axis. For on-resonance spin-locking $\beta = \pi/2$. The relaxation orthogonal to the quantization axis is termed $T_{2\rho}$ equivalent to the definition of T_2.

8.1 Spin-Lattice Relaxation in the Weak Collision Limit

In the weak collision limit the fluctuation of the spin interactions due to lattice motion characterized by the correlation time τ is much more rapid than the variation $\delta\omega$ in the spin interactions, i.e. $1/\tau \gg \delta\omega$. The relaxation times T_1, $T_{1\rho}$ and T_2 can be readily calculated in this limit for the standard types of spin interactions like direct homonuclear dipole-dipole interaction (D), indirect homonuclear interaction (J), quadrupole interaction (Q), chemical shift interaction (CS) and indirect and direct heteronuclear interaction (IS).

The following relations were derived in Appendix I (see also Ref. [3]).

(a) indirect homonuclear interaction

$$\frac{1}{T_{1\rho}} = \tfrac{1}{3}I(I+1)\{\sin^2\beta\, j_{10}(\omega_e) + (1+\cos^2\beta)j_{11}(\omega_I)$$

$$+ 3\sin^2\beta\cos^2\beta\, j_{20}(\omega_e) + 3\sin^4\beta\, j_{20}(2\omega_e)\} \tag{8.2}$$

$$\frac{1}{T_1}=\tfrac{2}{3}I(I+1)\left\{j_{11}(\omega_I)+j_{21}(\omega_I)+4j_{22}(2\omega_I)\right\} \tag{8.3}$$

$$\frac{1}{T_2}=\tfrac{1}{3}I(I+1)\left\{j_{10}(0)+j_{11}(\omega_I)+3j_{20}(0)+5j_{21}(\omega_I)+2j_{22}(2\omega_I)\right\}. \tag{8.4}$$

Under isotropic motion, all the $j_{kq}(\omega)$ for the same k are equal. If we invoke the fast correlation limit $(\omega_I\tau\ll1)$, i.e. $j_{kq}(0)=j_{kq}(\omega_I)$ we obtain $T_1=T_2$. This relation also holds for the following cases (b) and (c).

(b) homonuclear dipole-dipole interaction

$$\frac{1}{T_{1\rho}}=\tfrac{1}{3}I(I+1)\left\{3\sin^2\beta\cos^2\beta\, j_{20}(\omega_e)+3\sin^4\beta\, j_{20}(2\omega_e)\right.$$
$$\left.+(5-3\cos^2\beta)j_{21}(\omega_I)+(6\cos^2\beta+2)j_{22}(2\omega_I)\right\} \tag{8.5}$$

$$\frac{1}{T_1}=\tfrac{2}{3}I(I+1)\left\{j_{21}(\omega_I)+4j_{22}(2\omega_I)\right\} \tag{8.6}$$

$$\frac{1}{T_2}=\tfrac{1}{3}I(I+1)\left\{3j_{20}(0)+5j_{21}(\omega_I)+2j_{22}(2\omega_I)\right\} \tag{8.7}$$

(c) quadrupole interaction

$$\frac{1}{T_{1\rho}}=\tfrac{1}{10}(2I-1)(2I+3)\left\{3\sin^2\beta\cos^2\beta\, j_{20}(\omega_e)+3\sin^4\beta\, j_{20}(2\omega_e)\right.$$
$$\left.+(5-3\cos^2\beta)j_{21}(\omega_I)+(6\cos^2\beta+2)j_{22}(2\omega_I)\right\} \tag{8.8}$$

$$\frac{1}{T_1}=\tfrac{1}{5}(2I-1)(2I+3)\left\{j_{21}(\omega_I)+4j_{22}(2\omega_I)\right\} \tag{8.9}$$

$$\frac{1}{T_2}=\tfrac{1}{10}(2I-1)(2I+3)\left\{3j_{20}(0)+5j_{21}(\omega_I)+2j_{22}(2\omega_I)\right\} \tag{8.10}$$

(d) chemical shift interaction

$$\frac{1}{T_{1\rho}}=\omega_I^2\left\{\tfrac{1}{3}\sin^2\beta\, j_{00}(\omega_e)+\tfrac{1}{2}(1+\cos^2\beta)j_{11}(\omega_I)\right.$$
$$\left.+\tfrac{2}{3}\sin^2\beta\, j_{20}(\omega_e)+\tfrac{1}{2}(1+\cos^2\beta)j_{21}(\omega_I)\right\} \tag{8.11}$$

$$\frac{1}{T_1}=\omega_I^2\left\{j_{11}(\omega_I)+j_{21}(\omega_I)\right\} \tag{8.12}$$

$$\frac{1}{T_2}=\omega_I^2\left\{\tfrac{1}{3}j_{00}(0)+\tfrac{1}{2}j_{11}(\omega_I)+\tfrac{2}{3}j_{20}(0)+\tfrac{1}{2}j_{21}(\omega_I)\right\}. \tag{8.13}$$

If we now assume isotropic motion in the fast correlation limit $(\omega_I\tau\ll1)$ and assume that fluctuations in the isotropic shift vanish, i.e. $j_{00}(0)=0$ which holds e.g. in the case of rapid isotropic rotation, we obtain

$$T_1/T_2=[3+4/(\alpha+1)]/6 \tag{8.14}$$

where $\alpha=j_{11}(\omega)/j_{2q}(\omega)$ is the anti-symmetry parameter. If there is no anti-symmetrical chemical shift tensor

$$\alpha=0:\ T_1/T_2=7/6. \tag{8.15}$$

In the other extreme, when the antisymmetric part is as large as the symmetric part

$$\alpha = 1: \quad T_1/T_2 = 5/6. \tag{8.16}$$

It might be possible to determine the antisymmetric part of the chemical shift tensor this way.

(e) indirect heteronuclear interaction (observation of I spins)

$$\frac{1}{T_{1\rho}} = \tfrac{1}{3}S(S+1)\,\{\sin^2\beta[\tfrac{1}{3}j_{00}(\omega_e) + j_{11}(\omega_S) + \tfrac{2}{3}j_{20}(\omega_e) + j_{21}(\omega_S)$$
$$+ (1+\cos^2\beta)\,[\tfrac{1}{3}j_{00}(\omega_S-\omega_I) + \tfrac{1}{2}j_{10}(\omega_S-\omega_I) + \tfrac{1}{2}j_{11}(\omega_I)$$
$$+ \tfrac{1}{6}j_{20}(\omega_S-\omega_I) + \tfrac{1}{2}j_{21}(\omega_I) + j_{22}(\omega_S+\omega_I)]\} \tag{8.17}$$

$$\frac{1}{T_1} = \tfrac{1}{3}S(S+1)\,\{\tfrac{2}{3}j_{00}(\omega_S-\omega_I) + j_{10}(\omega_S-\omega_I) + j_{11}(\omega_I)$$
$$+ \tfrac{1}{3}j_{20}(\omega_S-\omega_I) + j_{21}(\omega_I) + 2j_{22}(\omega_S+\omega_I)\} \tag{8.18}$$

$$\frac{1}{T_2} = \tfrac{1}{3}S(S+1)\,\{\tfrac{1}{3}j_{00}(0) + \tfrac{1}{3}j_{00}(\omega_S-\omega_I) + \tfrac{1}{2}j_{10}(\omega_S-\omega_I)$$
$$+ j_{11}(\omega_S) + \tfrac{1}{2}j_{11}(\omega_I) + \tfrac{2}{3}j_{20}(0) + \tfrac{1}{6}j_{20}(\omega_S-\omega_I) + j_{21}(\omega_S)$$
$$+ \tfrac{1}{2}j_{21}(\omega_I) + j_{22}(\omega_S+\omega_I)\}. \tag{8.19}$$

In the fast correlation limit we obtain as in the homonuclear case $T_1 = T_2$.

(f) heteronuclear dipole-dipole interaction

$$\frac{1}{T_{1\rho}} = \tfrac{1}{3}S(S+1)\,\{\sin^2\beta[\tfrac{2}{3}j_{20}(\omega_e) + j_{21}(\omega_S)]$$
$$+ (1+\cos^2\beta)\,[\tfrac{1}{6}j_{20}(\omega_S-\omega_I) + \tfrac{1}{2}j_{21}(\omega_I) + j_{22}(\omega_S+\omega_I)]\} \tag{8.20}$$

$$\frac{1}{T_1} = \tfrac{1}{3}S(S+1)\,\{\tfrac{1}{3}j_{20}(\omega_S-\omega_I) + j_{21}(\omega_I) + 2j_{22}(\omega_S+\omega_I)\} \tag{8.21}$$

$$\frac{1}{T_2} = \tfrac{1}{3}S(S+1)\,\{\tfrac{2}{3}j_{20}(0) + \tfrac{1}{6}j_{20}(\omega_S-\omega_I) + j_{21}(\omega_S)$$
$$+ \tfrac{1}{2}j_{21}(\omega_I) + j_{22}(\omega_S+\omega_I)\}. \tag{8.22}$$

The generalized spectral functions $j_{kq}(\omega)$ are defined as

$$j_{kq}(\omega) = \mathrm{Re}\int_0^\infty dt(-1)^q\,\langle \delta A_{kq}(0)\,\delta A_{k-q}(t)\rangle\,e^{-i\omega t} \tag{8.23}$$

where

$$\langle \delta A_{kq}(0)\,\delta A_{k-q}(t)\rangle = \langle A_{kq}(0)\,A_{k-q}(t)\rangle - \langle A_{kq}\rangle\langle A_{k-q}\rangle. \tag{8.24}$$

According to Appendix I the $j_{kq}(\omega)$ may be rewritten in terms of the spectral densities $J_k(\omega)$, i.e.

$$j_{kq}(\omega) = \langle|\delta A_{kq}|^2\rangle\,J_k(\omega) \tag{8.25}$$

and

$$\langle|\delta A_{kq}|^2\rangle = (-1)^q\,[\langle A_{kq}A_{k-q}\rangle - \langle A_{kq}\rangle\langle A_{k-q}\rangle] \tag{8.26}$$

where

$$J_k(\omega) = \text{Re} \int_0^\infty dt \, g_k(t) \, e^{-i\omega t} \tag{8.27}$$

with

$$g_k(t) = \frac{\langle \delta A_{kq}(0) \, \delta A_{k-q}(t) \rangle}{\langle \delta A_{kq} \, \delta A_{k-q} \rangle}. \tag{8.28}$$

In most cases $g_k(t)$ may be approximated by [1–3]

$$g_k(t) = e^{-t/\tau_k} \tag{8.29}$$

with

$$J_k(\omega) = \frac{\tau_k}{1 + \omega^2 \tau_k^2}. \tag{8.30}$$

The orientation dependence of T_1, $T_{1\rho}$ and T_2 with respect to the external field is fully contained in the $\langle |\delta A_{kq}|^2 \rangle$. The corresponding A_{kq} for the different spin interactions can be found in Table 2.2 of Chap. 2. Similar expressions for the different relaxation times have been derived by Blicharski [4] and by Spiess [3].

8.2 Spin Lattice Relaxation in Multiple-Pulse Experiments

In multiple pulse experiments, which we want to discuss first, one expects a relaxation in the rotating frame with a time constant of about $T_{1\rho}$. Let us recall, that $T_{1\rho}$ is the spin lattice relaxation in an effective field B_1 in the rotating frame. If we assume a Gaussian-Markoff process with the correlation time τ_c to cause a "random isotropic modulation" of the dipole Hamiltonian, we arrive at [1–7] (see Eq. (8.5) with $\beta = \pi/2$)

$$\frac{1}{T_{1\rho}} = M_2 \left[\frac{\tau_c}{1 + 4\omega_1^2 \tau_c^2} + \frac{5}{3} \frac{\tau_c}{1 + \omega_0^2 \tau_c^2} + \frac{2}{3} \frac{\tau_c}{1 + 4\omega_0^2 \tau_c^2} \right] \tag{8.31}$$

where

$$\omega_1 = \gamma B_1 \quad \text{and} \quad \omega_0 = \gamma B_0$$

and with

$$M_2 = \frac{3}{5} \frac{\gamma_I^4 \hbar^2}{r^6} I(I+1) \tag{8.32}$$

being the variation of the second moment of two spins I with the gyromagnetic ratio γ_I, separated by the distance r under isotropic rotation. Notice, that the maximum relaxation rate is reached for $\omega_1 \tau_c = 1/2$, which is considered as being a "slow motion", since ω_1 is of the order of several kHz.

We obtain for this maximum rate

$$\left(\frac{1}{T_{1\rho}} \right)_{max} = \tfrac{1}{2} M_2 \tau_c. \tag{8.33}$$

This is of course the "worst" case in terms of resolution, i.e. the ultimate resolution in multiple-pulse experiments is limited by spin lattice relaxation

[8]. Supposing for a while, that the effective rotating frame relaxation $T_{1\rho}$ in a multiple-pulse experiment is on the order of the spin locking $T_{1\rho}$, we want to make an estimate of the ultimate resolution which could be achieved in the presence of maximum spin lattice relaxation effect. Let us call δ_0 the "rigid line width" (for a more subtle discussion see Sect. 2.5 and Appendix C) with

$$\delta_0 \cong M_2^{1/2}$$

and

$$\delta = 1/T_{1\rho}$$

the "ultimate line width" caused by spin lattice relaxation. We then arrive under the condition of "maximum relaxation effect" according to Eq. (8.33) at

$$\delta_0/\delta = \frac{2}{\tau_c M_2^{1/2}} \tag{8.34}$$

or with $\omega_1 \tau_c = 1/2$

$$\delta/\delta_0 = 4\omega_1/M_2^{1/2}. \tag{8.35}$$

Inserting $B_1 = 10\,G$ and $(M_2)^{1/2} = 1\,G$ into Eq. (8.34) results in a factor of 40 for the ultimate resolution compared with the rigid line width. This may seriously deteriorate the line narrowing efficiency in a well chosen multiple pulse experiment i.e., we do have to worry about spin lattice relaxation in multiple pulse experiments if we demand highest resolution [9]. On the other hand, we shall see that spin lattice relaxation can be investigated by multiple pulse experiments in a very appealing way, especially in the regime of "slow motions" [8–10].

Haeberlen and Waugh [8] have discussed extensively spin lattice relaxation in multiple-pulse experiments and here we want to review some of their results briefly. Starting from the master equation for the spin density matrix ρ [1]

$$\frac{d\rho}{dt} = -\int_0^\infty dt' \, \langle [\mathscr{H}(t), [\mathscr{H}(t-t'), \rho]] \rangle_{av} \tag{8.36}$$

where $\mathscr{H}$ is the suitably transformed dipolar Hamiltonian of a pair of spins I coupled to each other and subjected to a random motion. Haeberlen and Waugh [8] arrive under the following conditions to decoupled Bloch equations for the different components of the magnetization.

The conditions are:
(i)   A frame of reference has to be chosen in which no external field, static or oscillatory, appears explicitly.
(ii) Restriction to the secular part of the dipolar Hamiltonian.
(iii) Only spins $I = 1/2$ are considered, or else higher order tensor operator averages like $\langle T_{3m} \rangle$ are neglected.

If we denote the relaxation time of the z-component of the magnetization in the suitably transformed rotating frame $T_{1\rho}$ and the corresponding relaxation time of the components orthogonal to the z-axis $T_{2\rho}$, we arrive at the following equations [8]

$$\frac{1}{T_{1\rho}} = \frac{8\pi}{3} M_2 [-\mathrm{Re}\{\mathscr{F}^1(\omega)\} + 4\,\mathrm{Re}\{\mathscr{F}^2(2\omega)\}] \tag{8.37}$$

and

$$\frac{1}{T_{2\rho}} = \frac{4\pi}{3} M_2 [3 \operatorname{Re}\{\mathscr{F}^0(0)\} + 2 \operatorname{Re}\{\mathscr{F}^2(2\omega)\} - 5 \operatorname{Re}\{\mathscr{F}^1(\omega)\}] \tag{8.38}$$

where a random stationary motion has been assumed, i.e.

$$\mathscr{F}^{-m}(-\omega) = \mathscr{F}^{m*}(\omega).$$

As the motion is assumed to be due to fluctuations in the dipolar interaction, we can write

$$\mathscr{F}^m(m\omega) = (-1)^m \int_0^\infty dt' \langle Y_{2m}^*(t) \, Y_{2m}(t-t') \rangle \, e^{im\omega t} \tag{8.39}$$

where $\langle \, \rangle$ denotes an ensemble average to obtain the correlation function of the randomly modulated spherical harmonics. For a Gauss-Markoff process with the correlation time τ_c

$$\operatorname{Re}\{\mathscr{F}^m(m\omega)\} = (-1)^m \frac{1}{4\pi} \frac{\tau_c}{1+(m\omega\tau_c)^2}$$

the familiar expression for T_1 and T_2 can be obtained [1]:

$$\frac{1}{T_1} = \frac{2}{3} M_2 \left[\frac{\tau_c}{1+\omega_0^2\tau_c^2} + \frac{4\tau_c}{1+4\omega_0^2\tau_c^2} \right]$$

$$\frac{1}{T_2} = \frac{1}{3} M_2 \left[3\tau_c + \frac{5\tau_c}{1+\omega_0^2\tau_c^2} + \frac{2\tau_c}{1+4\omega_0^2\tau_c^2} \right]$$

where T_1 represents the relaxation in the direction of the quantization axis, whereas T_2 describes the relaxation orthogonal to the quantization axis.

Let us discuss some representative experiments

a) Spinning Sample Experiment

In this case we have to replace Y_{2m} by [8]

$$d_{mm'}^{(2)}(\vartheta_r) \, Y_{2m'}(t) \, e^{im'\omega_r t}$$

where ϑ_r is the angle of the rotation axis with the magnetic field, and

$$d_{mm'}^{(2)}(\vartheta_r) = \langle 2m| \, e^{-i\vartheta_r I_y} \, |2m')$$

is the Wigner rotation matrix. ω_r is the rotation frequency. This results for a Gauss-Markoff process in [8]

$$\mathscr{F}^m(m\omega) = (-1)^m \frac{1}{4\pi} \sum_{m'} |d_{mm'}^{(2)}|^2 \frac{\tau_c}{1+(m\omega_0 - m'\omega_r)^2 \tau_c^2}. \tag{8.40}$$

Inserting Eq. (8.40) into Eq. (8.37) results in the spin lattice relaxation time T_1 in the spinning sample experiment. No dramatic change compared with the ordinary T_1 is expected, since $\omega_0 \gg \omega_r$ in practice. A more dramatic change is induced into the "transverse" spin lattice relaxation time T_2 as discussed by Haeberlen and Waugh [8] and also by Andrew and Jasinski [11].

If $\omega_0 \tau_c \gg 1$ is assumed and Eq. (8.40) is inserted into Eq. (8.38) we obtain under the "magic angle" condition $\vartheta_r = 54° 44'$

$$\frac{1}{T_2} = \frac{1}{3} M_2 \left[\frac{2\tau_c}{1+\omega_r^2 \tau_c^2} + \frac{\tau_c}{1+4\omega_r^2 \tau_c^2} \right]. \tag{8.41}$$

We notice, that the line-width in a "magic angle" spinning sample experiment may well be determined by spin lattice relaxation if $\omega_r \tau_c \cong 1$ and may cause a failure of such an experiment in terms of not achieving enough resolution. We turn now to the rotating frame relaxations, assuming throughout $\omega_0 \tau_c \gg 1$ and thus emphasizing the contribution of "slow motions".

b) Spin-Locking [12, 13] *and Lee-Goldburg Experiment* [14]

The relaxation takes place in an effective field $B_e = \omega_e/\gamma$ tilted by an angle ϑ with respect to the static magnetic field. With the restrictions we made so far we obtain [8]

$$\mathrm{Re}\{\mathscr{F}^m(m\omega_e)\} = (-1)^m |d_{0m}^{(2)}(\vartheta)|^2 \cdot \frac{1}{4\pi} \frac{\tau_c}{1+(m\omega_e\tau_c)^2}. \tag{8.42}$$

Notice at this stage already, that replacing ϑ by $-\vartheta$ or by $\vartheta+\pi$ in Eq. (8.42) does not affect the spectral distribution function i.e., phase alternation of the effective field or the rf field does not influence the relaxation. In other words, not the "average" field is the important parameter in spin lattice relaxation, but rather the modulus of this field i.e., the rate at which it samples the stochastic motion.

In the resonance spin-locking case ($\vartheta = \pi/2$) with

$$|d_{0\pm2}^{(2)}|^2 = \tfrac{3}{8}; \quad |d_{0\pm1}^{(0)}|^2 = 0; \quad |d_{00}^{(2)}|^2 = 1/4$$

we obtain for the relaxation rate along the applied rf field

$$\frac{1}{T_{1\rho}} = M_2 \frac{\tau_c}{1+4\omega_e^2 \tau_c^2} \tag{8.43}$$

which is the so-called spin-lattice relaxation in the rotating frame as can be derived from Eq. (8.31) under the condition $\omega_0 \tau_c \gg 1$. Similar the transverse relaxation orthogonal to the applied rf field

$$\frac{1}{T_{2\rho}} = \frac{1}{4} M_2 \left(\tau_c + \frac{\tau_c}{1+4\omega_e^2 \tau_c^2} \right) \tag{8.44}$$

is obtained, which might be observed in a rotary echo experiment.

Notice that:

(i) $\quad T_{2\rho} = 2 \cdot T_{1\rho} \quad$ for $\omega_e \tau_c \ll 1 \quad \omega_0 \tau_c \gg 1$

(ii) $\quad T_{1\rho} > T_{2\rho} \quad$ for $\omega_e \tau_c \gg 1$.

For $\omega_e \tau_c \gg 1$ the correlation time τ_c approaches the flip-flop correlation time of the I spins in the rigid lattice i.e. $T_{2\rho} \to T_2$.

Lee-Goldburg-Experiment ($\tan \vartheta = \sqrt{2}$) [8]

$$|d_{0\pm2}^{(2)}|^2 = \tfrac{1}{6}; \quad |d_{0\pm1}^{(2)}|^2 = \tfrac{1}{3}; \quad |d_{00}^{(2)}|^2 = 0.$$

Relaxation rate along the quantization axis (111) in the rotating frame:

$$\frac{1}{T_{1\rho}}=\frac{2}{9}M_2\left[\frac{\tau_c}{1+\omega_e^2\tau_c^2}+\frac{2\tau_c}{1+4\omega_e^2\tau_c^2}\right]. \tag{8.45}$$

Relaxation rate orthogonal to the (111) direction:

$$\frac{1}{T_{2\rho}}=\frac{1}{9}M_2\left[\frac{5\tau_c}{1+\omega_e^2\tau_c^2}+\frac{\tau_c}{1+4\omega_e^2\tau_c^2}\right]. \tag{8.46}$$

Haeberlen and Waugh [8] have discussed the interference which might occur if sample spinning combined with a Lee-Goldburg experiment is performed. In this case the spectral distribution function can be expressed as [8]

$$\mathrm{Re}\{\mathscr{F}^m(m\omega_0)\}=(-1)^m\,|d_{0m}^{(2)}(\vartheta)|^2\sum_{m'}|d_{0m'}^{(2)}(\vartheta_r)|^2\,\frac{1}{4\pi}\,\frac{\tau_c}{1+(m\omega_e-m'\omega_r)^2\tau_c^2}. \tag{8.47}$$

For the magic angle condition $\tan\vartheta=\tan\vartheta_r=\sqrt{2}$.

$\mathscr{F}^{(0)}$ in $1/T_{2\rho}$ drops out, because of $|d_{00}^{(2)}(\vartheta)|=|d_{00}^{(2)}(\vartheta_r)|=0$

and terms with $(m\omega_e-m'\omega_r)$ become important, where m, $m'=-2,\ldots+2$. In the case $\omega_e\cong\omega_r$ several of those terms vanish, leading to $1/T_{2\rho}\sim\tau_c$.

If the correlation time τ_c approaches the flip-flop rate of the rigid lattice, $(T_{2\rho}=T_2)$ a broad "solid" line is observed rather than a "narrowed" line. In this sense spinning of the sample counterbalances the rotation in spin space and destroys the line narrowing efficiency of each of those experiments alone [8]. It is highly recommended to make $\omega_e\neq\omega_r$, usual $\omega_e\gg\omega_r$.

c) Multiple-Pulse Experiments

Haeberlen and Waugh [8] and Mansfield [5] have discussed spin lattice relaxation in different multiple-pulse experiments, namely

MW-2 $P_y-(\tau-P_x-2\tau-P_{\bar{x}}-\tau)_n$

WHH-4 $(\tau-P_x-2\tau-P_{\bar{x}}-\tau-P_{\bar{y}}-2\tau-P_y-\tau)_n$

where P_i denotes a 90° rf pulse in the i direction.

It should be noted, that MW-4 namely $P_y-(\tau-P_x-2\tau-P_x-2\tau-P_x-2\tau-P_x-\tau)_n$ is governed by the same relaxation time as MW-2. Remember, that the periodic structure of the external fields is the important feature which samples the spectral distribution of the fluctuation, rather than the "average" effective field. Notice, that the "average" effective field in the MW-2 and the WHH-4 experiment performed on-resonance, is zero. Suppose off-resonance effects are disregarded and the same restrictions (i)-(iii) are applied as before with $\omega_0\tau_c\gg1$. Haeberlen and Waugh [8] arrive after some manipulation at the following relaxation rate along the quantization axis, namely the (111) direction in the rotating frame

$$\text{WHH-4:}\quad\frac{1}{T_{1\rho}}=\frac{6}{\pi^2}M_2\sum_{n=1}^{\infty}\frac{1}{n^2}\frac{\tau_c}{1+(n\omega_c\tau_c)^2} \tag{8.48a}$$

and for the relaxation rate orthogonal to the (111) direction at

$$\text{WHH-4:} \quad \frac{1}{T_{2\rho}} = \frac{9}{2\pi^2} M_2 \left\{ 3 \sum_k \frac{1}{k^2} \frac{\tau_c}{1+(k\omega_c\tau_c)^2} + \frac{1}{3} \sum_n \frac{1}{n^2} \frac{\tau_c}{1+(n\omega_c\tau_c)^2} \right\} \quad (8.48\,\text{b})$$

where $\omega_c = 2\pi/t_c$ is the "cycle frequency" as defined by the cycle time t_c. Here $k, n > 0$ and not a multiple of 3 and where k is even and n is odd i.e. $k = 2, 4, 8 \ldots$; $n = 1, 5, 7 \ldots$ Mansfield [5] has calculated $T_{1\rho}$ for the MW-2 experiment, following along those lines and obtained in the direction of the applied rf field

$$\text{MW-2:} \quad \frac{1}{T_{1\rho}} = \frac{8}{\pi^2} M_2 \sum_{n=1}^{\infty} \frac{1}{n^2} \frac{\tau_c}{1+(n\omega_c\tau_c)^2} \quad (8.49)$$

where the sum is now restricted to n odd. The factor 4 in Mansfields [5] calculation has to be replaced by 8 according to Wilsch et al. [15]. The main features which are displayed by Eqs. (8.48) and (8.49) are:

(i) $T_{1\rho}$ in the multiple pulse experiment is always larger, that $T_{1\rho}$ in the spin locking experiment.

(ii) The maximum relaxation rate is determined by $\omega_c\tau_c \cong 1$.

For further details we refer the reader to the papers of Haeberlen and Waugh [8] and by Mansfield [5]. The approach taken can be criticized on different grounds, questioning the validity of the master-equation Eq. (8.36) in this regime ($\omega_0\tau_c \gg 1$, $\omega_c\tau_c \cong 1$) and the different approximations which are made.

Gründer et al. [9, 10] have attempted a more rigorous treatment by calculating directly the time development of the magnetization in a suitable direction, say the x direction for the MW-2 cycle. We shall call this decay time of the magnetization in the x-direction of the rotating frame T_2^*.

Let us start with

$$\langle I_x(t) \rangle = \frac{1}{\text{Tr}\{I_x^2\}} \text{Tr}\{U(t)\, I_x\, U^{-1}(t)\, I_x\} \quad (8.50)$$

where

$$U(t) = \exp\left\{ -i \sum_{n=0}^{\infty} \mathbf{F}_n(t) \right\}$$

is the time evolution operator due to spin lattice relaxation in a suitably defined reference frame, expressed by a Magnus expansion (see Sect. 3.3). Restriction to the first term in the Magnus expansion and assuming a Gauss-Markoff process leads to

$$\langle I_x(t) \rangle = \exp\left\{ \frac{1}{2\text{Tr}\{I_x^2\}} \langle \text{Tr}\{[\mathbf{F}_0(t), I_x]^2\} \rangle_{av} \right\} \quad (8.51)$$

where $\langle \ \rangle_{av}$ means that an ensemble average has to be performed. Gründer et al. [9, 10] obtained after some algebra

$$\langle I_x(t) \rangle = \exp[-t R_1(\alpha) + \tau_c(1 - e^{-t/\tau_c}) R_2(\alpha)] \quad (8.52)$$

in the case of multiple pulse experiments, where $R_1(\alpha)$ and $R_2(\alpha)$ are functions of the parameter $\alpha = \tau/\tau_c$ (2τ pulse spacing, τ_c correlation time of the Gauss-

Markoff process), depending on the type of multiple pulse experiment, which is considered. Two limiting cases will be discussed, namely

(i) $t > \tau_c$, neglecting the 2. term in Eq. (8.52)

$$\langle I_x(t) \rangle = \exp(-t \cdot R_1(\alpha) \equiv \exp(-t/T_{1e}) \tag{8.53}$$

(ii) $t \ll \tau_c$ series expansion of the 2. term in Eq. (8.52)

$$\langle I_x(t) \rangle = \exp\left[-t(R_1(\alpha) - R_2(\alpha)) - \frac{t^2}{2\tau_c} R_2(\alpha) \right]$$
$$\equiv \exp[-t/T_{1e} + M_{2\,\text{eff}}\, t^2/2] \tag{8.54}$$

where

$$\frac{1}{T_{1e}} = R_1(\alpha) - R_2(\alpha) \quad \text{and} \quad M_{2\,\text{eff}} = -\frac{R_2(\alpha)}{\tau_c}.$$

However, Eq. (8.53) is more generally applicable than it seems to be, since

in case $\tau_c \ll \tau$ and $t \gg \tau_c$ Eq. (8.53) is valid

and

in case $\tau_c \gg \tau$ and $R_2(\alpha) \ll R_1(\alpha)$ Eq. (8.53) is again valid.

The relaxation rates following this scheme are in the case of
MW-2 *and* MW-4 *experiment* [9, 10]

$$\frac{1}{T_{1e}} = M_2 \tau_c \left(1 - \frac{\tanh \alpha}{\alpha} \right); \quad \alpha = \tau/\tau_c \tag{8.55}$$

and for the
WHH-4 *and* HW-8 *experiment* [9, 10]

$$\frac{1}{T_{1e}} = \frac{2}{3} M_2 \tau_c \left[1 - \frac{(5 \cosh \alpha - 2) \sinh^2 \alpha}{\alpha \sinh 3\alpha} \right] \tag{8.56}$$

where T_{1e} is the relaxation of the x magnetization, rather than the "longitudinal" magnetization along the (111) direction. In Table 8.1 we compare the individual T_{1e} values for different experiments, namely, spin locking, MW-2 and MW-4, WHH-4 and HW-8 for different values of α. Figure 8.1 gives a pictorial representation of the normalized relaxation times in the different experiments versus τ_c/τ. Notice, that $T_{1\rho}$ in the spin locking experiment is always shorter as compared with T_{1e} in the MW or WHH experiments. The T_{1e}-minimum occurs always at $\alpha = \tau/\tau_c = \pi/2$, from which the correlation time can be determined very accurately. The range of correlation times, which can be determined with the multiple-pulse method, corresponds to that of the spin-locking $T_{1\rho}$ experiment. However, the advantage of the multiple pulse method is, that the full relaxation curve can be observed in one shot. Other advantages of the multiple pulse method, connected with the quenching of spin diffusion will be discussed later. The MW-4 sequence was used to test the applicability of multiple-pulse experiments to the investigation of slow motions [9, 10].

So far we have assumed $\pi/2$ (or π)-pulses in the δ-pulse limit are applied in the different multiple-pulse experiments. Rhim et al. [16] and Vega and Vaug-

Table 8.1. Decay time constant T_{1e} of magnetization in the x, y-plane of the rotating frame on resonance for different multiple-pulse experiments (δ pulses assumed) due to spin lattice relaxation with correlation time τ_c. 2τ, pulse spacing; $\alpha=\tau/\tau_c$; M_2 "rigid" second moment; t_w, pulse width; $\omega_1=\gamma B_{1x}$; suggested preparation pulse in front of each sequence $(\frac{\pi}{2})_y$. (Gründer et al. [9, 10])

Multiple pulse experiment				$T_{1e}(T_{1\rho})$			
a) Homo-nuclear dipolar interaction	type	sequence	cycle time t_c	$\tau_c\ll\tau$	$\tau_c\ll\tau$ —	Minimum $\tau=\frac{\pi}{2}\tau_c$	$\tau_c\gg\tau$
	MW-2	$[\tau-(\frac{\pi}{2})_x-2\tau-(\frac{\pi}{2})_{\bar{x}}-\tau]$	4τ	$\dfrac{1}{T_{1e}}=M_2\tau\dfrac{1}{\alpha}\left(1-\dfrac{\tanh\alpha}{\alpha}\right)$	$T_{1e}=\dfrac{1}{M_2\tau_c}$	$T_{1e\,\min}=\dfrac{3.8}{M_2\tau}$	$T_{1e}=\dfrac{3\tau_c}{M_2\tau^2}$
	MW-4	$[\tau-(\frac{\pi}{2})_x-2\tau-(\frac{\pi}{2})_x-\tau]_2$	8τ				
	WHH-4	$[\tau-(\frac{\pi}{2})_{\bar{x}}-2\tau-(\frac{\pi}{2})_x-\tau-(\frac{\pi}{2})_y-2\tau-(\frac{\pi}{2})_{\bar{y}}]$	6τ	$\dfrac{1}{T_{1e}}=M_2\tau\cdot\dfrac{2}{3\alpha}$	$T_{1e}=\dfrac{3}{2M_2\tau_c}$	$T_{1e\,\min}=\dfrac{6.4}{M_2\tau}$	$T_{1e}=\dfrac{9\tau_c}{2M_2\tau^2}$
	HW-8	WHH-4$+(-$WHH-4$)$	6τ	$\cdot\left[1-\dfrac{(5\cosh\alpha-2)\sinh^2\alpha}{\alpha\sinh 3\alpha}\right]$		$\left(\tau=\dfrac{\pi}{4}\tau_c\right)$	
b) Hetero-nuclear dipolar interaction	MW-4	$[\tau-(\frac{\pi}{2})_x-2\tau-(\frac{\pi}{2})_x-\tau]_2$	8τ	$\dfrac{1}{T_{1e}}=M_2\tau\dfrac{1}{\alpha}\left[\dfrac{1-\tanh 2\alpha}{2\alpha}\right]$	$T_{1e}=\dfrac{1}{M_2\tau_c}$	$T_{1e\,\min}=\dfrac{1.9}{M_2\tau}$	$T_{1e}=\dfrac{3\tau_c}{4M_2\tau^2}$
	MW-4	$[\tau-(\pi)_x-2\tau-(\pi)_x-\tau]$	4τ	$\dfrac{1}{T_{1e}}=M_2\tau\dfrac{1}{\alpha}\left[\dfrac{1-\tanh\alpha}{\alpha}\right]$	$T_{1e}=\dfrac{1}{M_2\tau_c}$	$T_{1e}=\dfrac{3.8}{M_2\tau}$	$T_{1e}=\dfrac{3\tau_c}{M_2\tau^2}$

Similar relations have been obtained by Wilsch et al. [15]

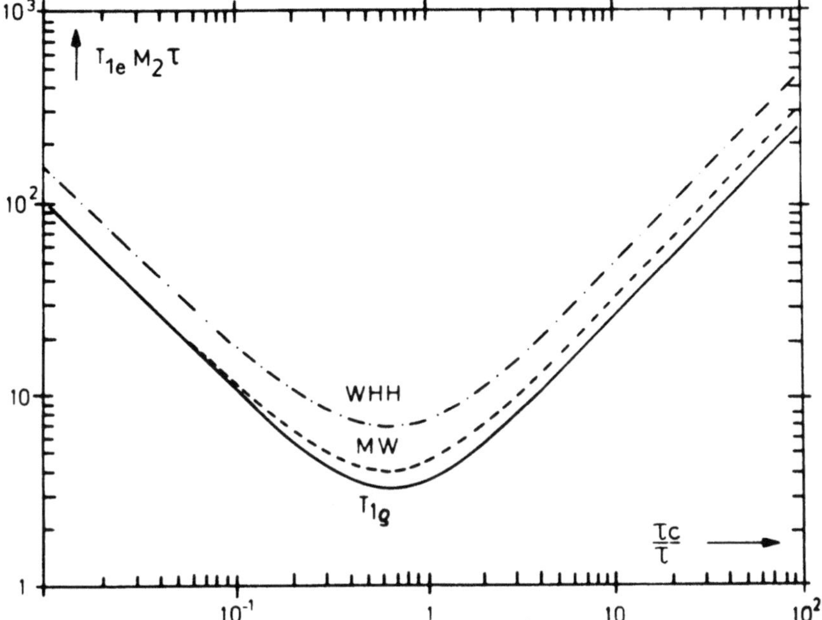

Fig. 8.1. Comparison of the normalized decay time T_{1e} due to spin lattice relaxation in different multiple pulse experiments versus the normalized correlation time τ_c of the motion. (W. Gründer et al. [9, 10])

han [17] have analysed more recently the situation, where these constraints are relaxed. This is an important extension, since Rhim et al. [18] have demonstrated in a series of papers that spin heating causes a destruction of the magnetization in multiple pulse experiments too, limiting the applicability of these sequences for relaxation measurements. Spin heating, however, can be reduced extensively, as shown by Rhim et al. [18] when Θ-pulses with $\Theta \ll \pi/2$ are used. It is therefore desirable to have formulae at hand which cover this situation.

The multiple-pulse experiment to be discussed contains a $\pi/2$-prepulse in the y-direction followed after a time τ_1 by a string of Θ-pulses phased in the x-direction of the rotating frame. The Θ-pulses of width t_w are separated by 2τ leaving a "window" of $2\tau_1 = 2\tau - t_w$ between the pulses. Following the procedure of Gründer et al. [9, 10] and Wilsch et al. [15], Rhim et al. [16] arrived in the case of homonuclear dipolar interaction at the expression

$$\frac{1}{T_{1e}} = \frac{M_2^{II} \tau}{2\alpha^2} \left([2\alpha_1 - \sinh(2\alpha_1) + D \sinh(2\alpha)] + \frac{1}{A_+} \{\beta + 2\sinh(2\alpha_1) \right.$$

$$- 2D[\sinh(2\alpha) - \omega_1 \tau_c \sin(2\Theta)]\}$$

$$\left. - \frac{A_-}{A_+^2} \{\sinh(2\alpha_1) - D[(\sinh(2\alpha) - 2\omega_1 \tau_c \sin(2\Theta)]\} \right) \tag{8.57}$$

where

$$\alpha = \tau/\tau_c; \quad \alpha_1 = \tau_1/\tau_c; \quad \beta = t_w/\tau_c; \quad A_\pm = 1 \pm 4\omega_1^2 \tau_c^2$$

and

$$D = \frac{\cosh(2\alpha_1) - 1}{\cosh(2\alpha) - \cos(2\Theta)}.$$

The following two limiting cases are immediately obtained:

(i) Ostroff-Waugh [12] and Mansfield [13] sequence (MW-2, MW-4) with $\Theta = \pi/2$ and δ-pulses ($\beta = 0$, $\alpha = \alpha_1$)

$$\frac{1}{T_{1e}} = M_2^{II}\,\tau\,\frac{1}{\alpha}\left[1 - \frac{\tanh\alpha}{\alpha}\right] \tag{8.58}$$

which is the same result as listed in Table 6.1 derived by Gründer et al. [9, 10].

(ii) Spin-locking case where $\tau_1 = 0$ ($2\alpha = \beta$), i.e.

$$\frac{1}{T_{1e}} = M_2^{II}\,\frac{\tau_c}{1 + 4\omega_1^2\tau_c^2} \tag{8.59}$$

which is identical to the results obtained by Jones [7] and Look and Lowe [6].

In the heteronuclear case, Rhim et al. [16] derived in the δ-pulse approximation

$$\frac{1}{T_{1e}} = M_2^{IS}\,\tau_c\left\{1 - \frac{\sinh 2\alpha(1 - \cos\Theta)}{2\alpha(\cosh 2\alpha - \cos\Theta)}\right\} \tag{8.60}$$

with $\Theta = \pi/2$ this leads to the same result obtained by Gründer et al. [9, 10] (see Table 8.1).

These expressions are useful, because they open the possibility of using small angles $\Theta \ll \pi/2$ to study relaxation processes in the slow motion regime.

8.3 Application of Multiple-Pulse Experiments to the Investigation of Spin-Lattice Relaxation

Roeder and Douglass [19] have measured $T_{1\rho}$ in cyclohexane (C_6H_{12}) at different temperatures using spin locking. The MW-4 sequence was applied to this substance by Gründer et al. [9, 10] and Fig. 8.2 represents the different T_{1e} data versus inverse temperature for different pulse spacing 2τ.

The T_{1e}-minima are expected according to Eq. (8.55) and Table 8.1 at

$$T_{1e\,\text{min}} = 3.8/M_2\tau \tag{8.61}$$

or with $\tau/\tau_c = \pi/2$

$$T_{1e\,\text{min}} = 7.6/(\pi M_2\tau_c). \tag{8.62}$$

Equation (8.57) allows an independent check of the theory, using no adjustable parameter. Using the value of $M_2 = 7.8 \cdot 10^8\,\text{s}^{-2}$ as calculated by Roeder and Douglass [19] the calculated values of $T_{2\,\text{min}}^*$ according to Eq. (8.57) are listed in the second column of Table 8.2 and are to be compared with the measured values as listed in the third column of the same table.

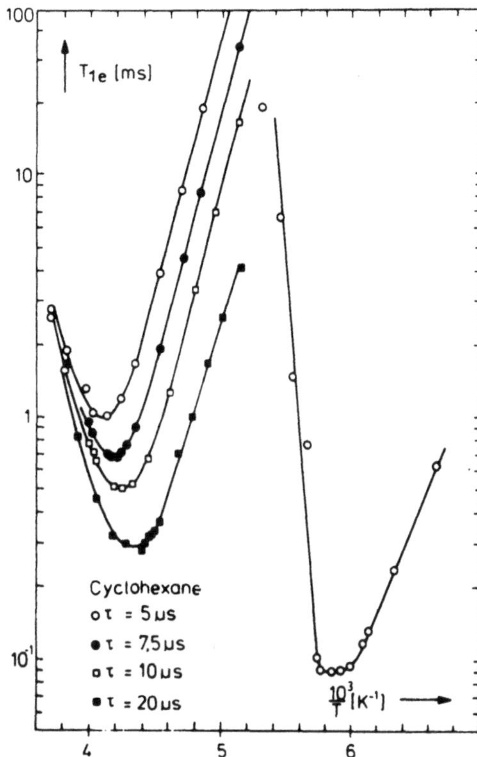

Fig. 8.2. Measured T_{1e} values in a MW-4 experiment on protons in cyclohexane versus inverse temperature for different pulse spacing τ, according to W. Gründer [10]

Table 8.2. Comparison of theoretical (Eq. (8.57), $M_2 = 7.8 \cdot 10^8\,s^{-2}$) and experimental $T_{1e\,min}$ values of cyclohexane for different values of the pulse spacing τ in a MW-4 experiment (see Fig. 8.2) according to Gründer [10]

$\tau(\mu s)$	$T_{1e\,min}$(theor.)	$T_{1e\,min}$(exptl.)
5	0.98	0.98
7.5	0.65	0.67
10	0.49	0.51
20	0.24	0.28

The agreement is quite convincing. The correlation time τ_c as determined by this method for varying temperature T is plotted in Fig. 8.3 on a logarithmic scale. The data can be fitted by an Arrhenius equation [10] to

$$\tau_c = 5.5 \cdot 10^{-15} \exp(9.6 \cdot 10^3/RT)$$

where the activation energy $\Delta E = 9.6\,kcal/Mol$ agrees well with the value of $\Delta E = 9.9\,kcal/Mol$ as determined by Roeder and Douglass [19] by means of a spin locking experiment. Further aspects, expecially the influence of pulse width are discussed by Gründer [10].

We are now going to discuss another interesting application of multiple-

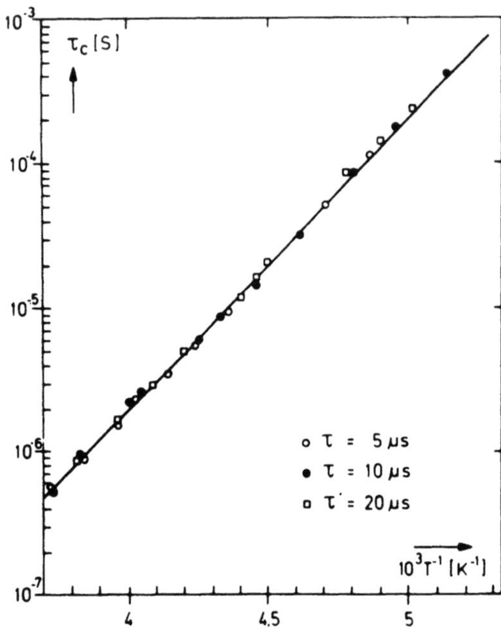

Fig. 8.3. Correlation time τ_c versus inverse temperature obtained from Fig. 8.2 by W. Gründer [10] for different pulse spacings in a MW-4 experiment

pulse experiments for the investigation of the orientation dependence of T_1. Suppose, the shielding tensor of a nuclear site is known, this knowledge may be exploited to monitor the orientation dependence of some nuclear quantity with respect to the static field in case of a powder sample. We have applied this method to ^{19}F in CF$_3$COOAg where the complete ^{19}F shielding tensor is known [20, 21]. At a high enough temperature this tensor is axially symmetric with the symmetry axis being the C$_3$-axis, which is also the "rotation" axis [21]. If we apply a π pulse to the spins a time τ prior to the application of a high resolution multiple-pulse burst, the spectrum observed after Fourier transformation of the multiple-pulse decay will be partially relaxed i.e., it will be inverted if $\tau \ll T_1$, it will be an ordinary spectrum if $\tau > T_1$ and it will be intermediate in between. This is demonstrated in Fig. 8.4, where the partially relaxed powder spectra of ^{19}F in a CF$_3$COOAg powder sample are shown for different values of τ.

By plotting the amplitude of the signal versus τ the relaxation function $R(\tau)$ can be obtained. This recovery signal will be in the case of exponential relaxation proportional to $1 - \exp(-\tau/T_1)$.

As clearly seen from Fig. 8.4 the relaxation is anisotropic, since for $\tau = 1.1$ s the unique axis component i.e., C$_3 \parallel B_0$ relaxes faster than the component with C$_3 \perp B_0$. The complete evaluation of the relaxation function $R(\tau)$ renders a non-exponential process which is most pronounced for C$_3 \parallel B_0$ i.e., $\beta = 0$. This has been demonstrated earlier in the case of a single crystal relaxation study. The interested reader is referred to Ref. [22].

Evaluating only the initial slope of the relaxation function $R(\tau)$ leads to the relaxation rate $1/T_1$, which is plotted versus the angle β of the C$_3$ axis with respect to the static field B_0 in Fig. 8.5 for two different temperatures. The

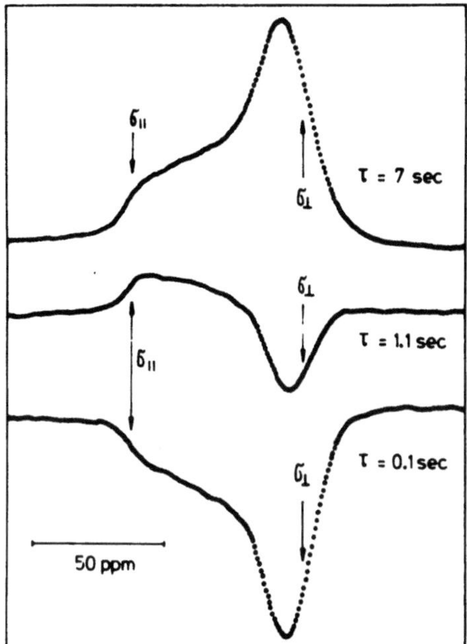

Fig. 8.4. Partially relaxed powder spectra of ^{19}F in CF$_3$COOAg at room temperature [20]. Notice, that the $\sigma_\parallel$ component relaxes faster than the $\sigma_\perp$ component, i.e., the spin lattice relaxation is much faster when the methyl C$_3$ axis is parallel to the static magnetic field B_0

temperatures are chosen so as to fulfill the conditions $\omega_0\tau_c \ll 1$ ($T=22°$ C) in one case and $\omega_0\tau_c \gg 1$, in the other case but with $\omega_c\tau_c \ll 1$ ($T=-85°$ C), where ω_c is the "cycle frequency". The relaxation of a 3-spin group has been calculated previously [23]

$$\frac{1}{T_1^{(3)}} = \frac{9}{16}\frac{\gamma^4\hbar^2}{r^6}\left[(1-\cos^4\beta)\frac{\tau_c}{1+\omega_0^2\tau_c^2}\right.$$
$$\left. + (1+6\cos^2\beta+\cos^4\beta)\frac{\tau_c}{1+(2\omega_0\tau_c)^2}\right]. \tag{8.63}$$

If we call the total relaxation rate $1/T_1 = 1/T_1^{(3)}+A$ to account for intermolecular relaxation and other types of relaxation mechanisms, such as spin-rotation interaction and shift anisotropies (those contributions can be shown to be less than 3 % [24]), we obtain

$$\frac{1}{T_1} = \frac{9}{16}T_0^{-1}(5+3\cos 2\beta)+A_1 \quad \text{for} \quad \omega_0\tau_c \ll 1 \tag{8.64}$$

and

$$\frac{1}{T_1} = \frac{9}{4}T_0^{-1}(\omega_0\tau_c)^{-2}(5+6\cos^2\beta-3\cos^4\beta)+A_2 \quad \text{for} \quad \omega_0\tau_c \gg 1 \tag{8.65}$$

where

$$T_0^{-1} = \gamma^4\hbar^2\tau_c/r^{-6}.$$

The solid lines in Fig. 8.5 are obtained by fitting Eqs. (8.64) and (8.65) to the experimental data. This results in $A_1 = 0.29\,\text{s}^{-1}$ and $\tau_c = 4.10^{-11}\,\text{s}$ at $T=22°$ C and $A_2 = 0.181\,\text{s}^{-1}$ and $\tau_c = 3.1\cdot 10^{-8}\,\text{s}$ at $T=-85\,°\text{C}$. The agreement between the experimental data and the solid lines in Fig. 8.5 is quite convincing.

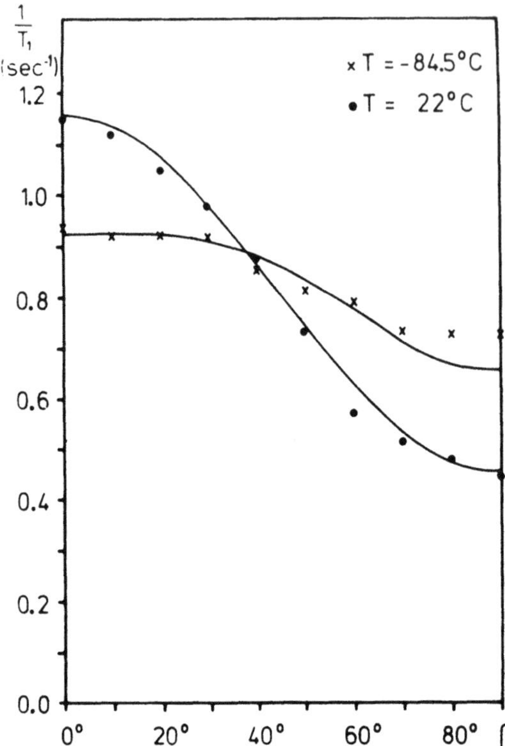

Fig. 8.5. Orientation dependence of the spin lattice relaxation rate $1/T_1$ of ^{19}F in CF_3COOAg for two different temperatures (above and below the T_1 minimum), obtained from powder samples by evaluating spectra as shown in Fig. 8.4 [20]. β is the angle between the C_3-axis of the methyl group and the magnetic field B_0. The solid lines correspond to theoretical expressions discussed in the text [Eqs. (8.64, 8.65)]

The example chosen is, however, a very fortitious one, because all the CF_3-groups in the sample are magnetically equivalent (see Fig. 8.6) i.e., the C_3 axes of all CF_3-groups are parallel [21]. If differently oriented CF_3-groups would be present in the unit cell, the whole beauty of this experiment would have been spoiled by spin diffusion. This is why there is so little orientation dependence of T_1 observed in methyl napthalene [25]. There is, of course, a way around this, by applying a locking field at the magic angle and thus quenching the spin diffusion process.

A multiple-pulse experiment, which serves the same goal, namely, the quenching of spin diffusion has been applied to the relaxation of ^{19}F spins by paramagnetic centers. Hartmann et al. [26] have applied a spin locking burst at the magic angle to serve the same goal. The relaxation of nuclear spins during the application of multiple-pulse experiments was studied earlier in this chapter, neglecting completely any contribution due to spin diffusion among different magnetically inequivalent nuclei. In the relaxation of nuclear spins via electronic paramagnetic centers, the process of spin diffusion becomes very important for remote nuclei. The direct relaxation rate of these nuclei is usually very small since $1/T_1 \sim r^{-6}$ where r is the distance to the paramagnetic center. General aspects of relaxation by paramagnetic centers may be found in Refs. [27, 28].

We only remark here, that "remote" nuclei usually communicate via spin diffusion (by the nuclear dipole reservoir or the electronic spin dipole reservoir

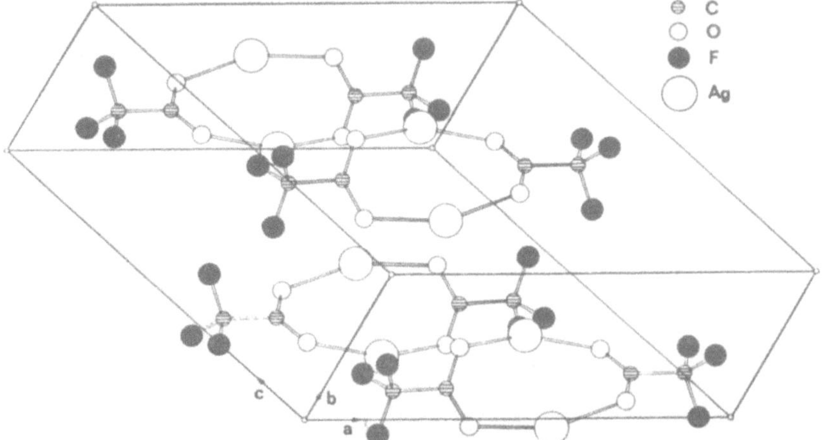

Fig. 8.6. Unit cell of CF_3COOAg according to the space group Cc. Internuclear distances and angles may be found in Ref. [21b]. (Courtesy of D. Suwelack)

[28]) with "closer" nuclei, resulting in a more or less exponential relaxation of the nuclei. It was realized, however, that in the spin diffusion vanishing case, the relaxation should follow $\exp[-(t/\tau_1)^{1/2}]$ [26, 27].

Suppose we have quenched spin diffusion by some sort of experiment, the relaxation rate of an individual nuclear spin i will be

$$\frac{1}{T_1^{(i)}} = C \cdot r_i^{-6}. \tag{8.66}$$

The total decay of the magnetization of the sample will follow

$$R(t) = \frac{1}{N} \sum_i \exp(-t/T_1^{(i)}) \tag{8.67}$$

excluding all the nuclei which are in a critical volume V_c around the paramagnetic center i.e., nuclei which are shifted by more than the resonance line width are excluded. The sum in Eq. (8.67) can be casted into an integral as

$$R(t) = \frac{1}{V} \int_{V_c}^{V} dV \exp(-t/T_1(r)). \tag{8.68}$$

Hartmann and co-workers [26] have evaluated this integral for different cases. All these cases may be expressed by [26–29]

$$R(t) = \exp[-(t/\tau_1)^{1/2} \{\psi_1(t) + \psi_2(t)\}] \tag{8.69a}$$

with

$$\omega_0 \tau_c \ll 1 : \frac{1}{\tau_1^{1/2}} = 1.3 \frac{2}{3} \pi^{3/2} N_p(\gamma_p \gamma_n \hbar)[S(S+1)\tau_c]^{1/2} \tag{8.69b}$$

and

$$\omega_0 \tau_c \gg 1 : \frac{1}{\tau_1^{1/2}} = \frac{16}{9\sqrt{3}} \pi^{3/2} N_p C_e^{1/2} \tag{8.69c}$$

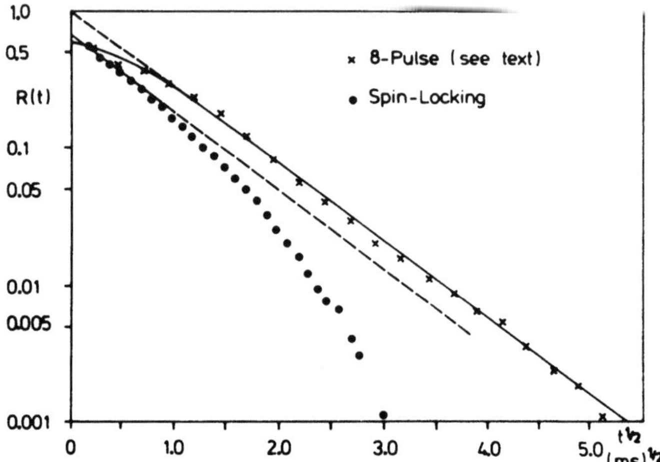

Fig. 8.7. Decay of magnetization of ^{19}F in CaF_2 (0.05 mol.-% Eu^{2+}) versus $t^{1/2}$. In the case of spin locking exponential relaxation is observed, whereas, in a multiple-pulse experiment (8-pulse), which quenches spin diffusion among the ^{19}F spins, the magnetization decays for large times as $\exp\left[-\left(\dfrac{t}{\tau_1}\right)^{1/2}\right]$. The solid line corresponds to Eq. (8.69) (see also Ref. [20])

where

$$C_e = \frac{2}{9} S(S+1)(\gamma_p \gamma_n \hbar)^2 \frac{\tau_c}{1+\omega_e^2 \tau_c^2}.$$

Here N_p is the number of electronic spins, S their quantum number and γ_p their gyromagnetic ratio, whereas γ_n is the gyromagnetic ratio of the nuclei. τ_c is the correlation time of the electronic spins.

$\psi_1(t)$ and $\psi_2(t)$ are "modifying functions" which depend on nuclear spin and electronic spin parameters. We remark, that $\psi_1(t) \to 0$ and $\psi_2(t) \to 1$ for $t \to \infty$. Plotting $\ln R(t)$ versus $t^{1/2}$ should therefore result in a straight line for large t. To demonstrate this behavior an experiment, which quenches spin diffusion was performed on ^{19}F in CaF_2 doped with 0.05 mol.-% Eu^{2+} (see Fig. 8.7) [20]. In order to eliminate the oscillations due to the resonance offset in an ordinary WHH-4 experiment, in every second WHH-4 cycle all phases were inverted, leading to an 8-pulse cycle. This eight pulse cycle has the property of quenching spin diffusion on one hand and eliminating resonance offset oscillations on the other hand. By shifting the spectrometer frequency by an amount of $\Delta\omega$ away from resonance, however, an effective field $B_{\text{eff}} = \frac{1}{\sqrt{3}} \Delta\omega/\gamma$ can be applied to the spins. Since the local field is virtually zero, extremely slow motions which are not accessible by ordinary $T_{1\rho}$ and T_{1D} measurements can be studied by this method.

Coming back to the relaxation by paramagnetic impurities (Fig. 8.7) we notice a linearity between $\ln R(t)$ versus $t^{1/2}$ over several decades for large t. The behavior of the relaxation function is completely different for the ordinary $T_{1\rho}$ as measured by spin locking. The interested reader is referred, for a further discussion, to Refs. [26–29].

8.4 Spin-Lattice Relaxation in Dilute Spin Systems

(i)   spin-lattice relaxation in the laboratory frame

A scheme for studying the spin lattice relaxation of dilute spins is drawn schematically in Fig. 8.8.

Suppose we are dealing with a powder sample of benzene (C_6H_6) and are observing the spectrum of the dilute ^{13}C in the sample by applying the CP experiment [31]. The observed shielding tensor is axially symmetric at the applied temperatures with the unique axis being the 6-fold rotation axis. We can therefore use the high resolution ^{13}C powder spectrum to monitor any anisotropic relaxation process such as spin lattice relaxation and the corresponding Overhauser effect [30].

Following along the lines of Gibby et al. [30] we assume that the dominant contribution to the relaxation of the ^{13}C spins (S spins) comes from the proton (I spins) directly bonded to it. The molecular motion is governed by random jumps about the 6-fold axis with a correlation time τ_c (Gauss-Markoff process). Coupled rate equations can be obtained for the I and S magnetization. Averaging over the six equivalent orientations of the benzene ring results in the following relaxation rates [30]

$$\frac{1}{T_1^{SS}} = \frac{3}{16} \frac{\gamma_S^2 \gamma_I^2 \hbar^2 I(I+1)}{r^6} \left[y_0(\beta) \frac{\tau_c}{1+(\omega_S-\omega_I)^2 \tau_c^2} + y_1(\beta) \frac{\tau_c}{1+\omega_S^2 \tau_c^2} \right.$$
$$\left. + y_2 \frac{\tau_c}{1+(\omega_S+\omega_I)^2 \tau_c^2} \right] \tag{8.70}$$

and

$$\frac{1}{T_1^{IS}} = \frac{3}{16} \frac{\gamma_S^2 \gamma_I^2 \hbar^2 S(S+1)}{r^6} \left[y_2(\beta) \frac{\tau_c}{1+(\omega_S+\omega_I)^2 \tau_c^2} \right.$$
$$\left. - y_0(\beta) \frac{\tau_c}{1+(\omega_S-\omega_I)^2 \tau_c^2} \right] \tag{8.71}$$

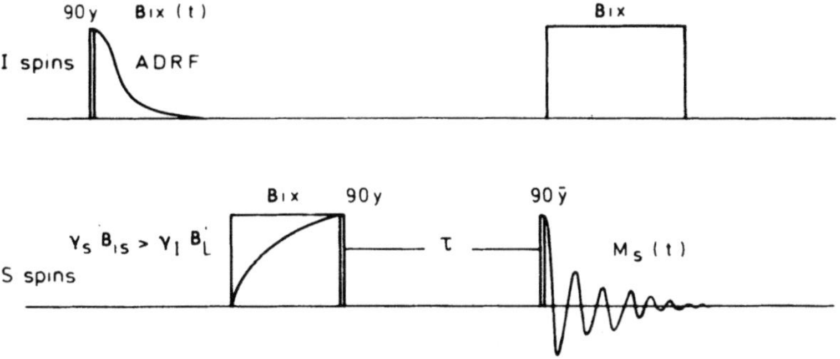

Fig. 8.8. Scheme for measuring spin lattice relaxation of dilute spins as proposed by Pines et al. [30, 31]

where

$$y_0(\beta) = \sin^4 \beta$$
$$y_1(\beta) = 2 \sin^2 \beta (1 + \cos^2 \beta)$$
$$y_2(\beta) = 1 + 6 \cos^2 \beta + \cos^4 \beta$$

and β is the angle between the C_6 axis and the static magnetic field B_0.

Two interesting special cases will be discussed

(i)   The S spins (^{13}C) will be disturbed e.g. by a π pulse which does not affect the $\langle I_z \rangle$ polarization. A relaxation of the S spins with T_1^{SS} will be observed,

(ii) The I spins (^{1}H) are saturated, $\langle I_z \rangle = 0$ and the S spin magnetization will reach a steady state with the time constant T_1^{SS}, leading to an Overhauser enhancement $1 + \eta$ in the S spin polarization, with

$$\eta = \frac{\gamma_I}{\gamma_S} \frac{T_1^{SS}}{T_1^{IS}}. \tag{8.72}$$

Since both T_1^{SS} and T_1^{IS} show strong angular dependence, we also expect η to vary strongly with the angle β. This angular dependence can be studied in a powder sample as demonstrated in Figs. 8.10 and 8.11. Instead of the angle β one may find it convenient to use the parameter

$$\cos^2 \beta = (\sigma - \sigma_\perp)/(\sigma_\parallel - \sigma_\perp) \tag{8.73}$$

to describe the angular dependence of T_1^{SS} and η. From Eq. (8.70) it is expected, that the anisotropy of T_1^{SS} changes sign by going from one side of the T_1 minimum to the other i.e., at high temperature the right hand side of the powder spectrum relaxes more rapidly than the left hand side, whereas at low temperatures the reverse is true. Evaluating partially relaxed powder spectra as in Fig. 8.9 leads to the T_1^{SS} anisotropy as shown in Fig. 8.10. The temperature dependence of the correlation time τ_c has been determined previously by deuteron relaxation as

$$\tau_c = 9.2 \cdot 10^{-15} \exp(-4.2 \cdot 10^3 / RT).$$

At $T = -80°$ C, which is above the T_1 minimum, a value of $\omega_S \tau_c = 0.075$ is predicted in excellent agreement with Fig. 8.10. A similar agreement can be obtained for the nuclear Overhauser effect for temperatures above and below the T_1 minimum as demonstrated in Fig. 8.11.

Thus the study of direct spin lattice relaxation anisotropies of dilute spins can give valuable information on the molecular processes involved. In a similar investigation it has been shown by GPW [30] in adamantane, that those ^{13}C nuclei which have two rather than one directly bound proton relax more rapidly than the others.

These relaxation anisotropies should not be confused, however, with anisotropies of the cross relaxation rates $1/T_{IS}$ which have been discussed in Chap. 4 and which can of course result in distorted line shapes too. Another distinction has to be made concerning line shape changes due to anisotropic molecular reorientation, which show up when $\Delta\omega\tau_c \cong 1$, where $\Delta\omega$ is the size of the anisotropy (see Sect. 2.8).

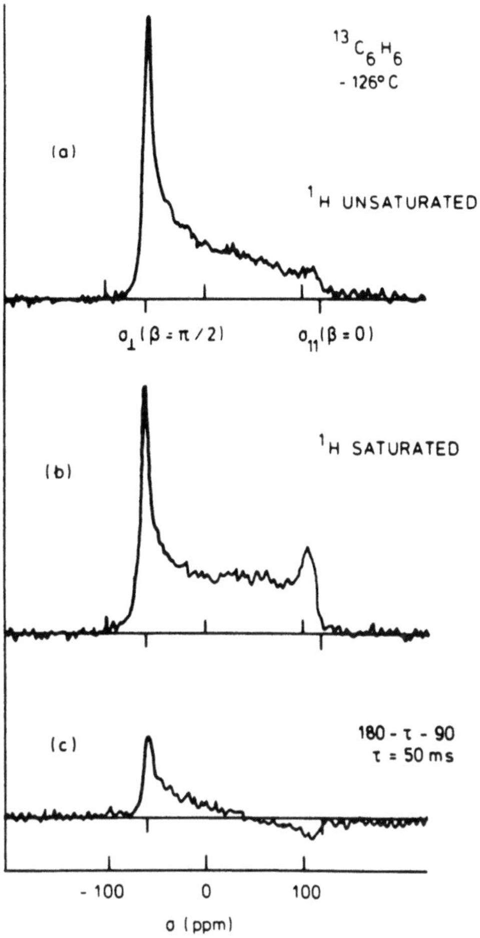

Fig. 8.9a–c. Relaxation anisotropy of ^{13}C spins due to anisotropic molecular motion in solid benzene according to Gibby, Pines and Waugh [30]. **a** Proton decoupled ^{13}C spectrum of benzene with 4 % abundance of ^{13}C. **b** Same as **a** but with a 1 G saturation field applied at the protons, demonstrating the anisotropic nuclear Overhauser enhancement. The powder pattern shown is expected from theory and depends strongly on temperature (see also Fig. 8.11). **c** Anisotropic T_1 relaxation after a recovery time of 50 ms. Molecules with $\beta = \frac{\pi}{2}$ $(\sigma = \sigma_\perp)$ show a seven times faster relaxation than those with $\beta = 0$ $(\sigma = \sigma_\parallel)$ (see also Fig. 8.10)

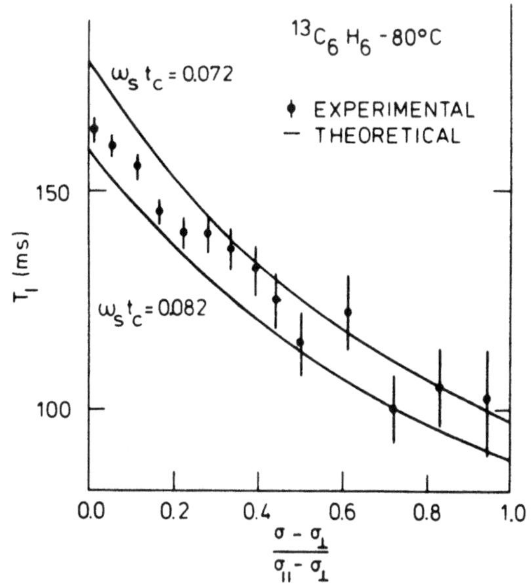

Fig. 8.10. Orientation dependence of the spin lattice relaxation time T_1 of ^{13}C in benzene according to Gibby, Pines and Waugh [30]. The experimental points were obtained by the analysis of spectra like in Fig. 8.9c. The theoretical curves correspond to Eq. (8.70) using $r = 1.085$ Å

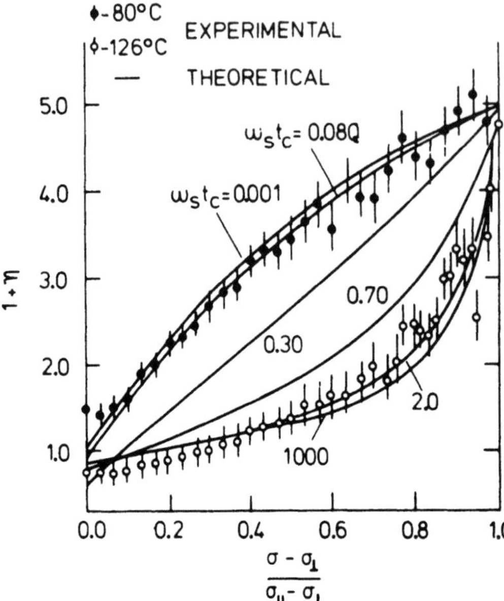

Fig. 8.11. Orientation dependence of the ^{13}C nuclear Overhauser enhancement $1+\eta$ in benzene according to Gibby, Pines and Waugh [30] for two different temperatures. The experimental points were obtained from spectra like Fig. 8.9b, whereas the theoretical curves are derived from Eqs. (8.70–8.73)

(ii) spin-lattice relaxation in the rotating frame

If slow motion governs the molecular dynamics spin-lattice relaxation in the rotating frame T_{1x} is more sensitive to the lattice motion than T_1 [32–34]. By T_{1x} we label the usual $T_{1\rho}$ relaxation in the rf field $\omega_1 = \gamma \cdot B_1$ as well as the relaxation time of dipolar order T_{1D} ($\omega_1 = 0$). Different pulse schemes for observing these relaxation processes have been used [32–34] and some are shown in Fig. 8.12.

Let us discuss the relaxation processes involved in the standard CP experiment shown in Fig. 8.12a. The equation of motion for the inverse temperatures β_I and β_S of the I and S spins was already discussed in Chap. 4. The general solution of Eq. (4.66) can be expressed as (see also [32a])

$$\beta_S(t) = e^{-a_- t/T_{IS}}[(\varepsilon\alpha^2 + T_{IS}/T_{1x}^{(I)} - a_-)\beta_{S0} + \beta_{I0}]/(a_+ - a_-)$$
$$- e^{-a_+ t/T_{IS}}[(\varepsilon\alpha^2 + T_{IS}/T_{1x}^{(I)} - a_+)\beta_{S0} + \beta_{I0}]/(a_+ - a_-) \qquad (8.74a)$$

and

$$\beta_I(t) = e^{-a_- t/T_{IS}}[(1 + T_{IS}/T_{1x}^{(S)} - a_-)\beta_{I0} + \varepsilon\alpha^2 \beta_{S0}]/(a_+ - a_-)$$
$$- e^{-a_+ t/T_{IS}}[(1 + T_{IS}/T_{1x}^{(S)} - a_+)\beta_{I0} + \varepsilon\alpha^2 \beta_{S0}]/(a_+ - a_-) \qquad (8.74b)$$

where $\beta_{I0} = \beta_I(t=0)$ and $\beta_{S0} = \beta_S(t=0)$ are the initial conditions and T_{IS} is the transfer time between I and S spins. The individual relaxation times of I and S spins in the rotating frame are labelled $T_{1x}^{(I)}$ and $T_{1x}^{(S)}$ respectively.

Equation (8.74) holds also in the I spin ADRF case the only difference being that T_{IS} differs from the one in the spin-locking case and $T_{1x}^{(I)} = T_{1D}^{(I)}$.

The other parameters are the same as defined in Chap. 4, namely

$$a_\pm = a_0[1 \pm (1 - b/a_0^2)^{1/2}] \qquad (8.75a)$$

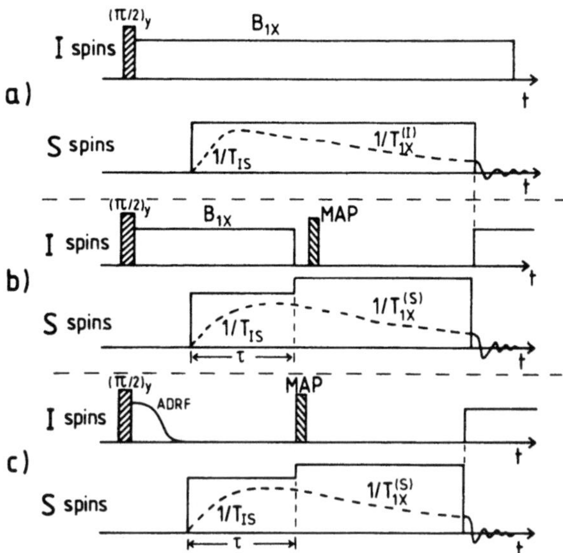

Fig. 8.12a–c. Pulse schemes for measuring spin-lattice relaxation of dilute spins (S) in the rotating frame. Cross-polarization occurs during time τ with the rate $1/T_{IS}$. In the ordinary CP experiment (**a**) the S spin magnetization relaxes with $T_{1x}^{(I)}$ which is mainly governed by $T_{1\rho}$ of the I spins. Pulse schemes (**b**) and (**c**) allow $T_{1x}^{(S)}$ to be measured if $T_{IS} \gg T_{1x}^{(S)}$ which may be achieved by making $\gamma_S B_{1S} = \omega_{1S} \gg \omega_{LI}$ (local field of the I spins). MAP is the Emid [35] magic angle pulse which eliminates dipolar order

with
$$a_0 = \tfrac{1}{2}(1 + \varepsilon\alpha^2 + T_{IS}/T_{1x}^{(I)} + T_{IS}/T_{1x}^{(S)}) \tag{8.75b}$$
and
$$b = \frac{T_{IS}}{T_{1x}^{(I)}} \left(1 + \frac{T_{IS}}{T_{1x}^{(S)}}\right) + \varepsilon\alpha^2 \frac{T_{IS}}{T_{1x}^{(S)}}. \tag{8.75c}$$

The Hartmann-Hahn mismatch parameter $\alpha = \omega_{eS}/\omega_{eI}$ is usually of the order one and $\varepsilon = (N_S S(S+1))/(N_I I(I+1))$ is usually very small (about 10^{-2}).

With the initial condition $\beta_{S0} = 0$ as in the standard CP experiment $\beta_S(t)$ rises to a maximum and decays again for longer times as was shown in Sect. 4.3. Both spin systems (I and S) are still coupled (T_{IS}) and decay jointly as

$$t \gg T_{IS}: \quad \beta_S(t) \cong \beta_I(t) \sim e^{-t/T_{1x}}$$

where

$$\frac{1}{T_{1x}} = \frac{1}{2} \frac{b}{a_0 T_{IS}}$$

which can be derived from Eq. (8.74) under the condition $T_{IS} \ll T_{1x}^{(I)}, T_{1x}^{(S)}$.

By inserting the expressions for a_0 and b from Eq. (8.75) one arrives at $T_{1x} \cong T_{1x}^{(I)}$; i.e. the S spin decay is governed by the I spin relaxation.

In order to measure $T_{1x}^{(S)}$; i.e. the S spin relaxation in the rotating frame the pulse schemes of Fig. 8.12(b) and (c) have to be used. There I and S spins are decoupled during relaxation by a strong rf field applied to the S spins ($\omega_{eS} \gg \omega_{LI}$). Moreover, any leftover dipolar order of the I spins is eliminated by a magic angle pulse (MAP [35]). This way we invoke the initial condition β_{I0} $=0$ when the relaxation decay begins. Under the condition $T_{IS} \gg T_{1x}^{(I)}$, $T_{1x}^{(S)}$. Equation (8.62) can be brought into the form

$$\beta_S(t) = \beta_{S0}\, e^{-t/T_{1x}^{(S)}} + \beta_I(t) \tag{8.76a}$$

and

$$\beta_I(t) = \beta_{S0}\, \varepsilon\alpha^2 (e^{-t/T_{1x}^{(S)}} - e^{-t/T_{1x}^{(I)}})/(T_{IS}/T_{1x}^{(I)} - T_{IS}/T_{1x}^{(S)}). \tag{8.76b}$$

If $T_{1x}^{(S)} = T_{1x}^{(I)}$, $\beta_I(t)$ vanishes. In general, however, the β_I-correction to $\beta_S(t)$ has to be taken into account. If T_{IS} is still much larger than $T_{1x}^{(I)}$, $T_{1x}^{(S)}$ this correction will be minute. In the more general case where $T_{IS} \cong T_{1x}^{(I)}$, $T_{1x}^{(S)}$ however, the time evolution is quite complicated and corresponds to the general solution (Eq. (8.74)) with the initial condition $\beta_I = 0$. The decay will be bi-exponential in this case. In fact for slow spin-lattice relaxation $T_{1x}^{(S)} \ll T_{IS}$ the decay of S spin magnetization will be entirely governed by the cross-relaxation (T_{IS}). This will be especially the case when ω_{es} is not much larger than ω_{LI}. An estimate on how large ω_{es} should be can be obtained from the approximate expression for T_{IS} in the ADRF case (Eq. (4.105b))

$$\frac{1}{T_{IS}} = \frac{1}{2}\sin^2\vartheta_S\, M_2^{IS}\, \tau_c \exp(-\omega_{es}\tau_c)$$

where M_2^{IS} is the dipolar second moment of the $I\text{-}S$ interaction and τ_L is the flip-flop correlation time (on the order of 10^{-4} s for most $^{13}C\text{-}^1H$ systems) as discussed in Sect. 4.3.

If CP-NMR is combined with magic angle sample spinning (MAS) another contribution to relaxation comes from the interference of lattice motion with the sample rotation. This amounts to a linebroadening mechanism or T_2 process, which is most pronounced, when $\omega_r\tau \cong 1$, where τ is the correlation time of the lattice motion and ω_r is the sample rotation frequency [36]. A similar interference effect results between the decoupling field ω_1 and lattice motion when $\omega_1\tau \cong 1$ [32a].

8.5 Selective Excitation and Spectral Diffusion

In the previous sections we have always prepared the initial state to be homogeneous over the spectral distribution of the spins, i.e. the saturation pulses or the π-pulse in a T_1-experiment are always assumed to cover the full spectral width. The relaxation of the initial state towards thermal equilibrium is usually exponential in this case. In this section we want to discuss the reverse situation where the preparation of the initial state occurs selectively, i.e. at a particular frequency affecting only the spin packet at that frequency and leaving other spins unaffected. Consider the case of a multi-line spectrum e.g. in the high resolution spectrum of a single crystal. Saturation of a particular line and its relaxation may depend on the other lines in the spectrum, if they are connected with the saturated line, e.g. by spin interactions or lattice motion (heterogeneous line, see Appendix D). In this case a non-exponential relaxation or better multi-exponential relaxation will occur.

In order to be specific we treat the case of ^{125}Te atomic motion in tellurium. Günther et al. [37] have shown recently that ultra slow atomic motion can be detected by observing the multi-exponential spin-lattice relaxation after selective excitation. There are three magnetically non-equivalent sites in the unit cell of tellurium, resulting in three different lines in the ^{125}Te spectrum. No special technique for resolving these lines is necessary since the chemical shift anisotropies are on the order of 1000 ppm and the dipole-dipole interactions are weak. Let us suppose, that we irradiate selectively only, e.g. line 1 of the spectrum. Direct as well as indirect phonon processes, and contributions from conduction electrons will cause a "local" relaxation rate $1/T_1$ [37]. Moreover, at sufficiently large temperatures hopping of the nuclei via translational atomic diffusion occurs with a mean rate κ. The hopping process does reduce the magnetization $m_{z\,1}(t)$ of the line 1 since magnetization is carried from line 1 to lines 2 and 3. The following set of differential equations results [37]

$$\frac{d}{dt}\begin{pmatrix} m_{z\,1}(t) \\ m_{z\,2}(t) \\ m_{z\,3}(t) \end{pmatrix} = \begin{pmatrix} -\dfrac{1}{T_1}-2\kappa & \kappa & \kappa \\ \kappa & -\dfrac{1}{T_1}-2\kappa & \kappa \\ \kappa & \kappa & -\dfrac{1}{T_1}-2\kappa \end{pmatrix} \begin{pmatrix} m_{z\,1}(t) \\ m_{z\,2}(t) \\ m_{z\,3}(t) \end{pmatrix} \tag{8.77}$$

with the solution

$$\begin{pmatrix} m_{z\,1}(t) \\ m_{z\,2}(t) \\ m_{z\,3}(t) \end{pmatrix} = \exp(-t/T_1)\begin{pmatrix} a & b & b \\ b & a & b \\ b & b & a \end{pmatrix} \begin{pmatrix} m_{z\,1}(0) \\ m_{z\,2}(0) \\ m_{z\,3}(0) \end{pmatrix} \tag{8.78}$$

where

$$a=\tfrac{1}{3}(1+2e^{-3\kappa t}) \quad \text{and} \quad b=\tfrac{1}{3}(1-e^{-3\kappa t}).$$

Two limiting cases may be considered:

(i)   All spectral lines are prepared in the same initial state, e.g. $m_{z\,1}(0)=m_{z\,2}(0)$ $=m_{z\,3}(0)=1$, which leads to

$$m_{z\,j}(t)=\exp(-t/T_1)$$
$$j=1,2,3. \tag{8.79}$$

(ii) Selective excitation of line 1, e.g.

$$m_{z\,1}(0)=1; \quad m_{z\,2}(0)=m_{z\,3}(0)=0$$

leads to non-exponential decay as

$$m_{z\,1}(t)=\exp(-t/T_1)\,[1+2\exp(-3\kappa t)]/3 \tag{8.80a}$$

$$m_{z\,2}(t)=m_{z\,3}(t)=\exp(-t/T_1)\,[1-\exp(-3\kappa t)]/3. \tag{8.80b}$$

It is obvious, that the final magnetization is 1/3 of the initial magnetization if $1/T_1=0$. In the case of slow motion $(1/T_1\ll\kappa\ll1/T_2)$ this non-exponential relaxation behaviour is most pronounced and the mean hopping rate κ can be determined directly from the experimental data without knowledge of the different kinds of spin-interactions involved. Figure 8.13 shows as an example the timedependent magnetization $m_{z\,1}(t)$ for ^{125}Te in a tellurium single crystal

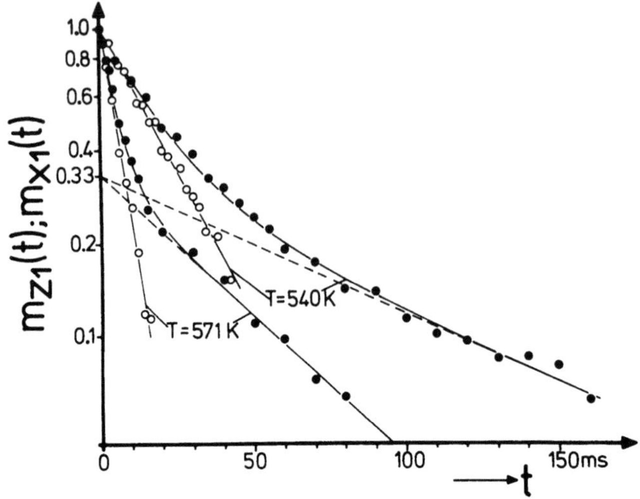

Fig. 8.13. Non-exponential decay of the magnetization $m_{z1}(t)$ (●●●) and $m_{x2}(t)$ (○○○) of ^{125}Te in tellurium at two different temperatures according to Günther et al. [37]

for two different temperatures. Note, that irradiation of all three lines, i.e. non-selective excitation leads to the usual exponential behaviour governed by the background relaxation $1/T_1$. This relaxation is caused in the present case by charged vacancies. The relaxation rate due to the fluctuating chemical shift tensor with rate κ, however, is very small, i.e. about 10^{-6} for a chemical shift anisotropy of about 1000 ppm.

It is readily seen that the rotating frame magnetization $m_{xj}(t)$ $(j=1,2,3)$ in a spin-locking experiment, however, decays exponentially. Only spins leaving the spectral line j with a rate 2κ lead to a destruction of m_{xj} whereas spins entering line j do not contribute since they appear with a random phase and cannot be locked in the locking-field B_1. The same argument holds, by the way in spin echo experiments. Starting from a similar differential equation as before we arrive at [37]

$$m_{xj}(t) = \exp[-t(1/T_{1\rho} + 2\kappa)]\, m_{xj}(0) \tag{8.81}$$

leading to an exponential decay with decay rate $1/T_{1\rho} + 2\kappa$. This behaviour is also shown in Fig. 8.13.

The cross-relaxation rates $1/T_{1CR} = 3\kappa$ and $1/T_{1\rho CR} = 2\kappa$ according to Eqs. (8.80) and (8.81) were determined by Günther et al. [37] for ^{125}Te in a tellurium single crystal from selective excitation experiments and are plotted in Fig. 8.14, versus temperature. The data in Fig. 8.14 illustrate the proposed ratio $T_{1CR}/T_{1\rho CR} = 2/3$. Note, that hopping rates of less than $1\,\mathrm{s}^{-1}$ can be detected. The connection of the hopping rates κ with the atomic diffusion constant in comparison with tracer experiments is discussed in more detail by Günther et al. [37].

Extension of the method of selective excitation outlined here to other systems with more hopping sites is straightforward. It opens the possibility to investigate ultra slow motion directly resorting to spin dynamics. Consider e.g.

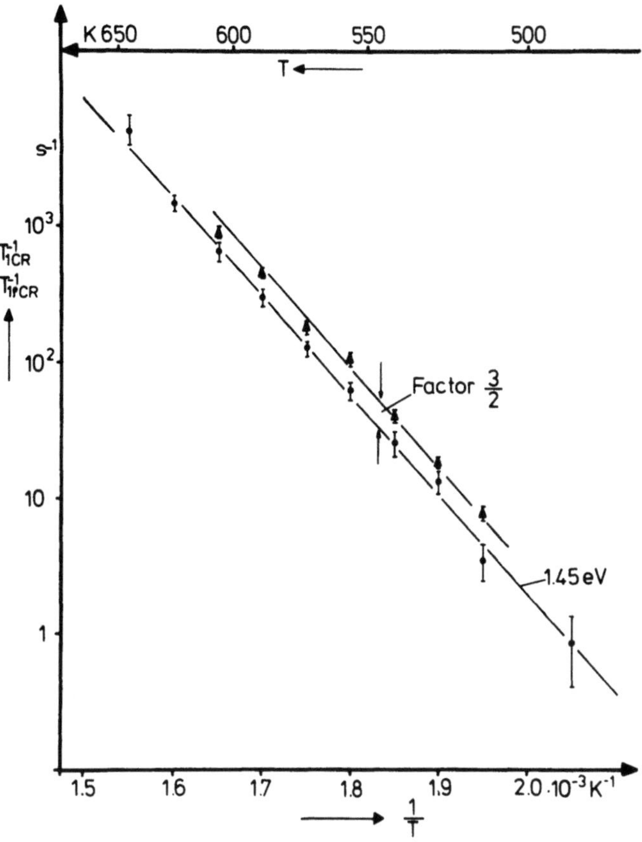

Fig. 8.14. Cross-relaxation rates $T^{-1}_{1CR}=3\kappa$ (▲▲▲) and $T^{-1}_{1\rho CR}=2\kappa$ (●●●) obtained from the initial slope of the $m_{z1}(t)$ and $m_{x1}(t)$ curves, respectively, versus inverse temperature [37]. An activation energy of 1.45 eV is determined from the slope. Note the theoretically expected 3/2 ratio of the two rates

molecular rotation in solids where a high resolution NMR spectrum can be obtained by the methods discussed in the preceding chapters. In the ultra slow motional regime at low temperatures the background relaxation might be minutes or hours allowing hopping rates down to that order of magnitude to be determined. In fact this approach is related to what is called "saturation transfer" [38] proposed in *ESR* and to a recently proposed 2D-NMR scheme by Jeener et al. [39] applied to site exchange.

An extension of Eqs. (8.77) and (8.78) to an N-site exchange problem is trivial and results in

$$\mathbf{m}_z(t)=\exp(-t/T_1)\,\tilde{\mathbf{M}}\mathbf{m}_z(0) \tag{8.82}$$

where $\mathbf{m}_z(t)$ is an N-dimensional column vector and the $N\times N$ matrix $\tilde{\mathbf{M}}$ contains diagonal elements a and off-diagonal elements b similar to Eq. (8.78) with

$$a=\frac{1}{N}[1+(N-1)\,e^{-N\kappa t}]; \quad b=\frac{1}{N}(1-e^{-N\kappa t}). \tag{8.83}$$

9 Appendix

A Irreducible Tensor Representation of Spin Interactions

Although Cartesian tensors lead to a direct physical apprehension in terms of direction cosines it might be more convenient to express tensors in a spherical basis, when rotations are involved. We therefore introduce a spherical representation of second rank tensors here.

The nine components T_{ij} $(i,j=x,y,z)$ of a Cartesian tensor of second rank can be decomposed into a [1]

scalar

$$\mathbf{T}_0 = \tfrac{1}{3}\mathrm{Tr}\{T_{ij}\} = \tfrac{1}{3}\sum_i T_{ii} \tag{A.1}$$

an *antisymmetric* tensor of first rank

$$\mathbf{T}_1: \ T'_{ij} = \tfrac{1}{2}(T_{ij} - T_{ji}) \tag{A.2}$$

with three components and zero trace, and into a tracless second rank tensor, which is

symmetric

$$\mathbf{T}_2: \ T''_{ij} = \tfrac{1}{2}(T_{ij} + T_{ji}) - \tfrac{1}{3}\mathrm{Tr}\{T_{ij}\} \tag{A.3}$$

with five components. This leads to a sum of three irreducible tensors as

$$T_{ij} = \tfrac{1}{3}\mathrm{Tr}\{T_{ij}\} + T'_{ij} + T''_{ij}. \tag{A.4}$$

The components of the three quantities $\mathbf{T}_0$, $\mathbf{T}_1$, and $\mathbf{T}_2$ transform in the same way as the spherical harmonics of order zero, one and two, respectively.

When tensors are to be expressed in a new coordinate system (primed axes) obtained by a rotation of the old (unprimed) coordinate system about the Euler angles (α, β, γ) it is found convenient to use a spherical representation. In this spherical representation the irreducible spherical tensor $\mathbf{T}_k$ of rank k with $2k+1$ components $\mathbf{T}_{kq}$ transforms according to the irreducible representation D_k of the rotation group

$$T'_{kq} = \mathbf{R}(\alpha\beta\gamma)\, T_{kq}\, \mathbf{R}^{-1}(\alpha\beta\gamma) = \sum_{p=-k}^{+k} T_{kp}\, D^k_{pq}(\alpha\beta\gamma). \tag{A.5}$$

With the spherical unit vectors

$$\mathbf{e}_{10} = \mathbf{e}_z; \ \mathbf{e}_{1\pm 1} = \mp(1/\sqrt{2})(\mathbf{e}_x \pm i\mathbf{e}_y) \tag{A.6}$$

the components of a first rank irreducible tensor T_{1q} in terms of the Cartesian components (T_x, T_y, T_z) are given by [2]

$$T_{1q} = \mathbf{e}_{1q} \cdot \mathbf{T} = \mathbf{e}_{iq} \cdot (\mathbf{e}_x \, T_x + \mathbf{e}_y \, T_y + \mathbf{e}_z \, T_z) \tag{A.7}$$

or

$$T_{10} = T_z; \quad T_{1 \pm 1} = \mp (1/\sqrt{2})(T_x \pm i T_y). \tag{A.8}$$

A second rank tensor is decomposed into the three irreducible spherical tensors in a similar fashion. First we have to find the corresponding spherical unit vectors $\mathbf{e}_{kq}$. These may be contructed from the rule for the product of two irreducible tensor operators [1, 2]:

$$\mathbf{e}_{kq} = [\mathbf{e}_1 \times \mathbf{e}_1]_{kq} = (2k+1)^{1/2} \sum_{q_1 q_2} (-1)^q \begin{pmatrix} 1 & 1 & k \\ q_1 & q_2 & -q \end{pmatrix} \mathbf{e}_{1q_1} \mathbf{e}_{1q_2}$$
$$(k = 0, 1, 2). \tag{A.9}$$

The irreducible spherical tensors in terms of the Cartesian tensors are obtained by

$$T_{kq} = \mathbf{e}_{kq} \cdot \mathbf{T} = (2k+1)^{1/2} \sum_{q_1 q_2} (-1)^q \begin{pmatrix} 1 & 1 & k \\ q_1 & q_2 & -q \end{pmatrix} T_{1q_1} T_{1q_2} \tag{A.10}$$

where

$$\mathbf{T} = \sum_{ij} \mathbf{e}_i \mathbf{e}_j T_{ij} \quad (i, j = x, y, z) \tag{A.11}$$

as [2]

$$\begin{aligned}
T_{00} &= -(1/\sqrt{3})\,[T_{xx} + T_{yy} + T_{zz}] \\
T_{10} &= -(i/\sqrt{2})\,[T_{xy} - T_{yx}] \\
T_{1 \pm 1} &= -\tfrac{1}{2}[T_{zx} - T_{xz} \pm i(T_{zy} - T_{yz})] \\
T_{20} &= (1/\sqrt{6})\,[3 T_{zz} - (T_{xx} + T_{yy} + T_{zz})] \\
T_{2 \pm 1} &= \mp \tfrac{1}{2}[T_{xz} + T_{zx} \pm i(T_{yz} + T_{zy})] \\
T_{2 \pm 2} &= \tfrac{1}{2}[T_{xx} - T_{yy} \pm i(T_{xy} + T_{yx})].
\end{aligned} \tag{A.12}$$

In cases where the Cartesian tensor is symmetric and traceless only the second rank irreducible tensor is non-vanishing, e.g. in the case of dipole-dipole interaction and quadrupole interaction.

Hamiltonians are usually expressible as scalar products of tensors. The scalar product of two irreducible tensors $\mathbf{A}_k$ and $\mathbf{T}_k$ with the components A_{kq} and T_{kp}, respectively, is defined as [1, 2]

$$\mathbf{A}_k \cdot \mathbf{T}_k = \sum_{q=-k}^{+k} (-1)^q A_{kq} T_{k-q} = \sum_{q=-k}^{+k} (-1)^q A_{k-q} T_{kq}. \tag{A.13}$$

In Sect. 2.1 we had expressed spin interactions by second rank tensors as

$$\mathcal{H} = \mathbf{X} \cdot \tilde{\mathbf{A}} \cdot \mathbf{Y} = \sum_{i, j} A_{ij} X_i Y_j \tag{A.14}$$

where $\mathbf{X}$ and $\mathbf{Y}$ are vectors and $\tilde{\mathbf{A}}$ is a matrix.

In terms of irreducible spherical tensors, we may express the spin interaction Hamiltonian as

$$\mathcal{H} = \sum_{k=0}^{2} \sum_{q=-k}^{+k} (-1)^q A_{kq} T_{k-q} \tag{A.15}$$

where the spherical tensor components A_{kq} and T_{kq} can be expressed by the Cartesian tensor components A_{ij} and T_{ij} according to Eq. (A.12).

Note, however, that in order to form T_{ij} as a dyadic product from X_j and Y_i their position has to be exchanged, i.e.

$$T_{ij} = Y_i X_j \tag{A.16}$$

Applying this to the shift interaction $S_{ij} = \gamma_I \sigma_{ij}$ yields:

$$\mathcal{H}_S = \mathbf{I} \cdot \mathbf{S} \cdot \mathbf{B}_0 = \sum_{i,j} S_{ij} I_i B_{0j} \qquad (i,j = x, y, z)$$

or

$$\mathcal{H}_S = \sum_{k=0}^{2} \sum_{q=-k}^{+k} (-1)^q A_{kq} T_{k-q}$$

where

$$\begin{aligned}
A_{00} &= -(1/\sqrt{3})[S_{xx} + S_{yy} + S_{zz}] = -(1/\sqrt{3}) \operatorname{Tr}\{S_{ij}\}\\
A_{10} &= -(i/\sqrt{2})[S_{xy} - S_{yx}]\\
A_{1\pm1} &= -\tfrac{1}{2}[S_{zx} - S_{xz} \pm i(S_{zy} - S_{yz})]\\
A_{20} &= (1/\sqrt{6})[3S_{zz} - \operatorname{Tr}\{S_{ij}\}]\\
A_{2\pm1} &= \mp\tfrac{1}{2}[S_{xz} + S_{zx} \pm i(S_{yz} + S_{zy})]\\
A_{2\pm2} &= \tfrac{1}{2}[S_{xx} - S_{yy} \pm i(S_{xy} + S_{yx})]
\end{aligned} \tag{A.17}$$

and by obeying $T_{ij} = B_{0i} T_j$

$$\begin{aligned}
T_{00} &= -(1/\sqrt{3})[I_x B_{0x} + I_y B_{0y} + I_z B_{0z}]\\
T_{10} &= (i/\sqrt{2})[I_x B_{0y} - I_y B_{0x}]\\
T_{1\pm1} &= \tfrac{1}{2}[I_z B_{0x} - I_x B_{0z} \pm i(I_z B_{0y} - I_y B_{0z})]\\
T_{20} &= (1/\sqrt{6})[3 I_z B_{0z} - (I_x B_{0x} + I_y B_{0y} + I_z B_{0z})]\\
T_{2\pm1} &= \mp\tfrac{1}{2}[I_x B_{0z} + I_z B_{0x} \pm i(I_y B_{0z} + I_z B_{0y})]\\
T_{2\pm2} &= \tfrac{1}{2}[I_x B_{0x} - I_y B_{0y} \pm i(I_x B_{0y} + I_y B_{0x})].
\end{aligned} \tag{A.18}$$

This is readily simplified with $\mathbf{B}_0 = (0, 0, B_0)$ as

$$\begin{aligned}
T_{00} &= -(1/\sqrt{3}) I_z B_0\\
T_{10} &= 0; \quad T_{1\pm1} = -\tfrac{1}{2}(I_x \pm iI_y) B_0\\
T_{20} &= (2/\sqrt{6}) I_z B_0\\
T_{2\pm1} &= \mp\tfrac{1}{2}(I_x \pm iI_y) B_0\\
T_{2\pm2} &= 0
\end{aligned} \tag{A.19}$$

from which follows

$$\mathcal{H}_S = A_{00} T_{00} - (A_{11} T_{1-1} + A_{1-1} T_{11}) + A_{20} T_{20}$$
$$- (A_{2-1} T_{21} + A_{21} T_{2-1}). \tag{A.20}$$

In case S_{ij} is symmetric i.e., $S_{ij} = S_{ji}$, in addition $A_{1\pm1}$ vanishes and we are left with

$$\mathcal{H}_S = A_{00} T_0 + A_{20} T_{20} - (A_{2-1} T_{21} + A_{21} T_{2-1}) \tag{A.21}$$

where A_{00} and A_{20} are given by Eq. (A.17) and $A_{2\pm1}$ is further simplified

$$A_{2\pm1} = \mp (S_{xz} \pm iS_{yz}). \tag{A.22}$$

When rotations are applied to the symmetric part of $\mathcal{H}_S$ in real space, only the components A_{2q} have to be transformed. The separation of the shielding Hamiltonian $\mathcal{H}_S$ into the second rank spherical tensor products is not as convenient as it appears to be, since rotations are usually performed, whether in real space or in spin space. For rotations in real space the former representation is fine, but rotations in spin space cannot be applied to the T_{kq}, since they also contain B_0.

We find it therefore more convenient to separate strictly geometrical and spin variables

$$\mathcal{H}_S = \mathbf{I} \cdot \tilde{\mathbf{S}} \cdot \mathbf{B}_0 = \mathbf{I} \cdot \mathbf{B}_S \tag{A.23}$$

where

$$\mathbf{B}_S = \tilde{\mathbf{S}} \cdot \mathbf{B}_0.$$

It follows

$$\mathcal{H}_S = \sum_{q=-1}^{1} (-1)^q A_{1q} T_{1-q} \tag{A.24}$$

where

$$A_{10} = B_{Sz}; \quad A_{1\pm1} = \mp (1/\sqrt{2})(B_{Sx} + iB_{Sy}) \tag{A.25}$$

and

$$T_{10} = I_z; \quad T_{1\pm1} = \mp (1/\sqrt{2})(I_x \pm iI_y). \tag{A.26}$$

With $\mathbf{B}_0 = (0, 0, B_0)$ we obtain further

$$B_{Sx} = S_{xz} B_0; \quad B_{Sy} = S_{yz} B_0; \quad B_{Sz} = S_{zz} B_0$$

or

$$\mathcal{H}_S = \gamma_I B_0 (\sigma_{xz} I_x + \sigma_{yz} I_y + \sigma_{zz} I_z). \tag{A.27}$$

Spin-spin interactions which are bilinear in the spin variable are conveniently expressed by irreducible spherical tensor products as

$$\mathcal{H}_{IS} = \mathbf{I} \cdot \tilde{\mathbf{D}} \cdot \mathbf{S} = \sum_{ij} D_{ij} I_i S_j$$

$$\mathcal{H}_{IS} = \sum_{k=0}^{2} \sum_{q=-k}^{k} (-1)^q A_{kq} T_{k-q}$$

where $\mathbf{D}$ represents the direct as well as the indirect spin-spin interaction. If we assume $\mathbf{I}\equiv\mathbf{S}$, even the quadrupole interaction with $\mathbf{D}=\mathbf{Q}$ and $k=2$ can be represented in this fashion. The spherical tensor components A_{kq} and T_{kq} can be readily obtained from Eq. (A.12) in terms of their Cartesian counterparts. Now spin and geometrical variable are well separated, since the A_{kq} represent only geometrical variables, whereas the T_{kq} contain only spin variables.

If only symmetric components have to be considered we obtain

$$
\begin{aligned}
A_{00} &= -(1/\sqrt{3})\,\mathrm{Tr}\{D_{ij}\} \\
A_{10} &= A_{1\pm1}=0 \\
A_{20} &= (3/\sqrt{6})\,D_{zz} \\
A_{2\pm1} &= \mp(D_{xz}\pm iD_{yz}) \\
A_{2\pm2} &= \tfrac{1}{2}(D_{xx}-D_{yy}\pm 2iD_{xy})
\end{aligned}
\tag{A.28}
$$

and

$$
\begin{aligned}
T_{20} &= (1/\sqrt{6})(3I_z S_z - \mathbf{I}\cdot\mathbf{S}) \\
T_{2\pm1} &= \mp\tfrac{1}{2}(I_z S_\pm + I_\pm S_z) \\
T_{2\pm2} &= \tfrac{1}{2}I_\pm S_\pm
\end{aligned}
\tag{A.29}
$$

As a specific example let us consider the dipolar interaction. With $\mathrm{Tr}\{D_{ij}\}=0$ and

$$
D_{ij}=\frac{\gamma_I\gamma_S\hbar}{r^3}(\delta_{ij}-3\mathbf{e}_i\cdot\mathbf{e}_j)\quad (i,j=x,y,z)
\tag{A.30}
$$

with $\mathbf{e}_i (i=x,y,z)$ are the x, y and z components of a unit vector pointing from one spin to the other we obtain

$$
A_{2q}=-\sqrt{6}\frac{\gamma_I\gamma_S\hbar}{r^3}\sqrt{\frac{4\pi}{5}}\,Y_{2q}
\tag{A.31}
$$

where Y_{kq} are spherical harmonics. Sometimes it might be convenient to use the modified spherical harmonics.

$$
C_{kq}=\left[\frac{4\pi}{2k+1}\right]^{1/2} Y_{kq}.
\tag{A.32}
$$

B Rotations

We want to summarize some general rotation properties for the convenience of the reader. A more rigorous treatment can be found in standard texts. We adopt here the sign conventions and the Euler angle definition of Rose [1a]. If a positive rotation is to be applied to a frame (x,y,z) about the Euler angles (α,β,γ) as

$$
\mathbf{R}(\alpha\beta\gamma)=\mathbf{R}_{z''}(\gamma)\,\mathbf{R}_{y'}(\beta)\,\mathbf{R}_z(\alpha)
\tag{B.1}
$$

where α is a rotation about the original z axis, β is about the new y axis, and γ is about the final z axis, the product of the three rotation matrices in Eq. (B.1) leads to

$$
\mathbf{R}(\alpha\beta\gamma) = \left\{
\begin{array}{ll}
\cos\alpha\cos\beta\cos\gamma - \sin\alpha\sin\gamma & \\
-\cos\alpha\cos\beta\sin\gamma - \sin\alpha\cos\gamma & \\
\cos\alpha\sin\beta &
\end{array}
\right.
$$

$$
\left.
\begin{array}{ll}
\sin\alpha\cos\beta\cos\gamma + \cos\alpha\sin\gamma & -\sin\beta\cos\gamma \\
-\sin\alpha\cos\beta\sin\gamma + \cos\alpha\cos\gamma & \sin\beta\sin\gamma \\
\sin\alpha\sin\beta & \cos\beta
\end{array}
\right\} \tag{B.2}
$$

The product of rotations in Eq. (B.1) may be readily expressed in terms of rotations about the original axes (x, y, z) as:

$$
\mathbf{R}(\alpha\beta\gamma) = \mathbf{R}_z(\alpha)\,\mathbf{R}_y(\beta)\,\mathbf{R}_z(\gamma). \tag{B.3}
$$

These rotations in Cartesian coordinates can be transformed to spherical coordinates as discussed similarily in Appendix A. We come to this later.

Let us now summarize some positive rotations (counter clockwise) in spin space:

$$
e^{-i\alpha I_x}
\begin{Bmatrix} I_x \\ I_y \\ I_z \end{Bmatrix}
e^{i\alpha I_x} =
\begin{Bmatrix} I_x \\ I_y\cos\alpha + I_z\sin\alpha \\ I_z\cos\alpha - I_y\sin\alpha \end{Bmatrix}
$$

$$
e^{-i\alpha I_y}
\begin{Bmatrix} I_x \\ I_y \\ I_z \end{Bmatrix}
e^{i\alpha I_y} =
\begin{Bmatrix} I_x\cos\alpha - I_z\sin\alpha \\ I_y \\ I_z\cos\alpha + I_x\sin\alpha \end{Bmatrix} \tag{B.4}
$$

$$
e^{-i\alpha I_z}
\begin{Bmatrix} I_x \\ I_y \\ I_z \end{Bmatrix}
e^{i\alpha I_z} =
\begin{Bmatrix} I_x\cos\alpha + I_y\sin\alpha \\ I_y\cos\alpha - I_x\sin\alpha \\ I_z \end{Bmatrix}
$$

From these relations, combined rotations can be constructed e.g.

$$
e^{-i\alpha I_x} = e^{i\frac{\pi}{2}I_z}\, e^{-i\alpha I_y}\, e^{-i\frac{\pi}{2}I_z} \tag{B.5}
$$

or

$$
e^{-i\alpha I_x} = e^{-i\frac{\pi}{2}I_y}\, e^{-i\alpha I_z}\, e^{i\frac{\pi}{2}I_y} \tag{B.6}
$$

and in a similar way

$$
e^{-i\alpha(I_x + I_y + I_z)/\sqrt{3}} = e^{-i\frac{\pi}{4}I_z}\, e^{-i\beta_m I_y}\, e^{-i\alpha I_z}\, e^{i\beta_m I_y}\, e^{i\frac{\pi}{4}I_z} \tag{B.7}
$$

where

$$
\tan\beta_m = \sqrt{2}.
$$

Since in the definition of the Euler angles [Eq. (B.3)] only rotations about the z and y axis are allowed, we may write any x-rotation according to Eqs. (B.5, B.6) as

$$
e^{-i\alpha I_x} = \mathbf{R}\left(-\frac{\pi}{2}\,\alpha\,\frac{\pi}{2}\right) \tag{B.8}
$$

Table B.1. Expressions of $d^j_{mm'}(\beta)$ for $j = \frac{1}{2}$, 1, $\frac{3}{2}$, and 2 (Brink and Satchler [1b])

$$d^{1/2}_{1/2\,1/2} = d^{1/2}_{-1/2\,-1/2} = \cos\left(\frac{\beta}{2}\right)$$

$$d^{1/2}_{-1/2\,1/2} = -d^{1/2}_{1/2\,-1/2} = \sin\left(\frac{\beta}{2}\right)$$

$$d^{1}_{11} = d^{1}_{-1-1} = \cos^2\left(\frac{\beta}{2}\right)$$

$$d^{1}_{1-1} = d^{1}_{-11} = \sin^2\left(\frac{\beta}{2}\right)$$

$$d^{1}_{01} = d^{1}_{-10} = -d^{1}_{0-1} = -d^{1}_{10} = \sin\beta/\sqrt{2}$$

$$d^{1}_{00} = \cos\beta$$

$$d^{3/2}_{3/2\,3/2} = d^{3/2}_{-3/2\,-3/2} = \cos^3\left(\frac{\beta}{2}\right)$$

$$d^{3/2}_{3/2\,1/2} = d^{3/2}_{-1/2\,-3/2} = -d^{3/2}_{1/2\,3/2}$$
$$= -d^{3/2}_{-3/2\,-1/2} = -\sqrt{3}\cos^2\left(\frac{\beta}{2}\right)\sin\left(\frac{\beta}{2}\right)$$

$$d^{3/2}_{3/2\,-1/2} = d^{3/2}_{-1/2\,3/2} = d^{3/2}_{1/2\,-3/2}$$
$$= d^{3/2}_{-3/2\,1/2} = \sqrt{3}\cos\left(\frac{\beta}{2}\right)\sin^2\left(\frac{\beta}{2}\right)$$

$$d^{3/2}_{3/2\,-3/2} = -d^{3/2}_{-3/2\,3/2} = -\sin^3\left(\frac{\beta}{2}\right)$$

$$d^{3/2}_{1/2\,1/2} = d^{3/2}_{-1/2\,-1/2} = \cos\left(\frac{\beta}{2}\right)\left[3\cos^2\left(\frac{\beta}{2}\right) - 2\right]$$

$$d^{3/2}_{1/2\,-1/2} = -d^{3/2}_{-1/2\,1/2} = \sin\left(\frac{\beta}{2}\right)\left[3\sin^2\left(\frac{\beta}{2}\right) - 2\right]$$

$$d^{2}_{22} = d^{2}_{-2-2}\cos^4\left(\frac{\beta}{2}\right)$$

$$d^{2}_{21} = -d^{2}_{12} = -d^{2}_{-2-1}$$
$$= d^{2}_{-1-2} = -\tfrac{1}{2}\sin\beta(1 + \cos\beta)$$

$$d^{2}_{20} = d^{2}_{02} = d^{2}_{-20} = d^{2}_{0-2} = \sqrt{\tfrac{3}{8}}\sin^2\beta$$

$$d^{2}_{2-1} = d^{2}_{1-2} = -d^{2}_{-21} = -d^{2}_{-12} = \tfrac{1}{2}\sin\beta(\cos\beta - 1)$$

$$d^{2}_{2-2} = d^{2}_{-22} = \sin^4\left(\frac{\beta}{2}\right)$$

$$d^{2}_{11} = d^{2}_{-1-1} = \tfrac{1}{2}(2\cos\beta - 1)(\cos\beta + 1)$$

$$d^{2}_{1-1} = d^{2}_{-11} = \tfrac{1}{2}(2\cos\beta + 1)(1 - \cos\beta)$$

$$d^{2}_{10} = d^{2}_{0-1} = -d^{2}_{01} = -d^{2}_{-10} = -\sqrt{\tfrac{3}{2}}\sin\beta\cos\beta$$

$$d^{2}_{00} = \tfrac{1}{2}(3\cos^2\beta - 1)$$

References for additional tables for $d^j_{mm'}(\beta)$ are as follows:
$j = 2, 4, 6$: Buckmaster, H.A., Can. J. Phys. **42**, 386 (1964)
$j = 1, 3, 5$: ibid. **44**, 2525 (1966)
$j = 3$ Ying-Nan Chiu, J. Chem. Phys. **45**, 2969 (1966)

or

$$e^{-i\alpha I_x} = \mathbf{R}\left(0\,\frac{\pi}{2}\,0\right)\mathbf{R}_z(\alpha)\,\mathbf{R}\left(0\,-\frac{\pi}{2}\,0\right) \tag{B.9}$$

and in a similar way

$$e^{-i\alpha(I_x + I_y + I_z)/\sqrt{3}} = \mathbf{R}\left(\frac{\pi}{4}\,\beta_m\,0\right)\mathbf{R}_z(\alpha)\,\mathbf{R}\left(0\,-\beta_m\,-\frac{\pi}{4}\right). \tag{B.10}$$

We turn now to rotations of irreducible tensor operators, by writing

$$T'_{kq} = \mathbf{R}(\alpha\beta\gamma)\,T_{kq}\,\mathbf{R}^{-1}(\alpha\beta\gamma) = \sum_{p=-k}^{+k} T_{kp}\,D^k_{pq}(\alpha\beta\gamma) \tag{B.11}$$

with the Wigner rotation matrices

$$D^k_{pq}(\alpha\beta\gamma) = \langle kp|\,e^{-i\alpha I_z}\,e^{-i\beta I_y}\,e^{-i\gamma I_z}\,|kq\rangle \tag{B.12}$$

or

$$D^k_{pq}(\alpha\beta\gamma) = e^{-i\alpha p}\, d^k_{pq}(\beta)\, e^{-i\gamma q}. \tag{B.13}$$

The reduced rotation matrices $d^k_{pq}(\beta)$ are real and are expressed explicitly for the values $k = 1/2, 1, 3/2, 2$ in Table B.1.

As a representative example let us consider

$$I'_z = e^{-i\alpha I_x}\, I_z\, e^{+i\alpha I_x}$$

or

$$T'_{10} = \mathbf{R}\left(-\frac{\pi}{2}\,\beta\,\frac{\pi}{2}\right) T_{10}\, \mathbf{R}^{-1}\left(-\frac{\pi}{2}\,\beta\,\frac{\pi}{2}\right)$$

$$T'_{10} = \sum_q T_{1q}\, D^1_{q0}\left(-\frac{\pi}{2}\,\beta\,\frac{\pi}{2}\right)$$

$$T'_{10} = \sum_q T_{1q}\, e^{iq\frac{\pi}{2}}\, d^1_{q0}(\beta).$$

With the help of Table B.1 and $T_{11} + T_{1-1} = -i\sqrt{2}\, I_y$, this leads to

$$T'_{10} = I_z \cos\alpha - I_y \sin\alpha$$

as in Eq. (B.4).

C General Line Shape Theory

In this context we consider the "line-shape" in a general sense, covering the static resonance line shape of the NMR spectrum as well as spectral distribution functions which are involved in relaxation, cross-relaxation, cross-polarization as well as spin decoupling processes, thus providing the basic formalism for Sects. 2.5, 3.8, 4.3, 4.4. Only the "non-trivial" case of many body interactions, such as dipolar interaction is considered [3].

Let us first define a Liouville space by considering the ordinary operators of quantum mechanics to be state vectors in this Liouville space i.e., if the Hilbert space of state vectors ψ has the dimension n, the corresponding Liouville space will have the dimension $n \times n$.

The Liouville-v. Neumann equation of motion for the density matrix ρ may now be expressed in Liouville space as [4]

$$\frac{d}{dt}|\rho) = -i\hat{\mathscr{H}}|\rho) \tag{C.1}$$

with the formal solution $|\rho(t)) = T\exp\left[-i\int_0^t dt'\,\hat{\mathscr{H}}(t)\right]|\rho(0))$ where $\hat{\mathscr{H}}$ is a superoperator or Liouville operator (Liouvillian) acting on the state vector $|\rho)$ in Liouville space and T the usual Dyson time ordering operator.

$\hat{\mathscr{H}}$ is defined by

$$\hat{\mathscr{H}}|\mathbf{A}) = |[\mathscr{H},\mathbf{A}]) \tag{C.2}$$

correspondingly

$$\hat{\mathscr{H}}^n |A) = |[\mathscr{H},[\mathscr{H},...,[\mathscr{H},A]]...]_n) \tag{C.3}$$

and

$$e^{\alpha\hat{A}} |B) = |e^{\alpha A} B e^{-\alpha A}). \tag{C.4}$$

A scalar product between two Liouville state vectors is defined as [4]

$$(A|B) \underset{\text{def}}{=} \text{Tr}\{A^+ B\} \tag{C.5}$$

where A^+ is the Hermitian adjoint of A. This definition [Eq. (C.5)] is appropriate for finite dimensional Liouville spaces.

Let us summarize some more relations

$$(A|B)^* = (B|A)$$
$$(A|\beta B) = \beta(A|B) \quad \beta \text{ complex number}$$
$$(A|B+C) = (A|B)+(A|C) \tag{C.6}$$
$$(A|A) \geqq 0 \quad \text{equals zero only if } A = 0.$$

We define the expectation value of an operator Q as

$$\langle Q \rangle = (Q|\rho) \tag{C.7}$$

or

$$\langle Q(t) \rangle = (Q| \, T \exp\left\{-i \int_0^t dt' \, \hat{\mathscr{H}}(t')\right\} |\rho(0)). \tag{C.8}$$

Let us always consider the case where an appropriate interaction representation has been chosen to give the time evolution operator a simple form. Moreover we shall often represent it by an average Hamiltonian by means of a Magnus expansion [5, 6] (see also Appendix F)

$$T \exp\left[-i \int_0^t dt' \, \mathscr{H}(t')\right] = \exp[-it(\bar{\mathscr{H}}^{(0)} + \bar{\mathscr{H}}^{(1)} + ...)]. \tag{C.9}$$

In this case we arrive at a simple form for Eq. (C.8) as

$$\langle Q(t) \rangle = (Q| \exp[-it\hat{\bar{\mathscr{H}}}] |\rho(0)) \tag{C.10}$$

with

$$\bar{\mathscr{H}} = \bar{\mathscr{H}}^{(0)} + \bar{\mathscr{H}}^{(1)} + ...$$

according to a Magnus expansion which keeps $\bar{\mathscr{H}}$ Hermitian at every stage of approximation.

The corresponding line shape is formally defined as

$$J(\omega) = \int_{-\infty}^{+\infty} dt \langle Q(t) \rangle \, e^{i\omega t} = (Q| \, \delta(\omega - \hat{\bar{\mathscr{H}}}) |\rho(0)) \tag{C.11}$$

where $\delta(\omega - \hat{\bar{\mathscr{H}}})$ is a δ function operator.

As an example we treat the Bloch decay in solid where spin species I couple to their neighbors via the dipolar interaction. The zeroth order Hamiltonian $\mathscr{H}^{(0)}$ is in this case the truncated or secular dipolar Hamiltonian $\mathscr{H}'_D$, which produces a Bloch decay of the e.g. x-component of the magnetization as

$$G(t) = \frac{(I_x | \exp[-it\,\hat{\mathcal{H}}_D'] \,|\rho(0))}{(I_x|\rho(0))} \tag{C.12}$$

where the exponential operator may be expanded to yield

$$G(t) = \sum_{n=0}^{\infty} \frac{(-it)^n}{n!} M_n \tag{C.13}$$

with the n-th moment M_n defined as

$$M_n = \frac{(I_x | \hat{\mathcal{H}}_D'^n |\rho(0))}{(I_x|\rho(0))}. \tag{C.14}$$

The moment expansion, however, is not rapidly converging and the calculation of higher order moments (higher than the fourth moment) becomes a formidable task. Different approximations have therefore been devised to calculate the Bloch decay due to the many body dipolar interaction more accurately.

We are now going to discuss a very convenient way of approximation which is based on the Mori formalism [7b–10]. First the equation of motion for an operator is rewritten in a different form, which is still exact, but more susceptible to approximations i.e., the hierarchy of approximations is always pertained.

Let us expand the density matrix into a set of m "orthogonal" operators $\mathbf{Q}_k$ $k = 1, \dots, m$ as

$$|\rho) = Z^{-1}[1 + \beta_1 \mathbf{Q}_1 + \beta_2 \mathbf{Q}_2 + \dots \beta_m \mathbf{Q}_m + \mathbf{O}_{\text{residual}}] \tag{C.15}$$

where "orthogonal" means

$$(\mathbf{Q}_i|\mathbf{Q}_j) = 0 \quad \text{for } i \neq j \tag{C.16}$$

and

$$(\mathbf{Q}_i|\mathbf{O}_{\text{residual}}) = 0.$$

We do not consider the system to be at equilibrium explicitly, thus avoiding the concept of a spin temperature. We may, however, define thermodynamic coordinates (which are not temperatures necessarily) by

$$\beta_k = \frac{(\rho|\mathbf{Q}_k)}{(\mathbf{Q}_k|\mathbf{Q}_k)} = \frac{\langle \mathbf{Q}_k \rangle}{\text{Tr}\{\mathbf{Q}_k^+ \mathbf{Q}_k\}}. \tag{C.17}$$

Nevertheless β_k is small in NMR.

A suitable projection operator is defined which projects the "relevant" part of the density matrix onto the individual operators $\mathbf{Q}_k$ by

$$P \underset{\text{def}}{=} \sum_{k=1}^{m} \frac{|\mathbf{Q}_k)(\mathbf{Q}_k|}{(\mathbf{Q}_k|\mathbf{Q}_k)} \tag{C.18}$$

with $P^2 = P$ (idempotent).

The density matrix is now properly separated into the "relevant" part and the "irrelevant" part by

$$|\rho) = P\,|\rho) + (1 - P)\,|\rho) \tag{C.19}$$

which results in a Liouville-v. Neumann equation as

$$\frac{d}{dt}|\rho) = -i\hat{\mathscr{H}}P\,|\rho) - i\hat{\mathscr{H}}(1-P)\,|\rho) \tag{C.20}$$

and a corresponding equation of motion for the expectation value $\langle \mathbf{Q}_k \rangle$ of the operator $\mathbf{Q}_k$ as

$$\frac{d}{dt}(\mathbf{Q}_k|\rho) = -i(\mathbf{Q}_k|\,\hat{\mathscr{H}}P\,|\rho) - i(\mathbf{Q}_k|\,\hat{\mathscr{H}}(1-P)\,|\rho). \tag{C.21}$$

The first term represents the direct or trivial contribution to the motion of $\langle \mathbf{Q}_k(t) \rangle$, whereas the second term contains "memory" effects as will be seen later. In order to calculate the second term we write

$$\frac{d}{dt}(1-P)\,|\rho) = -i(1-P)\,\hat{\mathscr{H}}P\,|\rho) - i(1-P)\,\hat{\mathscr{H}}(1-P)\,|\rho) \tag{C.22}$$

which has the formal solution

$$(1-P)\,|\rho(t)) = \mathbf{S}(t,0)\,(1-P)\,|\rho(0)) - i\int_0^t dt'\,\mathbf{S}(t,t')\,(1-P)\,\hat{\mathscr{H}}(t')\,P\,|\rho(t)) \tag{C.23}$$

where

$$\mathbf{S}(t,t') = T\exp\left[-i\int_{t'}^t d\tau(1-P)\,\hat{\mathscr{H}}(\tau)\right]. \tag{C.24}$$

We insert Eq. (C.23) into Eq. (C.21) to obtain

$$\frac{d}{dt}\langle \mathbf{Q}_k(t) \rangle = \mathbf{K} + \mathbf{L} + \mathbf{M} \tag{C.25}$$

where

$$\mathbf{K} = -i(\mathbf{Q}_k|\,\hat{\mathscr{H}}P\,|\rho) = -i\sum_{j=1}^m \frac{(\mathbf{Q}_k|\,\hat{\mathscr{H}}(t)\,|\mathbf{Q}_j)}{(\mathbf{Q}_j|\mathbf{Q}_j)}\,\langle \mathbf{Q}_j(t) \rangle. \tag{C.26}$$

$$\mathbf{L} = -i(\mathbf{Q}_k|\,\hat{\mathscr{H}}\,\mathbf{S}(t,0)\,(1-P)\,|\rho(0)). \tag{C.27}$$

$$\mathbf{M} = -\sum_{j=1}^m \int_0^t dt' \frac{(\mathbf{Q}_k|\,\hat{\mathscr{H}}(t)\,\mathbf{S}(t,t')\,(1-P)\,\hat{\mathscr{H}}(t')\,|\mathbf{Q}_j)}{(\mathbf{Q}_j|\mathbf{Q}_j)}\,\langle \mathbf{Q}_j(t') \rangle. \tag{C.28}$$

If the complete Liouville space is exhausted by the m operators $\mathbf{Q}_k$ i.e. dim $\mathscr{L}$ $= m$, $\mathbf{L}$ and $\mathbf{M}$ vanish and Eq. (C.25) reduces to the Liouville equation again. In other words there is no irreversibility if dim $\mathscr{L} = m$. Irreversibility, therefore, can occur only if $m < \dim \mathscr{L}$ i.e., the m orthogonal operators $\mathbf{Q}_k$ do not span the complete Liouville space. The term $\mathbf{L}$ represents the initial condition and this term is, in general, not zero. However, in most experiments the initial condition can be prepared so as to let $\mathbf{L}$ vanish [8]. The first term $\mathbf{K}$ as defined in Eq. (C.26) can be readily shown to vanish if $[\mathbf{Q}_k, \mathbf{Q}_j] = 0$, which leads to $P\hat{\mathscr{H}}P = 0$.

We always assume in the following that the conditions for $\mathbf{K} = \mathbf{L} = 0$ are fulfilled (by proper transformations and/or initial conditions) and summarize

[10]

$$\frac{d}{dt} \langle \mathbf{Q}_k(t) \rangle = -\int_0^t dt' \sum_{j=1}^m K_{kj}(t,t') \langle \mathbf{Q}_j(t') \rangle \qquad \text{(C.29)}$$

with

$$K_{kj}(t,t') = \frac{(\mathbf{Q}_k| \,\hat{\mathscr{H}}(t)\, S(t,t')\,(1-P)\,\hat{\mathscr{H}}(t')\, |\mathbf{Q}_j)}{(\mathbf{Q}_j|\mathbf{Q}_j)} \qquad \text{(C.30)}$$

The Kernel $K_{kj}(t,t')$ represents the "memory" effects and takes the history of earlier times into account. $K_{kj}(t,t')$ is therefore called the "memory function". We note that Eq. (C.29) is still exact, but is of course by no means easier to evaluate than the formal solution Eq. (C.8). However, approximations may be applied to the "memory function" $K_{kj}(t,t')$ which amounts to moving a step upwards in the hierarchy of approximations.

Let us introduce further simplifications. We assume, that a convenient interaction representation has been chosen, such that the interaction Hamiltonian can be considered time independent i.e., to a good approximation the Hamiltonian is represented by an average Hamiltonian.

Following Eq. (C.29), where we suppose $[\mathbf{Q}_k, \mathbf{Q}_j] = 0$ i.e. $P\hat{\mathscr{H}}P = 0$, we obtain

$$K_{kj}(\tau) = \frac{(\mathbf{Q}_k| \,\hat{\mathscr{H}} \exp[-i\tau(1-P)\hat{\mathscr{H}}]\,\hat{\mathscr{H}}\, |\mathbf{Q}_j)}{(\mathbf{Q}_j|\mathbf{Q}_j)} \qquad \text{(C.31)}$$

with

$$K_{kj}(0) = \frac{(\mathbf{Q}_k| \,\hat{\mathscr{H}}^2\, |\mathbf{Q}_j)}{(\mathbf{Q}_j|\mathbf{Q}_j)} = M_{2kj}. \qquad \text{(C.32)}$$

We may expand $K_{kj}(\tau)$ as

$$K_{kj}(\tau) = \sum_{n=0}^{\infty} \frac{(-i\tau)^n}{n!} K_n$$

with

$$K_n = \frac{(\mathbf{Q}_k| \,\hat{\mathscr{H}} \{(1-P)\hat{\mathscr{H}}\}^n \,\hat{\mathscr{H}}\, |\mathbf{Q}_j)}{(\mathbf{Q}_j|\mathbf{Q}_j)}. \qquad \text{(C.33)}$$

We specialize now in the case, where the corresponding spectral distribution function is an even function with frequency i.e., $J(\omega) = J(-\omega)$ and consequently all odd order moments in $K_{kj}(\tau)$ vanish

$$K_{kj}(\tau) = K_{kj}(0) \left[1 - \frac{N_2}{2!} \tau^2 + \frac{N_4}{4!} \tau^4 - + \dots \right]$$

with $K_{kj}(0)$ as before, [Eq. (C.32)] and $[\mathbf{Q}_k, \mathbf{Q}_j] = 0$ (implying $P\hat{\mathscr{H}}P = 0$), we obtain:

$$N_2 \cdot K_{kj}(0) = \frac{(\mathbf{Q}_k| \,\hat{\mathscr{H}}^4\, |\mathbf{Q}_j)}{(\mathbf{Q}_j|\mathbf{Q}_j)} - \frac{(\mathbf{Q}_k| \,\hat{\mathscr{H}}^2 P \hat{\mathscr{H}}^2\, |\mathbf{Q}_j)}{(\mathbf{Q}_j|\mathbf{Q}_j)}. \qquad \text{(C.34)}$$

$$N_4 K_{kj}(0) = \frac{(\mathbf{Q}_k| \,\hat{\mathscr{H}}^6 - \hat{\mathscr{H}}^4 P \hat{\mathscr{H}}^2 - \hat{\mathscr{H}}^2 P \hat{\mathscr{H}}^4 + \hat{\mathscr{H}}^2 P \hat{\mathscr{H}}^2 P \hat{\mathscr{H}}^2\, |\mathbf{Q}_j)}{(\mathbf{Q}_j|\mathbf{Q}_j)} \qquad \text{(C.35)}$$

If we restrict ourselves to only one observable, we finally obtain

$$\frac{d}{dt}\langle \mathbf{Q}|(t)\rangle = -\int_0^t dt'\, K(t-t')\,\langle \mathbf{Q}(t')\rangle \tag{C.36}$$

with

$$
\begin{aligned}
K(0) &= M_2 \\
N_2 &= M_2(\mu - 1) \\
N_4 &= M_2^2(v - 2\mu + 1)
\end{aligned}
\tag{C.37}
$$

where

$$M_n = \frac{(\mathbf{Q}|\,\hat{\mathscr{H}}^n\,|\mathbf{Q})}{(\mathbf{Q}|\mathbf{Q})} \tag{C.38}$$

and

$$\mu = M_4/M_2^2\,;\quad v = M_6/M_2^3. \tag{C.39}$$

The corresponding line shape $I(\omega)$ can be obtained from Eq. (C.36) with

$$I(\omega) = \int_0^\infty dt\, \langle \mathbf{Q}(t)\rangle \cos \omega t \tag{C.40}$$

by formal integration as [11]

$$I(\omega) = \langle \mathbf{Q}(0)\rangle \cdot \frac{K'(\omega)}{[\omega - K''(\omega)]^2 + [K'(\omega)]^2} \tag{C.41}$$

where

$$K'(\omega) = \int_0^\infty d\tau\, K(\tau) \cos \omega \tau \tag{C.42a}$$

and

$$K''(\omega) = \int_0^\infty d\tau\, K(\tau) \sin \omega \tau. \tag{C.42b}$$

The half width at half height of the line shape $I(\omega)$ according to Eq. (C.41) can be obtained by iteration from

$$\delta = K''(\delta) + [2K'(0)\,K'(\delta) - K'^2(\delta)]^{1/2} \tag{C.43a}$$

Note, that δ is connected with the full width at half height (FWHH) in frequency units Δv by $\Delta v = \delta/\pi$ in general.

So far no approximation other than the average Hamiltonian approximation in a suitably chosen interaction representation has been used. The iteration of Eq. (C.43a) can in many cases be avoided by using as a first order approximation

$$\delta \simeq K'(0) \quad \text{or} \quad \Delta v = K'(0)/\pi \tag{C.43b}$$

In order to calculate actual line shapes, we will have to approximate the memory function $K(\tau)$. A very convenient approximation to $K(\tau)$ would be a Gaussian

$$K(\tau) = K(0) \exp\left[-\frac{N_2}{2}\tau^2\right] \tag{C.44}$$

since only the second moment of $K(\tau)$ is involved.

In fact, it has been shown by several authors, that this approximation is indeed widely applicable and that it covers a wealth of different physical situations (see also Sects. 2.5, 3.8, 4.3, 4.4).

The lineshape $I(\omega)$ according to Eq. (C.41) is readily obtained by insertion of the sine- and cosine-transform of $K(\tau)$ according to Eq. (C.44) as [10]

$$K'(\omega) = \sqrt{\frac{\pi}{2}} \frac{K(0)}{\sqrt{N_2}} \exp(-\omega^2/2N_2). \tag{C.45a}$$

$$K''(\omega) = \frac{K(0)\,\omega}{N_2} \exp(-\omega^2/2N_2) \cdot F(\tfrac{1}{2}; \tfrac{3}{2}; \omega^2/2N_2) \tag{C.45b}$$

where $K(0)$ and N_2 are given by Eq. (C.37). $F(\tfrac{1}{2}; \tfrac{3}{2}; \omega^2/2N_2)$ is the confluent hypergeometric function, which is tabulated and can be calculated numerically. Notice that no adjustable parameter is involved at all and the only approximation being the functional form of the memory function.

The linewidth δ which is a suitable experimental parameter can now be expressed according to Eqs. (C.43) and (C.45) as

$$\delta = \sqrt{\frac{\pi}{2}} \left(\frac{K^2(0)}{N_2}\right)^{1/2} \cdot f(x) \tag{C.46}$$

with

$$f(x) = [2 - e^{-x^2}]^{1/2} \cdot e^{-x^2/2} + \frac{2x}{\sqrt{\pi}} \cdot e^{-x^2} \cdot F(\tfrac{1}{2}; \tfrac{3}{2}; x^2) \tag{C.47}$$

where $f(x)$ is a function close to one with

$$x = \delta/(2N_2)^{1/2}.$$

Another relation between $\mu = M_4/M_2^2$ and $f(x)$ is readily found with

$$\mu - 1 = \frac{\sqrt{\pi}}{2} \frac{f(x)}{x}. \tag{C.48}$$

This allows $f(\mu)$ to be expressed by μ as is shown in Fig. C.1. For large values of μ the function $f(\mu)$ approaches the value 1 and the linewidth δ may be expressed according to Eqs. (C.43, C.45, C.46) as

$$\delta = \sqrt{\frac{\pi}{2}} \left[\frac{M_2}{\mu - 1}\right]^{1/2}. \tag{C.49}$$

The lineshape $I(\omega)$ is shown to be Lorentzian in this case. This linewidth may be contrasted with the cut-off Lorentzian linewidth[3]. A better approximation to $f(\mu)$ for smaller values of μ is expressed as

$$f(\mu) = \left(\frac{\mu - 1}{\mu - 2}\right)^{1/2} g(\mu)$$

where now $g(\mu)$ is always close to one. Along these lines we propose the linewidth δ_G with a Gaussian memory function approximation to be

$$\delta_G = \sqrt{\frac{\pi}{2}} \left[\frac{M_2}{\mu - 2}\right]^{1/2}. \tag{C.50a}$$

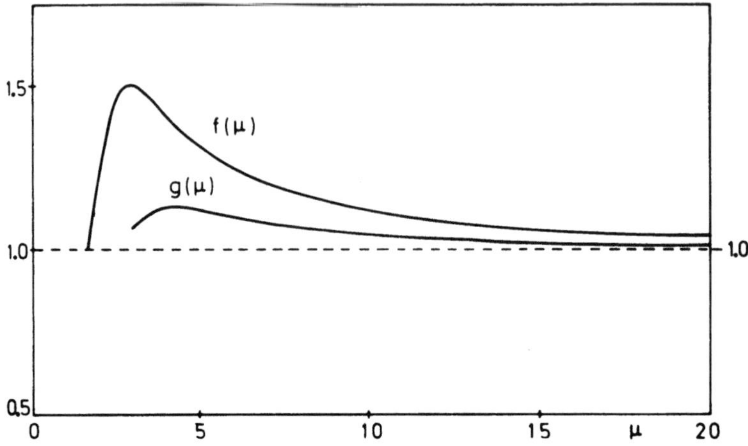

Fig. C.1. Line width correction functions $g(\mu)$ and $f(\mu)$ versus moment ratio $\mu = M_4/M_2^2$ (see text)

In order to obtain the correct linewidth in the case of a Gaussian ($\mu = 3$) and a close enough approximation for Lorentzians ($\mu > 3$) we suggest to use as a universal linewidth

$$\delta = \sqrt{\frac{\pi}{2}} \left[\frac{M_2}{\mu - 1.87} \right]^{1/2} \quad \text{or} \quad \Delta v = \frac{1}{\sqrt{2\pi}} \left[\frac{M_2}{\mu - 1.87} \right]^{1/2} \tag{C.50b}$$

with $\mu = M_4/M_2^2$ and Δv being the FWHH in frequency units.

We finally remark that higher order approximations of the memory function $K(\tau)$ are readily obtained by writing down an integro-differential Eq. (C.36) for $K(\tau)$ itself, followed by an approximation of the corresponding memory function of $K(\tau)$. This extension is straightforward and will not be discussed any further here.

Another approximation of $K(\tau)$, does not assume a particular functional form of $K_{kj}(\tau)$, but rather assumes the so-called "short correlation limit" to hold, where the correlation time of $K_{kj}(\tau)$ is assumed to be very short compared with any change of $\langle Q_j(t) \rangle$. This is usually the case in relaxation and also in line narrowing multiple-pulse experiments (see Sect. 3.8). In this limit $\langle Q_j(t) \rangle$ can be separated from the integral in Eq. (C.29) and the integration limit is extended to infinity

$$\frac{d}{dt} \langle Q_k(t) \rangle = - \int_0^\infty d\tau \sum_{j=1}^m K_{kj}(\tau) \cdot \langle Q_j(t) \rangle \tag{C.51}$$

which results in a sum of exponentials or corresponding Lorentzian line shape functions. The individual relaxation rates are defined as

$$\delta = \frac{1}{T_{kj}} = \int_0^\infty d\tau \, K_{kj}(\tau) = K'_{kj}(\omega = 0) \tag{C.52}$$

with $\Delta v = 1/(\pi T_{kj}) = K'_{kj}(0)/\pi$.

Other functional forms of $K(\tau)$ may be appropriate for special interactions, as was employed in Sect. 4.4.

The approximations of the memory function $K(t)$ discussed sofar assumed always a monotonic decrease of $K(t)$ with increasing time. If, however, the spins are driven by a periodic field e.g. by an rf field of strength $\omega_1 = \gamma B_1$ or a pulse sequence, $K(t)$ will in general be no monotonic function of time, but may be periodic itself. In the simplest approximation we may choose

$$K(t) = K_0(t) \cos \omega_1 t \tag{C.53}$$

where $K_0(t)$ now is a monotonic function in time and the cosinefunction takes the precessional motion with frequency ω_1 into account. In the case of spin-decoupling, discussed in Chap. 4, this is the correct form. In multiple-pulse experiments, however, the correct periodic function should be inserted or as a first order approximation $\cos \omega_c t$ may be used, where $\omega_c = 2\pi/t_c$ is the cycle frequency.

The consequences of this periodic motion for the calculation of the linewidth δ (or Δv) will now be discussed. The Fouriertransform of $K(t)$ now takes the form

$$K'(\omega) = \tfrac{1}{2} K'(\omega - \omega_1) + \tfrac{1}{2} K'(\omega + \omega_1) \tag{C.54}$$

We shall use here only the approximation Eqs. (C.43b; C.52), namely

$$\delta = K'(\omega = 0) = K'(\omega_1) \tag{C.55a}$$

In the case of Gaussian memory function this leads to

$$\delta = \sqrt{\frac{\pi}{2}} \left[\frac{M_2}{\mu - 1} \right]^{1/2} \exp[-\omega_1^2/2M_2(\mu - 1)] \tag{C.56}$$

In order to make contact with the universal linewidth formula Eq. (C.50b) we are tempted to write

$$\Delta v = \frac{1}{\sqrt{2\pi}} \left[\frac{M_2}{\mu - 1.87} \right]^{1/2} \exp[-\omega_1^2/2M_2(\mu - 1)] \tag{C.57}$$

for the full width at half height in frequency units, when periodic motion of the spins with frequency ω_1 modifies the lineshape (see Eq. (4.145)).

D Homogeneous, Inhomogeneous and Heterogeneous Lineshapes

Three types of spectral lineshapes, namely homogeneous, inhomogeneous and heterogeneous lineshapes may be distinguished. They are drawn schematically in Fig. D.1.

The *homogeneous line* is a sum of individual lines having all the same lifetime broadening $1/T_2$ and no shift with respect to each other.

This may be the case in NMR e.g. in a liquid where all the spins are decoupled from each other, but where coherence is relaxing with time T_2 or in a solid, where dipole-dipole interaction among the spins makes the spins

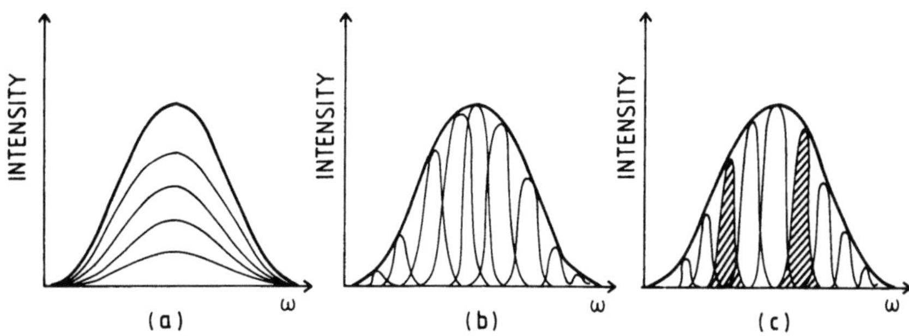

Fig. D.1a–c. Homogeneous (**a**), inhomogeneous (**b**) and heterogeneous (**c**) line shapes. The hatched area in (**c**) indicates that the corresponding spin packets are coupled. In the heterogeneous line different spin packets are coupled to each other whereas in the inhomogeneous line (**b**) all the spin packets are uncoupled

"indistinguishable" in the frequency domain and where spin flip-flop processes relax all the spins equally. Selective irradiation at a particular frequency of the line causes a saturation of the whole line (no hole burning).

In contrast the *inhomogeneous line* is a sum of non-overlapping individual lines. The lineshape is determined by a distribution of shifts. Although there may be some individual lifetime broadening connected with each spectral packet, no coupling between them exists. Such a situation may be encountered in NMR e.g. when a liquid is placed in a very inhomogeneous field. Selective irradiation at a particular frequency of the line causes a saturation of only that part of the line (hole burning). Since coupling to other spectral elements is excluded, no other parts of the line are affected.

The *heterogeneous line* plays an intermediate role between the homogeneous and inhomogeneous lineshape. It is made up of individual spectral packets, which are distributed in frequency and which are coupled to each other. The typical case where this behaviour is found in NMR, is in the case of heteronuclear spin coupling. There dilute S spins are coupled to a more or less isotropic "sea" of I spins (mostly protons) which in turn are strongly coupled via dipole-dipole interaction. The flip-flop processes of this dipole-dipole interaction among the I spins causes exchange between the different spectral parts of the S spins. This is depicted in Fig. D by the hatched areas in order to emphasize coupling between them. If the flip-flop processes are quenched, which may be achieved to a good extend e.g. by magic angle irradiation or suitable multiple-pulse experiments (see Chap. 3) the S spin line becomes inhomogeneous. If on the other hand the flip-flop rate could be enhanced (or more easily the I-S interaction is scaled to a small value) a homogeneous line would result.

Hole burning of a heterogeneous line becomes very intriguing since several parts of the spectrum may become saturated and additional spectral diffusion may occur. Note, that the notion of a heterogeneous line is not restricted the heteronuclear coupling, but can be extended to the more general situation where spectral elements of a line are coupled via a reservoir.

E Lineshape and Relaxation due to Fluctuating Chemical Shift Tensors

In order to present a simple example of the lineshape theory derived in Appendix C, we consider the lineshape and relaxation of spins which are subject to a fluctuating chemical shift tensor. The total Hamiltonian of the system is

$$\mathscr{H} = \mathscr{H}_I + \mathscr{H}_L \tag{E.1}$$

where the spin Hamiltonian

$$\mathscr{H}_I = \mathscr{H}_0 + \mathscr{H}_S \tag{E.2}$$

contains the Zeeman Hamiltonian $\mathscr{H}_0 = -\omega_0 I_z$ and the shift Hamiltonian

$$\mathscr{H}_S = \omega_x I_x + \omega_y I_y + \omega_z I_z \tag{E.3}$$

where

$$\omega_j = \omega_0 \sigma_{jz} \quad j = x, y, z$$

as derived e.g. in Appendix A. The lattice Hamiltonian $\mathscr{H}_L$ causes transitions between different sites, leading to a fluctuating chemical shift tensor, i.e. $[\mathscr{H}_L, \mathscr{H}_I] \neq 0$. $\mathscr{H}_L$ does operate, however, only on the lattice part of $\mathscr{H}_I$, namely on ω_j. Since $\mathscr{H}_0$ causes a precession about the z-axis with frequency ω_0 we are interested in the timedependence of the magnetization $\langle I_j(t) \rangle$; $j = x, y, z$, in the "rotating frame". The total Hamiltonian in this frame becomes timedependent

$$\mathscr{H}^* = \mathscr{H}_S(t) + \mathscr{H}_L \tag{E.4}$$

where

$$\mathscr{H}_S(t) = e^{-i\omega_0 t I_z} \mathscr{H}_S \, e^{i\omega_0 t I_z} \tag{E.5}$$

We are interested in the timedependence of the spin column vector

$$|\mathbf{I}) \equiv \begin{pmatrix} I_x \\ I_y \\ I_z \end{pmatrix} \tag{E.6}$$

in the rotating frame. Following along the lines of Appendix C we use the projector

$$P = \frac{|I_x)(I_x|}{(I_x|I_x)} + \frac{|I_y)(I_y|}{(I_y|I_y)} + \frac{|I_z)(I_z|}{(I_z|I_z)} \tag{E.7}$$

and derive the equation of motion

$$\frac{d}{dt}|\mathbf{I}(t)) = -i\tilde{\Omega}\,|\mathbf{I}(t)) - \int_0^t dt' \, \tilde{K}(t,t')\,|\mathbf{I}(t')) \tag{E.8}$$

where

$$\Omega_{jk} \equiv \frac{(I_j|\,\hat{\mathscr{H}}\,|I_k)}{(I_k|I_k)} = \frac{(I_j|\,\hat{\mathscr{H}}_S(t)\,|I_k)}{(I_k|I_k)} \quad (j, k = x, y, z) \tag{E.9}$$

and

$$K_{jk}(t,t') \equiv \frac{(I_j|\, \hat{\mathscr{H}}^*(t)\, S(t,t')\, (1-P)\, \hat{\mathscr{H}}^*(t')\, |I_k)}{(I_k|I_k)} \tag{E.10}$$

$$S(t,t') \equiv T \exp\left[-i\int_{t'}^{t} dt_1 (1-P)(\hat{\mathscr{H}}_S(t_1)+\hat{\mathscr{H}}_L)\right] \tag{E.11}$$

Once the frequency matrix $\tilde{\Omega}(t)$ and the memory function matrix $\tilde{K}(t,t')$ are know, Eq. (E.8) may be solved. So far Eq. (E.8) is rigorous, i.e. no approximation is involved. In order to solve Eq. (E.8) one has to use some sort of approximation, however. Since ω_0 is so much larger than all other spin interactions it is not a severe "mistake" to neglect oscillatory parts of the form $\cos\omega_0 t$, $\sin\omega_0 t$ in the rotating frame. Neglecting non-secular terms now simplifies the matrix

$$\tilde{\Omega}' = i\langle\omega_z\rangle \begin{pmatrix} 0 & -1 & 0 \\ 1 & 0 & 0 \\ 0 & 0 & 0 \end{pmatrix} \tag{E.12}$$

where $\langle\omega_z\rangle$ is the expectation value of $\omega_z=\omega_0\sigma_{zz}$, averaged over the lattice motion. Here we have used that

$$(A|\, \hat{B}\, |C) \equiv \mathrm{Tr}_I\,\mathrm{Tr}_L\{A^+[B,C]\} \tag{E.13}$$

where the trace is taken over the spinvariables and the lattice variables. If $\tilde{K}(t,t')$ would vanish, Eq. (E.8) together with Eq. (E.12) would desribe the precession of the spins around the z axis with the frequency $\langle\omega_z\rangle$. It is evident, that this must be the limit of extremely fast lattice fluctuations. In general $K(t,t')$ will be a decaying function connected with some correlation time τ. If we want to calculate $K(t,t')$ the main difficulty arises in how to approximate $S(t,t')$. A rigorous solution is exceedingly difficult. In the first step of approximation we apply average Hamiltonian theory, i.e. we neglect the non-secular terms in $\mathscr{H}_S(t_1)$ in Eq. (E.11)

$$S(t-t') = \exp[-i(t-t')(1-P)(\omega_z\hat{I}_z + \hat{\mathscr{H}}_L)] \tag{E.14}$$

or

$$S(t-t') = e^{-i(t-t')\hat{\mathscr{H}}_L}\, T e^{-i\int_{0}^{t-t'} dt_2(1-P)\omega_z(t_2)\hat{I}_z} \tag{E.15}$$

Now the timedependence of $\omega_z(t_2)$ is caused by the lattice fluctuations.

Two limits may be considered.

(a) Weak collision limit [3]. Since the lattice fluctuation rate $1/\tau$ is much larger than the variation in the spin interaction, i.e. $1/\tau \gg \delta\omega_z$ in this case, the spin dependent part in $S(t-t')$ may be neglected all together

$$S(t-t') \simeq e^{-i(t-t')\hat{\mathscr{H}}_L} \tag{E.16}$$

(b) Strong collision limit [11, 12]. Here the fluctuations due to the lattice interactions are very slow, i.e. $1/\tau \ll \delta\omega_z$ leading to

$$S(t-t') \simeq e^{-i(1-P)\hat{I}_z \int_{0}^{t-t'} dt_2\,\omega_z(t_2)} \tag{E.17}$$

where $\omega_z(t_2)$ is caused by the lattice fluctuations. Further approximations are usually imposed to simplify Eq. (E.17).

Here we want to restrict ourselves to the "weak collision limit" which we will consider in the following. We can express the memory function matrix $\tilde{K}(t,t')$ in this case as

$$K_{jk}(t,t') = \frac{(I_j|\,\tilde{\mathscr{H}}_S(t)\,e^{-i(t-t')\hat{\mathscr{H}}_L}\,\hat{\mathscr{H}}_S(t')\,|I_k)}{(I_k|I_k)}$$
$$-\frac{1}{(I_k|I_k)}\sum_{l=1}^{3}(I_j|\,\hat{\mathscr{H}}_S(t)\,|I_e)\,(I_e|\,\hat{\mathscr{H}}_S(t')\,|I_k) \tag{E.18}$$

$\tilde{K}(t,t')$ in this form still contains non-secular terms of the form $\cos\omega_0 t$, $\sin\omega_0 t$ etc. which will be neglected. We do, however, keep terms of the form $\cos\omega_0(t-t')$, $\sin\omega_0(t-t')$ which belong to the secular part, leading to

$$\tilde{\mathbf{K}}(\theta) = \begin{pmatrix} K_{xx} & K_{xy} & 0 \\ -K_{xy} & K_{yy} & 0 \\ 0 & 0 & K_{zz} \end{pmatrix} \quad \text{where} \quad \theta = t - t' \tag{E.19}$$

and

$$K_{zz}(\theta) = [g_{xx}(\theta) + g_{yy}(\theta)]\cos\omega_0\theta + [g_{xy}(\theta) - g_{yx}(\theta)]\sin\omega_0\theta \tag{E.20}$$

$$K_{xx}(\theta) = K_{yy}(\theta) = g_{zz}(\theta) + \tfrac{1}{2}K_{zz}(\theta) \tag{E.21}$$

$$K_{xy}(\theta) = \tfrac{1}{2}[g_{xy}(\theta) - g_{yx}(\theta)]\cos\omega_0\theta - \tfrac{1}{2}[g_{xx}(\theta) + g_{yy}(\theta)]\sin\omega_0\theta \tag{E.22}$$

with

$$g_{jk}(\theta) = \langle\omega_j(\theta)\,\omega_k(0)\rangle - \langle\omega_j\rangle\langle\omega_k\rangle \tag{E.23}$$

Using the symmetry $g_{xy}(\theta) = g_{yx}(\theta)$ results in

$$K_{zz}(\theta) = [g_{xx}(\theta) + g_{yy}(\theta)]\cos\omega_0\theta$$
$$K_{xy}(\theta) = -\tfrac{1}{2}[g_{xx}(\theta) + g_{yy}(\theta)]\sin\omega_0\theta \tag{E.24}$$

Having determined the frequency matrix $\tilde{\boldsymbol{\Omega}}$ Eq. (E.12) and the memory function matrix $\tilde{\mathbf{K}}(t-t')$ Eqs. (E.19-E.24) we can solve the equation of motion Eq. (E.8) rigorously by Laplace transform

$$|\mathbf{I}(s)) = [s\tilde{\mathbb{1}} + i\tilde{\boldsymbol{\Omega}} + \tilde{\mathbf{K}}(s)]^{-1}\,|\mathbf{I}(t=0)) \tag{E.25}$$

where $\mathbb{1}$ is the unit matrix and

$$\mathbf{I}(s) = \int_0^\infty dt\,\mathbf{I}(t)\,e^{-st} \tag{E.26}$$

It is, however, more convenient to separate first the z-component in the time domain

$$\frac{d}{dt}|I_z(t)) = -\int_0^t dt'\,K_{zz}(t-t')\,|I_z(t')) \tag{E.27}$$

Invoking, that $K_{zz}(\theta)$ decays much more rapidly than $I_z(t)$ (Markoff approximation) we may write

$$\frac{d}{dt}|I_z(t)) = -\frac{1}{T_1}|I_z(t))$$

where

$$\frac{1}{T_1} = \int\limits_0^\infty dt\, K_{zz}(t) \tag{E.28}$$

which leads to

$$\frac{1}{T_1} = j_{xx}(\omega_0) + j_{yy}(\omega_0) \tag{E.29}$$

with

$$j_{kl}(\omega_0) = \int\limits_0^\infty dt\, g_{kl}(t) \cos\omega_0 t \tag{E.30}$$

Following the same arguments one obtains for $1/T_2$

$$\frac{1}{T_2} = \int\limits_0^\infty dt\, K_{xx}(t) = j_{zz}(0) + \frac{1}{2T_1} \tag{E.31}$$

The Markoff approximation applied to Eq. (E.27) can, of course be avoided and a rigorous solution is immediately obtained by Laplace transform. In general this leads to a non-exponential decay. However, in the weak collision limit this non-exponential behaviour is negligible.

Using the Laplace transform we may calculate the spectrum

$$\langle I_x(\omega)\rangle = \frac{i\omega + K_{xx}(\omega)}{[i\omega + K_{xx}(\omega)]^2 + [\langle\omega_z\rangle - K_{xy}(\omega)]^2} \tag{E.32}$$

where

$$K(\omega) = \int\limits_0^\infty dt\, K(t)\, e^{-i\omega t} \tag{E.33}$$

In order to make contact with the general relaxation theory (Appendix I) we express the frequencies ω_j, $j = x, y, z$ by the spherical tensor components A_{kg} according to Appendix A as

$$\omega_z = \omega_0 \sigma_{zz} = \omega_0 \left[\sqrt{\tfrac{2}{3}}\, A_{20} - \frac{1}{\sqrt{3}}\, A_{00}\right] \tag{E.34}$$

$$\omega_x \pm i\omega_y = \omega_0 [A_{1\pm1} \pm A_{2\pm1}] \tag{E.35}$$

We therefore derive the following relations

$$j_{zz}(\omega) = \omega_0^2 [\tfrac{2}{3} j_{20}(\omega) + \tfrac{1}{3} j_{00}(\omega)] \tag{E.36}$$

and

$$j_{xx}(\omega) + j_{yy}(\omega) = \omega_0^2 [j_{11}(\omega) + j_{21}(\omega)] \tag{E.37}$$

where

$$j_{kq}(\omega) = \int\limits_0^\infty dt\, \{\langle A_{kq}(t)\, A_{k-q}(0)\rangle - |\langle A_{kq}\rangle|^2\}\, e^{-i\omega t} \tag{E.38}$$

which results in

$$\frac{1}{T_1} = \omega_0^2 [j_{11}(\omega_0) + j_{21}(\omega_0)] \tag{E.39}$$

and

$$\frac{1}{T_2} = \omega_0^2 \left[\tfrac{2}{3} j_{20}(0) + \tfrac{1}{3} j_{00}(0) \right] + \frac{1}{2T_1} \tag{E.40}$$

These results are identical to those of Appendix I Eq. (I.33) where spherical tensor notation has been used throughout.

F Time Evolution and Magnus Expansion

The general equation of motion for the density matrix $\rho(t)$ [7a]

$$\frac{d}{dt} \rho(t) = -i \hat{\mathscr{H}}(t) \rho(t) \tag{F.1}$$

where we have used the Liouville space notation (see Appendix C) with

$$\hat{\mathscr{H}}(t) \rho(t) \equiv [\mathscr{H}(t), \rho(t)].$$

Equation (F.1) is equivalent to the following integral equation [13]

$$\rho(t) = \rho(0) - i \int_0^t dt_1 \, \hat{\mathscr{H}}(t_1) \rho(t_1) \tag{F.2}$$

which has the formal iterative solution [13]

$$\rho(t) = \left[1 - i \sum_{n=1}^{\infty} \int_0^t dt_n \ldots \int_0^{t_2} dt_1 \, \hat{\mathscr{H}}(t_n) \ldots \hat{\mathscr{H}}(t_1) \right] \rho(0) \tag{F.3}$$

and can be expressed in short form as

$$\rho(t) = \mathbf{L}(t) \rho(0) \tag{F.4}$$

The disadvantages of this formula are twofold: Approximations obtained by truncation of the series are (i) not unitary and (ii) accurate only for small t or small $\|\mathscr{H}\|$. The evolution operator $L(t)$ may be formally expressed as [5, 14, 15]

$$\mathbf{L}(t) = T \exp\left[-i \int_0^t dt' \, \hat{\mathscr{H}}(t') \right] \tag{F.5}$$

in form of Dyson's time-ordered exponential. Magnus [5a] has solved the problem of finding an expression of the form

$$\mathbf{L}(t) = \exp[-i \hat{A}(t)] \tag{F.6}$$

where $A(t)$ is a function of multiple time integrals over products of $\mathscr{H}(t)$. $A(t)$ can be expressed by an infinite sum $(A(t) = A_1(t) + A_2(t) + \ldots$, called the Magnus-expansion. The first few terms in the Magnus-expansion were obtained by Magnus himself and others [13–15], namely

$$A_1(t) = \int_0^t dt_1 \, B_1 \tag{F.7}$$

$$A_2(t) = -\frac{i}{2} \int_0^t dt_2 \int_0^{t_2} dt_1 [B_2, B_1] \tag{F.8}$$

$$A_3(t) = -\frac{1}{6} \int_0^t dt_3 \int_0^{t_3} dt_2 \int_0^{t_2} dt_1 \{[B_3,[B_2,B_1]] + [B_1,[B_2,B_3]]\} \tag{F.9}$$

$$A_4(t) = \frac{i}{12} \int_0^t dt_4 \int_0^{t_4} dt_3 \int_0^{t_3} dt_2 \int_0^{t_2} dt_1 \{[B_4,[B_3,[B_2,B_1]]] + [B_1,[B_2,[B_4,B_3]]]$$

$$+ [B_4,[B_1,[B_2,B_3]]] + [B_3,[B_2,[B_1,B_4]]]\} \tag{F.10}$$

where $B_j = \mathscr{H}(t_j)$.

In fact the whole series has been derived by Bialynicki-Birula et al. [13] and can be expressed in compact form by

$$A_n(t) = (-i)^{n-1} \int_0^t dt_n \ldots \int_0^{t_2} dt_1 \, F(B_n \ldots B_1) \tag{F.11}$$

where

$$F(B_n \ldots B_1) = \frac{1}{n!} \sum_\pi (-1)^{n-k-1} k!(n-k-1)! \frac{1}{n} \{B_n \ldots B_1\} \tag{F.12}$$

Here the sum is to be formed over all permutations π of $B_n \ldots B_1$ and k is the total number of chronologically ordered neighbors (with $t_{j+1} > t_j$) in the permutation π and where $\{\ldots\}$ is the multiple commutator.

G Coherent Versus Secular Averaging Theory

By utilizing the Magnus-expansion, Haeberlen and Waugh [5d] have proposed to express the time evolution operator $L(t)$ at integer intervals nt_c, when the timedependence of $\mathscr{H}(t)$ is cyclic, with a cycle time t_c, i.e.

$$\mathscr{H}(t) = \mathscr{H}(t + nt_c) \tag{G.1}$$

leading to

$$L(nt_c) = \exp[-int_c \bar{\mathscr{H}}] \tag{G.2}$$

where

$$\bar{\mathscr{H}} = \bar{\mathscr{H}}^{(0)} + \bar{\mathscr{H}}^{(1)} + \bar{\mathscr{H}}^{(2)} + \ldots \tag{G.3}$$

and

$$\bar{\mathscr{H}}^{(n)} = \frac{1}{t_c} A_{n+1}(t_c)$$

with the $A(t)$ as derived in Appendix F Eqs. (F.7–F.10). This corresponds to a "stroboscopic" description of the time-evolution of the density matrix. What one is interested in, however, is not the behaviour of $\rho(t)$ at times nt_c but rather one is interested in the "slow motion" evolution of $\rho(t)$ at all times t. In other words one wants to smooth out all non-secular behaviour. H. Grupp [16] pointed out that such an approach was applied by Buishvili and Menabde [17], who used the averaging procedure of Krylov, Bogoljubov and Mitropolskij [18].

Here we follow a similar route based, however, on successive approximations in the frequency domain. In NMR we usually encounter the situation of periodic excitation. The time-dependence of the Hamiltonian is therefore conveniently expressed by a Fourier series

$$\mathcal{H}(t) = \sum_{n=-\infty}^{+\infty} H_n \, e^{i\omega_n t} \tag{G.4}$$

with

$$H_n = \frac{1}{t_c} \int_0^{t_c} dt \, \mathcal{H}(t) \, e^{-i\omega_n t}; \quad \omega_n = \frac{2\pi}{t_c} n \tag{G.5}$$

Correspondingly we express the density matrix $\rho(t)$ as

$$\rho(t) = \sum_m \rho_m(t) \, e^{i\omega_m t} \tag{G.6}$$

where $\rho_m(t)$ represents the slow motion evolution. Inserting $\mathcal{H}(t)$ and $\rho(t)$ into the equation of motion

$$\frac{d}{dt}\rho(t) = -i\hat{\mathcal{H}}(t)\,\rho(t) \tag{G.7}$$

leads to

$$\left\{ \frac{d}{dt}\rho_m(t) + i\omega_m \, \rho_m(t) + i\sum_k \hat{H}_{m-k}\,\rho_k(t) \right\} e^{i\omega_m t} = 0 \tag{G.8}$$

the slowly moving part of $\rho(t)$ we are looking for is $\rho_0(t)$, whoose equation of motion is readily obtained from Eq. (G.8)

$$\frac{d}{dt}\rho_0(t) + i\hat{H}_0\,\rho_0(t) + i\sum_{k\neq 0} \hat{H}_{-k}\,\rho_k(t) = 0 \tag{G.9}$$

In a zeroth order approximation we neglect the sum over $k \neq 0$ and obtain

$$\rho_0(t) = e^{-it\hat{\mathcal{H}}_0}\,\rho_0(t=0) \tag{G.10}$$

where

$$H_0 = \frac{1}{t_c} \int_0^{t_c} dt \, \mathcal{H}(t) \tag{}$$

according to Eq. (G.5), i.e. the same result as in coherent averaging is obtained (see Chap. 3.3 and reference [5d]). The deviation, however, occurs in the higher order corrections. In order to obtain these it is convenient to Laplace transform Eqs. (G.8) and (G.9)

$$(s + i\omega_m)\,\rho_m(s) + i\hat{H}_m\,\rho_0(s) + i\sum_{k\neq 0}\hat{H}_{m-k}\,\rho_k(s) = 0 \quad \text{for } m\neq 0 \tag{G.11}$$

and

$$s\rho_0(s) + i\hat{H}_0\,\rho_0(s) + i\sum_{k\neq 0}\hat{H}_{-k}\,\rho_k(s) = \rho(0) \tag{G.12}$$

where

$$\rho_m(s) = \int_0^\infty dt \, \rho_m(t) \, e^{-st}$$

and where the initial condition $\rho_{m\neq 0}(t=0)=0; \; \rho_0(t=0)=\rho(0)$.

We are looking for an expression of the form

$$(s+i\hat{\bar{H}})\,\rho_0(s)=\rho(0) \tag{G.13}$$

where

$$\hat{\bar{H}}=\hat{\bar{H}}^{(0)}+\hat{\bar{H}}^{(1)}+\hat{\bar{H}}^{(2)}+\ldots$$

which has the solution

$$\rho_0(t)=e^{-it\hat{\bar{H}}}\,\rho(0)$$

for the time evolution of the "secular" part of the density matrix. We therefore call this procedure "secular averaging". In order to obtain the higher order terms one has to solve Eq. (G.11) for $\rho_m(s)$ and insert it redefined as $\rho_k(s)$ into Eq. (G.12). Since only the secular motion is required ($|s|\ll\omega_m$) $s+i\omega_m$ is replaced by $i\omega_m$. The following series for $\hat{\bar{H}}$ is readily obtained

$$\hat{\bar{H}}^{(1)}=-\sum_{k\neq 0}\frac{1}{\omega_k}\hat{H}_{-k}\hat{H}_k$$

$$\hat{\bar{H}}^{(2)}=\sum_{\substack{k\neq 0\\m\neq 0}}\frac{1}{\omega_k\omega_m}\hat{H}_{-k}\hat{H}_{k-m}\hat{H}_m \tag{G.14}$$

$$\vdots$$

$$\hat{\bar{H}}^{(n)}=(-1)^n\sum_{k_1\ldots k_n}\frac{1}{\omega_{k_1}\ldots\omega_{k_n}}\hat{H}_{-k_1}H_{k_1-k_2}\ldots\hat{H}_{k_{n+1}-k_n}\hat{H}_{k_n}$$

Note, that

$$\hat{A}\hat{B}\ldots\hat{C}X\equiv[A,[B,[\ldots[C,X]]\ldots] \tag{G.15}$$

In order to obtain useful expressions for the average Hamiltonian we have to rearrange the commutators into the form $[[A,[B,\ldots,C]]\ldots],X]$. For the first two terms of the average Hamiltonian we obtain

$$\bar{H}^{(0)}=H_0$$

$$\bar{H}^{(1)}=-\sum_{k>0}\frac{1}{\omega_k}[H_{-k},H_k] \tag{G.16}$$

This is the expression obtained by Buishvili and Menabde [17]. However, Eq. (G.14) allows the calculation of higher order approximations.

Let us restrict ourselves here to the comparison of the secular averaged Hamiltonian H_1 with the coherently averaged Hamiltonian $\mathscr{H}^{(1)}$. In order to facilitate comparison we express both in the same notation

$$\bar{\mathscr{H}}^{(1)}=-\frac{i}{2t_c}\int_0^{t_c}dt_2\int_0^{t_2}dt_1[\mathscr{H}(t_2),\mathscr{H}(t_1)] \tag{G.17}$$

$$\bar{\mathscr{H}}^{(1)}=\tfrac{1}{2}\sum_{k\neq 0}\frac{1}{\omega_k}\{[H_{-k},H_k]-2[H_{-k},H_0]\} \tag{G.18}$$

$$\bar{H}^{(1)}=\tfrac{1}{2}\sum_{k\neq 0}\frac{1}{\omega_k}[H_{-k},H_k] \tag{G.19}$$

$$\bar{H}^{(1)}=-\frac{i}{2t_c}\int_0^{t_c}dt_2\int_0^{t_2}dt_1[\mathscr{H}(t_2)-\bar{\mathscr{H}}^{(0)},\mathscr{H}(t_1)-\bar{\mathscr{H}}^{(0)}] \tag{G.20}$$

It is evident, that both average Hamiltonian are equal if $\mathscr{H}^{(0)} = H_0 = 0$. If, however, $\bar{\mathscr{H}}^{(0)} = H_0 \neq 0$ there is a "non-secular" contribution to $\mathscr{H}^{(1)}$ which leads to improper results. It seems therefore advisable to subtract $\bar{\mathscr{H}}^{(0)}$ from $\mathscr{H}(t)$ before calculating $\bar{\mathscr{H}}^{(1)}$ following the coherent averaging approach. Examples of average Hamiltonian calculations can be found in the text and Appendix H.

H Applications of Average Hamiltonian Theory

(a) Contribution of non-secular shielding tensor elements to the resonance shift

The shift of the nuclear resonance line is given by the "size" of the shift Hamiltonian $\mathscr{H}_S$, namely

$$\|\mathscr{H}_S\| = \left[\frac{(\mathscr{H}_S | \mathscr{H}_S)}{(I_z | I_z)}\right]^{1/2} \tag{H.1}$$

In the laboratory frame we have

$$\mathscr{H}_S = \mathscr{H}_{Sz} + \mathscr{H}_S'$$

where

$$\mathscr{H}_{Sz} = \omega_0 \sigma_{zz} I_z$$

and

$$\mathscr{H}_S' = \omega_0(\sigma_{xz} I_x + \sigma_{yz} I_y). \tag{H.2}$$

In the rotating frame, where we perform our measurements we can write

$$\tilde{\mathscr{H}}_S(t) = e^{-i\omega_0 t I_z} \mathscr{H}_S e^{i\omega_0 t I_z} \tag{H.3}$$

where only the term $\omega_0 \sigma_{zz} I_z$ is stationary.

We are going to express the shielding Hamiltonian in the rotating frame by an secular averaged Hamiltonian $\bar{\mathscr{H}}_S$ according to Appendix F as

$$\bar{\mathscr{H}}_S = \bar{H}_S^{(0)} + \bar{H}_S^{(1)} + \ldots \tag{H.4}$$

where we use

$$|\bar{H}_S^{(0)}) = \omega_0 \sigma_{zz} |I_z) \tag{H.5}$$

$$|\bar{H}_S^{(1)}) = -\frac{i}{2t} \int_0^t dt_2 \int_0^{t_2} dt_1 \, e^{-i\omega_0 t_2 \hat{I}_z} \hat{\mathscr{H}}_S' e^{i(t_2 - t_1)\omega_0 \hat{I}_z} |\mathscr{H}_S'). \tag{H.6}$$

Using Eqs. (H.1) and (H.4) we write

$$\|\mathscr{H}_S\|^2 = \frac{(\bar{H}_S^{(0)} | \bar{H}_S^{(0)}) + 2(\bar{H}_S^{(0)} | \bar{H}_S^{(1)}) + (\bar{H}_S^{(1)} | \bar{H}_S^{(1)})}{(I_z | I_z)}. \tag{H.7}$$

We readily obtain

$$\frac{(\bar{H}_S^{(0)} | \bar{H}_S^{(0)})}{(I_z | I_z)} = \omega_0^2 \sigma_{zz}^2 \tag{H.8}$$

$$\frac{(\bar{H}_S^{(0)}|\bar{H}_S^{(1)})}{(I_z|I_z)} = -\frac{i}{2t}\int_0^t dt_2 \int_0^{t_2} dt_1\,\omega_0\sigma_{zz}\left\{\frac{(I_z|\,e^{-i\omega_0 t_2 \hat{I}_z}\,\hat{\mathcal{H}}_S'\,e^{i(t_2-t_1)\omega_0 \hat{I}_z}\,|\mathcal{H}_S')}{(I_z|I_z)}\right\}$$

$$= \frac{\omega_0\sigma_{zz}}{2t}\int_0^t dt_2 \int_0^{t_2} dt_1\,[-i\{\ \}] \tag{H.9}$$

with

$$-i\{\ \} = \omega_0\left[\sigma_{xz}\frac{(I_y|\,e^{i(t_2-t_1)\omega_0 \hat{I}_z}\,|\mathcal{H}_S')}{(I_z|I_z)} - \sigma_{yz}\frac{(I_x|\,e^{i(t_2-t_1)\omega_0 \hat{I}_z}\,|\mathcal{H}_S')}{(I_z|I_z)}\right] \tag{H.10}$$

or

$$i\{\ \} = \omega_0^2(\sigma_{xz}^2 + \sigma_{yz}^2)\sin\omega_0(t_2-t_1). \tag{H.11}$$

Integration over a "cycle" $\omega_0 t = 2\pi$ leads to

$$-2\cdot\frac{(\bar{H}_S^{(0)}|\bar{H}_S^{(1)})}{(I_z|I_z)} = \omega_0^2\sigma_{zz}(\sigma_{xz}^2 + \sigma_{yz}^2). \tag{H.12}$$

The contribution of $(\bar{H}_S^{(1)}|\bar{H}_S^{(1)})$ is next higher order and will be neglected here. It can be of course included straightforwardly. Inserting Eqs. (H.8, H.12) into Eq. (H.7) leads to

$$\|\mathcal{H}_S\| = \omega_0\sigma_{zz}\left[1 - \frac{\sigma_{xz}^2 + \sigma_{yz}^2}{\sigma_{zz}}\right]^{1/2}. \tag{H.13}$$

Since σ is a small number on the order of $\leq 10^{-4}$, the relative contribution of non-secular terms to the resonance shift $\omega_0\sigma_{zz}$ is of the same order of amgnitude. The same argument applies to the antisymmetric tensor elements, of course, since the σ_{xz}, σ_{yz} contain the symmetric and the antisymmetric constituents of the shielding tensor. Equation (H.13) justifies that we have neglected antisymmetric tensor contributions before. Equations (H.13) may be further approximated by [6]

$$\|\mathcal{H}_S\| = \omega_0\sigma_{zz}\left(1 - \frac{\sigma_{xz}^2 + \sigma_{yz}^2}{2\sigma_{zz}}\right) = \omega_0(\sigma_{zz} - \tfrac{1}{2}(\sigma_{xz}^2 + \sigma_{yz}^2)) \tag{H.14}$$

let us separate the shielding tensor σ_{ij} into a symmetric part $\sigma_{ij}^{(s)}$ and an antisymmetric part $\sigma_{ij}^{(a)}$

$$\sigma_{ij} = \sigma_{ij}^{(s)} + \sigma_{ij}^{(a)}. \tag{H.15}$$

The symmetric part may be diagonalized and is given in its principal axis frame $(1,2,3)$ by

$$\sigma_{ii}^{(s)} = \begin{pmatrix} \sigma_{11}^{(s)} & 0 & 0 \\ 0 & \sigma_{22}^{(s)} & 0 \\ 0 & 0 & \sigma_{33}^{(s)} \end{pmatrix} \tag{H.16}$$

whereas the antisymmetric part $\sigma_{ij}^{(a)}$ is given in the same frame as

$$\sigma_{ij}^{(a)} = \begin{pmatrix} 0 & \sigma_{12}^{(a)} & \sigma_{13}^{(a)} \\ -\sigma_{12}^{(a)} & 0 & \sigma_{23}^{(a)} \\ -\sigma_{13}^{(a)} & -\sigma_{23}^{(a)} & 0 \end{pmatrix}. \tag{H.17}$$

With the transformation $\mathbf{R}(\alpha, \beta, \gamma)$ we transform the shielding tensor from the principal axis frame into the magnetic field frame (lab.-frame) as

$$\sigma_{ij} = \mathbf{R}(\alpha\beta\gamma)\,\sigma_{ij}^{(1,2,3)}\,\mathbf{R}^{-1}(\alpha\beta\gamma), \tag{H.18}$$

$$\sigma_{ij} = \sum_{k,l} r_{ik}\,\sigma_{kl}^{(1,2,3)}\,r_{ji}. \tag{H.19}$$

We obtain

$$\sigma_{zz} = \sum_{k} r_{3k}^2\,\sigma_{kk}^{(s)} + \sum_{k,e} r_{3k}\,r_{3e}\,\sigma_{ke}^{(a)}. \tag{H.20}$$

Since $\sigma_{kl}^{(a)} = -\sigma_{lk}^{(a)}$ the last term in Eq. (H.20) vanishes, leaving

$$\sigma_{zz} = \sum_{k} r_{3k}^2\,\sigma_{kk}^{(s)}. \tag{H.21}$$

The non-secular components are correspondingly

$$\sigma_{xz} = \sum_{k,e} r_{1k}\,r_{3e}\,\sigma_{ke}, \tag{H.22}$$

$$\sigma_{yz} = \sum_{k,e} r_{2k}\,r_{3e}\,\sigma_{ke}. \tag{H.23}$$

This can be combined to obtain the non-secular contribution to the shift as

$$\tfrac{1}{2}(\sigma_{xz}^2 + \sigma_{yz}^2) = \sum_{i,j,k,e} r_{3j}\,r_{3e}\,\tfrac{1}{2}(r_{1i}r_{1k} + r_{2i}r_{2k})\,\sigma_{ij}\,\sigma_{ke}. \tag{H.24}$$

Equation (H.24) is quite a general formula, from which special cases can be obtained readily.

Let us now discuss some specific examples:

(i) The shielding tensor σ is purely symmetric

$$\sigma_{ij}^{(s)} = \sigma_{ii}^{(s)}\,\delta_{ij} \tag{H.25}$$

$$\tfrac{1}{2}(\sigma_{xz}^{(s)2} + \sigma_{yz}^{(s)2}) = \sum_{i,k} r_{3i}\,r_{3k}\,\tfrac{1}{2}(r_{1i}r_{1k} + r_{2i}r_{2k})\,\sigma_{ii}^{(s)}\,\sigma_{kk}^{(s)}.$$

(ii) The tensor σ is purely antisymmetric

$$\sigma_{ij}^{(a)} = -\sigma_{ji}^{(a)}(1 - \delta_{ij}) \tag{H.26}$$

$$\tfrac{1}{2}(\sigma_{xz}^{(a)2} + \sigma_{yz}^{(a)2}) = \sum_{i<j;k<e} \big[r_{3j}r_{3e}\tfrac{1}{2}(r_{1i}r_{1k} + r_{2i}r_{2k})$$
$$ - r_{3i}r_{3k}\tfrac{1}{2}(r_{1j}r_{1e} + r_{2j}r_{2e})\big]\,\sigma_{ij}^{(a)}\,\sigma_{ke}^{(a)}.$$

(iii) Special orientation dependence: $\alpha = \gamma = 0$

$$R = \begin{pmatrix} \cos\beta & 0 & -\sin\beta \\ 0 & 1 & 0 \\ \sin\beta & 0 & \cos\beta \end{pmatrix}$$

$$\sigma_{zz} = \sin^2\beta\,\sigma_{11}^{(s)} + \cos^2\beta\,\sigma_{33}^{(s)}$$

$$\tfrac{1}{2}(\sigma_{xz}^{(s)2} - \sigma_{yz}^{(s)2}) = \tfrac{1}{2}\sin^2\beta\,\cos^2\beta(\sigma_{33}^{(s)} - \sigma_{11}^{(s)})^2$$

$$\tfrac{1}{2}(\sigma_{xz}^{(a)2} + \sigma_{yz}^{(a)2}) = \tfrac{1}{2}\cos^2\beta\,\sigma_{23}^{(a)2} - \tfrac{1}{2}\sin^2\beta\,\sigma_{12}^{(a)2} + \tfrac{1}{2}(\cos^4\beta - \sin^4\beta)\,\sigma_{13}^{(a)2}. \tag{H.27}$$

Mixed terms have to be considered in the general case, if both symmetric and antisymmetric constituents are present. This can be done straightforwardly by starting from Eq. (H.24).

(b) Bloch-Siegert Shift [19]

Irradiation of a spin system by an rf field

$$\mathscr{H}_1 = -2\omega_1 I_x \cos \omega t \tag{H.28}$$

causes a rotation of the spins in a suitable reference frame. Suppose for example, the irradiation is performed

(i) on-resonance $\omega = \omega_0$ Larmor frequency

$\mathscr{H}_1$ is transformed into the rotating frame as

$$\mathscr{H}_1(t) = -[\omega_1 I_x + \omega_1 I_x \cos 2\omega_0 t + \omega_1 I_y \sin 2\omega_0 t] = -\omega_1 I_x + \tilde{\mathscr{H}}_1(t) \tag{H.29}$$

The secular part $-\omega_1 I_x$ causes a rotation about the x axis with the frequency ω_1. The non-secular term is due to the counter rotating part of the rf field and contributes to the rotation only in higher order. This non-secular contribution is called the "Bloch-Siegert shift" and can be usually neglected as follows.

Expressing $\mathscr{H}_1$ by an average Hamiltonian according to Appendix F as

$$\bar{H}_1 = \bar{H}_1^{(0)} + \bar{H}_1^{(1)} + \ldots \tag{H.30}$$

we arrive at $i = 1, 2.$

$$\bar{H}_1^{(0)} = -\omega_1 I_x \tag{H.31}$$

$$\bar{H}_1^{(1)} = \frac{-i}{2\omega_0^2 t} \int\limits_0^{\omega_0 t} d\alpha_2 \int\limits_0^{\alpha_2} d\alpha_1 [\tilde{\mathscr{H}}_1(\alpha_2), \tilde{\mathscr{H}}_1(\alpha_1)] \tag{H.32}$$

where $\alpha_i = \omega_0 t_i$ $i = 1, 2.$

If we define a cycle by $\omega_0 t = 2\pi$ ("stroboscopic observation") [6], we arrive at

$$\bar{H}_1^{(1)} = -\omega_1^{(1)} I_z \tag{H.33}$$

where

$$\omega_1^{(1)} = \omega_0 \left(\frac{\omega_1}{2\omega_0}\right)^2. \tag{H.34}$$

The relative Bloch-Siegert shift [6, 19]

$$\varepsilon = \omega_1^{(1)}/\omega_0 = \left[\frac{\omega_1}{2\omega_0}\right]^2 \tag{H.35}$$

is usually negligible.

However, in decoupling experiments, where a strong rf field is applied at a frequency usually much closer to the Larmor frequency of the spins to be observed, than the counter rotating component, remarkable resonance shifts can be expected.

(ii) *off-resonance* $|\omega - \omega_0| \gg \omega_1$

In the "resonance frame" rotating with the frequency ω_0, we can express $\tilde{\mathscr{H}}_1(t)$ as:

$$\tilde{\mathscr{H}}_1(t) = -\omega_1 I_x [\cos(\omega + \omega_0)t + \cos(\omega - \omega_0)t]$$
$$- \omega_1 I_y [\sin(\omega + \omega_0)t - \sin(\omega - \omega_0)t]. \tag{H.36}$$

We express now the propagator of $\tilde{\mathscr{H}}_1(t)$ by an average Hamiltonian

$$\bar{\mathscr{H}}_1 = \bar{\mathscr{H}}_1^{(0)} + \bar{\mathscr{H}}_1^{(1)} + \ldots$$

where we have to choose convenient cycles.

It can be shown, that a cycle time can be always defined with $\omega t_c = m2\pi$ and $\omega_0 t_c = n2\pi$ ("stroboscopic observation") [6].

Using this cycle property of $\tilde{\mathscr{H}}_1(t)$ we obtain

$$\bar{\mathscr{H}}_1^{(0)} = 0$$

$$\bar{\mathscr{H}}_1^{(1)} = \frac{-i}{2t_c} \int_0^{t_c} dt_2 \int_0^{t_2} dt_1 [\tilde{\mathscr{H}}_1(t_2), \tilde{\mathscr{H}}_1(t_1)]$$

and by choosing "proper" cycles

$$\bar{\mathscr{H}}_1^{(1)} = -\omega_1^{(1)} I_z$$

where

$$\omega_1^{(1)} = -\frac{1}{2}\omega_1^2 \left[\frac{1}{\omega_0 + \omega} + \frac{1}{\omega_0 - \omega} \right] \tag{H.37}$$

or

$$\omega_1^{(1)} = \omega_0 \left[\frac{\omega_1^2}{\omega^2 - \omega_0^2} \right] \tag{H.38}$$

with the relative shift

$$\varepsilon = \omega_1^{(1)}/\omega_0 = \frac{\omega_1^2}{\omega^2 - \omega_0^2} \tag{H.39}$$

Examples:

a) ^{1}H decoupled, ^{13}C observed

$$\nu_0(^{13}\text{C}) = 23 \text{ MHz}; \ \nu(^1\text{H}) = 90 \text{ MHz}; \ \nu_1(^1\text{H}) = 100 \text{ kHz } (23.5 \text{ G})$$

results in

$$\varepsilon = 1.32 \text{ ppm}$$

b) ^{1}H decoupled, ^{19}F observed

$$\nu_0(^{19}\text{F}) = 84.6 \text{ MHz}; \ \nu(^1\text{H}) = 90 \text{ MHz}; \ \nu_1(^1\text{H}) = 100 \text{ kHz } (23.5 \text{ G})$$

results in

$$\varepsilon = 10.6 \text{ ppm}.$$

Notice, that the sign of the Bloch-Siegert shift changes, according to Eq. (H.38) depending, whether irradiation is performed above $(+)$ or below $(-)$ resonance of the observed spins. In decoupling experiments this shift has to be taken into account. This is usually done by measuring the resonance frequency of a reference sample with the decoupling field on, i.e. reference signal is affected by the same shift and the frequency difference to the spectral line is about the same. Notice, however, that different lines in a spectrum are affected differently. A simple calculation shows, however, that this distinction is usually negligible.

Suppose two lines in the spectrum are separated by $\Delta\omega$. The change of this separation by the Bloch-Siegert shift is according to Eq. (H.38)

$$\Delta\omega_1^{(1)} \cong \Delta\omega \cdot \varepsilon$$

i.e. this absolute shift is of the order of some 10^{-12}. This is even in "Highest Resolution NMR" undetectable.

I Relaxation Theory

Here we outline a general relaxation theory which follows directly from the equation of motion for the density matrix $\rho(t)$, governed by the time-independent total Hamiltonian

$$\rho(t) = e^{-it\mathcal{H}} \rho(0) \tag{I.1}$$

with

$$\mathcal{H} = \mathcal{H}_z + \mathcal{H}_1 + \mathcal{H}_{int} + \mathcal{H}_L \tag{I.2}$$

which includes the spin-Hamiltonian

$$\mathcal{H}_z = \omega_I I_z + \omega_S S_z$$
$$\mathcal{H}_1 = \omega_1 I_x 2\cos\omega t$$
$$\mathcal{H}_{int} = \sum_{k=0}^{2} \sum_{q=-k}^{+k} (-1)^q A_{kq} T_{k-q}$$

and the lattice Hamiltonian $\mathcal{H}_L$. Note, that

$$[\mathcal{H}_I, \mathcal{H}_S] = [\mathcal{H}_L, \mathcal{H}_z] = [\mathcal{H}_L, \mathcal{H}_1] = 0$$

but

$$[\mathcal{H}_z, \mathcal{H}_{int}] \neq 0; \quad [\mathcal{H}_L, \mathcal{H}_{int}] \neq 0 \tag{I.3}$$

Therefore any explicit time-dependence in the Hamiltonian is only introduced by the following two steps:

(i) transformation into the rotating frame [3, 12]

$$\rho(t) = e^{-it(\omega I_z + \omega_S S_z)} T e^{-i\int_0^t dt'(\Delta\omega \hat{I}_z + \mathcal{H}_1(t') + \mathcal{H}_{int}(t') + \mathcal{H}_L)} \rho(0) \tag{I.4a}$$

or

$$\rho(t) = e^{-it(\omega I_z + \omega_S S_z)} \rho^*(t) \tag{I.4b}$$

where $\rho^*(t)$ is the density matrix in the rotating frame.

The Hamiltonian in the rotation frame

$$\mathscr{H}^*(t) = \Delta\omega I_z + e^{it(\omega I_z + \omega_S S_z)}(\mathscr{H}_1 + \mathscr{H}_{int})\, e^{-it(\omega I_z + \omega_S S_z)} + \mathscr{H}_L \tag{I.5}$$

where

$$\Delta\omega = \omega - \omega_I$$

becomes now explicitly *time-dependent*. One usually neglects the non-secular parts of $\mathscr{H}_1(t)$, leading to

$$\mathscr{H}^*(t) = \Delta\omega I_z + \omega_1 I_x + \mathscr{H}_{int}(t) + \mathscr{H}_L \tag{I.6}$$

and where

$$\mathscr{H}_{int}(t) = e^{it(\omega I_z + \omega_S S_z)}\, \mathscr{H}_{int}\, e^{-it(\omega I_z + \omega_S S_z)} \tag{I.7}$$

Since $\omega \simeq \omega_I$ we will replace ω by ω_I in the following, when applying Eq. (I.7).

In order to treat all the common spin-interactions equivalently, we shall use the abbreviation D: for homonuclear dipolar interaction, J: for J-coupling of homonuclear spins, Q: for quadrupole interaction, CS: for chemical shift interaction and IS: for heteronuclear interaction between I spins and S spins. According to Appendix A we can express now $\mathscr{H}_{int}(t)$ for any of those spin-interactions by

$$\mathscr{H}_{int}(t) = \sum_{k=0}^{2} \sum_{q=-k}^{+k} (-1)^q A_{kq}\, T_{k-q}(t) \tag{I.8}$$

where

$$T_{kq}(t) = \begin{cases} e^{itq\omega_I}\, T_{kq} & \text{for } D, J, Q \\ e^{it(q_1\omega_I + q_2\omega_S)}\, T_{kq} & \text{for } IS, CS\ (q_2 = 0) \end{cases} \tag{I.9}$$

with $q_1 + q_2 = q$.

The timedependence of $\mathscr{H}_{int}(t)$, however, is a coherent motion and does not lead to relaxation. In order to treat relaxation we have to include the influence of lattice fluctuation on the spin interactions due to $[\mathscr{H}_{int}, \mathscr{H}_L] \neq 0$. This is most directly done by using the lineshape theory outlined in Appendix C.

In order to treat T_1, $T_{1\rho}$ and T_2 by the same formalism, we are looking for the timeevolution of $\langle I_\rho(t) \rangle$ where

$$I_\rho = e^{-i\beta I_y} I_z\, e^{i\beta I_y} \tag{I.10}$$

and

$$\tan\beta = \omega_1/\Delta\omega; \quad \omega_e = (\omega_1^2 + \Delta\omega^2)^{1/2} \tag{I.11}$$

We expect to derive an equation of the form

$$\frac{d}{dt}\langle I_\rho(t) \rangle = -\frac{1}{T_{1\rho}}\langle I_\rho(t) \rangle \tag{I.12}$$

from where T_1 is obtained by setting $\beta = 0$; $\omega_e = 0$ and where T_2 corresponds to $\beta = \pi/2$; $\omega_e \to 0$. If we want to calculate $T_{2\rho}$, the relaxationtime orthogonal to the quantization axis in the rotating frame, we have to replace I_ρ by

$$I_\rho^{(\perp)} = e^{-i\beta I_y} I_x\, e^{i\beta I_y}$$

(ii) Transformation into the "lattice fluctuating frame"

In order to apply the general lineshape theory according to Appendix C the "relevant" part of the density matrix $\rho(t)$ is given by its projection onto I_ρ by

$$P = \frac{|I_\rho)(I_\rho|}{(I_\rho|I_\rho)} \tag{I.13}$$

Following the procedure outlined in Appendix C we obtain

$$\frac{d}{dt}\langle I_\rho(t)\rangle = -\int_0^t dt'\, K(t,t')\,\langle I_\rho(t')\rangle \tag{I.14}$$

where

$$K(t,t') = \frac{(I_\rho|\,\hat{\mathscr{H}}_{\mathrm{int}}(t)\,S(t,t')\,\hat{\mathscr{H}}_{\mathrm{int}}(t')\,|I_\rho)}{(I_z|I_z)} \tag{I.15}$$

with

$$S(t,t') = T e^{-i\int_{t'}^t dt_1\,\hat{\mathscr{H}}^*(t_1)} \tag{I.16}$$

where $\mathscr{H}^*(t)$ is given by Eq. (I.6). $S(t,t')$ may be separated into three parts

$$S(t,t') = S_1(t-t')\,S_L(t-t')\,S_{\mathrm{int}}(t,t') \tag{I.17}$$

with

$$S_1(t-t') = e^{-i(t-t')(\Delta\omega \hat{I}_z + \omega_1 \hat{I}_x)} = e^{-i\beta \hat{I}_y}\,e^{-i(t-t')\omega_e \hat{I}_z}\,e^{i\beta \hat{I}_y}$$

$$S_L(t-t') = e^{-i(t-t')\hat{\mathscr{H}}_L}$$

and

$$S_{\mathrm{int}}(t,t') = T e^{-i\int_{t'}^t dt_1\,\hat{\tilde{\mathscr{H}}}_{\mathrm{int}}(t_1)} \tag{I.18}$$

where

$$\hat{\tilde{\mathscr{H}}}_{\mathrm{int}}(t) = S_L(t-t')\,S_1(t-t')\,\hat{\mathscr{H}}_{\mathrm{int}}(t)\,S_1^{-1}(t-t')\,S_L^{-1}(t-t')$$

In the "weak collision limit" the fluctuations caused by $\mathscr{H}_L$, characterized by the correlationtime τ are rapid compared with $\|\mathscr{H}_{\mathrm{int}}\|$, i.e. $\tau \ll T_2$ or

$$S_{\mathrm{int}}(t,t') \simeq e^{-i(t-t')\bar{\mathscr{H}}_{\mathrm{int}}} \tag{I.19}$$

where $\bar{\mathscr{H}}_{\mathrm{int}}$ is the timeaveraged spin interaction. In the "strong collision limit", however, $\tau \gg T_2$, i.e. the timevariation of $\tilde{\mathscr{H}}_{\mathrm{int}}(t)$ due to the lattice motion has to be taken into account in the expression for $S_{\mathrm{int}}(t,t')$ [11, 12]. Here we restrict ourselves to the "weak collision limit". Moreover $S_{\mathrm{int}}(t,t')$ according to Eq. (I.19) does not contribute to the decay of $K(t,t')$ significantly in this case, since $\tau \ll T_2$ and we obtain in the "weak collision limit"

$$K(t,t') = \frac{(I_\rho|\,\hat{\mathscr{H}}_{\mathrm{int}}(t)\,S_1(t-t')\,S_L(t-t')\,\hat{\mathscr{H}}_{\mathrm{int}}(t')\,|I_\rho)}{(I_z|I_z)}. \tag{I.20}$$

Note, that the lattice Hamiltonian $\mathscr{H}_L$ operates only on the lattice part of $\mathscr{H}_{\mathrm{int}}$, namely on A_{kq}. Since we are interested in relaxation rather than lineshapes, we consider only the deviation of $A_{kq}(t)$ from its equilibrium value $\langle A_{kq}\rangle$, i.e. $A_{kq}(t)$ stands in the following for the expression $A_{kq}(t) - \langle A_{kq}\rangle$. Using the invariance of $K(t,t')$ under translation of time, we can express $K(t,t') = K(t-t')$

after some algebra as

$$K(t) = \sum_{k=0}^{2} \sum_{q=-k}^{+k} g_{kq}(t)\, e^{-it\tilde{q}\omega}\, X_{kq}(\beta, t) \tag{I.21}$$

where

$$g_{kq}(t) = (-1)^q \langle A_{kq}(0)\, A_{k-q}(t)\rangle$$
$$\equiv (-1)^q \mathrm{Tr}_L\{A_{kq}\, e^{-it\mathscr{H}_L}\, A_{k-q}\, e^{it\mathscr{H}_L}\} \tag{I.22}$$

$$\tilde{q}\omega \equiv \begin{cases} q\dot{\omega}_I & \text{for } D, J, Q \\ q_1\omega_I + q_2\omega_S & \text{for } IS, CS\,(q_2=0) \end{cases}$$

with $q_1 + q_2 = q$ and

$$X_{kq}(\beta, t) \equiv \frac{(I_\rho|\, \hat{T}_{k-q}\, \mathbf{S}_1(t)\, \hat{T}_{kq}\, |I_\rho)}{(I_z|I_z)}$$
$$= \frac{(T_{kq}|\, e^{-i\beta \hat{I}_y}\, \hat{I}_z\, e^{-it\omega_e \hat{I}_z}\, \hat{I}_z\, e^{i\beta \hat{I}_y}\, |T_{kq})}{(I_z|I_z)}. \tag{I.23}$$

For the calculation of $T_{2\rho}$ we have to replace I_z in Eq. (I.23) by I_x. Equation (I.14) can be solved rigorously by Laplace transform, leading to the spectral representation $\langle I_\rho(\omega)\rangle$ as shown in Appendix C, which after transformation into the timedomain causes a non-exponential behaviour of $\langle I_\rho(t)\rangle$. Here we neglect those "memory effects" for simplicity and assume in consistence with the "weak collision limit" that $K(t)$ decays much more rapidly than $\langle I_\rho(t)\rangle$ itself, i.e. [3, 12]

$$\frac{d}{dt}\langle I_\rho(t)\rangle = -\int_0^\infty dt'\, K(t')\, \langle I_\rho(t)\rangle \tag{I.24}$$

leading to an exponential relaxation with the relaxation rate

$$\frac{1}{T_{1\rho}} = \mathrm{Re} \int_0^\infty dt\, K(t) \tag{I.25}$$

where T_1 is obtained by setting $\beta=0$, $\omega_e=0$ and T_2 results from $\beta=\pi/2$, $\omega_e\to 0$. Inserting $K(t)$ according to Eq. (I.21) into Eq. (I.23) leads us to the general expression

$$\frac{1}{T_{1\rho}} = \mathrm{Re} \sum_{k=0}^{2} \sum_{q=-k}^{+k} \int_0^\infty dt\, g_{kq}(t)\, e^{-it\tilde{q}\omega}\, X_{kq}(\beta, t) \tag{I.26}$$

In order to be more specific, we treat now the different spin-interactions separately. It should be noted, that equivalent expressions for $1/T_1$ as Eq. (I.26) have been obtained in the calculation of spin-lattice relaxation in the weak collision limit before [3, 12, 20, 21].

(a) D, J, Q

In this case $X_{kq}(\beta, t)$ according to Eq. (I.23) can be expressed as [20]

$$X_{kq}(\beta) = \lambda \sum_{q_1=-k}^{+k} q_1^2 (d_{qq_1}^{(k)}(\beta))^2\, e^{-itq_1\omega_e} \tag{I.27}$$

where $\lambda = I(I+1)/3$ for D and J, whereas $\lambda = (2I-1)(2I+3)/10$ in the case of Q.

Introducing the spectral density function

$$j_{kq}(q\omega) = \mathrm{Re} \int_0^\infty dt\, g_{kq}(t)\, e^{-itq\omega} \tag{I.28}$$

with

$$j_{kq}(\pm q\omega) = j_{kq}(q\omega)$$

because of the symmetry of $g_{kq}(t)$, we derive from Eq. (I.26)

$$\frac{1}{T_{1\rho}} = \lambda \sum_{k=0}^{2} \sum_{q=-k}^{+k} \sum_{q_1=-k}^{+k} q_1^2 [d_{qq_1}^{(k)}(\beta)]^2 j_{kq}(q\omega_I + q_1\omega_e) \tag{I.29}$$

Inserting the $d_{q,q_1}^{(k)}(\beta)$ from Appendix B and using $\omega_I \gg \omega_e$, i.e. we obtain [19, 20]

$$j_{kq}(q\omega_I + q_1\omega_e) \simeq j_{kq}(q\omega_I) \quad \text{if} \quad q \neq 0$$

$$\frac{1}{T_{1\rho}} = \lambda\{\sin^2\beta\, j_{10}(\omega_e) + (1 + \cos^2\beta) j_{11}(\omega_I)$$

$$+ 3\sin^2\beta\cos^2\beta\, j_{20}(\omega_e) + 3\sin^4\beta\, j_{20}(2\omega_e)$$

$$+ (5 - 3\cos^2\beta) j_{21}(\omega_I) + (6\cos^2\beta + 2) j_{22}(2\omega_I)\} \tag{I.30}$$

$$\beta = 0: \frac{1}{T_1} = 2\lambda\{j_{11}(\omega_I) + j_{21}(\omega_I) + 4 j_{22}(2\omega_I)\}$$

$$\beta = \tfrac{\pi}{2}:\ \underset{\omega_e \to 0}{\frac{1}{T_2}} = \lambda\{j_{10}(0) + j_{11}(\omega_I) + 3 j_{20}(0) + 5 j_{21}(\omega_I) + 2 j_{22}(2\omega_I)\}$$

where

$$\lambda = \tfrac{1}{3}I(I+1) \quad \text{for } D, J$$

and

$$\lambda = \tfrac{1}{10}(2I-1)(2I+3) \quad \text{for } Q$$

Moreover the terms $j_{kq}(\omega)$ with $k=1$ vanish in the case of D and Q.

(b) CS

In order to calculate $X_{kq}(\beta, t)$ according to Eq. (I.23) we have to insert the appropriate expressions for T_{kq}. These may be obtained from Appendix A resulting in

$$\frac{1}{T_{1\rho}} = \gamma_I^2 B_0^2 \sum_{k=0}^{2} (2k+1) \sum_{q,q'=-1}^{+1} q'^2\, w_{kq} [d_{qq'}^{(1)}(\beta)]^2 j_{kq}(q\omega_I + q'\omega_e) \tag{I.31}$$

where the square of the 3j-symbol

$$w_{kq} = w_{k-q} = \begin{pmatrix} 1 & 1 & k \\ -q & 0 & q \end{pmatrix}^2 \tag{I.32}$$

has been used. Evaluation of Eq. (I.31) and using again

$$j_{kq}(q\omega_I + q'\omega_e) \simeq j_{kq}(q\omega_I) \quad \text{if} \quad q \neq 0$$

leads to [19, 20].

$$\frac{1}{T_{1\rho}}=\gamma_I\,B_0^2\{\tfrac{1}{3}\sin^2\beta\,j_{00}(\omega_e)+\tfrac{1}{2}(1+\cos^2\beta)\,j_{11}(\omega_I)$$

$$+\tfrac{2}{3}\sin^2\beta\,j_{20}(\omega_e)+\tfrac{1}{2}(1+\cos^2\beta)\,j_{21}(\omega_I)\}\tag{I.33}$$

$$\beta=0:\ \frac{1}{T_1}=\gamma_I^2\,B_0^2\{j_{11}(\omega_I)+j_{21}(\omega_I)\}$$

$$\beta=\tfrac{\pi}{2}:\ \frac{1}{T_2}=\gamma_I^2\,B_0^2\{\tfrac{1}{3}j_{00}(0)+\tfrac{1}{2}j_{11}(\omega_I)+\tfrac{2}{3}j_{20}(0)+\tfrac{1}{2}j_{21}(\omega_I)\}$$
$$\scriptstyle\omega_e\to0$$

Note, that the antisymmetric part of the chemical shift tensor, contained in $j_{11}(\omega_I)$ contributes in leading order to the relaxation, whereas it does not contribute to the spectrum in first order. This antisymmetric part may be measurable by investigating the orientation dependence of $1/T_1$ in a solid where the relaxation is dominated by chemical shift interactions.

(c) *IS*

Proceeding along the same lines as before, we obtain from Eq. (I.26)

$$\frac{1}{T_{1\rho}}=\tfrac{1}{3}S(S+1)\sum_{k=0}^{2}(2k+1)\sum_{q=-k}^{+k}\sum_{q_1,q_2=-1}^{+1}q_2^2\,w_{kq}(q_1)\,[d^{(1)}_{q_1q_2}(\beta)]^2$$

$$\cdot j_{kq}[(q-q_1)\,\omega_S+q_1\omega_I+q_2\omega_e]\tag{I.34}$$

where

$$w_{kq}(q_1)=w_{k-q}(-q_1)=\begin{pmatrix}1 & 1 & k\\ -q_1 & (q_1-q) & q\end{pmatrix}^2.$$

In Eq. (I.34) we have assumed, that the S spins relax slowly with respect to the correlation time τ, i.e. $\tau\ll T_{1S}$, T_{2S} consistent with the "weak collision limit". On the other hand we assume, that the S spins are either "decoupled" from the I spins e.g. by $T_{1S}\ll T_{1I}$ or are always in thermal equilibrium during the relaxation of the I spins. If these conditions are not fulfilled the coupled relaxation of I and S spins has to be treated e.g. as shown by Abragam [3].

Proceeding with Eq. (I.34) by writing the different sums explicitly we obtain

$$\frac{1}{T_{1\rho}}=\tfrac{1}{3}S(S+1)\,\{\sin^2\beta[\tfrac{1}{3}j_{00}(\omega_e)+j_{11}(\omega_S)+\tfrac{2}{3}j_{20}(\omega_e)+j_{21}(\omega_S)]$$

$$+(1+\cos^2\beta)\,[\tfrac{1}{3}j_{00}(\omega_S-\omega_I)+\tfrac{1}{2}j_{10}(\omega_S-\omega_I)+\tfrac{1}{2}j_{11}(\omega_I)$$

$$+\tfrac{1}{6}j_{20}(\omega_S-\omega_I)+\tfrac{1}{2}j_{21}(\omega_I)+j_{22}(\omega_S+\omega_I)]\}\tag{I.35}$$

$$\beta=0:\ \frac{1}{T_1}=\tfrac{1}{3}S(S+1)\,\{\tfrac{2}{3}j_{00}(\omega_S-\omega_I)+j_{10}(\omega_S-\omega_I)+j_{11}(\omega_I)$$

$$+\tfrac{1}{3}j_{20}(\omega_S-\omega_I)+j_{21}(\omega_I)+2j_{22}(\omega_S+\omega_I)\}$$

$$\beta=\tfrac{\pi}{2}:\ \frac{1}{T_2}=\tfrac{1}{3}S(S+1)\,\{\tfrac{1}{3}j_{00}(0)+\tfrac{1}{3}j_{00}(\omega_S-\omega_I)+\tfrac{1}{2}j_{10}(\omega_S-\omega_I)$$
$$\scriptstyle\omega_e\to0$$

$$+j_{11}(\omega_S)+\tfrac{1}{2}j_{11}(\omega_I)+\tfrac{2}{3}j_{20}(0)+\tfrac{1}{6}j_{20}(\omega_S-\omega_I)$$

$$+j_{21}(\omega_S)+\tfrac{1}{2}j_{21}(\omega_I)+j_{22}(\omega_S+\omega_I)\}$$

In the case of dipolar interaction all $j_{kq}(\omega)$ with $k=0,1$ vanish identically.

In order to utilize Eqs. (I.30, I.33 and I.35) the spectral functions $j_{kq}(\omega)$ have to be evaluated. Remember that

$$j_{kq}(\omega) = \mathrm{Re} \sum_{0}^{\infty} dt\,(-1)^q \langle \delta A_{kq}(0)\, \delta A_{k-q}(t) \rangle\, e^{-i\omega t}$$

where

$$\langle \delta A_{kq}(0)\, \delta A_{k-q}(t) \rangle \equiv \langle A_{kq}(0)\, A_{kq}(t) \rangle - \langle A_{kq} \rangle \langle A_{k-q} \rangle$$

It is convenient to express $j_{kq}(\omega)$ by the spectral densities $J_k(\omega)$ as

$$j_{kq}(\omega) = \langle |\delta A_{kq}|^2 \rangle\, J_k(\omega) \tag{I.36}$$

with

$$\langle |\delta A_{kq}|^2 \rangle = (-1)^q [\langle A_{kq} A_{k-q} \rangle - \langle A_{kq} \rangle \langle A_{k-q} \rangle]$$

where

$$J_k(\omega) = \mathrm{Re} \int_{0}^{\infty} dt\, g_k(t)\, e^{-i\omega t}$$

and

$$g_k(t) = \frac{\langle \delta A_{kq}(0)\, \delta A_{k-q}(t) \rangle}{\langle \delta A_{kq}\, \delta A_{k-q} \rangle}.$$

This definition of the correlation function $g_k(\omega)$ secures that $g_k(t=0)=1$ and $g(t \to \infty)=0$. In most cases $g_k(t)$ can be approximated by

$$g_k(t) = e^{-t/\tau_k} \qquad k = 0, 1, 2 \tag{I.37}$$

If there is only one type of motion one may even set $\tau_2 = \tau_1 = \tau_0 = \tau$ leading to

$$J(\omega) = \frac{\tau}{1 + \omega^2 \tau^2}. \tag{I.38}$$

The orientational dependence of $j_{kq}(\omega)$ with respect to the external magnetic field B_0 described by the Euler angles (θ, ϕ) is now contained in $\langle \delta A_{kq}^2 \rangle$ where

$$\delta A_{kq} = A_{kq}^{(0)} - \langle A_{kq} \rangle.$$

This leads to the simple relation

$$j_{kq}(\omega) = \langle |\delta A_{kq}|^2 \rangle\, \frac{\tau}{1 + \omega^2 \tau^2}. \tag{I.39}$$

The orientational dependence of $\langle |\delta A_{kq}|^2 \rangle$ is readily obtained from

$$\langle |\delta A_{kq}|^2 \rangle = (-1)^q \sum_{q_1, q_2 = -k}^{+k} \delta A_{kq_1}\, \delta A_{kq_2}\, e^{-i(q_1 + q_2)\phi}\, d_{q_1 q}^{(k)}(\theta)\, d_{q_2 - q}^{(k)}(\theta). \tag{I.40}$$

The δA_{kq} can be obtained for the different spin-interactions from Appendix A and Table 2.2, whereas the $d_{q,q'}^{(k)}(\theta)$ are given in Table B.1 in Appendix B. Similar results have been obtained by Spiess [20] and Blicharski [21].

10 References

Chapter 1

1. Abragam, A.: The Principles of Nuclear Magnetism. London: Oxford Univ. Press 1961
2. Friebolin, H.: NMR-Spektroskopie, p. 74. Weinheim: Physik-Verlag 1974
3. Andrew, E.R., Hinshaw, W.S., Jasinski, A.: Chem. Phys. Lett. *24*, 399 (1974)
4. a) Andrew, E.R., Eades, R.G.: Proc. Roy. Soc. London *A216*, 398 (1953)
   b) Andrew, E.R., Clough, S., Farnell, L.F., Gledhill, T.D., Roberts, I.: Phys. Lett. *19*, 6 (1966)
5. Lowe, I.J.: Phys. Rev. Lett. *2*, 285 (1959)
6. a) Andrew, E.R.: Prog. NMR Spectroscopy *8*, 1 (1971)
   b) Andrew, E.R.: Biennial Rev. Magn. Res. (ed. by C.A. McDowell, Vancouver) (1974)
7. Waugh, J.S., Huber, L.M., Haeberlen, U.: Phys. Rev. Lett. *20*, 180 (1968)
8. a) Ellett, J.D., Haeberlen, U., Waugh, J.S.: J. Am. Chem. Soc. *92*, 411 (1970)
   b) Ellett, J.D., Gibby, M.G., Haeberlen, U., Huber, L.M., Mehring, M., Pines, A.: Advan. Magn. Res. *5*, 117 (1971)
9. a) Mehring, M., Griffin, R.G., Waugh, J.S.: J. Am. Chem. Soc. *92*, 7222 (1970)
   b) Mehring, M., Griffin, R.G., Waugh, J.S.: Chem. Phys. *55*, 746 (1971)
10. Stacey, L.M., Vaughan, R.W., Elleman, D.D.: Phys. Rev. Lett. *26*, 1153 (1971)
11. Griffin, R.G., Ellett, J.D., Mehring, M., Bullitt, J.G., Waugh, J.S.: J. Chem. Phys. *57*, 2147 (1972)
12. Haeberlen, U., Kohlschuetter, U., Kempf, J., Spiess, H.W., Zimmermann, H.: Chem. Phys. *3*, 248 (1974)
13. Silberzyc, W.: Acta Phys. Satyr. *11*, 1111 (1970)
14. a) Pines, A., Gibby, M.G., Waugh, J.S.: J. Chem. Phys. *56*, 1776 (1972)
    b) Pines, A., Waugh, J.S.: J. Magn. Res. *8*, 354 (1972)
    c) Pines, A., Gibby, M.G., Waugh, J.S.: J. Chem. Phys. *59*, 569 (1973)
    d) Pines, A., Gibby, M.G., Waugh, J.S.: Chem. Phys. Lett. *15*, 373 (1972)
15. Pines, A., Schattuck, T.W.: J. Chem. Phys. *61*, 1255 (1974)
16. Schaefer, J., Stejskal, E.O., Buchdahl, R.: Macromolecules *10*, 384 (1977)
17. Jeener, J.: Ampere International Summer School II, Basko Polje, Yugoslavia, 1971
18. Aue, W.P., Bartholdi, E., Ernst, R.R.: J. Chem. Phys. *64*, 2229 (1976)
19. Freeman, R., Morris, G.A.: Bull. Magn. Res. *1*, 5 (1979)
20. a) Waugh, J.S.: Proc. Nat. Acad. Sci. USA, *73*, 1394 (1976)
    b) Hester, R.K., Ackerman, J.L., Neff, B., Waugh, J.S.: Phys. Rev. Lett. *36*, 1081 (1976)
    c) Rybaczewski, E.F., Neff, B.L., Waugh, J.S., Sherfinski, J.S.: J. Chem. Phys. *67*, 1231 (1977)
21. Warren, W.S., Sinton, S., Weitekamp, D.P., Pines, A.: Phys. Rev. Lett. *43*, 1791 (1979)
22. Mansfield, P.: Progr. NMR Spectroscopy *8*, 41 (1971)
23. Vaughan, R.W.: Annual Reviews in Materials Science, Vol. 4, (ed. by Huggins, R.A., Bube, H.R., Roberts, R.W.) (Annual Reviews, Palo Alto 1974)
24. Griffin, R.G.: Analytical Chemistry *49*, 951 (1977)
25. Haeberlen, U.: Advan. Magn. Res. supplement. New York: Academic Press 1976
26. Spiess, H.W.: Rotation of Molecules and Nuclear Spin Relaxation in NMR: Basic Principles and Progress, Vol. 15, p. 55. Berlin-Heidelberg-New York: Springer 1978

Chapter 2

1. Abragam, A.: The Principles of Nuclear Magnetism. London: Oxford Univ. Press 1961
2. a) Poole, C.P., Farach, H.A.: The Theory of Magnetic Resonance. New York: Wiley Interscience 1972

b) Slichter, C.P.: Principles of Magnetic Resonance, Berlin-Heidelberg-New York: Springer 1978

3. Tutunjian, P.N. and Waugh, J.S.: J. Chem. Phys. *76*, 1223 (1982), J. Magn. Res. *49*, 155 (1982)

4. a) Rose, M.E.: Elementary Theory of Angular Momentum. New York: John Wiley 1967
   b) Brink, D.M., Satchler, G.R.: Angular Momentum. Oxford: Clarendon Press 1968

5. Haeberlen, U., Waugh, J.S.: Phys. Rev. *185*, 420 (1969)

6. Haeberlen, U.: Advan. Magn. Res. (ed. by Waugh, J.S.) supplement (1976)

7. a) Lippmaa, E., Alla, M., Tuherm, T.: Proceedings of the XIXth Congress Ampere, p. 113, Groupement Ampere. Heidelberg-Geneva 1976
   b) Yarim-Agaev, Y., Tutunjian, P.N. and Waugh, J.S.: J. Chem. Phys. *56*, 1223 (1982)

8. Maricq, M.M., Waugh, J.S.: J. Chem. Phys. *70*, 3300 (1979)

9. Herzfeld, J., Berger, A.: J. Chem. Phys. *73*, 6021 (1980)

10. Griffin, R.G., Ellett, J.D., Mehring, M., Bullitt, J.G., Waugh, J.S.: J. Chem. Phys. *57*, 2147 (1972)

11. Bloembergen, N., Rowland, J.A.: Acta Met. *1*, 731 (1953)

12. Mehring, M., Griffin, R.G., Waugh, J.S.: J. Chem. Phys. *55*, 746 (1971)

13. a) Hentschel, R., Schlitter, J., Sillescu, H., Spiess, H.W.: J. Chem. Phys. *68*, 56 (1978)
    b) Hentschel, R., Sillescu, H., Spiess, H.W.: Polymer 1981 (in press)

14. Spiess, H.W.: Rotation of Molecules and Nuclear Spin Relaxation in NMR: Basic Principles and Progress, Vol. 15, p. 55. Berlin-Heidelberg-New York: Springer 1978

15. Opella, St.J., Waugh, J.S.: J. Chem. Phys. *66*, 4919 (1979)

16. Blinc, R., Juznic, S., Rutar, V., Seliger, J., Zumer, S.: Phys. Rev. Lett. *44*, 609 (1980)

17. Goldman, M.: Spin Temperature and Nuclear Magnetic Resonance in Solids. London: Oxford Univ. Press 1970

18. Haeberlen, U., Waugh, J.S.: Phys. Rev. *175*, 453 (1968)

19. Van Vleck, J.H.: Phys. Rev. *74*, 1168 (1948)

20. a) Zwanzig, R.: In: Lectures in Theoretical Physics. New York: Interscience 1961
    b) Mori, H.: Progr. Theor. Phys. Jpn. *34*, 399 (1965)
    c) Bosse, J.: (private communication)

21. a) Mori, H., Kawasaki, K.: Progr. Theor. Phys. (Kyoto) *27*, 529 (1962)
    b) Mori, H.: Progr. Theor. Phys. (Kyoto) *33*, 423 (1965)

22. Lado, F., Memory, J.D., Parker, G.W.: Phys. Rev. *B4*, 1406 (1971)

23. Tjon, J.A.: Phys. Rev. *143*, 259 (1966)

24. a) Robertson, B.: Phys. Rev. *144*, 151 (1966)
    b) Robertson, B.: Phys. Rev. *153*, 391 (1967)
    c) Andersson, P.W., Weiss, P.R.: Rev. Mod. Phys. *25*, 269 (1953)

25. Argyres, P.N., Kelley, P.L.: Phys. Rev. *134 A*, 93 (1964)

26. Kivelson, D., Ogan, K.: Advan. Magn. Res. 7, 71 (1974)

27. Parker, G.W., Lado, F.: Phys. Rev. *B8*, 3081 (1973)

28. Parker, G.W., Lado, F.: Phys. Rev. *B9*, 22 (1974)

29. Demco, D., Tegenfeldt, J., Waugh, J.S.: Phys. Rev. *B11*, 4133 (1975)

30. Mehring, M., Sinning, G.: Phys. Rev. *B15*, 2519 (1977)

31. Andrew, E.R., Bradbury, A., Eades, R.G.: Nature *182*, 1659 (1958); *183*, 1802 (1959)

32. Lowe, I.J.: Phys. Rev. Lett. *2*, 285 (1959)

33. Andrew, E.R.: Progr. NMR Spectroscopy *8*, 1 (1971)

34. Andrew, E.R.: Biennial Rev. Magn. Res. (ed. by C.A. McDowell, Vancouver) (1974)

35. Andrew, E.R., Farnell, L.F., Gledhill, T.D.: Phys. Rev. Lett. *19*, 6 (1967)

36. Andrew, E.R., Hinshaw, W.S., Riffen, R.S.: Phys. Lett. *46 A*, 57 (1973)

37. Andrew, E.R., Hinshaw, W.S.: Phys. Lett. *43 A*, 113 (1973)

38. Henriot, E.: Hugenard, E.: Compt. Rend. *180*, 1389 (1925)

39. a) Beams, J.W.: Rev. Sci. Instr. *8*, 795 (1937)
    b) Beams, J.W., Pickels, E.G.: Rev. Sci. Instr. *6*, 299 (1935)

40. Zilm, K.W., Alderman, D.W., Grant, D.M.: J. Magn. Res. *30*, 563 (1978)

41. Waugh, J.S., Maricq, P.M., Cantor, R.: J. Magn. Res. *29*, 183 (1978)

42. Griffin, R.G.: presented in Ref. [9]

43. Pines, A.: private communication

44.   Mehring, M., Raber, H.: Solid. State Commun. *13*, 1637 (1973)
45. a) Schnabel, B.: Wiss. Z. Univ. Jena *22*, p. 335 (1973)
    b) Schnabel, B., Taplick, T.: Phys. Lett. *27A*, 310 (1968)
46. a) Cohn, M., Kowalsky, A., Leigh, H., Maricić, S.: Magnetic Resonance in Biological System, p.45. New York: Pergamon Press Inc. (1967)
    b) Babka, J., Doskocilova, D., Pivcova, H., Ruzieka, Z., Schneider, B.: Proc. 16th Congress Ampere. Bucharest, Rumania 1970, 785 (Publ. House Academy R.S. Rumania)
47.   Schaefer, J., Stejskal, E.O., Buchdahl, R.: Macromolecules *10*, 384 (1977)
48.   Müller, R., Zachmann, H.: Colloid & Polymer Sci. *258*, 753 (1980)
49.   Stejskal, E.O., Schaefer, J., McKay, R.A.: J. Magn. Res. *25*, 569 (1977)
50.   Stejskal, E.O., Schaefer, J., Waugh, J.S.: J. Magn. Res. *28*, 105 (1977)
51.   Aue, W.P., Ruben, D.J., Griffin, R.G.: J. Magn. Res. *43*, 472 (1981)
52.   Dixon, W.: J. Magn. Res. *44*, 220 (1981)
53.   Garroway, A.N., Van der Hart, D.L., Earl, W.L.: Phil. Trans. Roy. Soc. London A, 299 (1981)
54.   Van der Hart, D.L., Earl, W.L., Garroway, A.N.: J. Magn. Res. *44*, 361 (1981)
55.   Suwelack, D., Rothwell, W.P., Waugh, J.S.: J. Chem. Phys. *73*, 2559 (1980)
56. a) Ackerman, J.L., Eckmann, R., Pines, A.: Chem. Phys. *42*, 423 (1979)
    b) Eckmann, R., Alla, M., Pines, A.: J. Magn. Res. *41*, 440 (1980)
    c) Eckmann, R., Müller, L., Pines, A.: Chem Phys. Lett. *74*, 376 (1980)
    d) Müller, L., Eckmann, R., Pines, A.: Chem. Phys. Lett. *76*, 149 (1980)
57. a) Gerstein, B.C., Pembleton, R.G., Wilson, R.C., Ryan, L.M.: J. Chem. Phys. *66*, 361 (1977)
    b) Ryan, L.M., Taylor, R.E., Paff, A.J., Gerstein, B.C.: J. Chem. Phys. *72*, 508 (1980)
58. a) Rosenberger, H., Schnabel, B.: Wiss. Z. Friedrich-Schiller Univ. Jena Math. Naturwiss. Reihe *27*, 257 (1978)
    b) Schnabel, B., Haubenreisser, U., Scheler, G., Müller, R.: "Proceedings, 20th Congress Ampere, Tallinn 1978", 106
59.   Scheler, G., Haubenreisser, U., Rosenberger, H.: J. Magn. Res. *44*, 134 (1981)
60.   Burum, D.P., Rhim, W.K.: J. Chem. Phys. *71*, 944 (1979)
61.   Oldfield, E., Meadows, M.: J. Magn. Res. *31*, 327 (1978)
62.   Bennett, J., Marsh, H.: 6th Internat. Gas Bearing Symp. A1-1 (1974)
63.   Balimann, G., Burgess, M.J.S., Harris, R.K., Oliver, A.G., Packer, K.J., Say, B.J., Tanner, St.F., Blackwell, R.W., Brown, L.W., Bunn, A., Cudby, M.E.A., Eldridge, J.W.: Chem. Phys. *46*, 469 (1980)
64.   Freed, J.H., Bruno, G.V., Polnaszek, C.: J. Phys. Chem. *75*, 3386 (1971)
65.   Freed, J.H., Fraenke, G.K.: J. Chem. Phys. *39*, 326 (1963)
66.   Sillescu, H., Kivelson, D.: J. Chem. Phys. *48*, 3493 (1968)
67.   Sillescu, H.: J. Chem. Phys. *54*, 2111 (1971)
68.   Sillescu, H.: Ber. Bunsenges. Phys. Chem. *75*, 283 (1971)
69.   Hensen, K., Riede, W.O., Sillescu, H., Wittgenstein, A.v.: J. Chem. Phys. *61*, 4365 (1974)
70.   Goldman, S.A., Bruno, G.V., Polnaszek, C.F., Freed, J.H.: J. Chem. Phys. *56*, 716 (1972)
71.   Spiess, H.W.: Chem. Phys. *6*, 217 (1974)
72.   Spiess, H.W., Grosescu, R., Haeberlen, U.: Chem. Phys. *6*, 226 (1974)
73. a) Alexander, S., Baram, A., Luz, Z.: Mol. Phys. *27*, 441 (1974)
    b) Baram, A., Luz, Z., Alexander, S.: J. Chem. Phys. *64*, 4321 (1976)
    c) Alexander, S., Luz, Z., Naor, Y., Poupko, R.: Mol. Phys. *33*, 1119 (1977)
74.   Wemmer, D.E., Ruben, D.J., Pines, A.: J. Am. Chem. Soc. *103*, 28 (1981)
75.   Gordon, R.G., McGinnis, R.P.: J. Chem. Phys. *49*, 2455 (1968)
76.   Wilkinson, J.H.: The Algebraic Eigenvalue Problem, Chap. 8. London: Oxford Univ. Press 1965
77.   Becker, H.J.: Diploma work. Dortmund 1975
78. a) Pschorn, U., Spiess, H.W.: J. Magn. Res. *39*, 217 (1980)
    b) Spiess, H.W., Sillescu, H.: J. Magn. Res. *42*, 381 (1981)
79.   Davis, J.H., Jeffrey, K.R., Bloom, M., Valic, M.I., Higgs, T.P.: Chem. Phys. Lett. *42*, 390 (1976)
80.   Seelig, A., Seelig, J.: Biochemistry *13*, 4839 (1974)
81.   Oldfield, E., Meadows, M., Rice, D., Jacobs, R.: Biochemistry *17*, 2727 (1978)

82. Huang, T.H., Skarjune, R.P., Wittebort, R.J., Griffin, R.G., Oldfield, E.: J. Am. Chem. Soc. *102*, 7377 (1980)
83. Gall, C.M., DiVerdi, J.A., Opella, J.S.: J. Am. Chem. Soc. *103*, 5039 (1981)
84. Wemmer, D.E.: Ph.D. Thesis. Univ. of California, Berkeley 1979
85. a) Wemmer, D.E., Ruben, D.J., Pines, A.: J. Am. Chem. Soc. *103*, 28 (1981)
    b) Wemmer, D.E., Pines, A.: J. Am. Chem. Soc. *103*, 34 (1981)
86. Rigny, P.: Physica *59*, 707 (1972)
87. Günther, B., Kanert, O., Mehring, M., Wolf, D.: Phys. Rev. *B24*, 6747 (1981)

Chapter 3

1. Hahn, E.L.: Phys. Rev. *80*, 580 (1950)
2. Carr, H.Y., Purcell, E.M.: Phys. Rev. *94*, 630 (1954)
3. Gill, D., Meiboom, S.: Rev. Sci. Instr. *29*, 688 (1958)
4. Ostroff, E.D., Waugh, J.S.: Phys. Rev. Lett. *16*, 1097 (1966)
5. Mansfield, P., Ware, D.: Phys. Lett. *22*, 133 (1966)
6. Waugh, J.S., Wang, C.H.: Phys. Rev. *162*, 209 (1967)
7. Mansfield, P., Ware, D.: Phys. Rev. *168*, 318 (1968)
8. Mansfield, P., Richards, K.H.B., Ware, D.: Phys. Rev. *B1*, 2048 (1970)
9. Waugh, J.S., Huber, L.M., Haeberlen, U.: Phys. Rev. Lett. *20*, 180 (1968)
10. Abragam, A.: The Principles of Nuclear Magnetism. London: Oxford Univ. Press 1961
11. Goldman, M.: Spin Temperature and Nuclear Magnetic Resonance in Solids. London: Oxford Univ. Press 1970
12. Haeberlen, U., Waugh, J.S.: Phys. Rev. *175*, 453 (1968)
13. Wilcox, R.M.: J. Math. Phys. *8*, 962 (1967)
14. a) Zwanzig, R.: In: Lectures in Theoretical Physics. New York: Interscience 1961
    b) Mori, H.: Progr. Theor. Phys. Jpn. *34*, 399 (1965)
    c) Bosse, J.: (private communication)
15. a) Mori, H., Kawasaki, K.: Progr. Theor. Phys. (Kyoto) *27*, 529 (1962)
    b) Mori, H.: Progr. Theor. Phys. (Kyoto) *33*, 423 (1965)
16. Lado, F., Memory, J.D., Parker, G.W.: Phys. Rev. *B4*, 1406 (1971)
17. Ellett, J.D., Jr., Waugh, J.S.: J. Chem. Phys. *51*, 2581 (1969)
18. Haeberlen, U.: Advan. Magn. Res. supplement. New York: Academic Press 1976
19. a) Mansfield, P.: J. Phys. C: Solid. State Phys. *4*, 1444 (1971)
    b) Mansfield, P., Orchard, M.J., Stalker, D.C., Richards, K.H.B.: Phys. Rev. *B7*, 90 (1973)
20. Stacey, L.M., Vaughan, R.W., Elleman, D.D.: Phys. Rev. Lett. *26*, 1153 (1971)
21. Griffin, R.G., Ellett, J.D., Mehring, M., Bullitt, J.G., Waugh, J.S.: J. Chem. Phys. *57*, 2147 (1972)
22. Mehring, M., Griffin, R.G., Waugh, J.S.: J. Am. Chem. Soc. *92*, 7222 (1970)
23. Mehring, M., Griffin, R.G., Waugh, J.S.: J. Chem. Phys. *55*, 746 (1971)
24. Mansfield, P., Orchard, M.J., Stalker, D.C., Richards, K.H.B.: Phys. Rev. *B7*, 90 (1973)
25. Griffin, R.G., Yeung, H.N., LaPrade, M.D., Waugh, J.S.: J. Chem. Phys. *59*, 777 (1973)
26. Rhim, W.-K., Elleman, D.D., Vaughan, R.W.: J. Chem. Phys. *59*, 3740 (1973)
27. a) Vaughan, R.W., Elleman, D.D., Rhim, W.-K., Stacey, L.M.: J. Chem. Phys. *57*, 5383 (1972)
    b) Vaughan, R.W.: Annual Review in Materials Science, Vol. 4, (ed. by Huggins, Bube, R.A., Roberts, R.W.) Annual Review. Palo Alto, 1974
28. Haeberlen, U., Kohlschütter, U.: Chem. Phys. *2*, 76 (1973)
29. Haeberlen, U., Kohlschütter, U., Kempf, J., Spiess, H.W., Zimmermann, H.: Chem. Phys. *3*, 248 (1974)
30. Grosescu, R., Achlama, A.M., Haeberlen, U., Spiess, H.W.: Chem. Phys. *5*, 119 (1974)
31. Achlama, A.M., Kohlschütter, U., Haeberlen, U.: Chem. Phys. *7*, 287 (1975)
32. Raber, H., Brünger, G., Mehring, M.: Chem. Phys. Lett. *23*, 400 (1973)
33. a) Haeberlen, U., Schnabel, B.: 18th Ampere Congress, p. 545. Nottingham 1974
    b) Dühler, H., Neubauer, R., Schnabel, B.: Phys. Stat. Sol. *B65*, K141 (1974)
34. Schreiber, L.B., Vaughan, R.W.: Chem. Phys. Lett. *28*, 586 (1974)
35. a) Van Hecke, P., Weaver, J.C., Neff, B.L., Waugh, J.S.: J. Chem. Phys. *60*, 1668 (1974)

b) Van Hecke, P., Weaver, J.C., Neff, B.L., Waugh, J.S.: Proc. First Specialised Colloque Ampere. Krakow, Poland 143 (1973)

36. a) Haubenreisser, U., Schnabel, B.: Proc. First Specialised Colloque Ampere. Krakow, Poland 140 (1973)
   b) Haubenreisser, U., Schnabel, B., Scheler, G., Burghoff, U., Müller, R., Wilsch, R.: Phys. Stat. Sol. *A20*, K45 (1973)
   c) Wilsch, R., Burghoff, U., Müller, R., Rosenberger, H., Scheler, G., Pettig, M., Schnabel, B.: 18th Ampere Congress, Vol. 2, p. 553. Nottingham 1974

37. Burghoff, U., Scheler, G., Müller, R.: Phys. Stat. Sol. *A25*, K31 (1974)
38. Mehring, M., Raber, H.: Solid. State Commun. *13*, 1637 (1973)
39. Kanert, O., Mehring, M.: NMR: Basic Principles and Progress, Vol. 3. Berlin-Heidelberg-New York: Springer 1971
40. a) Mehring, M., Kotzur, D., Kanert, O.: Phys. Stat. Sol. *B35*, K25 (1972)
   b) Kotzur, D., Mehring, M., Kanert, O.: Z. Naturforsch. *28A*, 1607 (1973)
41. Mehring, M., Pines, A., Rhim, W.-K., Waugh, J.S.: J. Chem. Phys. *54*, 3239 (1971)
42. a) Mehring, M., Raber, H., Sinning, G.: Proceedings of the 18th Congress Ampere, p. 35. Nottingham (1974)
   b) Ackermann, H., Fujara, F.: Spring Meeting of the DPG. Freudenstadt, Germany (1976)
   c) Fujara, F.: Diploma work. Heidelberg 1975
43. Bloch, F., Siegert, A.: Phys. Rev. *57*, 522 (1940)
44. Silberszyc, W.: Acta Phys. Satyr. *11*, 1111 (1970)
45. Wang, C.H., Ramshaw, J.D.: Phys. Rev. *B6*, 3253, 1972, also Wang, C.H.: J. Chem. Phys. *59*, 225 (1973)
46. Magnus, W.: Com. Pure Appl. Math. 7, 649 (1954)
47. Pechukas, P., Light, F.C.: J. Chem. Phys. *44*, 3897 (1966)
48. Evans, W.A.B.: Ann. Phys. (N.Y.) *48*, 72 (1968)
49. Bialnicky-Birula, Mielnik, B.: Ann. Phys. (N.Y.) *51*, 187 (1969)
50. Mehring, M., Waugh, J.S.: Phys. Rev. *B5*, 3459 (1972)
51. a) Rhim, W.-K., Pines, A., Waugh, J.S.: Phys. Rev. Lett. *25*, 218 (1970)
   b) Rhim, W.-K., Pines, A., Waugh, J.S.: Phys. Rev. *B3*, 684 (1971)
52. Garroway, A.N., Mansfield, P., Stalker, D.C.: Phys. Rev. *B11*, 121 (1975)
53. Rhim, W.-K., Elleman, D.D., Vaughan, R.W.: J. Chem. Phys. *58*, 1772 (1973)
54. Mansfield, P.: Proceedings of the 17th Congress Ampere. Turku 1972
55. Rhim, W.K., Elleman, D.D., Schreiber, L.B., Vaughan, R.W.: J. Chem. Phys. *60*, 4595 (1974)
56. Ryan, L.M., Wilson, R.C., Gerstein, B.C.: Chem. Phys. Lett. *52*, 341 (1977)
57. Burum, D.P., Rhim, W.K.: J. Chem. Phys. *59*, 944 (1979)
58. Burum, D.P., Rhim, W.K.: J. Chem. Phys. *70*, 3553 (1979)
59. Burum, D.P., Rhim, W.K.: J. Magn. Res. *34*, 241 (1979)
60. Rhim, W.K., Burum, D.P., Elleman, D.D.: J. Chem. Phys. *71*, 3139 (1979)
61. Burum, D.P., Linder, M., Ernst, R.R.: J. Magn. Res. *44*, 173 (1981)
62. Rose, M.E.: Elementary Theory of Angular Momentum, p. 77. New York: J. Wiley 1967
63. Brink, D.M., Satchler, G.R.: Angular Momentum, p. 48. London: Oxford Univ. Press 1968
64. a) Mehring, M.: Z. Naturforsch. *27A*, 1634 (1972), *28A*, 804 (1973)
   b) Mehring, M.: Rev. Sci. Instr. *44*, 64 (1973)
65. a) Ernst, H., Fenzke, D., Heink, W.: Colloq. Ampere. Krakow, Poland 1973
   b) Ernst, H., Fenzke, D., Heink, W.: Wiss. Z. Karl-Marx-Univ. Leipzig, Math.-Naturw. *R23*, 545 (1974)
   c) Haubenreisser, U., Schnabel, B.: 18th Ampere Congress, p. 551. Nottingham 1974
66. Haeberlen, U., Ellett, J.D., Jr., Waugh, J.S.: J. Chem. Phys. *55*, 53 (1971)
67. Pines, A., Waugh, J.S.: J. Magn. Res. *8*, 354 (1972)
68. Lan, N.Q., Pfeifer, H., Schmiedel, H.: Wiss. Z. Karl-Marx-Univ. Leipzig, Math.-Naturw. *R23*, 498 (1974)
69. Fenzke, D., Schmiedel, H.: Wiss. Z. Karl-Marx-Univ. Leipzig, Math.-Naturw. *R23*, 519 (1974)
70. Schmiedel, H.: Wiss. Z. Karl-Marx-Univ. Leipzig, Math.-Naturw. *R23*, 506 (1974)
71. Ernst, H., Fenzke, D., Heink, W.: Wiss. Z. Karl-Marx-Univ. Leipzig, Math.-Naturw. *R23*, 530 (1974)

72.　Lee, M., Goldberg, W.I.: Phys. Rev. *A140*, 1261 (1965)
73.　Yannoni, C.S., Vieth, H.M.: Phys. Rev. Lett. *37*, 1230 (1976)
74.　Ellett, D., Haeberlen, U., Waugh, J.S.: J. Polym. Sci., Polym. Lett. Ed., 7, 71 (1969)
75.　Garroway, A.N., Stalker, D.C., Mansfield, P.: Polymer *16*, 171 (1975)
76. a) English, A.D., Vega, A.J.: Macromolecules *12*, 353 (1979)
　　b) Vega, A.J., English, A.D.: Macromolecules *13*, 1635 (1980)
77.　Brandolini, A.J., Apple, Th.M., Dybowski, C., Pembleton, R.G.: Polymer *23*, 39 (1982)
78.　Hull, W.E., Sykes, B.D.: J. Mol. Biol. *98*, 121 (1975)
79. a) Gent, M.P.N., Ho, Ch.: Biochemistry *17*, 3023 (1978)
　　b) Gent, M.P.N., Cottam, P.F., Ho, Ch.: Proc. Natl. Acad. Sci. USA *75*, 630 (1978)
80.　Dybowski, C.R., Vaughan, R.W.: Proceedings of the 18th Colloque Ampere. Nottingham 1974
81.　Duncan, T.M., Dybowski, C.: Surface Sci. Rep. (in press)
82.　Blinc, R., Pirs, J., Zupancic, I.: Phys. Rev. Lett. *30*, 546 (1973)
83. a) Rhim, W.K., Burum, D.P., Elleman, D.D.: Phys. Rev. Lett. *37*, 1764 (1976)
　　b) Rhim, W.K., Burum, D.P., Elleman, D.D.: Phys. Lett. *62A*, 507 (1977)
84.　Burum, D.P., Elleman, D.D., Rhim, W.K.: J. Chem. Phys. *68*, 1164 (1978)
85.　Pembleton, R.G., Ryan, L.M., Gerstein, B.C.: Rev. Sci. Instr. *48*, 1286 (1977)
86.　Schnabel, B., Haubenreisser, U., Scheler, G., Müller, R.: In: Proceedings 19th Congress Ampere. Heidelberg 1976, p. 441
87. a) Rosenberger, H., Grimmer, A.R., Haubenreisser, U., Schnabel, B.: In: Proceedings 20th Congress Ampere. Tallin 1978, p. 106
　　b) Rosenberger, H., Grimmer, A.R.: Z. Anorg. Allg. Chemie *448*, 11 (1979)
88.　Ryan, L.M., Taylor, R.E., Paff, A.J., Gerstein, B.C.: J. Chem. Phys. *72*, 508 (1980)
89.　Scheler, G., Haubenreisser, U., Rosenberger, H.: J. Magn. Res. *44*, 134 (1981)
90.　Mansfield, P., Haeberlen, U.: Z. Naturforsch. *28A*, 1081 (1973)
91.　Rhim, W.K., Burum, D.P., Vaughan, R.W.: Rev. Sci. Instr. *47*, 720 (1976)

Chapter 4

1. a) Hartmann, S.R., Hahn, E.L.: Phys. Rev. *128*, 2042 (1962)
　　b) Jones, E.P., Hartmann, S.R.: Phys. Rev. *36*, 757 (1972)
2. a) Lurie, F.M., Slichter, C.P.: Phys. Rev. *133A*, 1108 (1964)
　　b) Slichter, C.P., Holton, W.C.: Phys. Rev. *122*, 1701 (1961)
3. a) Pines, A., Gibby, M.G., Waugh, J.S.: J. Chem. Phys. *56*, 1776 (1972)
　　b) Pines, A., Gibby, M.G., Waugh, J.S.: J. Chem. Phys. *59*, 569 (1973)
　　c) Pines, A., Gibby, M.G., Waugh, J.S.: Chem. Phys. Lett. *15*, 373 (1972)
4. a) Abragam, A.: The Principles of Nuclear Magnetism. London: Oxford Univ. Press 1961
　　b) Abragam, A., Proctor, W.G.: Phys. Rev. *109*, 1441 (1958)
5.　Redfield, A.G.: Phys. Rev. *98*, 1787 (1955)
6.　Goldman, M.: Spin Temperature and Nuclear Magn. Res. in Solids. London: Oxford Univ. Press 1970
7.　Solomon, I.: Compt. Rend. *248*, 92 (1950)
8.　Anderson, A.G., Hartmann, S.R.: Phys. Rev. *128*, 2023 (1960)
9. a) Jeener, J., Broekaert, P.: Phys. Rev. *157*, 232 (1967)
　　b) Jeener, J., Du Bois, R., Broekaert, P.: Phys. Rev. *139A*, 1959 (1965)
10.　McArthur, D.A., Hahn, E.L., Walstedt, R.E.: Phys. Rev. *188*, 609 (1969)
11.　Demco, D., Tegenfeldt, J., Waugh, J.S.: Phys. Rev. *B11*, 4133 (1975)
12.　Grannell, P.K., Mansfield, P.K., Whittaker, M.A.: Phys. Rev. *B8*, 4149 (1973)
13.　Ernst, H.: Wiss. Z. Karl-Marx-Univ. Leipzig, Math.-Naturw. *23*, 449 (1974)
14. a) Kanert, O., Mehring, M.: NMR: Basic Principles and Progress, Vol. 3. Berlin-Heidelberg-New York: Springer 1971
　　b) Schmid, D.: Springer Tracts in Modern Physics. Berlin-Heidelberg-New York: Springer 1973
15.　Walstedt, R.E., McArthur, D.A., Hahn, E.L.: Phys. Lett. *15*, 7 (1965)
16.　Bleich, H.E., Redfield, A.: J. Chem. Phys. *55*, 5406 (1971)

17. a) Yannoni, C.S., Bleich, H.E.: J. Chem. Phys. *55*, 5406 (1971)
    b) Yannoni, C.S.: J. Chem. Phys. *58*, 1773 (1973)
18.   Bleich, H.E., Redfield, A.G.: J. Chem. Phys. *67*, 5040 (1977)
19.   Stoll, M.E., Vega, A.J., Vaughan, R.W.: Phys. Rev. *A16*, 1521 (1977)
20. a) Maier, G., Haeberlen, U., Wolf, H.C., Hausser, K.H.: Phys. Lett. *25A*, 384 (1967)
    b) Maier, G., Wolf, H.C.: Z. Naturforsch. *23A*, 1068 (1968)
    c) Lau, P., Stehlik, D., Hausser, K.H.: J. Magn. Res. *15*, 270 (1970)
    d) Stehlik, D., Doehring, A., Colpa, J.P., Callaghan, E., Kesmarky, S.: Chem. Phys. *7*, 165 (1975)
    e) Hausser, K.H., Wolf, H.C.: Advan. Magn. Res. Vol. 8 (1976)
21. a) Goldman, M., Chapellier, M., Vu Huang, Chau, Abragam, A.: Phys. Rev. *B10*, 226 (1974)
    b) Jacquinot, J.F., Wenckenach, W.Th., Goldman, M., Abragam, A.: Phys. Rev. Lett. *32*, 1096 (1974)
22.   Demco, D., Kaplan, S., Pausak, S., Waugh, J.S.: Chem. Phys. Lett. *1*, 77 (1975)
23.   Goldman, M.: Compt. Rend. *246*, 1038 (1958)
24. a) Goldman, M., Landesmann, A.: Compt. Rend. *252*, 263 (1961)
    b) Goldman, M., Landesmann, A.: Phys. Rev. *132*, 610 (1963)
25.   Goldman, M.: Phys. Rev. *138A*, 1668 (1965)
26.   Schwab, M., Hahn, E.L.: J. Chem. Phys. *52*, 3152 (1970)
27. a) Sarles, L.R., Cotts, R.M.: Phys. Rev. *111*, 853 (1958)
    b) Bloch, F.: Phys. Rev. *111*, 841 (1958)
    c) Ackermann, H., Fujara, F.: Presented at the Spring Meeting of the DPG. Freudenstadt, Germany (1976)
    d) Fujara, F.: Diploma work. Heidelberg 1975
28. a) Mehring, M., Sinning, G., Pines, A., Shattuck, T.W.: XXVth IUPAC. Jerusalem 1975
    b) Mehring, M., Sinning, G., Pines, A., Shattuck, T.W.: Second Specialized Colloque Ampere. Budapest 1975
29.   Mehring, M., Pines, A., Rhim, W.-K., Waugh, J.S.: J. Chem. Phys. *54*, 3239 (1971)
30.   Mehring, M., Raber, H., Sinning, G.: 18th Congress Ampere, p. 35. Nottingham, England 1974
31.   Pines, A.: First Specialized Colloque Ampere. Krakow, Poland, August 1973
32.   Haeberlen, U. et al.: J. Magn. Res. *36*, 453 (1979)
33. a) Maudsley, A.A., Müller, L., Ernst, R.R.: J. Magn. Res. *28*, 463 (1977)
    b) Müller, L., Ernst, R.R.: Mol. Phys. *38*, 963 (1979)
    c) Burum, D.P., Ernst, R.R.: J. Magn. Res. *39*, 163 (1980)
34.   Morris, G.A., Freeman, R.: J. Am. Chem. Soc. *101*, 760 (1979)
35. a) Bertrand, R.D., Moniz, W.B., Garroway, A.N., Chingas, G.C.: J. Am. Chem. Soc. *100*, 5227 (1978)
    b) Bertrand, R.D., Moniz, W.B., Garroway, A.N., Chingas, G.C.: J. Magn. Res. *32*, 465 (1978)
36. a) Chingas, G.C., Garroway, A.N., Bertrand, R.D., Moniz, W.B.: J. Magn. Res. *35*, 283 (1979)
    b) Chingas, G.C., Garroway, A.N., Bertrand, R.D., Moniz, W.B.: J. Am. Chem. Soc. *102*, 2526 (1980)
37.   Garroway, A.N., Chingas, G.C.: J. Magn. Res. *38*, 179 (1980)
38.   Packer, K.J., Wright, K.M.: J. Magn. Res. *41*, 268 (1980)
39.   Pines, A., Shattuck, T.W.: J. Chem. Phys. *61*, 1255 (1974)
40. a) Mori, H., Kawasaki, K.: Progr. Theor. Phys. (Kyoto) *27*, 529 (1962)
    b) Mori, H.: Progr. Theor. Phys. (Kyoto) *33*, 423 (1965)
41. a) Lado, F., Memory, J.D., Parker, G.W.: Phys. Rev. *B4*, 1406 (1971)
    b) Parker, G.W., Lado, F.: Phys. Rev. *B9*, 22 (1974)
    c) Parker, G.W., Lado, F.: Phys. Rev. *B8*, 3081 (1973)
42.   Strombotne, R.L., Hahn, E.L.: Phys. Rev. *133A*, 1616 (1964)
43.   Müller, L., Kumar, A., Baumann, T., Ernst, R.R.: Phys. Rev. Lett. *32*, 1902 (1974)
44.   Hester, R.K., Ackermann, J.L., Cross, V.R., Waugh, J.S.: Phys. Rev. Lett. *34*, 993 (1975)
45.   Rybaczewski, E.F., Neff, B.L., Waugh, J.S., Sherfinski, J.S.: J. Chem. Phys. *67*, 1231 (1977)
46.   Abragam, A., Winter, J.: Compt. Rend. *249*, 1633 (1959)
47.   Walstedt, R.E.: Phys. Rev. *B5*, 41 (1972)
48. a) Mehring, M., Sinning, G., Pines, A.: Z. Phys. *B24*, 73 (1976)

b) Sinning, G., Mehring, M., Pines, A.: Chem. Phys. Lett. *43*, 382 (1976)

   c) Mehring, M., Sinning, G.: Phys. Rev. *15*, 2519 (1977)

49.   Van der Hart, D.L., Earl, W.L., Garroway, A.N.: J. Magn. Res. *44*, 361 (1981)

50.   Pines, A., Ruben, D.J., Vega, S., Mehring, M.: Phys. Rev. Lett. *36*, 110 (1976)

51. a) Snyder, L.C., Meiboom, S.: J. Chem. Phys. *58*, 5096 (1973)

   b) Hewitt, R.C., Meiboom, S., Snyder, L.C.: J. Chem. Phys. *58*, 5089 (1973)

52.   Van der Hart, D.L., Gutowsky, H.S.: J. Chem. Phys. *49*, 261 (1968)

53.   Kaplan, S., Pines, A., Griffin, R.G., Waugh, J.S.: Chem. Phys. Lett. *25*, 78 (1973)

54.   Ishol, L.M., Scott, R.A.: J. Magn. Res. *27*, 23 (1977)

55.   Grimmer, A.R., Peter, R., Fechner, E.Z.: Chem. *18*, 109 (1978)

56.   Zilm, K.W., Grant, D.M.: J. Am. Chem. Soc. *103*, 2913 (1981)

57.   Zilm, K.W., Conlun, R.T., Grant, D.M., Michl, J.: J. Am. Chem. Soc. *100*, 8038 (1978)

58. a) Pschorn, U., Spiess, H.W.: J. Magn. Res. *39*, 217 (1980)

   b) Hentschel, R., Sillescu, H., Spiess, H.W.: *42*, 381 (1981)

59. a) van Willigen, H., Griffin, R.G., Haberkorn, R.A.: J. Chem. Phys. *67*, 5855 (1977)

   b) Stark, R.E., Haberkorn, R.A., Griffin, R.G.: J. Chem. Phys. *68*, 1996 (1978)

   c) Haberkorn, R.A., Stark, R.E., van Willigen, H., Griffin, R.G.: J. Am. Chem. Soc. *103*, 2534 (1981)

60.   Seelig, A., Seelig, J.: Biochemistry *13*, 4839 (1974)

61.   Huang, T.H., Skarjune, R.P., Wittebort, R.J., Griffin, R.G., Oldfield, E.: J. Am. Chem. Soc. *102*, 7377 (1980)

62. a) Opella, S.J., Wise, W.B., DiVerdi, J.A.: Biochemistry *20*, 284 (1981)

   b) DiVerdi, J.A., Opella, S.J.: Biochemistry *20*, 280 (1981)

63. a) Hemminga, M.A., Veeman, W.S., Hilhorst, H.W.M., Schaafsma, T.J.: Biophys. J. *35*, 463 (1981)

   b) Nall, B.T., Rothwell, W.P., Waugh, J.S., Rupprecht, A.: Biochemistry *20*, 1881 (1981)

64.   Schaefer, J., Stejskal, E.O., Buchdahl, R.: Macromolecules *10*, 384 (1977)

65. a) Van der Hart, D.L., Garroway, A.N.: J. Chem. Phys. *71*, 2773 (1979)

   b) Earl, W.L., Van der Hart, D.L.: Macromolecules *12*, 762 (1979)

66.   Veeman, W.S., Menger, E.M., Ritchey, W., de Boer, E.: Macromolecules *12*, 924 (1979)

67.   Miknis, F.P., Bartuska, V.J., Maciel, G.E.: American Laboratory, Nov. (1979)

68.   Stejskal, E.O., Schaefer, J., Waugh, J.S.: J. Magn. Res. *28*, 105 (1977)

69.   Opella, St.J., Waugh, J.S.: J. Chem. Phys. *66*, 4919 (1977)

70.   Mehring, M., Weber, H., Müller, W., Wegener, G.: Solid. State Commun. (in press)

71. a) Bartuska, V.J., Maciel, G.E., Schaefer, G.E., Stejskal, E.O.: FUEL *56*, 354 (1977)

   b) Maciel, G.E., Bartuska, V.J., Miknis, F.P.: FUEL *57*, 505 (1978)

   c) Maciel, G.E., Bartuska, V.J., Miknis, F.P.: FUEL *58*, 391 (1979)

   d) Miknis, F.P., Maciel, G.E., Bartuska, V.J.: Org. Geochem. *1*, 169 (1979)

72.   Resing, H.A., Garroway, A.N., Hazlett, R.N.: FUEL *57*, 450 (1978)

73. a) Suwelack, D., Becker, J.D., Mehring, M.: Solid. State Commun. *22*, 597 (1977)

   b) Becker, J.D., Suwelack, D., Mehring, M.: Solid. State Commun. *25*, 1145 (1978)

   c) Mehring, M., Suwelack, D.: Phys. Rev. Lett. *42*, 317 (1979)

   d) Mehring, M., Becker, J.D.: Phys. Rev. Lett. *47*, 366 (1981)

74.   Blinc, R., Burgar, M., Rutar, V., Seeliger, J., Zupancic, J.: Phys. Rev. Lett. *38*, 92 (1977)

75. a) Garroway, A.N., Moniz, W.B., Resing, H.A.: ACS Preprints *36*, 133 (1976)

   b) Garroway, A.N., Moniz, W.B., Resing, H.A.: Chem. Soc. *13*, 63 (1979)

76.   Resing, A.H., Slotfeldt-Ellingsen, D., Garroway, A.N., Weber, D.C., Pinnavaia, T.J., Unger, K.: Magn. Res. in Coll. and Interface Sci. ed. J.P. Fraissard and H.A. Resing, D. Reidel Publishing Company 1980, p. 239

77.   Duncan, T.M., Dybowski, C.: Surface Science Rep. (in press)

78.   Balimann, G.E., Groombridge, Ch.J., Harris, R.K., Packer, K.J., Say, B.J., Tanner, St.F.: Phil. Trans. Roy. Soc. *299*, 643 (1981)

Chapter 5

1.   Jeener, J.: Ampere International Summer School II. Basko Polje, Yugoslavia 1971

2.   Kumar, A., Welti, D., Ernst, R.R.: J. Magn. Res. *18*, 69 (1975)

3.   Müller, L., Kumar, A., Ernst, R.R.: J. Chem. Phys. *63*, 5490 (1975)
4.   Aue, W.P., Bartholdi, E., Ernst, R.R.: J. Chem. Phys. *64*, 2229 (1976)
5. a) Bodenhausen, G., Freeman, R., Niedermeyer, R., Turner, D.L.: J. Magn. Res. *24*, 291 (1976)
   b) Bodenhausen, G., Freeman, R., Turner, D.L.: J. Chem. Phys. *65*, 839 (1976)
6. a) Bodenhausen, G., Freeman, R., Niedermeyer, R., Turner, D.L.,: J. Magn. Res. *26*, 133 (1977)
   b) Freeman, R., Morris, G.A., Turner, D.L.: J. Magn. Res. *26*, 373 (1977)
7. a) Hester, R.K., Ackerman, J.L., Cross, V.R., Waugh, J.S.: Phys. Rev. Lett. *34*, 993 (1975)
   b) Hester, R.K., Ackerman, J.L., Neff, B.L., Waugh, J.S.: Phys. Rev. Lett. *36*, 1081 (1976)
8.   Waugh, J.S.: Proc. Natl. Acad. Sci. (USA) *73*, 1394 (1976)
9.   Opella, S.J., Waugh, J.S.: J. Chem. Phys. *65*, 4919 (1977)
10.   Rybaczewski, E.F., Neff, B.L., Waugh, J.S., Sherfinski, J.S.: J. Chem. Phys. *67*, 1231 (1977)
11. a) Stoll, M.E., Vega, A.J., Vaughan, R.W.: J. Chem. Phys. *65*, 4093 (1976)
   b) Stoll, M.E., Vega, A.J., Vaughan, R.W.: J. Chem. Phys. *69*, 5458 (1978)
12.   Alla, M., Lippmaa, E.: Chem. Phys. Lett. *37*, 260 (1976)
13.   Freeman, R., Morris, G.A.: Bull. Magn. Res. *1*, 5 (1979)
14.   Aue, W.P., Bachmann, P., Wokaun, A., Ernst, R.R.: J. Magn. Res. *29*, 523 (1978)
15.   Hahn, E.L., Maxwell, D.E.: Phys. Rev. *88*, 1070 (1952)
16.   Aue, W.P.: Ph.D. Thesis, ETH Zürich 1979
17.   Ernst, R.R., Aue, W.P., Bachmann, P., Höhener, A., Linder, M., Meier, B.H., Müller, L., Wokaun, A.: Proceedings of the XXth Congress Ampere. Tallin 1978, p.15. Springer-Verlag
18.   Linder, M., Höhener, A., Ernst, R.R.: J. Chem. Phys. *73*, 4959 (1980)
19.   Nagayama, K., Wüthrich, K., Ernst, R.R.: Biochem. and Biophys. Res. Comm. *90*, 305 (1979)
20.   Nagayama, K., Kumar, A., Wüthrich, K., Ernst, R.R.: J. Magn. Res. *40*, 321 (1980)
21.   Maudsley, A.A., Ernst, R.R.: Chem. Phys. Lett. *50*, 368 (1977)
22.   Bodenhausen, G., Stark, R.E., Ruben, D.J., Griffin, R.G.: Chem. Phys. Lett. *67*, 424 (1979)
23.   Munowitz, M.G., Griffin, R.G., Bodenhausen, G., Huang, T.H.: J. Am. Chem. Soc. *103*, 2529 (1981)
24.   Aue, W.P., Ruben, D.J., Griffin, R.G.: J. Magn. Res. *43*, 472 (1981)
25.   Jeener, J., Meier, B.H., Bachmann, P., Ernst, R.R.: J. Chem. Phys. *71*, 4546 (1979)
26. a) Vega, S., Shattuck, T.W., Pines, A.: Phys. Rev. Lett. *37*, 43 (1976)
   b) Vega, S., Pines, A.: J. Chem. Phys. *66*, 5624 (1977)
27.   Wokaun, A., Ernst, R.R.: Chem. Phys. Lett. *52*, 407 (1977)

Chapter 6

1.   Goeppert-Mayer, M.: Ann. Phys. (Leipzig) *9*, 273 (1931)
2.   Walther, H.: In: "Laser Spectroscopy of Atoms and Molecules", Topics in Applied Physics 2, p. 1. Berlin-Heidelberg-New York: Springer 1976
3.   Koningstein, G.: "Introduction to the Theory of the Raman Effect". Reidel, Dordrecht (1972)
4.   Bloembergen, N., Levenson, M.D.: In: "High Resolution Laser Spectroscopy". K. Shimoda (ed.): Topics in Applied Physics 13, p. 315. Berlin-Heidelberg-New York: Springer 1976
5.   Letokhov, V.S., Chebotayev, V.P.: "Nonlinear Laser Spectroscopy". Springer Series in Optical Sciences 4. Berlin-Heidelberg-New York: Springer 1977
6.   Anderson, W.A.: Phys. Rev. *104*, 850 (1956)
7.   Kaplan, J.I., Meiboom, S.: Phys. Rev. *106*, 499 (1957)
8.   Yatsiv, S.: Phys. Rev. *113*, 1522 (1952)
9. a) Bucci, P., Martinelli, M., Santucci, S.: J. Chem. Phys. *52*, 4041 (1970)
   b) Bucci, P., Martinelli, M., Santucci, S.: J. Chem. Phys. *53*, 4524 (1970)
10. a) Vega, S., Pines, A.: J. Chem. Phys. *66*, 5624 (1977)
   b) Vega, S.: J. Chem. Phys. *68*, 5518 (1978)
11.   Wokaun, A., Ernst, R.R.: J. Chem. Phys. *67*, 1752 (1977)
12.   Warren, W.S., Sinton, S., Weitekamp, D.P., Pines, A.: Phys. Rev. Lett. *43*, 1791 (1979)
13. a) Snyder, L.C., Meiboom, S.: J. Chem. Phys. *58*, 5096 (1973)
   b) Hewitt, R.C., Meiboom, S., Snyder, L.C.: J. Chem. Phys. *58*, 5089 (1973)
14.   Pines, A., Ruben, D.J., Vega, S., Mehring, M.: Phys. Rev. Lett. *36*, 110 (1976)

15.   Pines, A., Vega, S., Mehring, M.: Phys. Rev. *B18*, 112 (1978)
16.   Suwelack, D., Mehring, M., Pines, A.: Phys. Rev. *B19*, 238 (1979)
17.   Abragam, A.: The Principles of Nuclear Magnetism. London: Oxford Univ. Press 1961
18.   Suwelack, D.: Doctoral Thesis. Dortmund 1979
19.   Brewer, R.G., Hahn, E.L.: Phys. Rev. *A11*, 1641 (1975)
20.   Grischkowsky, D., Loy, N.M.T., Liao, P.F.: Phys. Rev. *A12*, 2514 (1975)
21.   Chandra, S., Yen, W.M.: Phys. Lett. *57A*, 217 (1976)
22.   Flusberg, A., Hartmann, S.R.: Phys. Rev. *A14*, 813 (1976)
23.   Loy, N.M.T.: Phys. Rev. Lett. *36*, 5624 (1977)
24.   Hatanaka, H., Terao, T., Hashi, T.: J. Phys. Soc. Jpn. *39*, 835 (1975)
25.   Hatanaka, H., Hashi, T.: J. Phys. Soc. Jpn. *39*, 1139 (1975)
26.   Gold, D.G., Hahn, E.L.: Phys. Rev. *A16*, 324 (1977)
27.   Vega, S., Shattuck, T.W., Pines, A.: Phys. Rev. Lett. *37*, 43 (1976)
28.   Mehring, M., Wolff, E.K., Stoll, M.E.: J. Magn. Res. *37*, 475 (1980)
29.   Bodenhausen, G., Vold, R.L., Vold, R.R.: J. Magn. Res. *37*, 93 (1980)
30.   Wemmer, D.E., Wolff, E.K., Mehring, M.: J. Magn. Res. *42*, 460 (1981)
31. a) Stoll, M.E., Wolff, E.K., Mehring, M.: Phys. Rev. *A17*, 1561 (1978)
    b) Wolff, E.K., Mehring, M.: Phys. Lett. *20A*, 125 (1979)
32.   Stoll, M.E., Vega, A., Vaughan, R.W.: Phys. Rev. *A16*, 1521 (1977)
33.   Waugh, J.S.: private communication
34.   Haeberlen, U., Waugh, J.S.: Phys. Rev. *185*, 420 (1969)
35.   Pines, A., Vega, S., Ruben, D.J., Shattuck, T.W., Wemmer, D.E.: Proc. of the IVth Ampere
      Int. Summer School. Pula, Yugoslavia 1976, p. 127
36.   Vega, S., Pines, A.: Proc. of the XIXth Congress Ampere. Heidelberg 1976, p. 395
37.   Wokaun, A., Ernst, R.R.: Chem. Phys. Lett. *52*, 407 (1977)
38.   Aue, W.P., Bartholdi, E., Ernst, R.R.: J. Chem. Phys. *64*, 2229 (1976)
39.   Vega, S., Naor, Y.: J. Chem. Phys. *75*, 75 (1981)
40.   Warren, W.S.: Ph.D. Thesis. Berkeley 1980
41. a) Warren, W.S., Weitekamp, D.P., Pines, A.: J. Magn. Res. *40*, 581 (1980)
    b) Warren, W.S., Weitekamp, D.P., Pines, A.: J. Chem. Phys. *73*, 2084 (1980)
42.   Drobny, G., Pines, A., Sinton, S., Weitekamp, D.P., Wemmer, D.E.: Chem. Soc. Symp. *13*, 49
      (1979)
43. a) Warren, W.S., Pines, A.: J. Chem. Phys. *74*, 2808 (1981)
    b) Warren, W.S., Pines, A.: J. Am. Chem. Soc. *103*, 1613 (1981)
44.   Drobny, G., Pines, A., Sinton, S., Warren, W.S., Weitekamp, D.P.: Phil. Trans. Roy. Soc.
      London *A299*, 585 (1981)
45.   Wokaun, A., Ernst, R.R.: Mol. Phys. *36*, 317 (1978)
46.   Mehring, M., Waugh, J.S.: Phys. Rev. *B5*, 3459 (1972)
47.   Rhim, W.K., Pines, A., Waugh, J.S.: Phys. Rev. *B3*, 684 (1971)
48. a) Sinton, S., Pines, A.: Chem. Phys. Lett. *76*, 263 (1980)
    b) Tang, J., Pines, A.: J. Chem. Phys. *73*, 2512 (1980)
    c) Tang, F., Pines, A.: J. Chem. Phys. *72*, 3290 (1980)
    d) Tang, J., Pines, A.: J. Chem. Phys. *73*, 172 (1980)
    e) Weitekamp, D.P., Garbow, J.R., Murdoch, J.B., Pines, A.: J. Am. Chem. Soc. *103*, 3578
      (1981)
49. a) Shattuck, T.W.: Ph.D. Thesis, Univ. of California. Berkeley 1976
    b) Vega, S., Shattuck, T.W., Pines, A.: Phys. Rev. *A22*, 638 (1980)
50.   Brunner, P., Reinhold, M., Ernst, R.R.: J. Chem. Phys. *73*, 1086 (1980)
51.   Reinhold, M., Brunner, P., Ernst, R.R.: J. Chem. Phys. *74*, 184 (1981)
52.   Emid, S.: J. Magn. Res. *42*, 147 (1981)

Chapter 7

1.   Appleman, B.R., Dailey, B.P.: Advan. Magn. Res. *7*, 231 (1974)
2.   Emsley, J.W., Feeney, J., Sutcliff, L.H.: High Resolution NMR Spectroscopy. Oxford: Per-
     gamon 1965

3.   Pople, J.A., Scheider, W.G., Bernstein, H.J.: High Resolution Nuclear Magnetic Resonance. New York: McGraw Hill 1959
4.   Ramsey, N.F.: Phys. Rev. *78*, 699 (1950)
5.   Snyder, L.C., Parr, R.G.: J. Chem. Phys. *34*, 837 (1961)
6.   Karplus, M., Das, T.P.: J. Chem. Phys. *34*, 1683 (1961)
7.   Haeberlen, U.: Advan. Magn. Res. supplement. New York: Academic Press 1976
8.   Ditchfield, R., Miller, D.P., Pople, J.A.: J. Chem. Phys. *54*, 4186 (1971)
9.   Kolker, H.J., Karplus, M.: J. Chem. Phys. *41*, 1259 (1964)
10. Reid, R.V., May-Chu, A.H.: Phys. Rev. *A9*, 609 (1974)
11. Kempf, J., Spiess, H.W., Haeberlen, U., Zimmermann, H.: Chem. Phys. *4*, 269 (1974)
12. a) Gierke, T.D., Flygare, H.W.: J. Am. Chem. Soc. *94*, 7277 (1972)
    b) Gierke, T.D., Tigelaar, H.L., Flygare, H.W.: J. Am. Chem. Soc. *94*, 330 (1972)
13. Malli, G., Fraga, S.: Theoret. Chim. Acta *5*, 284 (1966)
14. Haeberlen, U., Kohlschütter, U., Kempf, J., Spiess, H.W., Zimmermann, H.: Chem. Phys. *3*, 248 (1974)
15. Grosescu, R., Achlama, A.M., Haeberlen, U., Spiess, H.W.: Chem. Phys. *5*, 119 (1974)
16. Achlama, A.M., Kohlschütter, U., Haeberlen, U.: Chem. Phys. *7*, 287 (1975)
17. Lau, K.F., Vaughan, R.W.: J. Chem. Phys. *65*, 4825 (1976)
18. a) Schmiedel, H.: Phys. Stat. Sol. *B67*, K27 (1975)
    b) Neubauer, R., Schnabel, B.: First Spec. Colloque Ampere. Krakow, Poland 1973
19. Haeberlen, U., Kohlschütter, U.: Chem. Phys. *2*, 76 (1973)
20. Raber, H., Brünger, G., Mehring, M.: Chem. Phys. Lett. *23*, 400 (1973)
21. Schreiber, L.B., Vaughan, R.W.: Chem. Phys. Lett. *28*, 586 (1974)
22. Rhim, W.-K., Elleman, D.D., Vaughan, R.W.: J. Chem. Phys. *59*, 3740 (1973)
23. Haddix, D.C., Lauterbuhr, C.C.: Natl. Bur. Std. (U.S.) Spec. Publ. No. 301, 403 (1969)
24. Haubenreisser, U., Schnabel, B.: Proc. 1st Spec. Colloque Ampere. Krakow 1973, p. 140
25. a) Burghoff, U., Scheler, G., Müller, R.: Phys. Stat. Sol. *25A*, K31 (1974)
    b) Terao, T., Hashi, T.: J. Phys. Soc. Jpn. *36*, 989 (1974)
26. Van Hecke, P., Weaver, J.C., Neff, B.L., Waugh, J.S.: J. Chem. Phys. *60*, 1668 (1974)
27. Kohlschütter, U.: Frühjahrstagung der DPG K In 1975 and Ph.D. Thesis. Heidelberg 1975
28. a) Van Hecke, P., Spiess, H.W., Haeberlen, U.: J. Magn. Res. *22*, 103 (1976)
    b) Van Hecke, P., Spiess, H.W., Haeberlen, U., Haussuehl, S.: J. Magn. Res. *22*, 93 (1976)
    c) Feucht, H., Haeberlen, U., Pollak-Stachura, M., Spiess, H.W.: Z. Naturforsch. *31A*, 1173 (1976)
29. Berglund, B., Carson, D.G., Vaughan, R.W.: J. Chem. Phys. *72*, 824 (1980)
30. Suwelack, D., Becker, J.D., Mehring, M.: Solid. State Commun. *22*, 597 (1977)
31. Dybowski, C.R., Gerstein, B.C., Vaughan, R.W.: J. Chem. Phys. *67*, 3412 (1977)
32. Pines, A., Ruben, D.J., Vega, S., Mehring, M.: Phys. Rev. Lett. *36*, 110 (1976)
33. Ryan, L.M., Wilson, R.C., Gerstein, B.C.: Chem. Phys. Lett. *52*, 341 (1977)
34. a) Rhim, W.K., Burum, D.P., Elleman, D.D.: J. Chem. Phys. *71*, 3139 (1979)
    b) Burum, D.P., Rhim, W.K.: J. Chem. Phys. *70*, 3553 (1979)
35. McKnett, C.L., Dybowski, C.R., Vaughan, R.W.: J. Chem. Phys. *63*, 4578 (1975)
36. Burum, D.P., Rhim, W.K.: J. Magn. Res. *34*, 241 (1979)
37. Spiess, H.W., Haeberlen, U., Zimmermann, H.: J. Magn. Res. *25*, 55 (1977)
38. Halstead, T.K., Spiess, H.W., Haeberlen, U.: Mol. Phys. *31*, 1569 (1976)
39. a) Sagnowski, S., Aravamudhan, S., Haeberlen, U.: J. Chem. Phys. *66*, 4697 (1977)
    b) Sagnowski, S., Aravamudhan, S., Haeberlen, U.: J. Magn. Res. *28*, 271 (1977)
40. Achlama, A.M., Post, H., Haeberlen, U.: Chem. Phys. *31*, 203 (1978)
41. Aravamudhan, S., Haeberlen, U., Irngartinger, H., Krieger, C.: Mol. Phys. *38*, 241 (1979)
42. Tegenfeldt, J., Feucht, H., Ruschityka, G., Haeberlen, U.: J. Magn. Res. *39*, 509 (1980)
43. a) Post, H., Haeberlen, U.: J. Magn. Res. *40*, 17 (1980)
    b) Wendliny, Th.: Diploma work. Heidelberg 1979
    c) Bittner, H.P.: Diploma work. Heidelberg 1981
    d) Haeberlen, U.: Private communication
44. Ernst, H., Fenzke, D., Heink, W.: Phys. Stat. Sol. (b) *57*, K103 (1973)
45. Voigtsberger, B., Rosenberger, H.: Phys. Stat. Sol. (a) *35*, K89 (1976)
46. a) Nicol, A.T., Vaughan, R.W.: J. Chem. Phys. *69*, 5211 (1978)
    b) Nicol, A.T., Vaughan, R.W.: J. Am. Chem. Soc. *101*, 583 (1979)

47.   Dubois-Murphy, P., Gerstein, B.C.: J. Chem. Phys. *70*, 4552 (1979)
48.   Waugh, J.S., Huber, L.M., Haeberlen, U.: Phys. Rev. Lett. *20*, 180 (1968)
49. a) Vaughan, R.W., Elleman, D.D., Rhim, W.-K., Stacey, L.M.: J. Chem. Phys. *57*, 5383 (1972)
    b) Stacey, L.M., Vaughan, R.W., Elleman, D.D.: Phys. Rev. Lett. *26*, 1153 (1971)
50.   Sears, R.E.: J. Chem. Phys. *59*, 5213 (1973)
51. a) Phillips, J.C.: Rev. Mod. Phys. *42*, 317 (1970)
    b) Gordy, W.: Phys. Rev. *69*, 604 (1946)
    c) Gordy, W., Thomas, W.J.O.: J. Chem. Phys. *24*, 439 (1956)
52.   Stoeckmann, H.J., Ackermann, H., Dubbers, D., Group, M., Heitjans, P.: Z. Phys. *269*, 47 (1974)
53.   Burum, D.P., Elleman, D.D., Rhim, W.K.: J. Chem. Phys. *68*, 1164 (1978)
54.   Sears, R.E.: J. Chem. Phys. *61*, 4368 (1974)
55.   Mehring, M., Pines, A., Rhim, W.-K., Waugh, J.S.: J. Chem. Phys. *54*, 3239 (1971)
56. a) Mehring, M., Griffin, G.R., Waugh, J.S.: J. Am. Chem. Soc. *92*, 7222 (1970)
    b) Mehring, M., Griffin, G.R., Waugh, J.S.: J. Chem. Phys. *55*, 746 (1971)
57.   Chan, S.I., Dubin, A.S.: J. Chem. Phys. *46*, 1745 (1967)
58.   Griffin, R.G., Ellett, J.D., Mehring, M., Bullitt, M., Waugh, J.S.: J. Chem. Phys. *57*, 2147 (1972)
59.   Raber, H., Mehring, M.: Chem. Phys. *26*, 123 (1977)
60. a) Nehring, J., Saupe, A.: J. Chem. Phys. *52*, 1307 (1970)
    b) Snyder, L.: J. Chem. Phys. *43*, 4041 (1965)
61.   Griffin, R.G., Yeung, H.N., LaPrade, M.D., Waugh, J.S.: J. Chem. Phys. *59*, 777 (1973)
62.   O'Reilly, D.E., Peterson, E.M., El Saffar, Z.M., Scheie, C.E.: Chem. Phys. Lett. *8*, 470 (1971)
63.   Carolan, J.L.: Chem. Pyhs. Lett. *12*, 389 (1971)
64. a) Hunt, E., Meyer, H.: J. Chem. Phys. *41*, 353 (1964)
    b) Brooks-Harris, A., Hunt, E., Meyer, H.: J. Chem. Phys. *42*, 2851 (1965)
65.   Andrew, E.R., Tunstall, D.P.: Proc. Phys. Soc. *81*, 986 (1963)
66.   Yannoni, C.S., Dailey, B.P., Ceasar, G.P.: J. Chem. Phys. *54*, 4020 (1971)
67.   Hull, W.E., Sykes, B.D.: J. Mol. Biol. *98*, 121 (1975)
68. a) Raber, H., Mehring, M.: DPG Frühjahrstagung München 1973
    b) Long, R.C., Goldstein, J.H.: J. Chem. Phys. *54*, 1563 (1971)
69.   Wilson III, C.W.: J. Polymer. Sci. *61*, 403 (1962)
70.   Blinc, R., Zupancic, I., Maricic, S., Veksli, Z.: J. Chem. Phys. *39*, 2109 (1963), *40*, 3739 (1964)
71.   Blinc, R., Pirkmayer, E., Slivnik, J., Zupancic, I.: J. Chem. Phys. *45*, 1488 (1966)
72.   Hendermann, D.K., Falconer, W.E.: J. Chem. Phys. *50*, 1203 (1969)
73.   Van der Hart, D.L., Gutowsky, H.S., Farrar, T.C.: J. Chem. Phys. *50*, 1058 (1969)
74. a) Lauterbur, P.C.: Phys. Rev. Lett. *1*, 343 (1958)
    b) Yannoni, C.S., Whipple, E.B.: J. Chem. Phys. *47*, 2508 (1967)
75.   Pines, A., Rhim, W.-K., Waugh, J.S.: J. Chem. Phys. *54*, 5438 (1971)
76.   Spiess, H.W., Schweitzer, D., Haeberlen, U., Hausser, K.H.: J. Magn. Res. *5*, 101 (1971)
77.   Spiess, H.W., Mahnke, H.: Z. Naturforsch. *27A*, 1536 (1972), Ber. Bunsen-Ges. Phys. Chem. *76*, 991 (1972)
78. a) Kempf, J., Spiess, H.W., Haeberlen, U., Zimmermann, H.: Chem. Phys. Lett. *17*, 39 (1972)
    b) Chem. Phys. *4*, 269 (1974)
79. a) Pines, A., Gibby, M.G., Waugh, J.S.: J. Chem. Phys. *56*, 1776 (1972)
    b) J. Chem. Phys. *59*, 569 (1973)
80.   Pausak, S., Pines, A., Waugh, J.S.: J. Chem. Phys. *59*, 591 (1973)
81. a) Pines, A., Gibby, M.G., Waugh, J.S.: Chem. Phys. Lett. *15*, 373 (1972)
    b) Waugh, J.S., Gibby, M.G., Pines, A., Kaplan, S.: Proc. of the 17th Colloque Ampere, p.13. Turku, Finland 1972
82.   Kaplan, S., Griffin, R.G., Waugh, J.S.: Chem. Phys. Lett. *25*, 78 (1974)
83. a) Chan, J.J., Griffin, R.G., Pines, A.: J. Chem. Phys. *60*, 2561 (1974)
    b) J. Chem. Phys. *62*, 4923 (1975)
    c) Pines, A., Abramson, E.: J. Chem. Phys. *60*, 5130 (1974)
84.   Pausak, S., Tegenfeldt, J., Waugh, J.S.: J. Chem. Phys. *61*, 1338 (1974)
85.   Pines, A., Chang, J.J., Griffin, R.G.: J. Chem. Phys. *61*, 1021 (1974)
86.   Ando, I., Nishioka, A., Kondo, M.: Chem. Phys. Lett. *25*, 212 (1974)
87.   Stoll, M.E., Vaughan, R.W., Saillant, R.B., Cole, T.: J. Chem. Phys. *61*, 2896 (1974)

88. Spiess, H.W., Grosescu, R., Haeberlen, U.: Chem. Phys. *6*, 226 (1974)
89. Spiess, H.W.: NMR: Basic Principles and Progress, Vol. 15. Berlin-Heidelberg-New York: Springer 1978, p. 55
90. Linder, M., Höhener, A., Ernst, R.R.: J. Magn. Res. *35*, 379 (1979)
91. van Dongen Torman, J., Veeman, W.S.: J. Chem. Phys. *68*, 3233 (1978)
92. van Dongen Torman, J.: Ph.D. Thesis. Nijmegen 1978
93. van Dongen Torman, J., Veeman, W.S., de Boer, E.: J. Magn. Res. *32*, 49 (1978)
94. Wemmer, D.E., Pines, A.: J. Am. Chem. Soc. *103*, 34 (1981)
95. Wolff, E.K., Griffin, R.G., Waugh, J.S.: J. Chem. Phys. *67*, 2387 (1977)
96. Haberkorn, R.A., Stark, R.E., van Willigen, H., Griffin, R.G.: J. Am. Chem. Soc. *103*, 2534 (1981)
97. Van der Hart, D.L.: J. Chem. Phys. *64*, 830 (1976)
98. Opella, S.J., Waugh, J.S.: J. Chem. Phys. *66*, 4919 (1977)
99. Earl, W.L., Van der Hart, D.L.: Macromolecules *12*, 762 (1979)
100. Van der Hart, D.L.: Macromolecules *12*, 1232 (1979)
101. Mehring, M., Becker, J.D.: Phys. Rev. Lett. *47*, 366 (1981)
102. Griffin, R.G., Pines, A., Pausak, S., Waugh, J.S.: J. Chem. Phys. *63*, 1267 (1975)
103. Griffin, R.G., Ruben, D.J.: J. Chem. Phys. *63*, 1272 (1975)
104. Cross, V.R., Waugh, J.S.: J. Magn. Res. *25*, 225 (1977)
105. Englert, G.: Z. Naturforsch. *27A*, 1536 (1972)
106. Resing, H.A., Slotfeldt-Ellingsen, D., Garroway, A.N., Pinnavaia, T.J., Unger, K.: Magnetic Resonance in Colloid and Interface Science, ed. J.P. Fraissard and H.A. Resing, Reidel Publishing Comp. 1980, p. 239
107. Mehring, M., Weber, H., Müller, W., Wegner, G.: Solid. State Commun. (in press)
108. Tegenfeldt, J., Feucht, H., Ruschitzka, G., Haeberlen, U.: J. Magn. Res. *39*, 509 (1980)
109. Pines, A., Chang, J.J., Griffin, R.G.: J. Chem. Phys. *61*, 1021 (1974)
110. Van der Hart, D.L.: J. Chem. Phys. *64*, 830 (1976)
111. a) Zilm, K.W., Grant, D.M., Conlin, R.T., Michl, J.: J. Am. Chem. Soc. *100*, 8038 (1978)
     b) Zilm, K.W., Grant, D.M.: J. Am. Chem. Soc. *103*, 2913 (1981)
112. Kaplan, S., Pines, A., Griffin, R.G., Waugh, J.S.: Chem. Phys. Lett. *25*, 78 (1974)
113. Schweitzer, D., Spiess, H.W.: J. Magn. Res. *16*, 243 (1974)
114. Gibby, M.G., Griffin, R.G., Pines, A., Waugh, J.S.: Chem. Phys. Lett. *17*, 80 (1972)
115. Bhattacharyya, P.K., Dailey, B.P.: J. Chem. Phys. *59*, 5820 (1973)
116. Schweitzer, D., Spiess, H.W.: J. Magn. Res. *15*, 529 (1974)
117. Gibby, M.G., Pines, A., Rhim, W.-K., Waugh, J.S.: J. Chem. Phys. *56*, 991 (1972)
118. Gibby, M.G., Pines, A., Waugh, J.S.: J. Am. Chem. Soc. *94*, 6231 (1972)
119. Koma, A., Tanaka, S.: Solid. State Commun. *10*, 823 (1972)
120. Koma, A.: Phys. Stat. Sol. *B57*, 299 (1973)
121. Terao, T., Hashi, T.: J. Phys. Soc. Jpn. *36*, 989 (1974)
122. Zumbulyadis, N., Dailey, B.P.: Chem. Phys. Lett. *26*, 273 (1974)
123. Lucken, E.A.C., Williams, D.F.: Mol. Phys. *16*, 17 (1969)
124. Bensoussan, M.: J. Phys. Chem. Sol. *28*, 1533 (1967)
125. Harbison, G., Herzfeld, J., Griffin, R.G.: J. Am. Chem. Soc. *103*, 4752 (1981)
126. Burghoff, U., Rosenberger, H., Zeisds, R., Müller, R., Rashkovich, L.N.: Phys. Stat. Sol. (a) *26*, K171 (1974)
127. Kohler, S.J., Ellett, J.D., Klein, M.P.: J. Chem. Phys. *64*, 4451 (1976)
128. Andrew, E.R., Hinshaw, W.S., Tiften, R.S.: J. Magn. Res. *15*, 191 (1974)
129. Nolle, A.: Z. Naturforsch. *33A*, 666 (1978)
130. a) Cheung, T.T.P., Worthington, L.E., Murphy, P. Du Bois, Gerstein, B.C.: J. Magn. Res. *41*, 158 (1980)
     b) Murphy, P. Du Bois, Gerstein, B.C.: J. Am. Chem. Soc. *103*, 3282 (1981)
131. a) Nolle, A.: Z. Naturforsch. *32A*, 964 (1977)
     b) Lutz, O., Nolle, A.: Z. Phys. *B36*, 323 (1980)

Chapter 8

1. Abragam, A.: The Principles of Nuclear Magn. London: Oxford Univ. Press 1961
2. Goldman, M.: Spin Temperature and Nuclear Magn. Res. in Solids. London: Oxford Univ.-Press 1970

3. Spiess, H.W.: Rotation of Molecules and Nuclear Spin-Relaxation in NMR: Basic Principles and Progress, Vol. 15, p. 55. Berlin-Heidelberg-New York: Springer 1978
4. a) Blicharski, J.S.: Acta Phys. Polon. *A41*, 223 (1972)
   b) Z. Naturforsch. *27A*, 1355 (1972)
   c) Z. Naturforsch. *27A*, 1456 (1972)
5. Mansfield, P.: Progr. NMR Spectroscopy *8*, 41 (1971)
6. Look, D.C., Lowe, I.J.: J. Chem. Phys. *44*, 2995 (1966)
7. Jones, G.P.: Phys. Rev. *148*, 332 (1966)
8. Haeberlen, U., Waugh, J.S.: Phys. Rev. *185*, 420 (1969)
9. Gründer, W., Schmiedel, H., Freude, D.: Ann. Phys. (London) *27*, 409 (1971)
10. Gründer, W.: Wiss. Z. Karl-Marx-Univ. Leipzig, Math.-Naturw. *23*, 466 (1974)
11. Andrew, E.R., Jasinski, A.: J. Phys. C. Sol. Stat. Phys. *4*, 391 (1971)
12. Ostroff, E.D., Waugh, J.S.: Phys. Rev. Lett. *16*, 1097 (1966)
13. Mansfield, P., Ware, D.: Phys. Lett. *22*, 133 (1966)
14. Lee, M., Goldberg, W.I.: Phys. Rev. *A410*, 1261 (1965)
15. a) Willsch, R., Müller, R., Scheler, G.: 2nd Spez. Colloque Ampere. Budapest (Hungary) 1975
    b) Müller, R., Willsch, R.: J. Magn. Res. *21*, 135 (1976)
16. Rhim, W.K., Burum, D.P., Elleman, D.D.: J. Chem. Phys. *68*, 692 (1978)
17. Vega, A.J., Vaughan, R.W.: J. Chem. Phys. *68*, 1958 (1978)
18. Rhim, W.K., Burum, D.P., Elleman, D.D.: Phys. Rev. Lett. *37*, 1764 (1976)
19. Roeder, St.B.W., Douglass, D.C.: J. Chem. Phys. *52*, 5525 (1970)
20. Mehring, M., Raber, H., Sinning, G.: Proceedings of the 18th Congress Ampere, p. 35. Nottingham 1974
21. Mehring, M., Griffin, R.G., Waugh, J.S.: J. Chem. Phys. *55*, 746 (1971)
22. Mehring, M., Raber, H.: J. Chem. Phys. *59*, 1116 (1973)
23. Hilt, R.L., Hubbard, P.S.: Phys. Rev. *134A*, 392 (1964)
24. Wolff, E.: Diploma work. Dortmund 1975
25. Schütz, J.U.v., Wolf, H.C.: Z. Naturforsch. *27A*, 42 (1972)
26. a) Tse, D., Hartmann, S.R.: Phys. Rev. Lett. *21*, 511 (1968)
    b) Lin, N.A., Hartmann, S.R.: Phys. Rev. Lett. *21*, 511 (1973)
27. a) Tse, D., Lowe, I.J.: Phys. Rev. *166*, 279 (1968)
    b) Lowe, I.J., Tse, D.: Phys. Rev. *166*, 292 (1968)
28. Wolfe, J.P., Markiewicz, R.S.: Phys. Rev. Lett. *28*, 1105 (1973)
29. Kaplan, J.I.: Phys. Rev. *B3*, 604 (1971)
30. Gibby, M.G., Pines, A., Waugh, J.S.: Chem. Phys. Lett. *16*, 296 (1972)
31. Pines, A., Gibby, M.G., Waugh, J.S.: J. Chem. Phys. *59*, 569 (1973)
32. a) Garroway, A.N.: J. Magn. Res. *34*, 283 (1979)
    b) Van der Hart, D.L., Garroway, A.N.: J. Chem. Phys. *71*, 2773 (1979)
33. a) Garroway, A.N., Van der Hart, D.L., Earl, W.L.: Phil. Trans. Roy. London *A299*, 609 (1981)
    b) Van der Hart, D.L., Earl, W.L., Garroway, A.N.: J. Magn. Res. *44*, 361 (1981)
34. a) Schaefer, J., Stejskal, E.O., Buchdahl, R.: Macromolecules *10*, 384 (1977)
    b) Schaefer, J., Stejskal, E.O., Steger, T.R., Sefcik, M.D., McKay, R.A.: Macromolecules *13*, 1121 (1980)
35. Emid, S.: J. Magn. Res. *42*, 147 (1981)
36. Suwelack, D., Rothwell, W.P., Waugh, J.S.: J. Chem. Phys. *73*, 2559 (1980)
37. Günther, B., Kanert, O., Mehring, M., Wolf, D.: Phys. Rev. *B24*, 6747 (1981)
38. Hyde, J.S., Dalton, L.R.: Chem. Phys. Lett. *16*, 568 (1972)
39. Jeener, J., Meier, B.H., Bachmann, P., Ernst, R.R.: J. Chem. Phys. *71*, 4546 (1979)

Chapter 9

1. a) Rose, M.E.: Elementary Theory of Angular Momentum. New York: J. Wiley 1967
   b) Brink, D.M., Satchler, G.R.: Angular Momentum. London: Oxford Univ. Press 1968
2. Cook, R.L., De Lucia, F.C.: Am. J. Phys. *39*, 1433 (1971)
3. Abragam, A.: Principles of Nuclear Magnetism, Chap. IV. London: Oxford Univ. Press 1961

4. a) Bosse, J.: DFG-Conference on Magnetic Relaxation. Hirschegg, Austria, Sept. 1974
   b) Mansfield, P.: Progr. Nucl. Magn. Res. *8*, 43 (1971)
5. a) Magnus, W.: Com. Pure Appl. Math. *7*, 649 (1954)
   b) Pechukas, P., Light, F.C.: J. Chem. Phys. *44*, 3897 (1966)
   c) Wilcox, R.M.: J. Math. Phys. *8*, 962 (1967)
   d) Haeberlen, U., Waugh, J.S.: Phys. Rev. *175*, 453 (1968)
6.   Haeberlen, U.: Advan. Magn. Res. supplement. New York: Academic Press 1976
7. a) Fano, U.: Rev. Mod. Phys. *29*, 74 (1957)
   b) Shimizu, T.: J. Phys. Soc. Jpn. *28*, 790 (1970); *28*, 811 (1970)
8.   Demco, D., Tegenfeldt, J., Waugh, J.S.: Phys. Rev. *B11*, 4133 (1975)
9. a) Zwanzig, R.: Lectures in Theoretical Physics. New York: Interscience 1961
   b) Mori, H.: Progr. Theor. Phys. Jpn. *34*, 399 (1965)
10. a) Lado, F., Memory, J.D., Parker, G.W.: Phys. Rev. *B4*, 1406 (1971)
    b) Parker, G.W., Lado, F.: Phys. Rev. *B8*, 3081 (1973)
11.   Slichter, C.P., Ailion, D.G.: Phys. Rev. *135A*, 1099 (1964)
12.   Wolf, D.: Spin-Temperature and Nuclear-Spin Relaxation in Matter. Clarendon Press: Oxford 1979
13.   Bialynicki-Birula, I., Mielnik, B., Plebanski, J.: Ann. of Phys. *51*, 187 (1969)
14.   Dyson, F.J.: Phys. Rev. *75*, 486 (1949)
15.   Schwinger, J.: Phys. Rev. *74*, 1439 (1948)
16.   Grupp, H.: German translation of reference 17 (private communication)
17.   Buishvili, L.L., Menabde, M.G.: Sov. Phys. JETP *50*, 1176 (1979)
18.   Bogolubov, N.N., Mitropolsky, Y.A.: Asymptotic methods in the theory of non-linear oscillations. New York: Gordon and Breach (1961)
19.   Bloch, F., Siegert, A.: Phys. Rev. *57*, 552 (1940)
20.   Spiess, H.W.: Rotation of molecules and nuclear spin relaxation, in NMR: Basic Principles and Progress, Vol. 15, p. 55. Berlin-Heidelberg-New York: Springer 1978
21. a) Blicharski, J.S.: Acta Phys. Polon. *A41*, 223 (1972)
    b) Z. Naturforsch. *27A*, 1355 (1972)
    c) Z. Naturforsch. *27A*, 1456 (1972)

Inorganic Chemistry Concepts

Editors: C. K. Jørgensen, M. F. Lappert,
S. J. Lippard, J. L. Margrave, K. Niedenzu, H. Nöth,
R. W. Parry, H. Yamatera

Volume 1
R. Reisfeld, C. K. Jørgensen

Lasers and Excited States of Rare Earths

1977. 9 figures, 26 tables. VIII, 226 pages
ISBN 3-540-08324-3

Volume 2
R. L. Carlin, A. J. van Duyneveldt

Magnetic Properties of Transition Metal Compounds

1977. 149 figures, 7 tables. XV, 264 pages
ISBN 3-540-08584-X

Volume 3
P. Gütlich, R. Link, A. Trautwein

Mössbauer Spectroscopy and Transition Metal Chemistry

1978. 160 figures, 19 tables, 1 folding plate.
X, 280 pages. ISBN 3-540-08671-4

"…The book is thus a remarkable source of information not only for aspiring research students but for any people concerned with physics and chemistry research in university and industry. It should remain an important reference for a long time."
Die Naturwissenschaften

Volume 4
Y. Saito

Inorganic Molecular Dissymmetry

1979. 107 figures, 28 tables. IX, 167 pages
ISBN 3-540-09176-9

"…The book is directed towards a general and synthetic understanding of chiral molecules, and their unique property of optical activity, in the field of transition metal chemistry. The level of treatment is suited to graduate or advanced undergraduate teaching. For these roles, and for library reference, the book is strongly recommended."
Nature

Springer-Verlag
Berlin Heidelberg New York

Volume 5
T. Tominaga, E. Tachikawa

Modern Hot-Atom Chemistry and Its Applications

1981. 57 figures, 34 tables. VIII, 154 pages
ISBN 3-540-10715-0

This book has long been awaited by students and researchers seeking a clear introduction to the concepts of modern hot atom chemistry. Various applications to inorganic, analytical, geochemical, biological, and energy-related studies are discussed with a view toward the promotion of interdisciplinary collaboration. Topics of current interest, such as NEET, laser isotope separation and mesic chemistry, are also described to expand the scope for future development in hot atom chemistry.

Volume 6
D. L. Kepert

Inorganic Stereochemistry

1982. 206 figures, 45 tables. XII, 227 pages
ISBN 3-540-10716-9

An important recent advance concerns the stereochemistry of molecules containing ring systems, which are extremely important throughout chemistry. Such molecules may not have stereochemistries corresponding to any of the usual polyhedra, but are intermediate between two different idealized polyhedra. The precise location of a particular molecule along this continuous range of stereochemistries depends upon the geometric design of the ring system, which includes the number of atoms in ring and the size of these atoms.
The simple techniques outlined in this work are the best way, and in most cases the only way, that such complicated structures with coordination numbers from four to twelve can be predicted.

Volume 7
H. Rickert

Electrochemistry of Solids

An Introduction
1982. 95 figures, 23 tables. XII, 240 pages
ISBN 3-540-11116-6

The electrochemistry of solids is of great current interest to research and development. The technical applications include batteries with solid electrolytes, high-temperature fuel cells, sensors for measuring partial pressures or activities, display units and, more recently, the growing field of chemotronic components. The science and technology of solid-state electrolytes is sometimes called solid-state ionics, analogous to the field of solid-state electronics. Only basic knowledge of physical chemistry and thermodynamics is required to read this book with utility. The chapters can be read independently from one another.

NMR

Basic Principles and Progress
Grundlagen und Fortschritte

Editors:
P. Diehl, E. Fluck, R. Kosfeld

Volume 12
B. Lindman, S. Forsén
Chlorine, Bromine and Iodine NMR
Physico-Chemical and Biological Applications
1976. 74 figures, 29 tables. XIV, 368 pages
ISBN 3-540-07725-1

Volume 13
Introductory Essays
Editor: **M. M. Pintar**
1976. 48 figures. XI, 154 pages
ISBN 3-540-07754-5

Volume 14
H. Nöth, B. Wrackmeyer
Nuclear Magnetic Resonance Spectroscopy of Boron Compounds
1978. 1 figure, 96 tables. XII, 461 pages
ISBN 3-540-08456-8

Volume 15
H. W. Spiess, A. Steigel
Dynamic NMR Spectroscopy
Corrected Reprint. 1982.
Approx. 76 figures. Approx. 220 pages
ISBN 3-540-08784-2

Volume 16
P. S. Pregosin, R. W. Kunz
^{31}P and ^{13}C NMR Transition Metal Phosphine Complexes
1979. 26 figures, 37 tables. IX, 156 pages
ISBN 3-540-09163-7

Volume 17
J.-P. Kintzinger, H. Marsmann
Oxygen–17 and Silicon–29
1981. 31 figures. V, 235 pages
ISBN 3-540-10414-3

Volume 18
G. J. Martin, M. L. Martin, J.-P. Gouesnard
^{15}N-NMR Spectroscopy
1981. 11 figures, 142 tables. VII, 382 pages
ISBN 3-540-10459-3

Volume 19
NMR in Medicine
Editor: **R. Damadian**
1981. 77 figures. V, 174 pages
ISBN 3-540-10460-7

Volume 20
G. Govil, R. V. Hosur
Conformation of Biological Molecules
New Results from NMR
1982. 92 figures. VIII, 216 pages
ISBN 3-540-10769-X

Springer-Verlag
Berlin
Heidelberg
New York

MIX
Papier aus verantwortungsvollen Quellen
Paper from responsible sources
FSC® C105338

If you have any concerns about our products,
you can contact us on
ProductSafety@springernature.com

In case Publisher is established outside the EU,
the EU authorized representative is:
Springer Nature Customer Service Center GmbH
Europaplatz 3, 69115 Heidelberg, Germany

Printed by Libri Plureos GmbH
in Hamburg, Germany